S0-BMG-661

MTP International Review of Science

Biochemistry
Series One

Consultant Editors
H. L. Kornberg, F.R.S. and
D. C. Phillips, F.R.S.

Publisher's Note

The MTP International Review of Science is an important new venture in scientific publishing, which is presented by Butterworths in association with MTP Medical and Technical Publishing Co. Ltd. and University Park Press, Baltimore. The basic concept of the Review is to provide regular authoritative reviews of entire disciplines. Chemistry was taken first as the problems of literature survey are probably more acute in this subject than in any other. Physiology and Biochemistry followed naturally. As a matter of policy, the authorship of the MTP Review of Science is international and distinguished, the subject coverage is extensive, systematic and critical, and most important of all, it is intended that new issues of the Review will be published at regular intervals.

In the MTP Review of Chemistry (Series One), Inorganic, Physical and Organic Chemistry are comprehensively reviewed in 33 text volumes and 3 index volumes. Physiology (Series One) consists of 8 volumes and Biochemistry (Series One) 12 volumes, each volume individually indexed. Details follow. In general, the Chemistry (Series One) reviews cover the period 1967 to 1971, and Physiology and Biochemistry (Series One) reviews up to 1972. It is planned to start in 1974 the MTP International Review of Science (Series Two), consisting of a similar set of volumes covering developments in a two year period.

The MTP International Review of Science has been conceived within a carefully organised editorial framework. The overall plan was drawn up, and the volume editors appointed by seven consultant editors. In turn, each volume editor planned the coverage of his field and appointed authors to write on subjects which were within the area of their own research experience. No geographical restriction was imposed. Hence the 500 or so contributions to the MTP Review of Science come from many countries of the world and provide an authoritative account of progress.

Butterworth & Co. (Publishers) Ltd.

BIOCHEMISTRY SERIES ONE

Consultant Editors
H. L. Kornberg, F.R.S.
Department of Biochemistry University of Leicester and
D. C. Phillips, F.R.S., *Department of Zoology, University of Oxford*

Volume titles and Editors

1 **CHEMISTRY OF MACRO-MOLECULES**
Professor H. Gutfreund, *University of Bristol*

2 **BIOCHEMISTRY OF CELL WALLS AND MEMBRANES**
Dr. C. F. Fox, *University of California*

3 **ENERGY TRANSDUCING MECHANISMS**
Professor E. Racker, *Cornell University, New York*

4 **BIOCHEMISTRY OF LIPIDS**
Professor T. W. Goodwin, F.R.S., *University of Liverpool*

5 **BIOCHEMISTRY OF CARBO-HYDRATES**
Professor W. J. Whelan. *University of Miami*

6 **BIOCHEMISTRY OF NUCLEIC ACIDS**
Professor K. Burton, F.R.S., *University of Newcastle upon Tyne*

7 **SYNTHESIS OF AMINO ACIDS AND PROTEINS**
Professor H. R. V. Arnstein, *King's College, University of London*

8 **BIOCHEMISTRY OF HORMONES**
Professor H. V. Rickenberg, *National Jewish Hospital & Research Center, Colorado*

9 **BIOCHEMISTRY OF CELL DIFFER-ENTIATION**
Dr. J. Paul, *The Beatson Institute for Cancer Research, Glasgow*

10 **DEFENCE AND RECOGNITION**
Professor R. R. Porter, F.R.S., *University of Oxford*

11 **PLANT BIOCHEMISTRY**
Professor D. H. Northcote, F.R.S., *University of Cambridge*

12 **PHYSIOLOGICAL AND PHARMACO-LOGICAL BIOCHEMISTRY**
Dr. H. F. K. Blaschko, F.R.S., *University of Oxford*

PHYSIOLOGY SERIES ONE

Consultant Editors
A. C. Guyton,
Department of Physiology and Biophysics, University of Mississippi Medical Center and
D. Horrobin,
Department of Medical Physiology, University College of Nairobi

Volume titles and Editors

1 **CARDIOVASCULAR PHYSIOLOGY**
Professor A. C. Guyton and Dr. C. E. Jones, *University of Mississippi Medical Center*

2 **RESPIRATORY PHYSIOLOGY**
Professor J. G. Widdicombe, *St. George's Hospital, London*

3 **NEUROPHYSIOLOGY**
Professor C. C. Hunt, *Washington University School of Medicine, St. Louis*

4 **GASTROINTESTINAL PHYSIOLOGY**
Professor E. D. Jacobson and Dr. L. L. Shanbour, *University of Texas Medical School*

5 **ENDOCRINE PHYSIOLOGY**
Professor S. M. McCann, *University of Texas*

6 **KIDNEY AND URINARY TRACT PHYSIOLOGY**
Professor K. Thurau, *University of Munich*

7 **ENVIRONMENTAL PHYSIOLOGY**
Professor D. Robertshaw, *University of Nairobi*

8 **REPRODUCTIVE PHYSIOLOGY**
Professor R. O. Greep, *Harvard Medical School*

INORGANIC CHEMISTRY SERIES ONE

Consultant Editor
H. J. Emeléus, F.R.S.
Department of Chemistry
University of Cambridge

Volume titles and Editors

1 **MAIN GROUP ELEMENTS—HYDROGEN AND GROUPS I-IV**
Professor M. F. Lappert, *University of Sussex*

2 **MAIN GROUP ELEMENTS—GROUPS V AND VI**
Professor C. C. Addison, F.R.S. and Dr. D. B. Sowerby, *University of Nottingham*

3 **MAIN GROUP ELEMENTS—GROUP VII AND NOBLE GASES**
Professor Viktor Gutmann, *Technical University of Vienna*

4 **ORGANOMETALLIC DERIVATIVES OF THE MAIN GROUP ELEMENTS**
Dr. B. J. Aylett, *Westfield College, University of London*

5 **TRANSITION METALS—PART 1**
Professor D. W. A. Sharp, *University of Glasgow*

6 **TRANSITION METALS—PART 2**
Dr. M. J. Mays, *University of Cambridge*

7 **LANTHANIDES AND ACTINIDES**
Professor K. W. Bagnall, *University of Manchester*

8 **RADIOCHEMISTRY**
Dr. A. G. Maddock, *University of Cambridge*

9 **REACTION MECHANISMS IN INORGANIC CHEMISTRY**
Professor M. L. Tobe, *University College, University of London*

10 **SOLID STATE CHEMISTRY**
Dr. L. E. J. Roberts, *Atomic Energy Research Establishment, Harwell*

INDEX VOLUME

PHYSICAL CHEMISTRY SERIES ONE

Consultant Editor
A. D. Buckingham
Department of Chemistry
University of Cambridge

Volume titles and Editors

1 **THEORETICAL CHEMISTRY**
Professor W. Byers Brown, *University of Manchester*

2 **MOLECULAR STRUCTURE AND PROPERTIES**
Professor G. Allen, *University of Manchester*

3 **SPECTROSCOPY**
Dr. D. A. Ramsay, F.R.S.C., *National Research Council of Canada*

4 **MAGNETIC RESONANCE**
Professor C. A. McDowell F.R.S.C., *University of British Columbia*

5 **MASS SPECTROMETRY**
Professor A. Maccoll, *University College, University of London*

6 **ELECTROCHEMISTRY**
Professor J. O'M Bockris, *University of Pennsylvania*

7 **SURFACE CHEMISTRY AND COLLOIDS**
Professor M. Kerker, *Clarkson College of Technology, New York*

8 **MACROMOLECULAR SCIENCE**
Professor C. E. H. Bawn, F.R.S., *University of Liverpool*

9 **CHEMICAL KINETICS**
Professor J. C. Polanyi, F.R.S., *University of Toronto*

10 **THERMOCHEMISTRY AND THERMO-DYNAMICS**
Dr. H. A. Skinner, *University of Manchester*

11 **CHEMICAL CRYSTALLOGRAPHY**
Professor J. Monteath Robertson, F.R.S., *University of Glasgow*

12 **ANALYTICAL CHEMISTRY—PART 1**
Professor T. S. West, *Imperial College, University of London*

13 **ANALYTICAL CHEMISTRY—PART 2**
Professor T. S. West, *Imperial College, University of London*

INDEX VOLUME

ORGANIC CHEMISTRY SERIES ONE

Consultant Editor
D. H. Hey, F.R.S.,
Department of Chemistry
King's College, University of London

Volume titles and Editors

1 **STRUCTURE DETERMINATION IN ORGANIC CHEMISTRY**
Professor W. D. Ollis, F.R.S., *University of Sheffield*

2 **ALIPHATIC COMPOUNDS**
Professor N. B. Chapman, *Hull University*

3 **AROMATIC COMPOUNDS**
Professor H. Zollinger, *Swiss Federal Institute of Technology*

4 **HETEROCYCLIC COMPOUNDS**
Dr. K. Schofield, *University of Exeter*

5 **ALICYCLIC COMPOUNDS**
Professor W. Parker, *University of Stirling*

6 **AMINO ACIDS, PEPTIDES AND RELATED COMPOUNDS**
Professor D. H. Hey, F.R.S., and Dr. D. I. John, *King's College, University of London*

7 **CARBOHYDRATES**
Professor G. O. Aspinall, *Trent University, Ontario*

8 **STEROIDS**
Dr. W. F. Johns, *G. D. Searle & Co., Chicago*

9 **ALKALOIDS**
Professor K. Wiesner, F.R.S., *University of New Brunswick*

10 **FREE RADICAL REACTIONS**
Professor W. A. Waters, F.R.S., *University of Oxford*

INDEX VOLUME

MTP International Review of Science

Biochemistry
Series One

Volume 9

Biochemistry of Cell Differentiation

Edited by **J. Paul**
The Beatson Institute for Cancer Research, Glasgow

Butterworths · London
University Park Press · Baltimore

THE BUTTERWORTH GROUP

ENGLAND
Butterworth & Co (Publishers) Ltd
London: 88 Kingsway, WC2B 6AB

AUSTRALIA
Butterworths Pty Ltd
Sydney: 586 Pacific Highway 2067
Melbourne: 343 Little Collins Street, 3000
Brisbane: 240 Queen Street, 4000

NEW ZEALAND
Butterworths of New Zealand Ltd
Wellington: 26–28 Waring Taylor Street, 1

SOUTH AFRICA
Butterworth & Co (South Africa) (Pty) Ltd
Durban: 152–154 Gale Street

ISBN 0 408 70503 5

UNIVERSITY PARK PRESS

U.S.A. and CANADA
University Park Press
Chamber of Commerce Building
Baltimore, Maryland, 21202

Library of Congress Cataloging in Publication Data

Paul, John, 1922–
Biochemistry of cell differentiation.

(Biochemistry, series one, v. 9) (MTP international review of science)
1. Cell differentiation. 2. Cytochemistry.
I. Title. II. Series. III. Series: MTP international review of science. [DNLM: 1. Biochemistry—Cell differentiation. W1B1633 ser. 1 v. 9 1974/ QH607 B615 1974]
QP501.B527 vol. 9 [QH607] 574.1′92′08s [574.8′761]
ISBN 0-8391-1048-0 74-6296

Typeset and printed in Great Britain by
REDWOOD BURN LIMITED
Trowbridge & Esher
and bound by R. J. Acford Ltd, Chichester, Sussex

Consultant Editors' Note

The MTP International Review of Science is designed to provide a comprehensive, critical and continuing survey of progress in research. Nowhere is such a survey needed as urgently as in those areas of knowledge that deal with the molecular aspects of biology. Both the volume of new information, and the pace at which it accrues, threaten to overwhelm the reader: it is becoming increasingly difficult for a practitioner of one branch of biochemistry to understand even the language used by specialists in another.

The present series of 12 volumes is intended to counteract this situation. It has been the aim of each Editor and the contributors to each volume not only to provide authoritative and up-to-date reviews but carefully to place these reviews into the context of existing knowledge, so that their significance to the overall advances in biochemical understanding can be understood also by advanced students and by non-specialist biochemists. It is particularly hoped that this series will benefit those colleagues to whom the whole range of scientific journals is not readily available. Inevitably, some of the information in these articles will already be out of date by the time these volumes appear: it is for that reason that further or revised volumes will be published as and when this is felt to be appropriate.

In order to give some kind of coherence to this series, we have viewed the description of biological processes in molecular terms as a progression from the properties of macromolecular cell components, through the functional interrelations of those components, to the manner in which cells, tissues and organisms respond biochemically to external changes. Although it is clear that many important topics have been ignored in a collection of articles chosen in this manner, we hope that the authority and distinction of the contributions will compensate for our shortcomings of thematic selection. We certainly welcome criticisms, and solicit suggestions for future reviews, from interested readers.

It is our pleasure to thank all who have collaborated to make this venture possible—the volume editors, the chapter authors, and the publishers.

Leicester H. L. Kornberg

Oxford D. C. Phillips

Preface

In recent years many biochemists have tended to consider problems of cell differentiation too biological to be respectable subjects for serious biochemical research. This attitude developed in the forties and fifties as a reaction to the disappointments which followed the heroic attempts of the twenties and thirties to apply biochemical methods to the analysis of development, attempts which were reflected in Needham's monumental work '*Chemical Embryology*'. At that time, some of the best minds of the day were engaged in the field—one can mention Child, Dalcq, Haldane, Hämmerling, Harrison, Holtfreter, J. and T. H. Huxley, Needham, Spemann, Waddington and Weiss—and it is not surprising that, in retrospect, it has been referred to as the molecular biology of the thirties. However, the complexities of the problems of development were not then fully understood and, in the absence of any knowledge of the genetic code, messenger RNA and protein synthesis, it was inevitable that progress should slow down and adventurous minds should move elsewhere. The flowering of classical molecular biology in the past two decades has again changed the situation and developmental biology bids fair to emerge once more as the molecular biology of the seventies. One of the main components of molecular biology is, of course, biochemistry and a series of volumes on contemporary biochemistry would clearly be incomplete without one dealing with biochemical aspects of development.

In planning a volume on the Biochemistry of Cell Differentiation, I have had very much in mind that the subject interests a wider group of biologists than those who are actively engaged in it. I have also been aware that not only the biochemistry but also the basic biology is very complex and biochemists, in general, often feel diffident about approaching the subject because they feel their knowledge of the biological background is inadequate. I have therefore tried to design this volume in such a way that it tells a continuous story. The contributions have been chosen to cover the field rather comprehensively although by no means completely and, where necessary, I have attempted to tie them together with an introduction and short comments between chapters; these are intended to fill in the biological background and to draw the readers' attention to some aspects of the field which are not covered by the articles themselves but are of importance. Following an introduction to the problems of developmental biology, there come six chapters dealing with particular experimental systems from slime moulds, through plants, sea urchins and amphibia to erythropoiesis and hormone action in birds and mammals. Each of these illustrates special problems and the ways in which they have been tackled; they lead on to the four remaining chapters which are concerned with mechanisms involved in gene expression in eukaryotes, such as gene masking, RNA synthesis and the metabolic stability of macromolecules.

This is a rapidly moving field and, even as the manuscripts go to press, new discoveries are being reported which one would like to see included. However, a line has to be drawn somewhere and this volume describes the scene as it was in mid-1973.

The chapters are all written by the acknowledged authorities in the field, with one exception. The original author of Chapter 3 failed to produce a manuscript and I, as editor, have, therefore, inserted a brief review to maintain the continuity of the text. I make no pretensions to being an expert on the subject of that chapter and ask the reader to exercise indulgence for its lack of authority.

Glasgow — J. Paul

Contents

Editor's Introduction

In multi-cellular organisms, a remarkably ordered programme of changes occurs during the development of the mature plant or animal from the fertilised egg: a great range of functions is simultaneously orchestrated to produce the harmonious pattern which is characteristic of normal development. In general terms, it is possible to recognise two classes of processes and these can conveniently be considered separately. The first is the emergence of functionally and morphologically distinct cell types; this is termed 'cell differentiation'. The other is the association of cells in precise patterns to form tissues and organs. This process is largely a mystery and biochemistry has not so far made any striking contributions to its understanding. For that reason, none of the contributions to this volume deals specifically with it. In contrast, many features of cell differentiation can be described quite precisely in biochemical terms. The mechanisms involved in cell differentiation are almost certainly of a biochemical nature and of a kind which can be investigated by extrapolation from existing knowledge. This volume is, therefore, devoted almost entirely to its consideration.

Cells which have differentiated along a specific pathway are characterised, among other things, by the spectrum of proteins they contain. Haemoglobin is typical of erythrocytes, actin and myosin of muscle cells and so on. Therefore, a major component of cell differentiation clearly has to do with the control of protein synthesis, about which we now have a rather detailed understanding. Hence, it is possible to investigate directly many of the mechanisms which are likely to be implicated.

It is now generally accepted that the genetic code applies universally in both prokaryotes (non-nucleated single-celled organisms like bacteria) and eukaryotes (organisms, often multi-cellular, in which the cells contain nuclei and usually chromosomes), and that nearly all the information for protein structure in eukaryotes is encoded in the DNA in chromosomes. It follows that the nucleus of the fertilised egg must contain information for the whole organism which develops from it. One possible mechanism for differentiation might be the partitioning of this information into different cells, for example by the asymmetric segregation of different chromosomes or parts of chromosomes into daughter cells during cell division. Indeed, since the early work of Boveri[2] and later Wilson[21] which demonstrated segregation of precisely this kind during the early development of Ascaris, this idea has been entertained. However, it is now fairly clear that cells can differentiate without any loss of genetic information. The most convincing evidence has been provided by nuclear-transplantation experiments (Gurdon[9]; Gurdon and Laskey[10]) which demonstrated that nuclei isolated from intestinal epithelial cells of tadpoles and also from cultured epithelial cells can contain all the information for the development of a complete frog.

Good evidence also came from the work of Hadorn[11]. He and his colleagues transplanted tissue from the imaginal discs of insect larvae into adult flies. In these circumstances, the disc cells never differentiated but divided repeatedly. However, if they were transplanted back into larvae, they differentiated into adult tissues during metamorphosis. It was observed that cells from specific discs, and even from specific areas within them, almost always gave rise to the same specific structures but, occasionally, transplanted imaginal disc tissue, which had invariably given rise to one structure on previous back transplantation, suddenly gave rise to a different structure (a phenomenon called 'trans-determination'). These findings showed that genetic information for the different structures was present in the same cells.

In plants, it is well-known that differentiation is more flexible than in animals; for example, it has been possible to grow isolated cells into complete plants (Steward *et al.*[18]), again demonstrating that the original cells had a full genetic complement. Convincing though these experiments are, there are too few of them yet for it to be possible to state categorically that cyto-differentiation generally occurs without a change in the genetic complement. Nevertheless, the evidence is sufficiently strong that it is almost universally assumed that the main regulatory events in cell differentiation are post-replicative.

Before dismissing changes in DNA in the course of differentiation, however, the phenomenon of 'gene amplification' must be mentioned briefly. In some flies with giant banded chromosomes it has been known for some time that DNA synthesis can occur in certain bands and not others (Pavan and de Cunha,[16]). Moreover, much information has accumulated to show that during certain stages of development, especially during maturation of the oocyte, in many organisms the ribosomal cistrons increase proportionally more than the rest of the genome. The phenomenon has been studied in particular in the maturing frog egg and is discussed by Dr Denis in Chapter 4.

One of the major questions which has arisen is whether gene amplification is a general mechanism in cell differentiation or whether it is confined to certain special cases. Too few studies have been done to permit generalisation but it seems that the silk fibroin gene in silkworms (Suzuki *et al.*[19]) and the globin gene in mammals (Harrison *et al.*[12]) are not 'amplified'. Hence, post-replicative processes command most interest in this field at present. Among these, the regulation of transcription so as to produce a characteristic spectrum of RNA molecules has received most attention; it will be repeatedly referred to in the individual chapters in this book.

Obviously analogies exist between regulation of enzyme activity in bacteria and transcriptional controls in eukaryotes but one of the major distinctions between eukaryotic and prokaryotic cells is in post-transcriptional events. The newly transcribed RNA is not immediately translated as it is in bacteria. First, it is processed, that is to say, some parts of the molecule are degraded specifically to leave smaller RNA molecules which have to migrate from the nucleus to the cytoplasm before they are involved in the protein synthesising machinery (Georgiev[7]). There is some suggestive evidence that selective control mechanisms may operate to determine which molecules enter the cytoplasm and which remain to be degraded in the nucleus.

Messenger RNA in eukaryotes is not rapidly degraded from the 5′–end, as is messenger RNA in prokaryotes. Measurements of messenger RNA half life in eukaryotic cells vary but some are quite long, of the order of days; variation in the stability of RNA itself may, therefore, be a factor in determining the spectra of protein synthesis as discussed by Drs Kafatos and Gelinas in Chapter 8.

Because of this high stability of mRNA, controls at the translational level assume particular importance in eukaryotic cells and these are referred to in several chapters. One particular example of translational control has turned out to be exceptionally interesting. During the very early stages of development, as discussed in Chapter 3, evidence has emerged that the egg contains preformed messenger RNA which is not translated until after fertilisation. The so called 'masked messenger' provides an extreme example of long-lived messengers with translational controls and, although other instances may emerge in future studies, this phenomenon seems to be a particular feature of early cell differentiation.

Finally, the product of protein synthesis, the protein itself, has to be degraded in eukaryotic cells. In rapidly growing bacteria, proteins can be very stable because, when their synthesis stops, they will be rapidly diluted out by cell growth. In eukaryotes, however, organs such as the liver may not show any gross growth for many years and yet enzyme activities can wax and wane in response to appropriate stimuli. In these organisms, therefore, for enzyme activity to diminish, the enzyme has to be inhibited, excreted or degraded. These important mechanisms are discussed in Chapter 7 by Dr Schimke.

Much of our thinking about the possible mechanisms of cyto-differentiation nevertheless derives from our knowledge of control mechanisms in prokaryotes, especially in *E. coli.* The work of Jacob and Monod[13] is, of course, particularly relevant but it is questionable whether some of the phenomena characteristic of cell differentiation occur in prokaryotes at all. This need not be considered surprising when it is recollected that eukaryates differ from prokaryotes not only in the characteristics outlined earlier but also in that they have much larger genomes and the chromosomes consist of nucleoprotein which contains large amounts of histones and nonhistone proteins besides DNA. One reason for suspecting that the behaviour of eukaryotic cells during cell differentiation is not exactly like that of bacterial cells during enzyme adaptation is that cell differentiation (or at least the 'determination' of phenotype) seems to be rather irreversible in many cells especially in animals. Reference has already been made to the work of Hadorn which demonstrates this fact. Moreover, in experiments in which cultured differentiated cells have been passaged serially, even by cloning (passage of single cells) the differentiated features have persisted (Konigsberg[15]; Cahn and Cahn[3], and Coon[4]). Accordingly the articles in this volume deal entirely with cell differentiation in eukaryotes.

Much research in developmental biology has been devoted to searching for suitable systems and many attempts have been made to find simple organisms which would differentiate but otherwise behave in most respects like bacteria. In this connection, the slime moulds have attracted particular attention. The acellular slime moulds such as *Physarum polycephalum* have been studied, for example by Rusch (Cummins and Rusch[5]). At some stages in development, these form large syncitia in which the nuclei divide synchronously. Accordingly, they have proved exceedingly useful in studying the control of nuclear division. Another major group of slime moulds, the cellular slime moulds, has particularly attracted developmental biologists because they go through a fascinating cycle during part of which they behave like a multi-cellular organism and during part of which they behave as a culture of amoebae. These are described by Dr Ashworth in Chapter 1. The next five chapters describe other systems, each of which has been found to have specific merits and each of which illustrates a specific problem. Most of these systems were originally the objects of non-biochemical studies but have been found useful for biochemical studies also.

A description of one particular system is notably absent from this volume but some mention must be made of it since it is the system which has provided many of the ideas which have led to subsequent biochemical investigations. The study of the development of Diptera such as *Drosophila* and *Chironomus* has proved remarkably fruitful from the points of view of cytogenetics and developmental genetics, but the organisms are so small that it is only relatively recently that useful biochemical information has been gleaned from them (Berendes and Beermann[1]; Daneholt *et al.*[6] and Grossbach[8]). The salivary glands and some other organs in the larvae of these insects contain the well known giant chromosomes. These chromosomes are polytenic, that is to say the single DNA thread of the chromosomes is replicated very many times to form several hundred threads which remain in parallel alignment and in perfect register. Hence, the giant chromosome is a ribbon-like structure, the bands of the ribbon representing chromomeres. There is now quite good evidence that each band plus interband represents a genetic complementation unit (Judd *et al.*[14]). Moreover, while most of the bands in any organ are apparently inactive in RNA synthesis, certain bands swell up to form puffs which are demonstrably active in RNA synthesis (Pelling[17]). The pattern of puffing is in many cases organ specific. These observations led to the idea that most genes might not be transcribed in a given organ, but only those which determined the specific spectrum of proteins to be found there. Biochemical studies based on these ideas emerge in Chapter 6 by Drs Means, and O'Malley, in Chapter 9 by Dr Rutter and in Chapter 10 by Drs MacGillivray and Rickwood.

Many readers will use individual chapters of this book for reference purposes but it has been arranged so that it forms a continuous account of the field. The first five chapters deal with descriptions of different experimental systems illustrating specific points. The sixth chapter is again a description of a system but it also leads into the detailed study of mechanisms with which the remaining chapters are entirely concerned—the role of stability and turnover of products in Chapters 7 and 8, of RNA polymerases in Chapter 9 and of chromosomal proteins in Chapter 10. At intervals I have introduced connecting statements which, like this Introduction, are intended to set the biochemical accounts in perspective within the biological framework of the phenomena of cell differentiation.

References

1. Berendes, H. and Beermann, W. (1969). *Handbook of Molecular Cytology*, 501 (A. Lima-de-Faria, editor) (Amsterdam and London: North Holland)
2. Boveri, T. (1899). Die Entwicklung von Ascaris megalocephales mit besonderer Rucksicht auf die Kernverthaltrusse. Festahr F.C. von Kupffer, Jena
3. Cahn, R. D. and Cahn, M. B. (1966). *Proc. Nat. Acad. Sci. USA*, **55,** 106
4. Coon, H. G. (1966). *Proc. Nat. Acad. Sci. USA*, **55,** 66
5. Cummins, J. E. and Rusch, H. P. (1967). *Biochim. Biophys. Acta*, **138,** 124
6. Daneholt, B., Edstrom, J. E., Egyhazy, E., Lambert, B. and Ringborg, V. (1969). *Chromosoma*, **28,** 379
7. Georgiev, G. P. (1972). *Current Topics Dev. Biol.*, **7,** 1
8. Grossbach, V. (1969). *Chromosoma*, **28,** 136
9. Gurdon, J. B. (1962). *J. Embryol. Exp. Morph.*, **10,** 622
10. Gurdon, J. B. and Laskey, R. H. (1970). *J. Embryol. Exp. Morph.*, **24,** 277
11. Hadorn, E. (1965). *Brookhaven Symp. Biol.*, **18,** 148
12. Harrison, P. R., Hell, A., Birnie, G. O. and Paul, J. (1972) in the press
13. Jacob, F. and Monod, J. (1961). *J. Mol. Biol.*, **3,** 318
14. Judd, B. H., Shen, M. W. and Kaufman, T. C. (1972). *Genetics*, **71,** 139
15. Konigsberg, I. (1963). *Science*, **140,** 1273
16. Pavan, C. and de Cunha, A. B. (1969). *Ann. Rev. Genet.*, **3,** 425
17. Pelling, C. (1964). *Chromosoma*, **15,** 71
18. Steward, F. C., Mapes, N. C. and Mean, K. (1958). *Amer. J. Botany*, **45,** 705
19. Suzuki, Y., Gage, L. P. and Brown, D. D. (1972). *J. Mol. Biol.*, **70,** 637
20. Weeks, D. P. and Marcus, A. (1971). *Biochim. Biophys. Acta*, **232,** 671
21. Wilson, E. B. (1925). 'The Cell in Development and Heredity', 3rd edition. (New York: Macmillan)

1
The Development of the Cellular Slime Moulds

J. W. ASHWORTH
University of Essex

1.1 INTRODUCTION

The cellular slime moulds or Acrasiales were first identified by Brefeld[1] in 1869 but, unfortunately, he mistakenly thought that the amoebae fused after aggregation to form a true plasmodium and thus he described his isolate as another member of the Myxomycetales or true slime moulds. This error was corrected in 1880 by van Tieghem[2] but by this time both the Acrasiales and Myxomycetales had acquired the trivial name of 'slime moulds' and the confusion between the two groups of organisms persists to this day. It is not uncommon to find textbooks of biology which contain an illustration of a member of the Acrasiales with a corresponding text description of a Myxomycete.

To add to the confusion three other groups of organism are often associated with the slime moulds on the grounds that they are all to some extent colonial and slimy and have both fungal and animal characteristics. These five groups are classified as the Mycetozoa thus:

Labrinthulales
Plasmodiophorales
Myxomycetales (or Myxomycetes or Myxogastrales)
Acrasiales (or Acrasina)
Prosteliales (or Prostelida)

Bonner[3] in his monograph on the Acrasiales describes these other orders briefly and the whole question of the relationships amongst this group of primitive organisms has been discussed recently by Olive[4] who has suggested an extensive revision of the usual classification. In practice the only likely confusion is between the cellular and the true slime moulds since these are the two groups which have features of their life cycles which render them peculiarly attractive to biochemists—whose cavalier attitude to the subtleties of taxonomy are well known.

1.2 LIFE CYCLE OF THE CELLULAR SLIME MOULDS

1.2.1 Dictyostelium discoideum

The cellular slime moulds with which I am most familiar and which most biochemical studies have used is *Dictyostelium discoideum* which was isolated by Raper[5] in 1935. This species, like all cellular slime moulds, is a soil organism and is found in slightly acid soils where the bacterial flora is plentiful—typically forest litter. When in the vegetative, or feeding, state the organism is in the form of an amoeboid cell which is very similar in appearance to any other soil amoeba (Figure 1.1). The amoeba (sometimes called a myxamoeba to distinguish it from true amoebae) feeds on bacteria which are

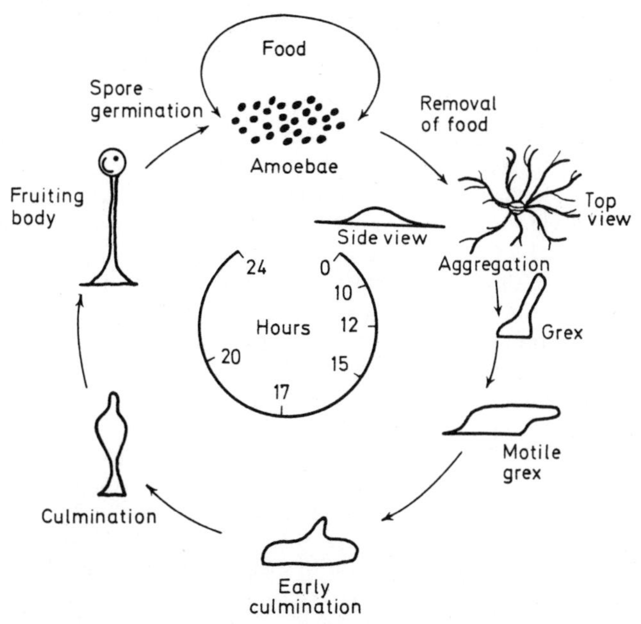

Figure 1.1 Life cycle of *D. discoideum*. The times refer to development on Millipore filters at 22°C using lightly buffered conditions[10]

engulfed by phagocytosis and digested in vacuoles which can be clearly seen in electron micrographs of thin sections of amoebae[6]. When about a thousand bacteria have been eaten, the amoeba divides by mitosis into two amoebae which in turn feed on more bacteria. This process continues indefinitely so long as the bacterial food supply continues. When the food supply is exhausted however, and if the amoebae are on a moist solid surface (as they would be on soil particles) they aggregate. This involves a considerable change in behaviour, for the feeding amoebae repel one another, and is mediated by a specific chemotactic system (*vide seq.*) involving a chemotactic agent called 'acrasin'[7] which is now believed to be 3′,5′-cyclic-AMP (or cAMP)[8]. Aggregation centres (Figure 1.1) secrete cAMP and non-feeding amoebae move up concentration gradients of cAMP thus forming aggregation streams (Figure 1.1). As the cells arrive at the aggregation centre they force those cells there originally off the substratum so forming a finger-like body called by Shaffer[9] a 'grex'. The grex is covered by a sheath of slimy material and can contain up to 40 000 individual cells. It is quite clear that at all times during aggregation, in the grex and during fruiting body formation, that the cells retain their individuality and do not fuse to form a syncitium or plasmodium. The production of a multinucleate plasmodium as a result of the fusion of amoeboid cells is characteristic of the Myxomycetales (the true or acellular moulds). As part of the confusion between these two groups of organism, the grex stage of the cellular slime moulds has been termed a 'pseudoplasmodium' but I will not adopt this terminology here.

The fate of the upright grex (Figure 1.1) depends on the environmental conditions[10]. If the underlying substratum has a low pH and a low ionic strength then the grex falls over on its side and, provided it is kept dark, will migrate indefinitely. The mode of migration has attracted much interest and is the subject of some controversy[11, 12]. It is established, however, that the slime sheath which covers the grex remains stationary on the substratum as the cells move through it and is left behind as a collapsed tube after the cells have passed. If the underlying substratum has a moderate ionic strength and a weak buffering capacity so that the pH increases during the course of development from slightly acid to near neutrality then the grex begins to migrate but, after a short time (which can be as little as an hour), the grex ceases to migrate, rounds up and begins fruiting body construction. If the underlying substratum has a high ionic strength and a weak buffering capacity then the migration stage can be omitted altogether and the upright grex will subside and immediately begin fruiting body construction. Newell, Telser and Sussman[10] have described the exact conditions necessary to elicit these three different modes of development in the NC-4 strain. I find that the exact conditions of ionic strength, buffering capacity and pH necessary vary from strain to strain and in the case of strain Ax-2 the conditions change depending on the nature of the medium used to grow the amoebae from which the grex was formed. However, the existence of these three modes of behaviour seems well established in all strains. The migrating grex moves along temperature, humidity and light gradients and there seems to be a clear ecological rationale for this behaviour since such responses would ensure that the grex moves towards the surface of the soil and away from the interstices where aggregation no doubt occurs. Light can act as a stimulus to fruiting body construction (although it is not essential) and the net effect of these complex responses on the part of the grex would be to lead to the construction of a fruiting body at the surface of the soil. The mechanism(s) whereby the grex receives and responds to these manifold physical changes in its environment are unknown and little investigated.

When the migrating grex stops moving and begins to construct a fruiting body, it has been shown using vital stains that it is the cells which were at the front of the grex which form the stalk and those which were at the back which form the spore cells[13] (*vide seq.*). As the cells which were at the front of the grex force their way through those that were at the back, they become vacuolated and encased in a cellulose cylinder up which the prespore cells move. Each stalk cell is in turn enclosed in a cellulose cell wall and thus it looks very much like a 'typical' plant cell. The turgor pressure of the stalk cells gives the newly formed stalk its rigidity but the stalk cells eventually become necrotic and non-viable.

The prespore cells synthesise a yellow, carotenoid[14] pigment as they differentiate and the final fruiting body consists of a yellow spore mass borne on a thin tapering cellulose stalk which is imbedded in a basal disc (Figure 1.1). The basal disc is composed of apparently undifferentiated cells derived from the back of the grex. The spores have a thick cell wall composed largely of cellulose and a dense, rather featureless cytoplasm. They stick to one another as a consequence of the synthesis of a mucopolysaccharide component which presumably forms the outermost layer of the spore wall[15].

The spores are resistant to many stresses which would be lethal to the amoebae (particularly desiccation) and remain viable after years of storage in a freeze-dried state. In the soil, raindrops would disrupt the spore mass and carry the spores to areas far removed from the site of construction of the fruiting body. In a favourable environment the spores germinate, each spore giving rise to one amoeba, and the life cycle is thus completed.

The life cycles of the other members of the genus *Dictyostelium* may be viewed (for convenience) as variations on that of *D. discoideum*.

1.2.2 D. mucoroides

This species differs from *D. discoideum* in that there is no clear separation of the migration phase from fruiting body construction. The grex formed from the initial aggregate immediately begins to form a stalk and then as the cell mass migrates it leaves behind it a stalk instead of a collapsed slime sheath (Figure 1.2). The tip shows a marked orientation towards light and will eventually turn upwards and the remaining cells then differentiate into spores similar in shape to those of *D. discoideum*. Unlike *D. discoideum* there is no basal disc at the bottom of the fruiting body.

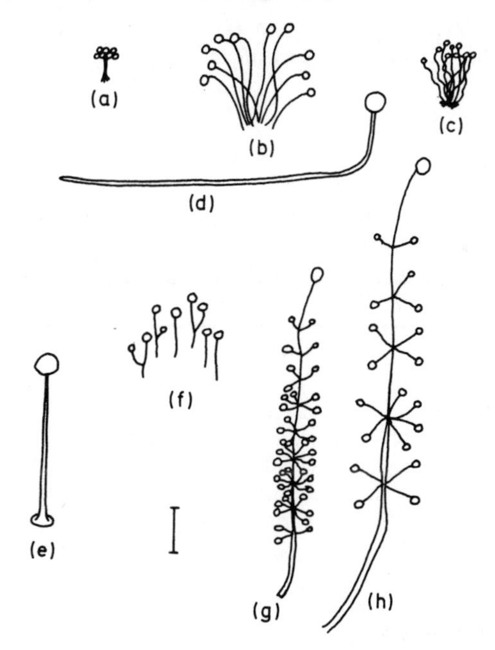

Figure 1.2 Semi-diagrammatic drawings of the fruiting bodies of the commoner species of the Acrasiales. The scale mark is approximately 1 mm. Redrawn from Ref. 119
(a) *D. polycephalum*, (b) *D. lacteum*, (c) *A. leptosomun*, (d) *D. mucoroides* and *D. purpureum*, (e) *D. discoideum*, (f) *D. minutum*, (g) *P. pallidum*, (h) *P. violaceum*

Some strains of *D. mucoroides* will form microcysts[16]. These are formed by single vegetative cells which round up and produce a thin protective wall of (probably) cellulose. In addition, the same authors describe strains which produce a macrocyst stage which consists of a condensed aggregate of cells derived from vegetative amoebae which is surrounded by a thickened wall. These various forms of resistant cyst are produced by amoebae when they are removed from food and are kept under water. Fruiting bodies only form on a solid surface and so the formation of these cysts presumably represents an adaptation to wet or aquatic environments. It may be significant that this is the only species that I have ever isolated from Leicestershire, a county well known for its damp and fogs! Nothing is known of the biochemical mechanism(s) which underly cyst formation or how the choice between the pathways of differentiation leading to the alternative resistant forms of spore, microcyst or macrocyst is made. Filosa and Dengler[17] have recently described the ultrastructure of the macrocyst and some of the stages preceding its formation.

1.2.3 D. purpureum

This species appears to be identical to *D. mucoroides* except that the spore mass is deep purple in colour instead of yellow.

1.2.4 D. lacteum

The characteristic feature of this species is that it has spherical instead of cylindrical spores. There appears to be no migrating grex stage, each aggregate breaking up into a number of little mounds each of which produces a small fruiting body.

1.2.5 D. minutum

As its name suggests this species also forms very small fruiting bodies. Again there appears to be no true migration stage and each aggregate gives rise to one or a number of very small fruiting bodies.

1.2.6 D. polycephalum

This species is unusual in several respects. When the amoebae aggregate, they do not form streams so much as sheets of cells which converge about the aggregation centre and this gives rise to a number of migrating grexes. The grexes are long and thin and show none of the usual tropisms. When they cease to move, the grex cells pile up into a mass from which numerous stalks arise which are cemented together for most of their length. The resulting fruiting body consists of several small fruiting bodies like those of *D. discoideum* (but without a basal disc) but bunched together (Figure 1.2).

1.2.7 Other cellular slime moulds

Acytostelium species are characterised by having spherical spores and a fruiting body with an acellular stalk. Hohl, Hamamoto and Hemmes[18] have described the detailed cytological structure of the cells.

Polysphodylium species characteristically have a whorl of branches which project from the main stem of the fruiting body—each branch being similar to a small *D. discoideum* fruiting body, except that the spore masses are either white (*P. pallidum*) or purple (*P. violaceum*). Like *D. mucoroides* the aggregate begins to form a stalk before the migration phase and during the migration the stalk is left behind (Figure 1.2).

Polysphondylium species (like *D. mucoroides*) can also form cyst-like resistant forms when incubated in shake cultures.

For a detailed account of these various organisms Bonner's[3] invaluable monograph should be consulted.

1.3 GENERAL BIOLOGY OF THE CELLULAR SLIME MOULDS

The stimulus to differentiate is in all cases removal of food and thus the formation of spores is in this, as in other groups of organism, a method of ensuring survival during unfavourable periods. The fact that this occurs after a long sequence of complex intercellular interactions is presumably related to the difficulty of dispersing the spores in a damp, but not aqueous, environment like soil. The world-wide distribution of the cellular slime moulds (Figure 1.3), and the fact that in especially favourable environments such as the Caribbean area[19] speciation seems to have run riot, testifies to the success of this strategy of dispersal.

In many organisms a process of genetic recombination is associated with the formation of the resistant stage of the life cycle. In the case of the Myxomycetales, for example, the amoebae are haploid and the plasmodium (which results from the fusion of amoebae of different 'mating' types) is diploid. When the plasmodium differentiates, meiosis precedes spore formation and the spores are thus haploid and represent genetic recombinants of the original haploid strains whose fusion resulted in the plasmodium. In the case of the Acrasiales the situation is quite different. Fruiting body formation and thus differentiation can occur within clones of amoebae and meiosis has never been reported. Instead there appears to be a 'para-sexual' system of genetic recombination[20, 21] involving processes analogous to those known to occur in *Aspergillus niger* and tissue culture cells (Figure 1.4).

Morphological[22, 23], drug resistant[24], temperature sensitive[25] and temporal[26, 27] mutants have been described but the formal genetic analysis[28] of *D. discoideum* has made slow progress. The lack of a simple system of genetic analysis (such as is now available for *E. coli K*12) represents perhaps the largest single technical obstacle to continued rapid progress in our understanding of the mechanisms of cell differentiation in this organism.

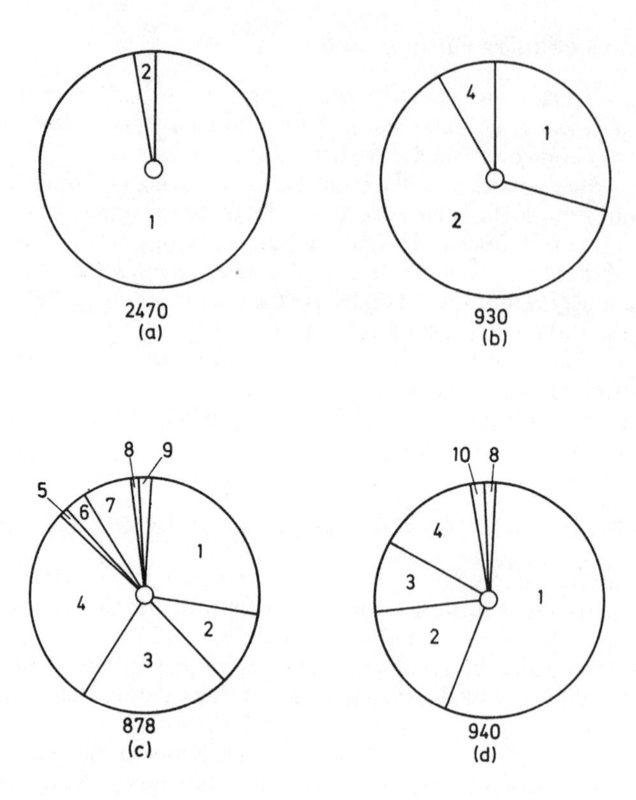

Figure 1.3 Distribution of the cellular slime moulds in different forest soils. The area of the wedge represents the frequency with which that species occurs in each location and the number represents the total number of clones of all types obtained from 1 g of soil[119-121]

1. *D. mucoroides*
2. *P. violaceum*
3. *P. pallidum*
4. *D. minutum*
5. *D. polycephalum*
6. *D. discoideum*
7. *D. lacteum*
8. *D. purpurem*
9. *A. leptosomum*
10. *D. mucoroides* (variant)

(a) Oak forest near Utrecht, Holland.
(b) Mixed deciduous forest near Trieste, Italy.
(c) Oak forest, Pine Buff, Wisconsin, U.S.A.
(d) Sub-tropical evergreen forest, Uganda.

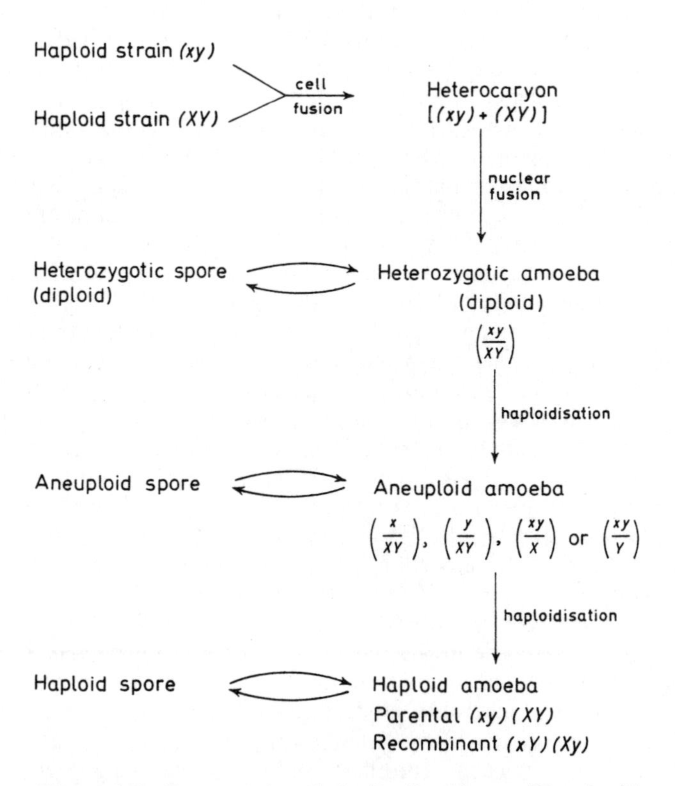

Figure 1.4 Para-sexual cycle in *D. discoideum*. x/X and y/Y represent two unlinked genes. The genotypes of the various kinds of nuclei are enclosed in parentheses[20]

1.4 METABOLISM OF GROWING AMOEBAE

1.4.1 Growth on bacteria

When an amoeba has eaten approximately 1000 bacteria, it will divide by binary fission and thus the problems involved with the growth of amoebae of *D. discoideum* are really associated with the growth of bacteria. Either the bacteria may be grown in media, harvested and then shaken in buffer (0.0167 M phosphate, pH 6.0) with amoebae[29], or bacteria and amoebae may be grown simultaneously in two-membered culture in (or on) a rich nutrient medium. The first method is rather cumbersome but provides the most reproducible growth situations whilst the second is technically far simpler but theoretically complicated since a predator/prey relationship exists between bacteria and the amoebae. After his extensive and detailed study of two-membered culture media Raper[30] concluded 'the following factors are of primary importance in governing the growth and development of *D. discoideum* (1) the composition of the culture medium, (2) the fermentations of the associated bacteria, (3) the pH and (4) the physical character of the bacterial growth.' These conclusions have stood the test of time and of the factors identified by Raper that of pH seems the most critical. Thus, in order to obtain a high yield of amoebae it is necessary to have luxuriant bacterial growth and thus a rich nutrient medium. However, many bacteria grow so fast on rich media that they will carry out fermentations even on the surface of Petri dishes and thus it is important to choose a balance of sugar, peptone and buffer that will enable the bacteria to grow well but not alter the pH far from the optimal 6.0 and not involve too high an ionic strength[10]. One simple solution to these problems is to use *Aerobacter aerogenes* (now more correctly designated *Klebsiella aerogenes*[31]) which produces rather few acid fermentation products and follow the procedures described by Sussman[32] using SM-agar medium which yield 1–10 $\times$ 10^8 amoebae per standard Petri dish.

Most workers have regarded the complexities outlined by Raper[30] as obstacles to be overcome in the pursuit of high and reproducible yields of amoebae but the two-membered culture system is a very convenient and experimentally very accessible predator/prey situation. The amoebae will eat a wide variety of bacterial species—only finding these with extensive extracellular polysaccharide capsules totally inedible—and there is some preliminary evidence that different species of slime mould have different food preferences[33]. There is also evidence from Raper's pioneer study that the interaction of bacteria and amoebae can be profoundly affected by the presence of a third, fungal, associate such as an *Aspergillus*. It would seem, therefore, that many of the ecological problems posed by higher organisms could, in principle, be modelled using cellular slime mould amoebae and various other microbial associates under controlled and highly reproducible conditions on agar plates.

Under optimal conditions doubling times of 3 h have been reported for the growth of *D. discoideum* on *K. aerogenes* at 22 °C. This implies that the amoebae capture, ingest and digest approximately five bacteria per min—an impressive performance. As might be expected therefore, these organisms have a very high level of a variety of digestive enzymes[34, 35] and large and extensive food vacuoles can be seen in electron micrographs of amoebae. Little is known of the way in which the amoebae ingest and capture their prey but recently Bonner and his colleagues[36] have shown that the feeding amoebae are very sensitive to folic acid and related pterins. They have suggested that the amoebae sense the presence of bacteria by detecting and following concentration gradients of folic acid and/or its degradation products. This behaviour resembles that shown by starving amoebae to cAMP (*vide seq.*) and Bonner suggests that as the amoebae become sensitive to cAMP gradients they lose their sensitivity to pterin gradients.

1.4.2 Growth in axenic culture

P. pallidum was the first cellular slime mould to be cultured axenically[37] and a fully defined medium has been reported for this strain[38]. Sussman[39] was also the first to obtain a strain of *D. discoideum* in axenic culture which he designated Ax-l. The medium used was rather complex and was soon simplified by Schwalb and Roth[40] and Watts and Ashworth[41]. The latter authors designated their strain Ax-2 and it is now maintained by the American Type Culture Collection (ATCC 24397). This strain is naturally the one with which I am most familiar and has the great advantage that it can be grown

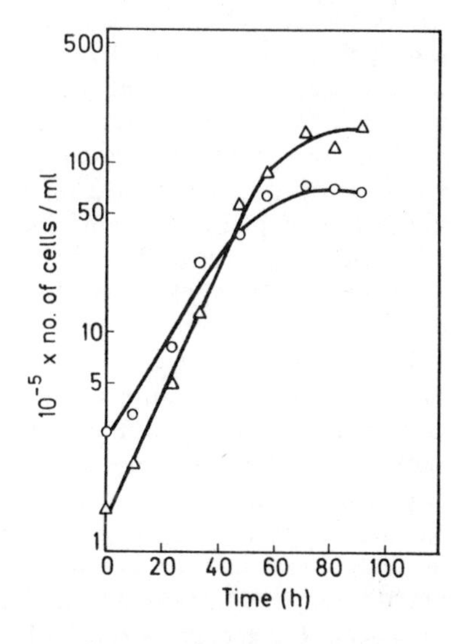

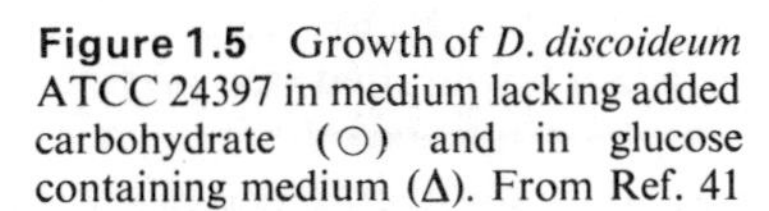
Figure 1.5 Growth of *D. discoideum* ATCC 24397 in medium lacking added carbohydrate (○) and in glucose containing medium (Δ). From Ref. 41

in medium (liquid or solid) in the absence of added carbohydrates but can utilise a number of different carbohydrates (such as glucose or maltose) if these are provided (Figure 1.5). *D. discoideum* has not yet been grown in a fully defined medium; the medium I routinely use for the growth of ATCC 24397 contains yeast extract, peptone and salts.

D. discoideum ATCC 24397 will also grow on bacteria and there are a number of differences between amoebae grown axenically in the presence or absence of added carbohydrate and those grown on bacteria (Table 1.1).

Table 1.1 Effect of changes in growth media on growth rate and composition of amoebae *D. discoideum* ATCC 24397

		Amount per 10^8 *amoebae harvested in exponential growth*			
Growth conditions	*Doubling time* (h)	DNA* (μg)	RNA† (μg *of ribose*)	*Protein*† (mg)	*Total carbohydrate*† (mg *glucose equiv*)
Axenic media	10	17	320	11	0.3
Axenic media + glucose (86 mM)	8	17	330	11	2.0
Bacteria	3–4	36	250	7	0.5

* data from 43
† data from 42

The difference in DNA content of axenically and bacterially grown amoebae was initially[42] thought to be due to the presence in the latter of bacterial DNA. However, later work has shown[43] that this can only account for 10–15% of the DNA content of bacterially grown cells and thus these cells must have twice the number of slime mould genome equivalents of DNA characteristic of axenically grown amoebae. The explanation advanced for this observation[43], namely that axenically grown cells are predominantly in the G1 phase of the cell cycle and that bacterially grown cells are predominantly in the G2 phase, has not been verified by any direct experimental test. If this explanation for the discrepancy in DNA contents is correct then, since bacterially and axenically grown amoebae differentiate in a similar fashion and retain their characteristic differences in DNA contents, the implication is that growth can cease and differentiation can be initiated at any point in the G1 and G2 phases of the cell cycle. This conclusion is very difficult to explain mechanistically and thus a direct test of the original explanation for the differences in DNA content is badly needed. So far, however, all attempts to synchronise the growth of populations of amoebae have failed.

The other differences in Table 1.1 present fewer theoretical difficulties. The differences in carbohydrate content are particularly interesting since, during the differentiation of the amoebae, there are dramatic changes in carbohydrate metabolism. The carbohydrate content of the amoebae can also be changed by altering the point in the growth cycle at which the amoebae are harvested—amoebae harvested in the stationary phase after growth in medium containing 86 mM glucose can have as much as 6 mg glycogen per 10^8 cells[44]. Correlated with these changes in chemical composition are changes in the specific activity of a number of enzymes[45] and changes in the physiological behaviour[42] of the amoebae.

1.5 AGGREGATION AND THE ROLE OF CYCLIC-AMP

When amoebae of the cellular slime moulds have exhausted their bacterial food supply, and if they find themselves on a moist, solid surface they aggregate. This process consists of the amoebae gathering together about a central point and there forming a finger-like projection—the grex—consisting of many thousands of cells encased in a slime sheath. The details of the process are different in different species and even in *D. discoideum* there exists a tremendous variation in the gross appearance of aggregating streams. Typically these consist of a centre on which converge branched streams which in turn consist of many hundreds of cells moving in a concerted and often pulsatile fashion[46]. Shaffer[9] has pointed out that whorls, stipples, flocks and clouds of cells (all of which look very different) are quite normally found amongst populations of aggregating cells. It seems as if the detailed form which an aggregating population of cells adopts is very sensitive to the exact nature of the surface and its temperature, humidity, ionic strength, etc. Little is known of how these environmental factors affect the morphogenesis. The process of aggregation is, however, known to be essentially a process of chemotaxis. The aggregating centre and its satellite streams secrete a substance ('acrasin') to which starving amoebae respond. In the case of *D. discoideum* this acrasin is known to be 3′,5′-cyclic-AMP[8] but, equally certainly, this is known not to be the acrasin for a number of other species such as *P. violaceum* and *P. pallidum* and a number of the Dictyosteliaceae (such as *D. minutum*, *D. lacteum* and *D. polycephalum*)[47]. The chemical nature of the acrasins for these species is unknown. In order for amoebae to respond to a cAMP signal, theory suggests that there must be some means of destroying the signalling molecule. Such an 'acrasinase' was first detected by Shaffer[48] and subsequently identified with an extracellular cAMP phosphodiesterase[49]. At least two extracellular phosphodiesterase activities have been reported to be excreted by aggregating amoebae, one with a high K_m for cAMP[50] and one with a low K_m[51]. It is possible that these two enzymes may be interconvertible forms of the same protein species[52] but it is clear from the cyclic-AMP concentration at which aggregation occurs (approximately 0.5 mM)[53] that only the low K_m enzyme can be physiologically significant. The situation is further complicated by the finding[54] that the amoebae also excrete a protein inhibitor of the extracellular, low K_m phosphodiesterase. This provides another possible mechanism whereby the extracellular cAMP concentration may be regulated. Phosphodiesterase activity has also been implicated[55] in the cAMP receptor sites which must occur on the cell membrane. Particle preparations made from such membranes have a phosphodiesterase activity clearly different from either of the two soluble, extracellular activities and which increases markedly in activity during the aggregation stage and then decreases. The existence of these numerous phosphodiesterases, which change markedly in activity at different times during the life cycle, is in marked contrast to the adenyl cyclase activity which remains constant throughout growth and development[56]. However, recent direct measurements[53] of the steady state level of cAMP during the growth and development phases of the life cycle have suggested that the *in vivo* activity of adenyl cyclase changes markedly during the life cycle even though the amount assayed *in vitro* remains apparently unchanging. The cAMP gradient to which the amoebae respond is thus the product of a complex interaction between the strength of the cyclic-AMP signal, the distance between the cell and the source of the signal and the net phosphodiesterase activity of the environment. Such complexities have attracted the attention of a number of mathematically inclined biologists and satisfactory explanations for many of Schaffer's observations[9] can now be offered[57,58,59].

Feeding amoebae do not respond to cAMP gradients as readily as do non-feeding amoebae. The reason for this is unknown; it might be that cAMP receptor sites not present during the feeding phase need to be elaborated, but it does not seem to be starvation, *per se*, which induces sensitivity to cAMP. Thus cells containing high or low levels of stored glycogen[41] aggregate in the same time and in a similar manner. Aggregating cells do seem, however, to differ from growing cells in their capacity to retain amino acids and Lee[60] has suggested that the resulting sudden efflux of amino acids at the end of growth triggers the aggregation response. The connection between the efflux of amino acids, the activation of adenyl cyclase activity and the changes in the net phosphodiesterase activity remain to be worked out, however.

Bonner[61] has shown that at very high concentrations (10^{-3} M) cAMP can induce isolated and unaggregated amoebae to form stalk-like cells. This observation suggests that cAMP might play a role in cell differentiation in addition to the role it plays in cell aggregation. This suggestion has received support from direct measurements of the extracellular cAMP concentration during differentiation of the amoebae which have shown[53] that two peaks in cAMP concentration occur, one at aggregation and the other during fruiting body construction.

1.6 CELL DIFFERENTIATION

1.6.1 Methods of approach

Two major problems are posed by the life cycle of *D. discoideum* for those interested in cell differentiation. These are, first, the nature of the events which lead any one amoeba to become either a spore or a stalk cell and, this choice having been made, the nature of the largely intracellular events which then lead to that cell's transformation or differentiation into a spore or stalk cell. The choice which each amoeba has to make (Figure 1.6) is made when that

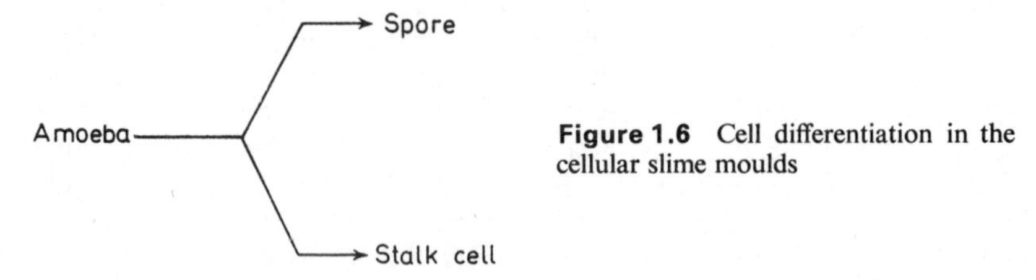

Figure 1.6 Cell differentiation in the cellular slime moulds

cell is a member of a cell population and thus must depend on cell–cell interactions in some way. In this respect the cell differentiations characteristic of the cellular slime moulds resemble, and are a model for, the differentiation events characteristic of higher organisms. The changes in the pattern of macromolecular synthesis which lead to the expression of that choice, however, probably involve regulatory devices (e.g. enzyme induction, repression, activation and inhibition) which have been well characterised in simpler organisms. In general terms, therefore, the attempts to account for the intracellular events have been rather more successful than the attempts to explain the nature of the cell–cell interactions.

However, those who work with the cellular slime moulds are divided as to whether it is the synthesis of specific developmental proteins which is the main point of intracellular control during differentiation or whether control of the activity of existing enzymes by alterations in the concentrations of various effectors[63] is the major control point. These viewpoints need not be mutually exclusive but in practice work has tended to become polarised in one direction or the other and reviews have tended to be rather polemical. Thus those authors impressed by the apparent analogies between enzyme induction in *E. coli K*12 and differentiation in *D. discoideum*[62,64,65] have stressed the importance of controls over transcription and translation which have been discounted by those[66,67] more impressed with the ability of computer programmes to simulate developmental events without the necessary invocation of such regulatory mechanisms. In part this situation has been due to the fact that until very recently most work in the field concentrated on the analysis and description of events which occur in bacterially-grown amoebae undergoing a highly synchronous and unstressed developmental programme. Differentiation is a complex process and thus it has been relatively easy, by judicious choice of experimental design, to find what one feels should be found. Recently, however, ways have been found to alter the developmental programme[10], stress the cells at intermediate stages[68] and alter the metabolic characteristics of the differentiating cells[42]. These new procedures should enable some of the theories developed on the basis of essentially descriptive studies to be put to experimental test and thus resolve some of disputes of the past few years.

1.6.2 Changes in cell composition

During differentiation of the amoebae there is little or no change in the DNA content of the cells[43]. There is, however, a marked (about 50%) decrease in the cell dry weight and cell volume and this is accounted for by the decrease in cellular content of RNA and, especially, protein[69]. If the amoebae are grown on bacteria there is little net change in the carbohydrate content of the cells[69] although there is a marked qualitative change in the nature of the carbohydrates present. Glycogen accounts for 95% or more of the carbohydrate present in bacterially grown amoebae[70] whereas little glycogen is present in the mature fruiting bodies. Instead the stalk cells contain largely cellulose[71] and the spores cellulose, a mucopolysaccharide[15], a little free glucose[69] and trehalose[72]. Development occurs in the absence of exogenous nutrients and so the deduction from these observations was that the energy needed for development was supplied by the oxidation of amino acids derived from protein degradation and, to a lesser extent, from the RNA which is also catabolised. The energy so provided was, it was suggested, used for the conversion of the amoebal glycogen into those carbohydrates characteristic of the fruiting bodies and, of course, for the maintenance of the vital functions of the differentiating cell. In apparent agreement with this idea were studies which showed that there was little glycolysis[66] or gluconeogenesis[73,74] and computer simulations which were based on these assumptions[66] 'worked' satisfactorily. However, it is important to know whether this description of the basic metabolic events during differentiation is a fundamental feature of the developmental programme or merely a trivial consequence of the fact that bacterially-grown amoebae always contain approximately the same amount of carbohydrate as do mature fruiting bodies (about 500 μg glucose equivalents per 10^8 cells). Amoebae can be grown in axenic media such that they contain anywhere between 46 and 5560 μg glucose equivalents of glycogen per 10^8 amoebae[75] and studies of the change in the macromolecular composition of such cells during their differentiation has shown that the basic carbohydrate metabolism of differentiating cells depends markedly on their glycogen concentration. Thus amoebae containing 46 μg glucose equivalents of glycogen per 10^8 amoebae form fruiting bodies with 550 μg glucose equivalents of carbohydrate material. In these cells there must, therefore, be extensive gluconeogenesis and isotope studies[76] have shown that amino acids provide the necessary carbon atoms. Amoebae containing between 1500 and 3620 μg glucose equivalents of glycogen per 10^8 amoebae form fruiting bodies with 650–750 μg glucose equivalents of carbohydrate material and so in these cases there must clearly be extensive glycolysis. Isotope experiments have confirmed this[76] and have shown that most of the amoebal glycogen is metabolised to carbon dioxide during differentiation. Amoebae with very high initial glycogen concentrations (5560 μg glucose equivalents) do not seem to be able to metabolise the excess glycogen at a rate sufficient to degrade it before fruiting body formation occurs and they form fruiting bodies with a significantly higher total carbohydrate content (1500 μg glucose equivalents) and glycogen content (500 μg glucose equivalents) than do other amoebae[75].

Studies of the rate of excretion of ammonia from differentiating amoebae have shown that only in cells containing a very high glycogen content is there any considerable decrease in the extent or rate of ammonia excretion. Assuming that ammonia excretion is a valid measure of the rate of amino acid oxidation this implies that during differentiation, unlike the situation in the growth phase[42], there is no marked sparing of amino acid oxidation by glycolysis. This conclusion is confirmed by studies of the rate and extent of protein loss during differentiation. Amoebae containing 71 μg glucose equivalents of glycogen lose 68% of their protein content during differentiation at a rate of 0.25 mg protein h^{-1} per 10^8 cells whereas amoebae containing 5560 μg glucose equivalents of glycogen initially lose 62% of their protein content at a rate of 0.21 mg protein h^{-1} per 10^8 cells during the same time[76]. Even in those cases where the high glycogen content of the cells leads to a decrease in the rate of ammonia excretion there is no alteration in the rate or extent of protein breakdown. The implication of these findings is that proteins are discarded during differentiation for reasons other than the necessity for energy provision and, as Sussman and Lovgren pointed out[77], the loss of an enzymic activity may be as important an event during cell differentiation as is the acquisition of an enzyme activity.

Similarly, we have found[76] that the amount and extent of RNA degradation during the differentiation of the amoebae is unaffected by their glycogen content which suggests, again, that during differentiation the cells discard RNA molecules for reasons other than, or in addition to, the necessity for provision of metabolic energy.

A comparison of the relative amounts of cellulose, mucopolysaccharide, glucose and trehalose formed by cells derived from amoebae containing, initially, different glycogen contents has also been very instructive[76]. Amoebae containing 5590 μg glucose equivalents of glycogen synthesise 30% more cell wall polysaccharide, 200% more mucopolysaccharide, 300% more free glucose and 400% more trehalose than do amoebae containing less than 300 μg glucose equivalents of glycogen during cell differentiation. Further, during the differentiation processes there are marked differences between these two populations of cells in the pool sizes of a number of low molecular weight metabolic intermediates such as glucose-6-phosphate (and thus presumably glucose-1-phosphate since the cells contain a very active phosphoglucomutase[74]) and UDP-glucose[78]. Thus although a qualitative description of differentiation can be given in terms of changes in the amounts of various carbohydrates formed (either transiently or not) it seems that these changes must be a consequence and not a cause of cell differentiation. It is difficult to see, for example, how the timing and spatial organisation of developmental events in *D. discoideum* can be determined by the utilisation of a specific amount of glycogen as has been suggested[79] or how the changes in the pool sizes of certain metabolites such as glucose-1-phosphate or UDP-glucose can be critical variables[65] since the same morphogenetic stage can be formed from cells with very different values for these parameters.

1.6.3 Changes in enzyme composition

It is widely held that the problems posed by cell differentiation are basically problems concerned with the way in which different genes can be active in the same cell at different times. Thus in its most simple form this differential gene activation theory assumes that if the right genes are activated at the right times then the cell metabolism will inevitably change appropriately. Seen in this light cell differentiation is the result of temporal changes in the specificity and/or activity of DNA-primed RNA polymerase(s) coupled with changes in the nature of the organisation of the DNA template itself. Much of the work with *D. discoideum* has been, or can be, interpreted in these terms.

If differential gene activity is, in fact, of importance then the amoebae should contain three sets of genes, one set concerned with events peculiar to the growth (amoebal) phase, one set concerned with events peculiar to the differentiation phase and the third set concerned with controlling events which must occur in all stages of the life cycle. Loomis[25] has shown that following chemical mutagenesis of the amoebae three classes of temperature sensitive (TS) mutant can be obtained, in agreement with this argument (Table 1.2). Curiously, of the nine temperature sensitive strains reported five were GTS (growth TS) (Table 1.2), two DTS (development TS) and two TS. These numbers are, of course, too low for serious analysis but I have also found it much easier to isolate GTS mutants than any other class. This is, I think, a little

Table 1.2 Temperature sensitive mutants of D. discoideum (From Loomis[25], by courtesy of *J. Bacteriol.*)

*Strain**	*Growth at* 27°C	*Development at* 27°C
Wild type	(+)	(+)
GTS	(−)	(+)
DTS	(+)	(−)
TS	(−)	(−)

* All strains grow and develop normally at 22°C

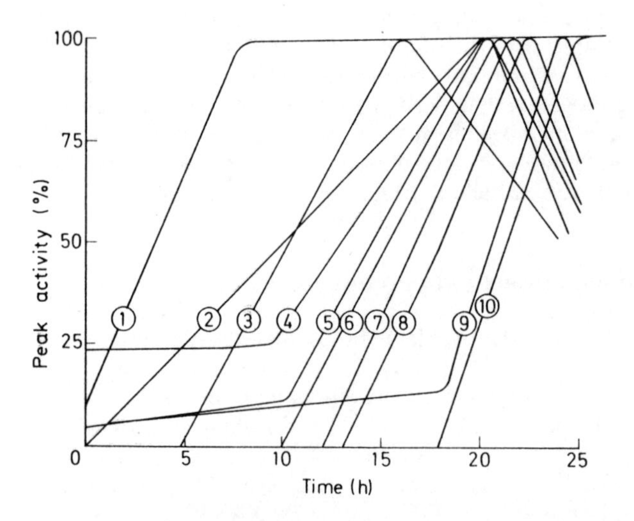

Figure 1.7 The appearance and disappearance of enzyme activities during the development of *D. discoideum* amoebae grown on bacteria. (1) *β-N*-acetyglucosaminidase (200)[81], (2) *α*-mannosidase (36)[82], (3) trehalose 6-phosphate synthase (18)[83], (4) tyrosine transaminase (45)[84], (5) UDP-glucose pyrophosphorylase (250)[85], (6) UDP galactose: polysaccharide transferase[86], (7) UDP-galactose epimerase (30)[87], (8) glycogen phosphorylase (24)[88], (9) alkaline phosphatase (40)[89], (10) *β*-glucosidase-2 (25)[90]. Peak specific activities have been normalised to 100 and the true value in nmoles substrate utilised (or product formed) min^{-1} mg^{-1} protein is given in parentheses in the key above where possible

unexpected since if all genes have an equal probability of being mutated the frequency of the different classes of mutant should reflect the relative abundances of the three classes of genes. *A priori* I would expect these to be in the order TS>GTS≃DTS. Hence it is possible that the mutagenic procedures adopted were such that one class of mutant (GTS) was formed preferentially. If this can be confirmed further analysis of such experiments might give valuable information about the nature of the organisation of different classes of genes in the amoebae and/or the nature of the action of different mutagens.

A mutant is usually temperature sensitive because it makes a protein molecule whose amino acid sequence differs from the wild type in such a way that its thermostability is affected. Loomis's work thus implies that the three classes of temperature sensitive mutant correspond to three classes of protein (enzyme). The appearance of novel enzyme activities (or enhanced amounts of previously present activities) during the differentiation phase has been well documented (Figure 1.7) but little is known about those enzymes which disappear during differentiation or which are present at all stages of the life cycle.

There has been considerable discussion in the literature about whether data such as those shown in Figure 1.7 can be interpreted as implying the differential synthesis of the appropriate enzyme protein and thus whether

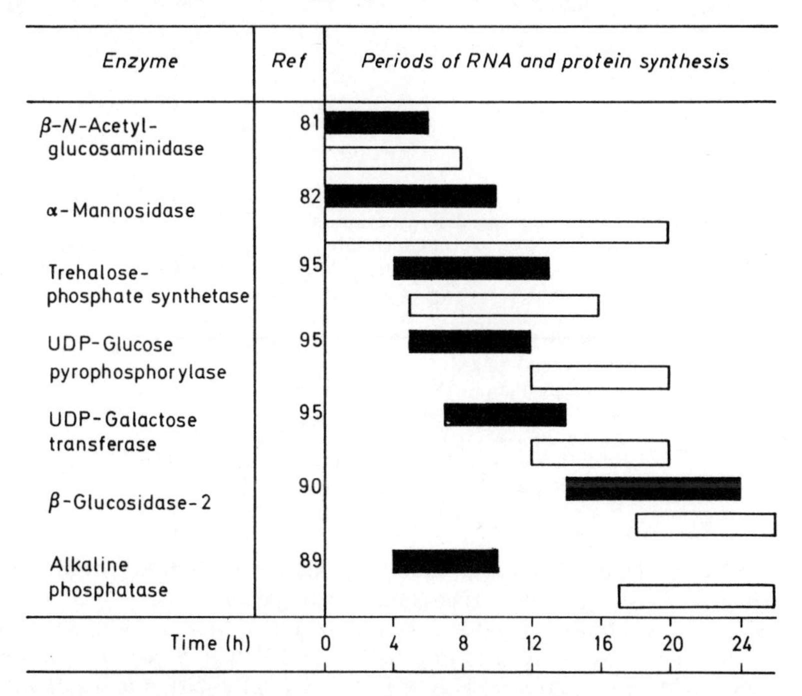

Figure 1.8 Periods of protein and RNA synthesis necessary for the appearance of various enzyme activities during cell differentiation in *D. discoideum.* The RNA synthetic period (■) was deduced by addition of actinomycin D at various times and measurement of the amount of enzyme subsequently formed. The protein synthetic period (□) was deduced from similar experiments using cycloheximide. (From Ashworth[65], by courtesy of Cambridge University Press)

studies of the effects of various drugs on the appearance/disappearance of enzyme activities can be interpreted in terms of the activity of the appropriate gene(s). There seems little point in repeating these, by now somewhat familiar, arguments and the reader is referred to the articles by Gustafson and Wright[66] and Wright[67] for a summary of the view that increases in enzyme specific activity cannot be taken to imply increases in the rate of differential protein synthesis and those of Sussman and Sussman[62], Newell[64] and Ashworth[65] who have taken the view that they can. Work published since these reviews

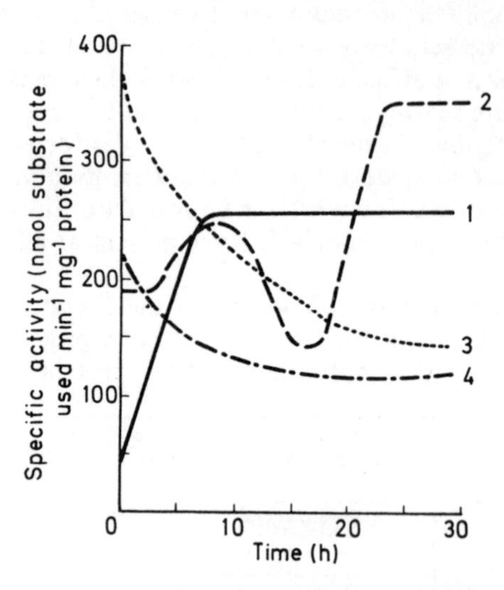

Figure 1.9 *β-N*-acetylglucosaminidase activities during the differentiation of amoebae of *D. discoideum* grown under different conditions but allowed to differentiate under identical conditions. (1) grown on bacteria, (2) grown on either axenic medium alone or axenic medium + 86 mM glucose and harvested when in the exponential phase of growth, (3) grown on axenic medium + 86 mM glucose and harvested when in the stationary phase of growth, (4) grown on axenic medium alone and harvested when in the stationary phase of growth. (From Ashworth and Wiener[122], by courtesy of North Holland Pub. Co.)

were written that bears on this question has included the demonstration by Franke and Sussman[91] that UDP-glucose pyrophosphorylase (enzyme 5, Figure 1.7) is synthesised *de novo* during differentiation and the apparently equally compelling demonstration by Killick and Wright[92] that trehalose 6-phosphate synthase (enzyme 3, Figure 1.7) is not. In addition it seems that the increase in the specific activity of *β-N*-acetylglucosaminidase (enzyme 1, Figure 1.7) is due, not to an increase in the differential rate of synthesis of this enzyme, but to its protection from proteolysis[93] and the description of *β-N*-acetylglucosaminidase and *α*-mannosidase (enzyme 2, Figure 1.7) as 'developmentally programmed' enzymes has been questioned[94]. On the other hand, qualitative changes in the pattern of mRNA synthesis during differentiation have been reported[80] and this, of course, implies that some of the changes in activity reported in Figure 1.7 represent changes in the differential rate of synthesis.

At the moment it seems that each enzyme should be considered separately in the expectation that different mechanisms may underly the changes in specific activity in Figure 1.7. Studies of the effect of actinomycin D on the enzyme synthetic pattern reinforce this conclusion. These studies are summarised in Figure 1.8. It is clear that the differences in the timing of what are referred to as the RNA and the protein synthetic periods imply differences in the mechanism whereby each alteration in specific activity is achieved. These differences might be in the way in which different mRNA molecules are translated (for which a specific suggestion has been made[96]) or they may refer, for example, to the synthesis of a specific ribosomal component or of an RNA molecule which codes for a specific protein activator of that enzyme rather than the synthesis of the mRNA for that enzyme itself.

The constancy and reproducibility of the changes in enzyme activity shown in Figure 1.7 is remarkable and has led to the idea that there exists a rigid sequence of changes in transcriptive events during differentiation which might be called a 'developmental programme'. However, just as in the case of the apparently similarly invariant pattern of changes in metabolism it is not clear whether this constancy is an inherent feature of the programme or whether it merely reflects the fact that amoebae grown on bacteria have a constant composition and therefore always need to make similar alterations in their enzyme composition during differentiation. Growth of the amoebae in different media has meant that the enzyme composition of the amoebae can be changed[45] and studies of the subsequent differentiation of such amoebae have shown that in the case of some enzyme activities very marked changes in the nature of the alterations occurring in specific activity during differentiation can be induced[94] (Figure 1.9). Thus it seems difficult to argue, for example, that part of the 'developmental programme' consists of the necessary transcription of the structural gene for β-*N*-acetylglucosaminidase[81].

1.6.4 Control of cell differentiation

Thus it seems that neither of the two ways of accounting for the biochemical changes which occur during the differentiation of the amoebae and which have been considered above are wholly satisfactory. Clearly there must be interactions between changes in metabolite pools and changes in the pattern of enzyme synthesis but formal attempts to accommodate these interactions have not been very successful[97]. It is possible that measurements of specific activity of enzymes *in vitro* give a wholly misleading estimate of relative enzyme concentrations and it is the latter which are regulated and of importance[65] and, of course, it is probable that metabolites are compartmentalised and thus measurements of the changes in the concentration of metabolites, which are expressed on a 'per cell' basis might also be grossly misleading. There is evidence, in fact, that the glycogen pool is compartmentalised[75] and that different parts of the glycogen pool might be undergoing very different reactions at the same time. However, although future work may refine our ideas about how regulation of enzyme activity and enzyme synthesis interact, I doubt that this will resolve the present confusion between causes and effects. These confusions might well be caused, I think, by our ignorance about how the controls which act at the cellular level interact with the intracellular controls which we have considered, thus far, as separate and distinct. Thus although it is true that during their differentiation slime mould cells make, say, trehalose and a mucopolysaccharide, it is also true that they only do this at an appropriate time and when in an appropriate position with respect to other cells. It is quite conceivable that the methods which the cells use to measure their position and/or the time[98] determine the way in which various subsequent biochemical events are controlled. Unfortunately we know very little about how cells measure position and time and certainly we are far from understanding how such measurements might be coupled with temporal and spatial changes in biochemical activity. However, several studies which have explored this area in *D. discoideum* have produced most unexpected results and suggest that this will, in the near future, be a most rewarding area of research.

1.7 CELL–CELL INTERACTIONS

1.7.1 Pattern formation

Pattern formation is concerned with the way in which differentiation is ordered in space and involves consideration of the mechanisms that ensure that cells have the correct spatial, temporal and proportional relationships to each other. Although manifestation of the cellular pattern—the actual appearance of stalk and spore cells—does not occur until fruiting body formation there is much evidence to suggest that the pattern arises during the grex stage of the life cycle. The grex clearly has a front (and the tip cells, in particular, look very different from the remainder under a microscope) and a back. Thus, if, for example, a grex is moving towards a light source which is suddenly moved through 180 degrees the grex also moves through 180 degrees with the tip cells leading. The classical experiments of Raper[13] suggested that the pattern of the grex was related to the pattern of the fruiting body. Thus when the tips of a normal, colourless grex and a red coloured grex (formed from amoebae grown on *Serratia marcesens*) were exchanged, Raper found that the cells of the front third of the grex became the stalk cells of the fruiting body and the cells of the back two-thirds of the grex became the spore cells. Since then a number of differences (summarised in Ref. 99) have been reported between front (pre-stalk) cells and back (pre-spore) cells. However, it is quite clear that these differences are readily reversible since if the slime sheath of the grex is broken whilst the grex is moving through a bacterial colony, the grex cells revert to the feeding, vegetative, stage[13]. Further, if the grex is cut into pieces those pieces derived from the back of the grex stop moving and construct normal (albeit small) fruiting bodies containing stalk as well as spore cells. The pieces derived from the front part of the grex will also construct fruiting bodies with the normal proportions of spore cells and stalk cells if they are allowed to migrate for 24 h, but the shorter the time between the cutting of the grex and the formation of the fruiting body the larger is the proportion of stalk in the fruiting body. This suggests that in the case of the front cells movement is necessary for reformation of the correct pattern in the grex. The grex thus poses, in a peculiarly simple form, the problem of how a linear pattern of cell differentiation arises and is maintained.

Takeuchi[100] has suggested that differences exist between the amoebae in the vegetative phase and that cells sort out during the migrating grex phase according to these differences. Cell sorting out during the grex stage has been reported[101, 102] to occur and support for Takeuchi's suggestion has recently come from work done on the fate of amoebae grown in different media[103]. A GTS (Table 1.2) mutant of *D. discoideum* ATCC 24397 was obtained and then wild type and mutant amoebae were grown in different media, mixed in all possible combinations and the mixtures allowed to form fruiting bodies. The ratio of wild type to mutant spores so formed was compared with the ratio in which the amoebae were mixed. It was found, for example, that amoebae grown in glucose containing media (G cells) when mixed with amoebae grown in media lacking glucose (NS cells) preferentially gave rise to spore cells and vice versa. Analysis of the tips and backs of grexes formed from such mixtures of cells showed that the G cells were at the back and the NS cells were at the front, i.e. sorting out had occurred at the grex stage. When the tip of a migrating grex formed from mutant NS cells was replaced (by surgery) with a tip from a migrating grex formed from wild type G cells, a composite grex was formed in which the cells were in the appropriate position with respect to their pre-stalk, pre-spore character but not with respect to their growth conditions. Analysis of the spores in the fruiting body formed by such composite grexes showed that the spores were predominantly of the wild type phenotype and thus that the cells had sorted out after the graft. This suggests that the cells of the grex do not use 'positional information' in the sense defined by Wolpert[104] to regulate their differentiation. Indeed it is far from clear what relevance the sorting out of the component cells of the grex has for fruiting body formation. The migrating grex is not an obligatory stage in the life cycle[10] and spherical balls of cells formed by incubating amoebae in roller cultures will give rise to fruiting bodies directly when placed at an air–water interface[105]. Further, G cells and NS cells will give rise to fruiting bodies[106] when incubated in isolation; hence although when mixed they show a preference for one mode of differentiation rather than another this suggests that superimposed on the pattern which results from the sorting out potential of the cells is another pattern forming mechanism which regulates the pre-spore/pre-stalk pattern. This may involve the slime sheath[107] and/or the tip[108, 109, 110] cells but these suggestions have yet to be convincingly proved. Thus there appear to be two pattern forming mechanisms operating in the grex, one leading to a sorting out of the cells according to their intrinsic biochemical activities and the other superimposed on the cells by some property of the cell population as a whole. It seems likely that these two mechanisms are to a certain extent interconnected since G cells and NS cells give rise to fruiting bodies with slightly different proportions of spore and stalk cells[106] but the nature of the interconnection remains obscure.

1.7.2 Morphogenesis

Morphogenesis refers to the development of the shape and form of the organism and its individual parts and is largely a problem of the co-ordination of cellular movements. The morphogenetic changes in the life cycle of the slime moulds have been reviewed[9,12,99,106]. The stage which has been most studied is aggregation since until very late in the aggregation process the cells remain on the substratum and can thus be clearly seen whereas the grex is a three dimensional cell mass in which it is impossible to follow individual cells for very long. Of the two important aspects of the aggregation process—chemotaxis and contact interaction of cells—chemotaxis has attracted most attention recently and has been discussed on p. 18. Shaffer[111] first drew attention to the importance of changes in the strength of cell–cell adhesions and cell–substrate adhesions during aggregation amd more recent work has confirmed, and emphasised the importance of, such changes[108,112]. Two classes of cell–cell adhesions have been recognised. One class appears to be inhibited by EDTA and to be present in all non-feeding cells whereas the other class appears specifically in aggregating cells and is not sensitive to EDTA. The development of what Gerisch has termed 'aggregation competence' seems to be associated with this latter class of adhesions and in an elegant use of immunological techniques[113] changes in the membrane have been correlated with changes in adhesion. Despite these changes in adhesions the cells remain able to move both with respect to one another and over a substratum and this presents serious problems of interpretation which have been reviewed[99].

Antigenic changes in the surface of cells have also been reported during fruiting body formation[114] but little is known of the relevance of these changes to the morphogenetic events. If, as has recently been suggested, the tip cells control the shape and the proportions of the fruiting body then, of course, changes in the cell–cell adhesions of this group of cells become of critical importance.

At a more strictly biochemical level Sussman and Newell and their co-workers have investigated the effect on enzyme specific activities of changes or disruptions in the morphogenetic sequence. They have shown that dramatic increases in the specific activity of UDP-glucose pyrophosphorylase can be induced in cells forced to remain for 30 h in the migating grex stage before forming fruiting bodies[115]. From a study of the effect of actinomycin D on the kinetics of appearance of these extra amounts of activity they have concluded they reflect extra rounds of transcription. Cell–cell interactions in developing aggregates may also be altered by disrupting the aggregate by trituration in EDTA and then redepositing the washed cells onto Millipore filters. If the disaggregation is done carefully, such cells rapidly recapitulate the normal morphogenetic sequence and within 2–3 h reach the stage at which they were disaggregated and then go on to form fruiting bodies if left undisturbed[68]. These procedures can lead, if disaggregation is done at the suitable time, to marked increases in the specific activity of UDP-glucose pyrophosphorylase, trehalose 6-phosphate synthetase, UDP-galactose 4-epimerase and UDP-galactose: polysaccharide transferase[116]. In these experiments, as in the earlier[115] studies, it was suggested that the increases were due to further rounds of transcription. The metabolic consequences of such additional quantities of enzymes being formed are obscure. Certainly there is little or no change in the total anthrone positive material of the fruiting bodies induced by disaggregation. Preliminary studies have shown, however, that in cells containing a very high initial content of glycogen[117] there can be some alteration in the relative amounts of the various carbohydrates found in the fruiting bodies as a result of disaggregation. Clearly it is now possible to alter, independently, both the enzyme content of a cell (by interfering with the morphogenetic sequence) and the metabolic flux through a number of relevant metabolic pools (by altering the growth conditions of the amoebae). This should enable the interrelationship between metabolic events and enzyme synthesis to be investigated as well as the dependance of both on morphogenesis.

Quantitative studies of the effects of alterations of the growth medium or of the morphogenetic sequence on various biochemical parameters[99] have shown that there is often a simple numerical relationship between the altered and unaltered value. This has been referred to as 'quantal control'[118]. This situation thus is very different from that in bacteria where the amount of, say, an enzyme formed is a linear function (within limits) of some parameter such as repressor or inducer concentration. In the slime mould such events seem to be discontinuous functions of unknown controlling agents.

Thus these essentially exploratory attempts to investigate the relationship between morphogenetic and biochemical events have already uncovered a number of surprising and unexpected effects and more surprises no doubt await us.

1.8 CONCLUSIONS

The cellular slime moulds have often been considered as 'model systems' for the study of cell differentiation in higher organisms. As such they possess a number of very attractive features which are attracting an ever increasing number of investigators. In particular this group of organisms seems ideally suited for investigations of the interactions between control mechanisms which operate at different levels of complexity, particularly the interaction of intracellular and intercellular controls. The different species of the Acrasiales seem to form fruiting bodies which represent variations on a theme and although *D. discoideum* has been the only species to have attracted intensive biochemical investigation so far there is no doubt that as investigations of the molecular basis of pattern formation and morphogenesis become more advanced the other species will become attractive as 'model systems' in their own right. In order for these studies to progress the techniques and ideas of the microbiologist/biochemist, which are particularly appropriate for studies of the amoebal phase, must merge with those of the cell biologist/embryologist, which are particularly appropriate to the developmental phase. This union is well under way and promises to lead, as a consequence of the well known principles of hybrid vigour, to exciting developments in which the Acrasiales will become, perhaps, something more than a 'model system', for, in the last resort, what is important about a 'model system' is not what it tells you about some other system but what it tells you about itself.

Note added in proof

It has recently been claimed that the macrocyst stage of the life cycle of *P. violaceum* represents a true diploid phase of the life cycle[1]. When the macrocyst germinates there must, therefore, be a meiotic division and formation of haploid amoebae. An extensive investigation of *Dictyostelium* and *Polysphondylium* species isolated from Nature has suggested that the ability to form macrocysts is genetically controlled by 'mating type' alleles and that in these species, therefore, like *P. violaceum*, the macrocyst represents a true diploid phase of the life cycle[124]. If these observations can be confirmed, they suggest that the cellular slime moulds have two possible pathways of differentiation, one associated with meiotic recombination and macrocyst formation involving two genetically distinct strains, and the other associated with parasexual recombination and fruiting body formation which need not involve more than one type of amoeba.

References

1. Brefeld, O. (1869). *Abh. Senckenberg. Naturforsch. Ges.*, **7,** 85
2. Tieghem, P. van (1880). *Bull. Soc. Bot. France*, **27,** 317
3. Bonner, J. T. (1967). *The Cellular Slime Moulds*, 4 (Princeton: Princeton University Press)
4. Olive, L. S. (1970). *Bot. Rev.*, **36,** 59
5. Raper, K. B. (1935). *J. Agr. Res.*, **50,** 135
6. Ashworth, J. M., Duncan, D. and Rowe, A. J. (1969). *Exptl. Cell. Res.*, **58,** 73
7. Bonner, J. T. (1947). *J. Exp. Zool.*, **106,** 1
8. Konijn, T. M., van de Meene, J. G., Bonner, J. T. and Barkley, D. S. (1967). *Proc. Nat. Acad. Sci. USA*, **58,** 1152
9. Shaffer, B. M. (1962). *Advances in Morphogenesis*, Vol. 2, 112 (M. Abercrombie and J. Brachet, editors) (New York: Academic Press)
10. Newell, P. C., Telser, A. and Sussman, M. (1969). *J. Bacteriol.*, **100,** 736
11. Garrod, D. R. (1969). *J. Cell. Sci.*, **4,** 781
12. Shaffer, B. M. (1964). *Primitive Motile Systems*, 387 (R. D. Allen and N. Kamiya, editors) (New York: Academic Press)
13. Raper, K. B. (1940). *J. Elisha Mitchell Sci. Soc.*, **56,** 241
14. Staples, S. O. and Gregg, J. H. (1967). *Biol. Bull.*, **132,** 413
15. White, G. J. and Sussman, M. (1963). *Biochim. Biophys. Acta*, **74,** 179
16. Blaskovics, J. C. and Raper, K. B. (1957). *Biol. Bull.*, **114,** 58
17. Filosa, M. F. and Dengler, R. E. (1972). *Develop. Biol.*, **29,** 1
18. Hohl, H. R., Hamamoto, S. T. and Hemmes, D. E. (1968). *Amer. J. of Botany*, **55,** 783
19. Cavender, J. C. (1970). *J. Gen. Microbiol.*, **62,** 113
20. Sinha, U. K. and Ashworth, J. M. (1969). *Proc. Roy. Soc. (London), Ser. B*, **173,** 531
21. Loomis, W. F. (1969). *J. Bacteriol.*, **97,** 1149
22. Sussman, R. R. and Sussman, M. (1953). *Ann. N.Y. Acad. Sci.*, **56,** 949
23. Loomis, W. F. and Ashworth, J. M. (1968). *J. Gen. Microbiol.*, **53,** 181
24. Fukui, Y. and Takeuchi, I. (1971). *J. Gen. Microbiol.*, **67,** 307
25. Loomis, W. F. (1969). *J. Bacteriol.*, **99,** 65
26. Sonneborn, D., White, G. J. and Sussman, M. (1963). *Develop. Biol.*, **7,** 79
27. Loomis, W. F. (1970). *Exp. Cell Res.*, **60,** 285
28. Katz, E. R. and Sussman, M. (1972). *Proc. Nat. Acad. Sci. USA*, **69,** 495
29. Gerisch, G. (1959). *Naturwissenschaften*, **46,** 654
30. Raper, K. B. (1939). *J. Agr. Res.*, **58,** 157
31. Bascomb, S., Lapage, S. P., Willcox, W. R. and Curtis, M. A. (1971). *J. Gen. Microbiol.*, **66,** 278

32. Sussman, M. (1966). *Methods in Cell Physiology*, Vol. 2, 397 (D. M. Prescott, editor) (New York: Academic Press)
33. Horn, E. G. (1968). *Bull. Ecol. Soc. Amer.*, **49,** 115
34. Wiener, E. and Ashworth, J. M. (1970). *Biochem. J.*, **118,** 505
35. Ferber, E., Munder, P. G., Fischer, H. and Gerisch, G. (1970). *Eur. J. Biochem.*, **14,** 253
36. Pan, P., Hall, E. M. and Bonner, J. T. (1972). *Nature New Biology*, **237,** 181
37. Sussman, M. (1963). *Science*, **139,** 338
38. Goldstone, E. M., Banerjee, S. D., Allen, J. R., Lee, J. J., Hutner, S. H., Bacchi, C. J. and Melville, J. F. (1966). *J. Protozool.*, **13,** 171
39. Sussman, M. and Sussman, R. R. (1967). *Biochem. Biophys. Res. Commun.*, **29,** 34
40. Schwalb, M. and Roth, R. (1970). *J. Gen. Microbiol.*, **60,** 283
41. Watts, D. J. and Ashworth, J. M. (1970). *Biochem. J.*, **119,** 175
42. Ashworth, J. M. and Watts, D. J. (1970). *Biochem. J.*, **119,** 174
43. Leach, C. K. and Ashworth, J. M. (1972). *J. Mol. Biol.*, **68,** 35
44. Weeks, G. and Ashworth, J. M. (1972). *Biochem. J.*, **126,** 617
45. Ashworth, J. M. and Quance, J. (1972). *Biochem. J.*, **126,** 601
46. Gerisch, G. (1971). *Naturwissenschaften*, **58,** 430
47. Konijn, T. M. (1972). *Advances in Cyclic Nucleotide Research*, Vol. 1, 17 (P. Greengard, G. A. Robinson, and R. Paoletti, editors) (New York: Raven Press)
48. Shaffer, B. M. (1956). *Science*, **123,** 1172
49. Goidl, E. A., Chassy, B. M., Love, L. and Krichevsky, M. I. (1972). *Proc. Nat. Acad. Sci. USA*, **69,** 1128
50. Chang, Y. Y. (1968). *Science*, **160,** 57
51. Riedel, V., Malchow, D., Gerisch, G. and Nägele, B. (1972). *Biochem. Biophys. Res. Commun.*, **46,** 279
52. Chassy, B. M. (1972). *Science*, **175,** 1016
53. Malkinson, A. and Ashworth, J. M. (1973). *Biochem. J.* in the press
54. Gerisch, G., Malchow, D., Riedel, V., Müller, E. and Every, M. (1972). *Nature New Biology*, **235,** 90
55. Malchow, D., Nägele, B., Schwarz, H. and Gerisch, G. (1972). *Eur. J. Biochem.*, **28,** 136
56. Rossomando, E. F. and Sussman, M. (1972). *Biochem. Biophys. Res. Commun.*, **47,** 604
57. Keller, E. F. and Segal, L. A. (1970). *J. Theor. Biol.*, **26,** 399
58. Keller, E. F. and Segal, L. A. (1970). *Nature (London)*, **227,** 1365
59. Cohen, M. H. and Robertson, A. (1971). *J. Theor. Biol.*, **31,** 101
60. Lee, K.-C. (1972). *J. Gen. Microbiol.*, **72,** 458
61. Bonner, J. T. (1970). *Proc. Nat. Acad. Sci., USA*, **65,** 110
62. Sussman, M. and Sussman, R. R. (1969). *Symp. Soc. Gen. Microbiol.*, Vol. 19, 403 (P. M. Meadow and S. J. Pirt, editors) (Cambridge: Cambridge University Press)
63. Wright, B. E. (1966). *Science*, **153,** 830
64. Newell, P. C. (1971). *Essays in Biochemistry*, Vol. 7, 87 (P. N. Campbell and F. Dickens, editors) (New York: Academic Press)
65. Ashworth, J. M. (1971). *Symp. Soc. Exp. Biol.*, Vol. 25, 27 (D. D. Davies and M. Balls, editors) (Cambridge: Cambridge University Press)
66. Gustafson, G. L. and Wright, B. E. (1972). *CRC Critical Reviews in Microbiology*, Vol. 1, 453 (A. I. Laskin and H. Lechebalier, editors) (Ohio: CRC Press)
67. Wright, B. E. (1968). *J. Cell. Physiol.*, **72,** Suppl. 1, 145
68. Newell, P. C., Longlands, M. and Sussman, M. (1971). *J. Mol. Biol.*, **58,** 541
69. White, G. T. and Sussman, M. (1961). *Biochim. Biophys. Acta*, **53,** 284
70. White, G. T. and Sussman, M. (1963). *Biochim. Biophys. Acta*, **74,** 173
71. Muhlethäler, K. (1956). *Amer. J. of Botany*, **43,** 673
72. Glegg, J. S. and Filosa, M. (1961). *Nature (London)*, **192,** 1077
73. Cleland, S. V. and Coe, E. L. (1969). *Biochim. Biophys. Acta*, **192,** 446
74. Cleland, S. V. and Coe, E. L. (1968). *Biochim. Biophys. Acta*, **156,** 44
75. Hames, B. D., Weeks, G. and Ashworth, J. M. (1972). *Biochem. J.*, **126,** 627
76. Hames, B. D. and Ashworth, J. M. (1973). *Biochem. J.* (in the press)
77. Sussman, M. and Lovgren, N. (1965). *Exp. Cell Res.*, **38,** 97
78. Hames, B. D. and Ashworth, J. M. (1973). quoted in ref. 99
79. Wright, B. E. (1967). *Arch. Mikrobiol.*, **59,** 335

80. Firtel, R. A., Jacobson, A. and Lodish, H. F. (1972). *Nature New Biology*, **239**, 225
81. Loomis, W. F., Jr. (1969). *J. Bacteriol.*, **97**, 1149
82. Loomis, W. F., Jr. (1970). *J. Bacteriol.*, **103**, 375
83. Roth, R. and Sussman, M. (1968). *J. Biol. Chem.*, **243**, 5081
84. Pong, S. S. and Loomis, W. F., Jr. (1971). *J. Biol. Chem.*, **246**, 4412
85. Ashworth, J. M. and Sussman, M. (1966). *J. Biol. Chem.*, **242**, 1696
86. Sussman, M. and Osborn, M. J. (1964). *Proc. Nat. Acad. Sci. USA*, **52**, 81
87. Telser, A. and Sussman, M. (1971). *J. Biol. Chem.*, **246**, 225
88. Jones, T. H. D. and Wright, B. E. (1970). *J. Bacteriol.*, **104**, 754
89. Loomis, W. F., Jr. (1969). *J. Bacteriol.*, **100**, 417
90. Coston, M. B. and Loomis, W. F., Jr. (1969). *J. Bacteriol.*, **100**, 1208
91. Franke, J. and Sussman, M. (1971). *J. Biol. Chem.*, **246**, 6381
92. Killick, K. A. and Wright, B. E. (1972). *J. Biol. Chem.*, **247**, 2967
93. Every, D. and Ashworth, J. M. (unpublished observations)
94. Quance, J. and Ashworth, J. M. (1972). *Biochem. J.*, **126**, 617
95. Roth, R., Ashworth, J. M. and Sussman, M. (1968). *Proc. Nat. Acad. Sci. USA*, **59**, 1235
96. Sussman, M. (1970). *Nature* (*London*), **225**, 1245
97. Francis, D. (1969). *Quart. Rev. Biol.*, **44**, 277
98. Cohen, M. (1971). *Symp. Soc. Exp. Biol.*, Vol. 25, 455 (D. D. Davies and M. Balls, editors) (Cambridge: Cambridge University Press)
99. Garrod, D. and Ashworth, J. M. (1973). *Symp. Soc. Gen. Microbiol.*, Vol. 23, 407 (J. M. Ashworth and J. E. Smith, editors) (Cambridge: Cambridge University Press)
100. Takeuchi, I. (1963). *Nucleic Acid Metabolism, Cell Differentiation and Cancer Growth*, 297 (E. V. Cowdry and S. Seno, editors) (Oxford: Pergamon Press)
101. Bonner, J. T. and Adams, M. S. (1958). *J. Embryol. Exp. Morphol.*, **6**, 346
102. Bonner, J. T., Sieja, T. W. and Hall, E. M. (1971). *J. Embryol. Exp. Morphol.*, **25**, 457
103. Leach, C. K., Ashworth, J. M. and Garrod, D. R. (1973). *J. Embryol. Exp. Morphol.* (in the press)
104. Wolpert, L. (1969). *J. Theor. Biol.*, **25**, 1
105. Gerisch, G. (1960). *Arch. Entwicklungsmech. Organismen*, **152**, 632
106. Garrod, D. R. and Ashworth, J. M. (1972). *J. Embryol. Exp. Morphol.*, **28**, 463
107. Loomis, W. F., Jr. (1972). *Nature* (*London*), **240**, 6
108. Gerisch, G. (1968). *Current Topics in Developmental Biology*, Vol. 3 (A. A. Moscona and A. Monroy, editors) (New York: Academic Press)
109. Farnsworth, P. A. (1973). *J. Embryol. Exp. Morphol.* (in the press)
110. Robertson, A. (1972). *Lepetit Colloquium on Biology and Medicine*, Vol. 3 (L. G. Silvestri, editor) (Amsterdam: North Holland)
111. Shaffer, B. M. (1957). *Quart. J. Microsc. Sci.*, **98**, 377
112. Garrod, D. R. (1972). *Exp. Cell Res.*, **72**, 588
113. Beng, H., Gerisch, G., Kempf, S., Riedel, V. and Cremer, G. (1970). *Exp. Cell Res.*, **63**, 147
114. Gregg, J. H. (1966). *The Fungi*, Vol. 2, 235 (G. C. Ainsworth and A. S. Sussman, editors) (New York: Academic Press)
115. Newell, P. C. and Sussman, M. (1970). *J. Mol. Biol.*, **49**, 627
116. Newell, P. C., Franke, J. and Sussman, M. (1972). *J. Mol. Biol.*, **63**, 373
117. Ashworth, J. M. (unpublished observations)
118. Sussman, M. and Newell, P. C. (1972). *Molecular Genetics and Developmental Biology* (in the press) (M. Sussman, editor) (New York: Prentice Hall)
119. Cavender, J. C. and Raper, K. B. (1965). *Amer. J. Bot.*, **52**, 302
120. Cavender, J. C. (1969). *Amer. J. Bot.*, **56**, 989
121. Cavender, J. C. (1969). *Amer. J. Bot.*, **56**, 973
122. Ashworth, J. M. and Wiener, E. (1973). *Lysosomes in Biology and Pathology*, Vol. 3 (in the press) (H. Fell and J. T. Dingle, editors) (Amsterdam: North Holland)
123. Erdos, G. W., Nickerson, A. W. and Raper, K. B. (1973). *Cytobiologie*, **6**, 351
124. Clark, M. A., Francis, D. and Eisenberg, R. (1973). *Biochem. Biophys. Res. Commun.*, **52**, 672

Editor's Comments

Dr Ashworth's description of the development of cellular slime moulds introduces many problems of development. It also raises many questions, the first of which, perhaps, are: What is the nature of the slime moulds? How are they related to other organisms and what relevance do observations on slime moulds have to these? The need to ask these questions is emphasised by the lack of general agreement about the classification of this group of organisms, which biochemists may be excused for finding particularly confusing. Dr Ashworth has classified the cellular slime moulds as belonging to the Mycetozoa which is a sub-class of the Class Rhizopoda commonly considered to belong to the Animal Kingdom. In a botanical context, however, they are included in the Acrasiales which are classified in the Phylum Myxophyta within the Plant Kingdom. Some other authorities prefer to classify them neither as animals nor as plants but as a separate Phylum, the Myxomycetes within the Protist Kingdom. No single one of these classifications is satisfactory since the properties of these organisms do not place them neatly in either the Animal or Plant Kingdom. Attention is drawn to this problem of classification of slime moulds simply to illustrate their somewhat unique nature.

The cellular slime moulds are unquestionably eukaryotes with a distinctive and classical nucleus. They have one of the smallest genomes among eukaryotes. It is only about five times larger than that of *E. coli* but the way in which these organisms produce and use messenger RNA is quite distinct from that in bacteria and seems to be identical with that in higher eukaryotes. For example, Lodish *et al.* (1973) have demonstrated that RNA is transcribed as a slightly larger molecule than the mRNA which appears in the cytoplasm. Within the nucleus about 20% of this precursor is degraded while a poly-adenylic sequence is added post-transcriptionally to the 3′ end. Only after this 'processing' does the mRNA become associated with cytoplasmic polysomes.

Hence, these organisms are likely to offer a very simple model of eukaryotic cells. Rather than devote space to considering the arguments concerning their classification it is, therefore, more profitable to ask how relevant observations made with them may be to developmental studies in plants and animals. To answer this it is necessary to recapitulate the major differences and similarities between plants and animals. Plants, in general, are immobile mainly because of their rigid cell walls, whereas animals are generally mobile. Although this difference has wider ramifications, it is not of high relevance to a discussion on cell differentiation. Secondly, plants on the whole have a more versatile metabolism than animals. Most of them are photosynthetic and very many are entirely autotrophic whereas animal cells depend extensively on nutrients synthesised by plants and other animals. Nevertheless, many basic metabolic pathways of animals and plant cells are very similar. In particular, the same genetic code is used by both animals and plants and the machinery for the synthesis of RNA and its translation into protein is virtually identical. Thirdly, development seems to be more labile in plants than in animals, as mentioned in the introduction. Whether this represents a real difference between the two kingdoms is not clear; it may simply be that the systems and phenomena studied most extensively in animals are of an inflexible nature whereas more attention has been directed to the study of labile phenomena in plants.

Looked at from the point of view of the biochemist or molecular biologist, the similarities between animals and plants are more striking than the differences. The cells have basically the same structure and, with the exception of plastids, the subcellular organelles are the same. The most striking difference is in the cell wall and this is not fundamental. In these terms, therefore, slime moulds are the same kinds of organisms as higher plants and animals. There are much more profound differences between all of these eukaryotic organisms and the prokaryotes, such as the bacteria.

From the point of view of cell differentiation, therefore, many of the phenomena described in the preceding and following chapters are likely to be relevant to both the animal and plant kingdoms. Nevertheless, a few functions are unique to the plant kingdom. A particular instance is the regulation of enzyme levels in plants by the photosensitive pigment, phytochrome. Not only does this represent a system unique for plant tissue but it presents an experimentally unique situation in the precision with which the input to the regulatory systems can be controlled and quantitated as described in the following essay by Professor Mohr.

References

Lodish, H. F., Firtel, R. A., Jacobson, A., Tuchman, J., Alton, T., Young, B. and Baxter, L. (1973). *Cold Spring Harbour Symp. Quant. Biol.*, **38** (in press)

2
The Role of Phytochrome in Controlling Enzyme Levels in Plants

H. MOHR
University of Freiburg

2.1 INTRODUCTION

2.1.1 Significance

This chapter deals with the photoregulation of enzyme levels in plants by the photochromic sensor pigment, phytochrome[1]. While photocontrol of enzymes (enzyme synthesis, degradation and activation) may be mediated by light absorption in other pigments as well (e.g. photosynthetic pigments[2,3]), the emphasis on phytochrome is justified for two reasons: firstly, phytochrome is the most important molecule in potentially autotrophic plants for the detection of photosignals from the environment and for making use of this information to regulate the orderly development of the living system; secondly, a detailed knowledge of the molecular mechanism(s) through which phytochrome exerts its control over the development of higher plants may serve as a useful model for understanding the control of development in general, including that in animals and man. Thus, our special topic 'photoregulation of enzyme levels in plants' will be treated within the framework of phytochrome-mediated photomorphogenesis[4], with particular emphasis on those phenomena where a description in 'molecular terms' can be envisaged.

2.1.2 Terminology

Originally, the terms 'induction' and 'repression' were used operationally to designate the appearance or lack of appearance (or disappearance) of an enzyme. In the present article the terms will be used in this original meaning, i.e. without *a priori* implications about the actual control mechanism(s). Thus, 'enzyme induction by phytochrome' means an increase of enzyme activity (or content) caused by phytochrome, whereas 'enzyme repression by phytochrome' means that an increase of enzyme activity (or content) is arrested by phytochrome. While the statement seems to be justified that, in spite of some differences in detail, all proteins are synthesised by the same basic mechanism (which involves messenger RNA and ribosomes)[5] our understanding of the *control* of protein synthesis, involving the action of hormones and light, is still very limited. Therefore it would be premature to base the terminology on any concept of regulation of protein synthesis and degradation. Rather, operational terms must be preferred at present.

Since no general answer can be given to the question of whether the difference in observed enzyme activity is due to a difference in the number of enzyme molecules (or to an activation of pre-existing molecules), the term 'synthesis' must still be used with great caution in the field of photoregulation of enzyme levels. While the use of the term 'synthesis' seems to be justified by the results of density or radioactive labelling[6,7] and of the usual inhibitor experiments[8,9], it must be realised that there is no *rigorous* proof in most instances that an increase of enzyme activity is paralleled by a corresponding increase in the number of enzyme molecules. Therefore the term 'apparent synthesis' is used in the present text for convenience to denote any increase in enzyme activity and is not *necessarily* meant to imply true *de novo* synthesis of a polypeptide chain.

2.1.3 Critical aspects of enzyme measurements

In general, it is assumed that the enzyme activity as measured in the *in vitro* assay represents the concentration of active enzyme molecules *in vivo*. The problems involved in this assumption with respect to enzyme levels in higher plants have recently been dealt with by Schopfer[10]. Therefore, a brief recapitulation of the major aspects of these will suffice:

(a) It must be ascertained that the total enzyme (or at least a constant percentage of it) can be extracted from the plant tissue. If part of the enzyme activity remains in the tissue residue it is necessary to check that its percentage is not changed by an exposure of the plants to light.

(b) The pH optimum (and the optimum ionic strength) of the enzyme assay is not necessarily identical with the pH optimum (and optimum ionic strength) for enzyme extraction.

(c) In the crude plant extract, 'factors' (in particular, inhibitors of phenolic nature) may be present which modify the enzyme activity in the *in vitro* assay. On the other hand, these substances are very probably not in contact with the enzyme *in situ*. Since the concentrations of inhibitors (e.g. phenolic substances) are likely to be controlled by light, procedures must be developed to remove or inactivate phenolics or other enzyme inhibitors in crude extracts[10].

In the work with the mustard seedling, adsorbants for phenolics such as Divergan (insoluble polyvinylpyrrolidone = PVP[11])[9,12,13], charcoal[8] or Sephadex gel chromatography[14] were used to purify the crude extract and to make sure that an apparent light effect which is due to a change of the *in vitro* enzyme activity by the presence of variable amounts of inhibitors in the crude extract does not come into play.

In the work with mustard seedlings mixing experiments were routinely carried out to check for the presence of inhibitors whose concentration is light-dependent. In these, extracts from dark-grown and light-grown plants are mixed. If the activities are strictly additive it is concluded that there is at least no differential inhibition or activation of the enzyme due to the light-mediated presence of 'factors' in the crude extract. Some examples of routine mixing experiments performed with enzyme extracts from mustard seedlings are shown in Table 2.1.

Table 2.1 Test for a differential influence by inhibitors or activators of enzyme activities in the *in vitro* enzyme assay. Extracts of mustard seedlings grown in the dark ('dark extract') or irradiated with continuous far-red light ('far-red extracts') were used. In the cases of amylase and phenylalanine ammonia lyase only the cotyledons were used. For amylase and lipoxygenase assays the crude extracts were treated with Divergan, and for phenylalanine ammonia-lyase assay with charcoal. (From Drumm, Oelze-Karow and Dittes[15], by courtesy of the authors.)

Enzyme	*Enzyme activity [relative units]*	
Amylase		
100% far-red extract	12.4	15.4
100% dark extract	3.1	
50% far-red extract + 50% dark extract	7.7 (× 2 = 15.4)	
Lipoxygenase		
100% far-red extract	97.4	254.7
100% dark extract	157.3	
50% far-red extract + 50% dark extract	127.1 (× 2 = 254.2)	
Phenylalanine ammonia-lyase		
100% far-red extract	0.38	0.38
100% dark extract	0.00	
50% far-red extract + 50% dark extract	0.19 (× 2 = 0.38)	

2.1.4 The phenomenon of photomorphogenesis[4]

All living systems on this planet depend on a narrow band in the electromagnetic spectrum which we call 'light' (*ca.* 390–760 nm in the case of vision; *ca.* 320–800 nm in the case of plants). Quanta of this spectral range have a relatively low energy of *ca.* 90–35 kcal einstein^{-1} (5.5–2.5 eV photon^{-1}), depending on the particular wavelengths. This energy is sufficient to alter the outer electronic energy levels of atoms or molecules but not sufficient to complete ionisation. Living systems have adapted themselves to this type of radiation in the course of evolution. Quanta of this size can be absorbed in a plant or animal by only a very few types of molecule which are characterised by extended π-electron systems, such as the chlorophylls or the carotenoids. Most molecules which occur in the cell (water, proteins, nucleic acids, lipids, carbohydrates and their metabolites) cannot absorb quanta in the spectral range of light to give an electronic excitation.

The term 'photomorphogenesis' is used to designate the fact that light can control development (that is, growth, differentiation and morphogenesis)

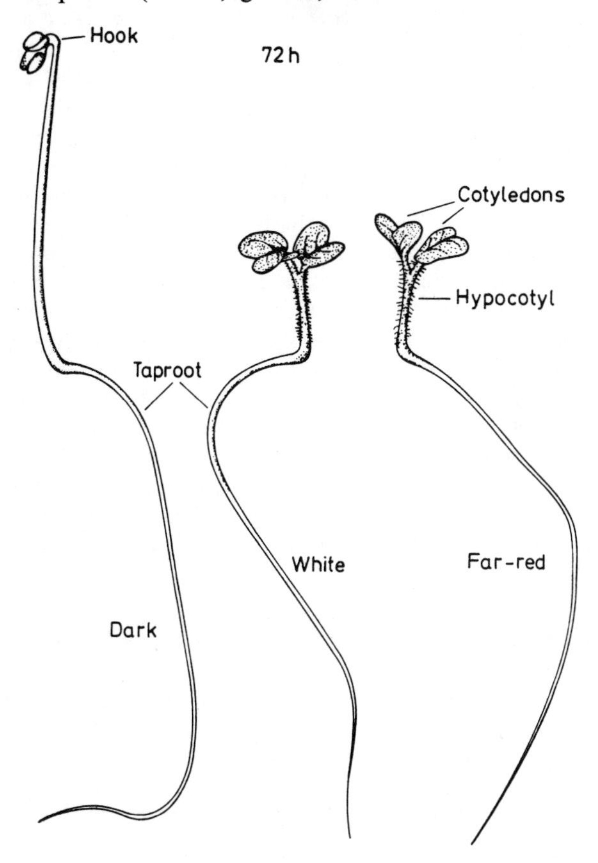

Figure 2.1 These seedlings of mustard (*Sinapis alba* L.) have the same chronological age (72 h after sowing at 25 °C) and are virtually identical genetically. The differences in development are due to the presence or absence of light. A characteristic of etiolation (dark-grown seedling, left) is that the axis grows rapidly while the leaves remain rudimentary. The biological significance of etiolation is obvious. As long as a plant has to grow in darkness it uses the limited supply of storage material predominantly for axis growth. In this way the probability is highest that the tip of the plant will reach the light before the storage material is exhausted. (From Mohr[4], by courtesy of Springer.)

of a plant independently of photosynthesis. While the specific development of every living system depends on its genetic information *and* on its environment, higher plants are particularly sensitive to the environmental factor of light. Naturally, light does not carry any *specific* information. However, light can be regarded as an 'elective factor' which deeply influences the manner in which those genes which are present in the particular organism are used.

Figure 2.1, showing seedlings of the mustard plant (*Sinapis alba* L.) illustrates the basic phenomena of photomorphogenesis. All three seedlings have virtually the same genes, the same chronological age (72 h after sowing at 25 °C), and all three were grown on the same medium. Light must be responsible for the obvious differences in morphogenesis of the etiolated and the light-grown seedlings. Since the development of the seedling under white light (which allows photosynthesis) does not significantly differ from the development of the seedling under continuous far-red light (which does not support photosynthesis), the effect of light on morphogenesis may not be considered a consequence of photosynthesis.

While photomorphogenesis is a general phenomenon in the ontogeny of higher plants, seedlings of dicotyledonous plants (that is, in the first stages of vegetative growth after seed germination) turned out to be particularly useful subjects for the causal analysis of the phenomenon, for the following reasons: The plant consists during this phase of only three parts, cotyledons, hypocotyl and radicle (taproot). The plumule is hardly developed (cf. Figure 2.1); the seedling (of *Sinapis alba* in the present case) contains so much storage material —mainly fat and protein—in the cotyledons that it is completely independent of photosynthesis or of an external supply of organic molecules for several days after seed germination (for at least 72 h at 25 °C); the seedling can be grown during this period on a medium which supplies only water. An

Table 2.2 Some phytochrome-mediated photoresponses of the mustard seedling, *Sinapis alba* L. (investigations carried out in our laboratory since 1957)

Inhibition of hypocotyl lengthening
Inhibition of translocation from the cotyledons
Enlargement of cotyledons
Unfolding of the lamina of the cotyledons
Hair formation along the hypocotyl
Opening of the hypocotylar ('plumular') hook
Formation of leaf primordia
Development of primary leaves
Increase of negative geotropic reactivity of the hypocotyl
Formation of tracheary elements
Differentiation of stomata in the epidermis of the cotyledons
Formation of plastids in the mesophyll of the cotyledons
Changes in the rate of cell respiration
Synthesis of anthocyanin
Increase in the rate of ascorbic acid synthesis
Increase in the rate of carotenoid synthesis
Increase in the rate of long-term protochlorophyll regeneration
Increase of RNA contents in the cotyledons
Decrease of RNA contents in the hypocotyl
Increase of protein synthesis in the cotlyedons
Changes in the rate of degradation of storage fat
Changes in the rate of degradation of storage protein
Elimination of the lag-phase of chlorophyll formation (in white light)
Increase of the rate of chlorophyll accumulation (in white light)

external supply of ions is not required for normal development. In comparing the dark-grown (etiolated) with the light-grown (normal) plant (cf. Figure 2.1) we realise that photomorphogenesis is a complex process. The integration of the different photoresponses of the different cells, tissues and organs must be precisely regulated in space and time. It is difficult to form a conception of this integration, even in the mustard seedling where more than 20 different phytochrome-mediated morphological, histological and biochemical photoresponses have been investigated over the years (Table 2.2).

At present we assume that development is primarily the consequence of an orderly sequence of changes in the enzyme complement of an organism. Therefore, the investigator of photomorphogenesis will primarily try to explore those phytochrome-mediated responses in which changes in enzyme levels have a well-defined causal role in well-defined developmental steps.

2.1.5 The phytochrome system[1, 4, 16]

Phytochrome is a bluish chromoprotein having two forms (species) which are interconvertible by light: P_r, with an absorption maximum in the red part of the spectrum at 660 nm and P_{fr}, with an absorption maximum in the far-red part of the spectrum at 730 nm. In a dark-grown seedling only P_r, the physiologically inactive species, is present. The physiologically active species P_{fr} (the 'effector molecule' of the phytochrome system) can only originate under the influence of light.

For the purpose of the present article the phytochrome system as it occurs in the mustard seedling at *ca.* 36 h after seed sowing at 25 °C can be described by a model which requires four elements and five rate constants (Figure 2.2).

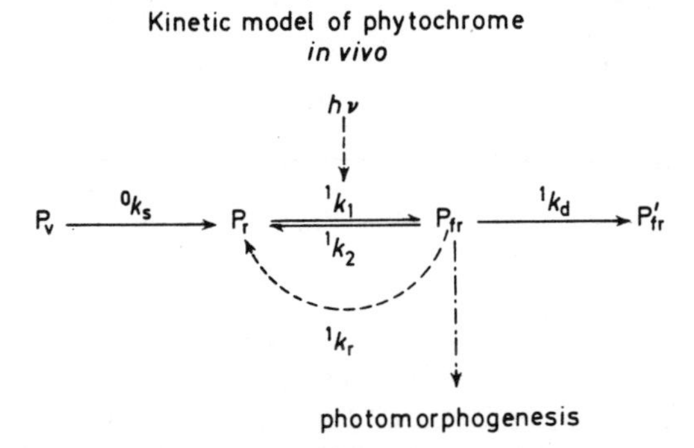

Figure 2.2 A model of the phytochrome system which includes *de novo* synthesis of P_r, dark reversion ($P_{fr} \dashrightarrow P_r$) and destruction ($P_{fr} \longrightarrow P'_{fr}$). (From Schäfer[17], by courtesy of the author.)

The model indicates that P_r is formed from a precursor, designated as P_v, through a zero order reaction (cf. Figure 2.6). The phototransformations ($P_r \longrightarrow P_{fr}$ and $P_{fr} \longrightarrow P_r$) follow first-order reactions in both directions. The dark reversion of P_{fr} to P_r follows first-order kinetics with a half-life of P_{fr} on the order of a few minutes. The dark reversion involves less than

20% of the total P_{fr}[18,19]; the remainder is subjected to an irreversible destruction, which is also a first-order process. The half-life of P_{fr} in this process is *ca.* 45 min at 25 °C in the cotyledons as well as in the hypocotyl hook of the mustard seedling[19].

Biochemically phytochrome is a chromoprotein consisting of a protein moiety and a chromophoric group. The phytochrome chromophore is an open chain tetrapyrrole (similar to the algal chromophore phycocyanobilin). A tentative model for the phytochrome chromophore and its binding to the protein moiety is shown in Figure 2.3. The model attempts, moreover, to

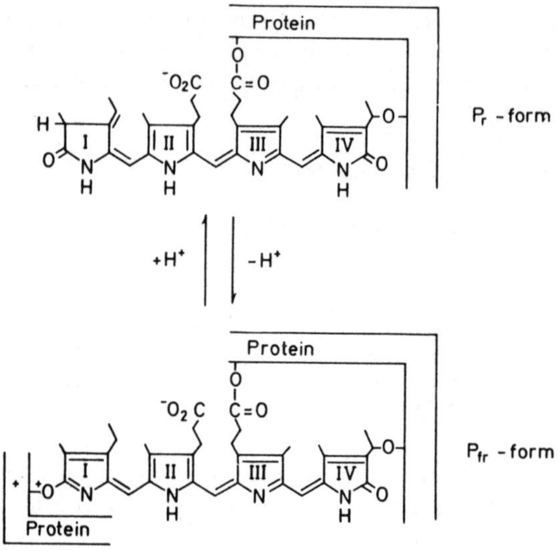

Figure 2.3 The most detailed structure proposed so far for the phytochrome chromophore and for the coupling between the chromophore and the protein moiety. The analysis was based on a sample of rye phytochrome which was denatured (From Rüdiger[20], by courtesy of Walter de Gruyter.)

describe the isomerisation of the chromophore during the transition from P_r to P_{fr} and vice versa. Irrespective of details which are still a matter of debate and intense experimentation[21] one may describe the photochemical transformations of the phytochrome system as a combination of isomerisation of the chromophore and a change in the conformation of the protein part. At physiological temperatures (e.g. at 25 °C), a protein conformation exists for each of the two chromophore isomers which gives a stable complex of the chromoprotein: P_r in the ground state, and P_{fr} in the ground state.

Biophysically, an important feature of the phytochrome system is that the absorption spectra of P_r and P_{fr} overlap throughout the visible range of the spectrum. This overlap is the reason that photostationary states are characteristic for the status of the phytochrome system in solution as well as in the cell, under conditions of saturating irradiations. Figure 2.4 gives some information on how much of the total phytochrome is present as P_{fr} if the photostationary state is established in the hypocotyl hook of the mustard seedling by monochromatic light of the indicated wavelengths. We see that *ca.* 80% at the most of the total phytochrome is available as P_{fr}, for example, if irradiation takes place with pure red light at *ca.* 660 nm. If the irradiation is with pure far-red at *ca.* 720 nm, only *ca.* 2.5% of the total phytochrome is present as P_{fr}. The photostationary state of the phytochrome system is rapidly established, at least in the red and far-red where the extinction coefficients of the phytochromes and the relative quantum efficiencies of the photochemical transformations are high. A minute or thereabouts of irradiation with medium quantum flux density is sufficient almost to establish the

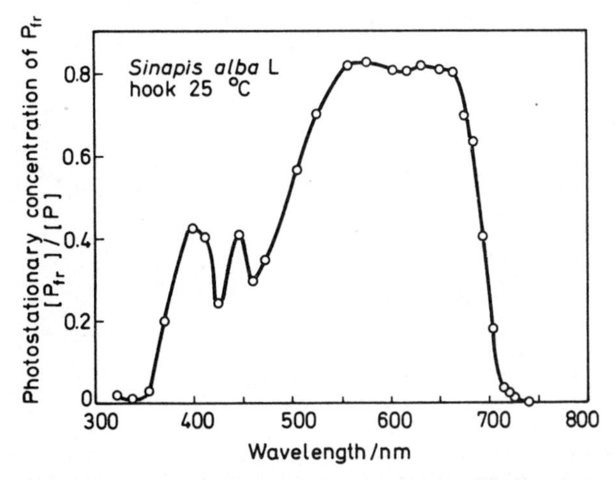

Figure 2.4 The fraction of P_{fr} at photoequilibrium as a function of wavelength (*in vivo* measurements at 25 °C with hypocotyl hooks of mustard seedlings). The data were obtained by K. M. Hartmann and C. J. P. Spruit. (From Hanke *et al.*[22], by courtesy of Springer.)

photostationary state. That is, only 'short-time irradiations' are required virtually to establish photostationary states in the phytochrome system *in vivo* as well as in the photochemically-active extract.

On the basis of this information the operational criteria for the involvement of phytochrome in a particular photoresponse can be defined as follows: a photoresponse can be induced by a brief irradiation (e.g. 5 min) with red light of medium quantum flux density; the induction by red light can be fully reversed by immediately following with a corresponding dose of far-red light; the extent of the response following the irradiation sequence red plus far-red must be identical with the extent of the response following a brief treatment with far-red light alone.

These operational criteria have been verified in many instances, classical examples being light-mediated seed germination in lettuce[23] and light-mediated anthocyanin synthesis in the mustard seedling (Figure 2.5). The dark-grown mustard seedling does not synthesise significant amounts of anthocyanin. Five minutes of red light (applied 36 h after sowing) will induce anthocyanin synthesis. The lag-phase, i.e. the duration of time between irradiation and the onset of anthocyanin synthesis, is *ca.* 3 h at 25 °C. The effect of the red light can be fully reversed by immediately following with 5 min of far-red light. The operational criteria for the involvement of phytochrome in controlling anthocyanin synthesis are thus clearly fulfilled.

In most quantitative experiments on the control of enzyme levels by phytochrome (that is, P_{fr}) an attempt must be made to maintain a stationary concentration of the effector molecule P_{fr} in the tissue, e.g. in the cotyledons or in the hypocotyl hook of the mustard seedling, over a considerable period of time. This means that conditions of long-term irradiation under which the

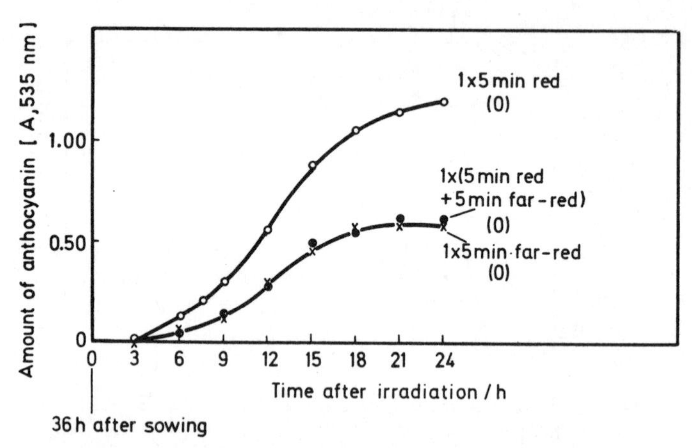

Figure 2.5 The time-courses of anthocyanin synthesis in the mustard seedling after brief irradiations with red and/or far-red light (5 min each). Irradiations were performed at time zero, i.e. 36 h after sowing. (After Lange *et al.*[24], by courtesy of the American Society of Plant Physiologists.)

deviations of the phytochrome system from a true photo-steady state are minimised, must be found empirically. In the course of time a 'standard far-red source[194] has been developed which satisfies these requirements[17, 19]. The theory behind it is briefly the following: according to the phytochrome model in Figure 2.2 any change of the total phytochrome ($P_{tot} = P_r + P_{fr}$) content in the mustard seedling can be described by

$$\frac{dP}{dt} - {}^0k_s - {}^1k_d[P_{fr}]$$

where $[P_{fr}]$ is the content of P_{fr}, ${}^0k_s$ is a zero-order rate constant of synthesis of P_r, and ${}^1k_d$ is a first-order rate constant for destruction of P_{fr}[19]. The rate constants ${}^0k_s$ and ${}^1k_d$ are independent of light at least within certain limits[105, 106]. The assumption of a zero-order rate constant of synthesis of P_r is justified by spectrophotometric data such as those in Figure 2.6. Moreover, using D_2O for density labelling of the protein moiety of phytochrome, Quail[104] has shown recently that the increase of the spectrophotometric signal in the apical hook and in the cotyledons of pumpkin seedlings must be attributed to *de novo* synthesis of phytochrome (P_r) protein in the dark-grown seedling, as well as in a seedling treated with light and later placed in the dark ('recovery increase' of P_r).

In the steady state, i.e. when $dP/d_t = 0$, then

$$^0k_s = {}^1k_d[P_{fr}] = {}^1k_d\,\varphi[P_{tot}]$$

whereby

$$\varphi = [P_{fr}]/[P_{tot}]$$

Hence, in the steady state the level of total phytochrome is a function of ^0k_s, ^1k_d and φ whereas the level of P_{fr} is only a function of the rate of synthesis of P_r and the rate of destruction of P_{fr}.

As an example, Figure 2.6 indicates that 'standard far-red light[194] which is equivalent (as far as the photo-equilibrium of phytochrome is concerned)

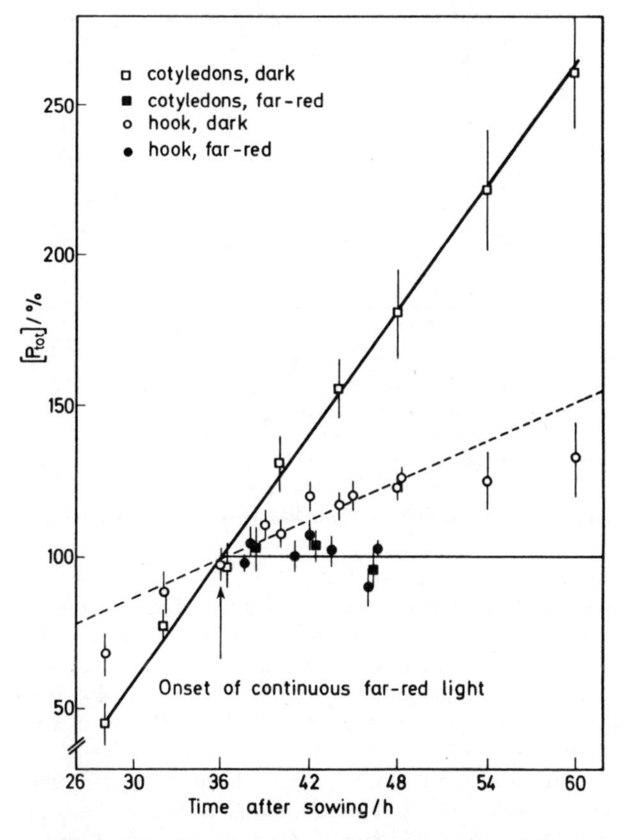

Figure 2.6 Time-courses of change of total phytochrome ($[P_{tot}] = [P_r] + [P_{fr}]$) in the cotyledons and in the hypocotyl hook of the mustard seedling in the dark and under continuous far-red light. Onset of light: 36 h after sowing. The deviations of the measured points from the regression line (hook) below 30 h and above 48 h can be accounted for by changes in the system of reference ('hook'). (From Schäfer *et al.*[19], by courtesy of the authors)

to the wavelength 718 nm[25] maintains a φ-value in cotyledons and hypocotyl hook of the mustard seedling which permits a long-term steady state of the phytochrome system (as defined by $[P_{tot}]$ = constant). Since the ^0k_s value differs by a factor of 3.2 in the two organs (cotyledons and hypocotyl hook) the φ-value must differ to the same degree ($\varphi_{fr,\,hook} = 0.023$; $\varphi_{fr,\,cotyledons} = 0.074$)[19, 25].

It has been mentioned that at least in dicotyledonous seedlings the destruction (decay) of P_{fr} follows first-order kinetics ($-dP_{fr}/dt = {}^1k_d[P_{fr}]$). This finding most likely implies that once a particular P_{fr} molecule is formed, its

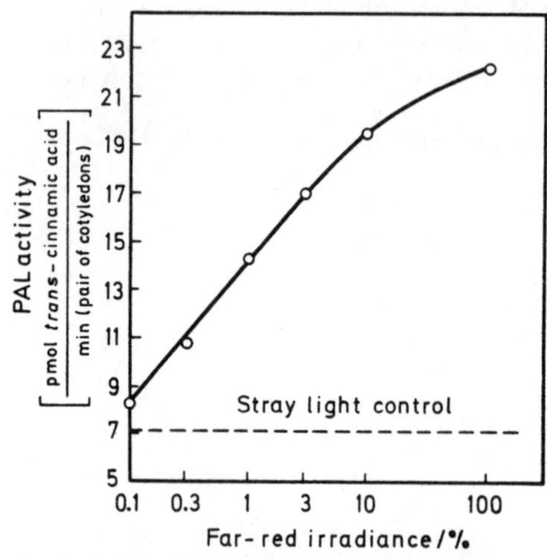

Figure 2.7 Dependence on irradiance of PAL synthesis. Irradiance of standard far-red light = 100%. Enzyme extraction: 3 h after onset of continuous far-red light (51 h after sowing). (From Schopfer and Mohr[14], by courtesy of the American Society of Plant Physiologists.)

chance of being destroyed as compared to the other molecules of the total P_{fr} population is random. The simplest model to account for this situation is one in which P_{fr} molecules are individually available to a degradative 'agent' which is present at all times in excess. While no detailed model is available, the temperature dependency of the decay of P_{fr} indicates an enzymic reaction. The Q_{10}-values between 14 and 34 °C are of the order of 3 (for details cf. Ref. 4).

There is one significant point which must be discussed briefly, the so-called 'high irradiance response' (HIR). This term was coined to designate photomorphogenic photoprocesses which are strongly dependent on irradiance, even under conditions of continuous long-term irradiation. For example, the extent of induction of phenylalanine ammonia-lyase (PAL) in the cotyledons of the mustard seedling is approximately a log-function of irradiance of the standard far-red light applied continuously (Figure 2.7). Hartmann[26-29] proposed that 'high irradiance responses' are also mediated by phytochrome and arise from the photochemical properties of the phytochrome system.

At any wavelength a photo-equilibrium is reached which is irradiance indedendent (assuming as a matter of course that the photochemical rate constants $^1k_1$ and $^1k_2$ are much higher than the ^1k_d rate constant for P_{fr} destruction and the ^1k_r rate constant for dark reversion). This assumption is only rigorously fulfilled if the irradiance is infinitely high. At any finite irradiance the actual 'photo-equilibrium' *in vivo* is less than the theoretical photo-equilibrum under the conditions $^1k_d = 0$ and $^1k_r = 0$. If ^1k_d and/or $^1k_r > 0$, the actual 'photo-equilibrium' is irradiance dependent. This point must be kept in mind in attempting to interpret the HIR[36]. In addition, however, cycling of the phytochrome molecules is a function of the irradiance. The larger the irradiance, the higher the rate of photoconversion of the phytochrome system.

Hartmann[26] attributes the irradiance dependency of continuous far-red light to an 'over-criticially activated P_{fr}'. Schopfer and Mohr[14] have introduced the symbol P_{fr}^* to designate this hypothetical P_{fr} species.

At the present state of the debate[1,4] one is led to the conclusion that a distinction must be made between $P_{fr(ground\ state)}$ and P_{fr}^* and that at least part of the effect of continuous standard far-red light must be attributed to the action of P_{fr}^*. Unfortunately, P_{fr}^* cannot be measured so far by physical means and the interpretation of P_{fr}^* in physical terms still remains completely unresolved. However, the *experimental* evidence that the strong action of continuous far-red light is due in some way to phytochrome has been convincingly presented by Hartmann[26-29]. The interpretation of $P_{fr(ground\ state)}$ is simpler. The established operational criteria for the involvement of phytochrome in a light-mediated response require that an induction effected by a brief irradiation with red light can be fully reversed by a subsequent pulse of far-red light (cf. Figure 2.5). Implicitly it is assumed that the 'effector molecule' is P_{fr} *in the ground state* (i.e. non-excited state) since P_{fr} is supposed to act in the dark. For theoretical reasons[30], the ground state of P_{fr} is very probably a $singlet_0$ state. The spectrophotometrically measurable P_{fr} represents P_{fr} in the ground state. In the case of the traditional light-pulse responses, the general function which relates the extent of a response Δm to the amount of spectrophotometrically detectable P_{fr}, must be written as

$$\Delta m = f([P_{fr}]_{ground\ state})$$

where $[P_{fr}]_{ground\ state}$ is the concentration of P_{fr} present immediately after termination of a brief irradiation. In principle this function can be elaborated empirically for every individual photoresponse.

2.1.6 Appropriate systems of reference

2.1.6.1 Biological units

The photochemical formation and maintenance of the chromoprotein P_{fr} in a plant will lead to a great number of photoresponses which together constitute the phenomenon of photomorphogenesis (cf. Figure 2.1, Table 2.2). It is hoped to understand 'photomorphogenesis' in molecular terms, in particular in terms of phytochrome-mediated enzyme induction and enzyme repression. The experimental material in many instances has been the dark-grown mustard seedling (*Sinapis alba* L.) between 36 and 72 h after sowing at 25 °C (cf. Figure 2.1). During this period there is no significant increase in DNA content (and very probably cell number) in either cotyledons or hypocotyl[31,32]. For this reason the biological unit (organ)—pair of cotyledons or hypocotyl—can be used as a system of reference for the enzyme data instead of 'cell' or 'unit DNA'. Other systems of reference such as 'total protein', 'soluble protein' or 'total RNA' would be misleading since these parameters change considerably under the influence of phytochrome[33-35]. One limitation, however, must always be kept in mind[10]: in most cases information about the compartmentalisation of a particular enzyme with respect to the organ (tissue) and with respect to the cell is not available. What can be achieved at best by applying extraction and *in vitro* assay procedures is a quantitative estimate of the *average* amount of active enzyme molecules in the biological unit (e.g., in a pair of cotyledons).

2.1.6.2 Phytochrome-insensitive enzyme levels

There are enzymes in the mustard seedling (possibly the majority of enzymes of the basic metabolism) whose temporal development, i.e. apparent synthesis and apparent decay, is not influenced by phytochrome even if dramatic photomorphogenic changes are induced[4]. Isocitrate-lyase (a key enzyme of the glyoxylate cycle) may serve as an example of those enzymes that have time-courses of enzyme levels which are totally independent of phytochrome (Figure 2.8). Neither light pulses ($P_{fr(ground\ state)}$) nor continuous far-red light

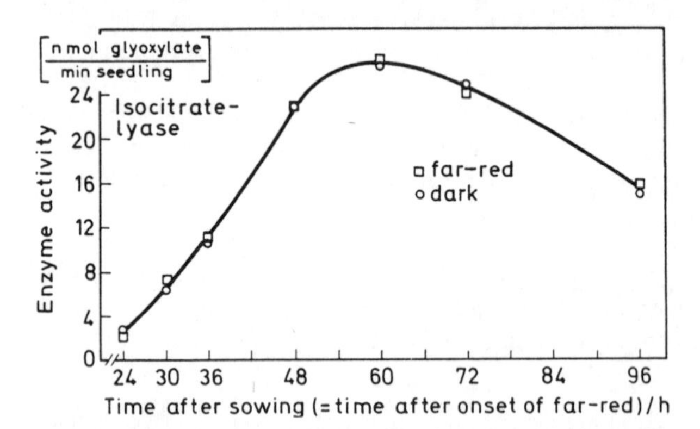

Figure 2.8 Time-course of isocitrate-lyase in the mustard seedling in darkness and under the influence of continuous far-red light. (From Karow and Mohr[12], by courtesy of Springer.)

(P^{*}_{fr}) have a significant influence on isocitrate-lyase, although the enzyme does show a 'grand period', i.e. a strong increase and a following decline of activity during the period of experimentation. If the onset of far-red light is only 36 h after sowing, the same result is obtained[12]. The fact that neither the temporal development nor the relative distribution of some enzymes within the seedling are significantly affected by the far-red light treatment[76] is amazing in so far as a seedling grown under far-red light for 36 h differs a great deal from a dark-grown seedling of the same age (cf. Figure 2.1): the far-red-grown seedling has large cotyledons and a short hypocotyl; the dark-grown seedling has small cotyledons and a long hypocotyl.

In conclusion, the occurrence of phytochrome-mediated photomorphogenesis is a specific phenomenon and does not automatically affect every aspect of metabolism and integration within the plant. The terms *differential* enzyme induction or *differential* enzyme repression have been used to describe the fact that in a given plant or plant organ some enzymes are induced or repressed by phytochrome while other enzyme levels are not affected.

2.2 PHYTOCHROME-MEDIATED ENZYME INDUCTION

Marcus[37] was the first to report a control of enzyme levels by phytochrome. He found an increase in NADP-dependent glyceraldehyde-3-phosphate

Table 2.3 Survey of plant enzymes controlled by phytochrome. The references refer to the first report in the literature dealing with the control by phytochrome as demonstrated by red-far-red reversibility and/or continuous far-red induction experiments. Supplemented from Schopfer[10], by courtesy of the author

Enzyme	*Plant*	*Author*
Glyceraldehyde 3-P dehydrogenase (NADP)	*Phaseolus*	Marcus[37]
Amino acid activating enzymes	*Pisum*	Henshall and Goodwin[41]
Phenylalanine ammonia-lyase	*Sinapis*	Durst and Mohr[39, 40]
	Pisum	Attridge and Smith[127]
Ribulose 1,5-diP carboxylase	*Secale*	Feierabend and Pirson[42]
Transketolase	*Secale*	Feierabend and Pirson[42]
Lipoxygenase	*Cucurbita*	Surrey[43]
Alkaline fructose 1,6-diphosphatase	*Pisum*	Graham *et al.*[44]
Glycollate oxidase	*Phaseolus*	Klein[45]
Ascorbate oxidase	*Sinapis*	van Poucke *et al.*[46]
Adenylate kinase	*Zea*	Butler and Bennett[47]
Inorganic pyrophosphatase	*Zea*	Butler and Bennett[47]
Phosphoenolpyruvate carboxylase	*Kalanchoe*	Queiroz[48]
Malic enzyme	*Kalanchoe*	Queiroz[48]
Malate dehydrogenase (NAD)	*Kalanchoe*	Queiroz[48]
NAD kinase	*Pharbitis*	Tezuka and Yamamoto[49]
Succinyl CoA synthetase	*Phaseolus*	Steer and Gibbs[50]
DNA-dependent RNA polymerase	*Pisum*	Bottomley[51]
Glyoxylate reductase	*Sinapis*	van Poucke *et al.*[52]
Amylase	*Sinapis*	Drumm *et al.*[9]
Cinnamic acid 4-hydroxylase*	*Pisum*	Russell[54]
Peroxidase	*Sinapis*	Schopfer and Plachy[55]
Glyceraldehyde 3-P dehydrogenase (NAD)	*Sinapis*	Cerff[56]
Ribonuclease	*Lupinus*	Acton[57]
Nitrate reductase	*Pisum*	Jones and Sheard[58]
Glutathione reductase	*Sinapis*	Drumm[59]

* In this case, fluorescent light pulses were found to be very effective but far-red reversibility was not tried.

dehydrogenase activity in the primary leaves of etiolated bean seedlings following short pulses of red light. The operational criteria for the involvement of phytochrome (cf. Figure 2.5) were found to be fulfilled. The participation of P_{fr} in this light-mediated response was thus clearly established. Margulies[38] confirmed these results and moreover found that a sequence of short pulses of red light could be replaced by continuous far-red light in producing the response. In 1966, the induction of phenylalanine ammonia-lyase by phytochrome was reported[39, 40] with particular emphasis on the lag-phase and *kinetics* (time-course) of the response. Since that time a considerable number of enzymes have been shown to be under the control of phytochrome. Table 2.3 includes only those enzymes for which the involvement of phytochrome in controlling the enzyme levels was clearly demonstrated.

There are many more enzymes whose levels have been found to be controlled by white light[10, 60]. However, unless the participation of phytochrome has been rigorously demonstrated by induction-reversion experiments (cf. Figure 2.5), other control mechanisms must be considered as well. As an example, induction of phenylalanine ammonia-lyase in disks of mature green *Xanthium* leaves[7, 61] and the induction of chymotrypsin inhibitor protein in potato[62-64] and tomato[63, 64] leaves when they are excised and placed in high irradiance white light, have been related to light absorption by chlorophyll and to products of photosynthesis rather than to the action of phytochrome[3].

In the following it is mainly the control by phytochrome of enzyme levels in the mustard seedling (cf. Figure 2.1). which is considered. The reasons for this restriction are that most of the *kinetic* data on phytochrome-mediated control of enzyme levels in plants were obtained with this system[4, 10]; that the spectrophotometric investigation of the phytochrome system is in an advanced state in this seedling[19]; that continuous standard far-red light maintains a true photo-steady state of the phytochrome system in cotyledons and hypocotyl hook (cf. Figure 2.6); and that suitable and simple systems of reference for the molecular data are available.

2.2.1 L-Phenylalanine ammonia-lyase (PAL) (E.C. 4.3.1.5)

2.2.1.1 Basic experiments

This enzyme catalyses the formation of *trans*-cinnamic acid from phenylalanine. In this function it is a key enzyme of the secondary plant metabolism including flavonoid biogenesis. Plants, unlike animals, produce and accumulate a large variety of compounds (secondary plant products) which are not essential for the basic metabolism common to all living systems. Control of biogenesis of secondary plant products (in particular, flavonoids) has often been used to study problems in genetics and developmental physiology. Since synthesis of anthocyanin in the mustard seedling (in particular in the mustard seedling cotyledons) can be induced by $P_{fr(ground\ state)}$ (cf. Figure 2.5), one could assume that phytochrome induces PAL synthesis. This is indeed the case (Figure 2.9). In the cotyledons of the dark-grown mustard seedling PAL activity can hardly be detected. However, the enzyme can rapidly be

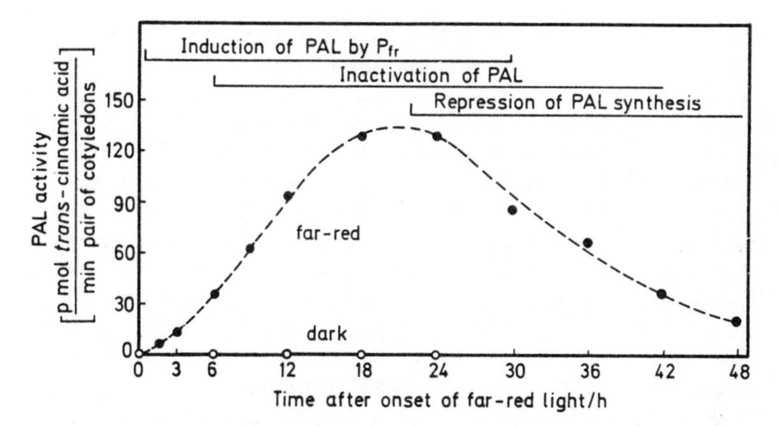

Figure 2.9 The influence of continuous far-red light on the basic kinetics of PAL levels in the cotyledons of the mustard seedling. Onset of far-red light: 36 h after sowing (25 °C). In the dark-grown mustard cotyledons PAL activity cannot be detected by the assay used. The basic kinetics can be explained by an interaction of the following three factors: enzyme induction, enzyme degradation (inactivation), repression of PAL synthesis. (From Dittes *et al.*[65], by courtesy of Verlag der Zeitschrift für Naturforschung.)

Table 2.4 Induction-reversion experiments demonstrating the involvement of phytochrome ($P_{fr(ground\ state)}$) in the light-mediated induction of PAL in the cotyledons of 48-h-old dark-grown mustard seedlings. With the very sensitive assay used in these experiments a dark level of PAL can be detected even in the cotyledons. (From Schopfer and Mohr[14], by courtesy of the American Society of Plant Physiologists.)

Treatment	*Enzyme activity* $\left[\frac{\text{pmol } \textit{trans-cinnamic acid}}{\text{min } \textit{pair of cotyledons}}\right]$
48 h dark	4.1 ± 0.3
52 h dark	5.2 ± 0.5
48 h dark + 5 min red + 4 h dark	13.6 ± 0.6
48 h dark + 5 min far-red + 4 h dark	10.8 ± 0.6
48 h dark + 5 min red + 5 min far-red + 4 h dark	10.4 ± 0.6
48 h dark + 5 min far-red + 5 min red + 4 h dark	14.3 ± 0.8

induced by continuous far-red light. The operational criteria for the involvement of phytochrome are clearly fulfilled (Table 2.4).

2.2.1.2 De novo *synthesis*

The usual experiments with the usual inhibitors of RNA and protein synthesis (Actinomycin D, Puromycin, Cycloheximide) suggest that the phytochrome-mediated increase of enzyme activity is due to an increase in the number of enzyme molecules[8, 65]. This conclusion has been supported by Schopfer and Hock[6] who used the *in vivo* density-labelling technique with deuterium oxide. The experiments show unambiguously that a true *de novo* synthesis of PAL occurs in the cotyledons of the mustard seedling. Even with improved methods of disk electrophoresis on polyacrylamide gel no indications were found of PAL isoenzymes in the organs (cotyledons, hypocotyl, taproot) of the mustard seedling. PAL moves as a single, homogeneous band under all conditions[66].

2.2.1.3 Interpretation of the 'basic kinetics'

The problem has been to understand the far-red-mediated kinetics of PAL ('basic kinetics') in the mustard cotyledons. Figure 2.9 indicates that at least three overlapping aspects are involved: induction (synthesis) of PAL; inactivation (degradation) of PAL; repression of PAL synthesis (i.e., for some reason or other the system no longer responds to phytochrome with the formation of PAL). The analysis of far-red⟶dark kinetics has contributed the most towards an interpretation (cf. Figure 2.12). The basis of these experiments is briefly that at the moment when the high irradiance far-red light is turned off the physiological effectiveness of a given P_{fr} pool drops instantaneously to a low level since cycling of the phytochrome system is arrested, and the species P^*_{fr} rapidly disappears. The drop of the PAL level following the withdrawal of the stimulus has been attributed to a degradation of the enzyme[40, 61, 67]. Changes in the amount of an enzyme in the cell (tissue) as a response to phytochrome can, of course, be due to a changing rate of synthesis, a changing rate of degradation, or to a combination of both. In the case of PAL the available information suggests that both rates change upon illumination[3]. The appearance of an 'inactivating principle' during PAL induction is thought to involve the induction of synthesis of a protein inactivator because inactivation of PAL can be prevented by the application of cycloheximide[61, 68, 69]. The data indicate that the 'degradative agent' (possibly a proteolytic enzyme) has a rapid turnover and needs to be constantly resynthesised .Unfortunately, nothing is known about the mechanism of PAL inactivation. This is part of the general problem that at present our knowledge concerning the modes of intracellular enzymatic enzyme degradation is very limited[70].

The interpretation of the basic kinetics of PAL in mustard cotyledons which involves induction of PAL synthesis, degradation of PAL and repression of PAL synthesis (by an unknown internal factor) has been supported by similar results obtained with other plant species. As an example, the decrease of PAL activity in the cotyledons of radish (*Raphanus sativus* L.) follows the kinetics of a first-order reaction after transfer of the seedlings from light to darkness[71]. The rate of decrease is the same under continuous far-red light after the stage of development is reached at which 'repression of PAL synthesis' comes into play (Figure 2.10). In gherkin hypocotyls (*Cucumis sativus* L.) Engelsma[72] has attributed this repression of PAL synthesis to the accumulation of 'end products' such as cinnamic acid and its derivatives. However, in

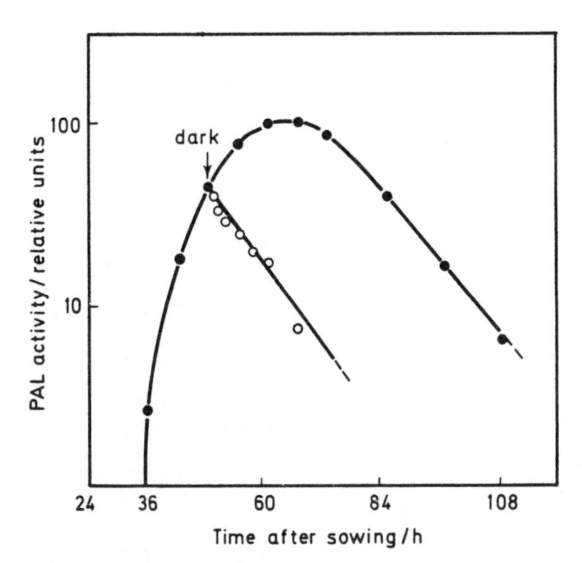

Figure 2.10 A comparison of the kinetics of PAL disappearance in the cotyledons of the radish seedling under continuous far-red light and in a far-red → dark experiment. The seedlings were transferred to darkness after 12 h of far-red light. (From Huault *et al.*[71], by courtesy of Acad. Sci., Paris.)

later experiments with the hypocotyl of red cabbage Engelsma[73] discovered that the effects of cinnamic acid and *p*-coumaric acid may be non-specific and due to damage of the segments. In any case, the problem of the 'mechanism of repression' is still unsolved.

2.2.1.4 Lag-phases

At 36 h or 48 h after sowing (at 25 °C) there is always a significant lag-phase, on the order of 45 min, before PAL synthesis becomes measurable after the onset of continuous far-red light or after a red or far-red light pulse (Figure 2.11). If, however, a seedling which has been preirradiated with 12 h of far-red light is kept in darkness for 6 h and is then re-irradiated with the same far-red light, no lag-phase for the action of the second irradiation can be measured. Enzyme increase is instantaneous and *linear*. Since the action of the second irradiation, as measured by increase of enzyme activity, can be completely inhibited by relatively low doses of Puromycin and Cycloheximide, it has been concluded that the reappearance of P_{fr} leads to rapid *de novo* synthesis of enzyme protein[8].

While in the mustard seedling a singificant secondary lag-phase (i.e., a lag-phase after the second onset of light in the programme light–dark–light) could not be detected in any response[4], Engelsma[73] has reported that in red cabbage (*Brassica oleracea* L.) the increase in PAL activity

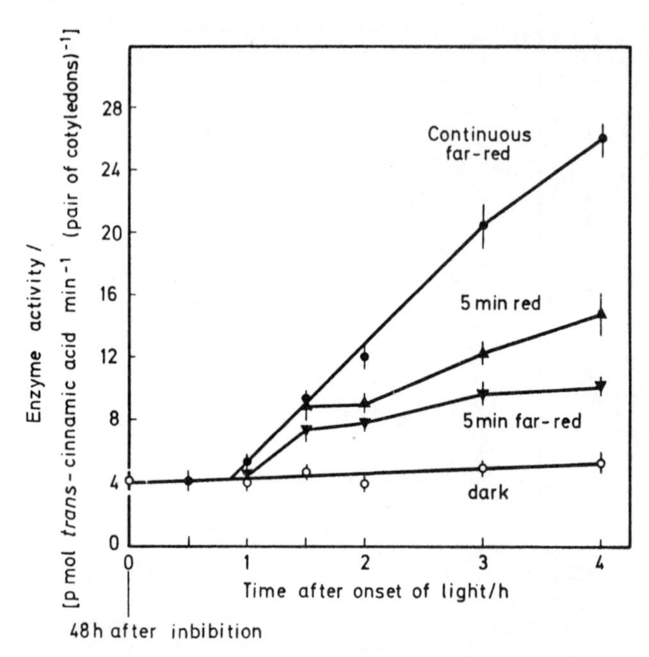

Figure 2.11 Increase of PAL level in mustard seedling cotyledons under continuous far-red light, or following a brief irradiation (5 min) with red or far-red light at time zero (48 h after sowing). The lag-phase is the same with all treatments. With the sensitive assay used in these experiments a dark level of PAL can be detected even in the cotyledons. (From Schopfer and Mohr[14], by courtesy of American Society of Plant Physiologists.)

induced by a second irradiation is preceded by a considerable lag-phase. To understand these differences a detailed knowledge of the action of phytochrome during the initial lag-phase is required. This aspect will be dealt with in connection with phytochrome-mediated anthocyanin synthesis.

2.2.1.5 Kinetics of PAL levels in different organs of the same plant (mustard seedling)

Figure 2.12 shows that in hypocotyl (as well as in the taproot) PAL appears in considerable amounts even in the dark. Furthermore, the basic kinetics of PAL (i.e. the time-course of PAL levels under continuous standard far-red light) in the hypocotyl are totally different as compared to the basic kinetics in the cotyledons. The preferred explanation of the hypocotyl data is based on the assumption that PAL synthesis in the dark and PAL synthesis mediated by phytochrome are totally independent phenomena. (However, there are no indications that PAL as a molecule is different in the two organs, according to studies made by Schopfer[66].) If the dark kinetics are subtracted from the far-red kinetics, a time-course of the 'remainder enzyme' is found which is

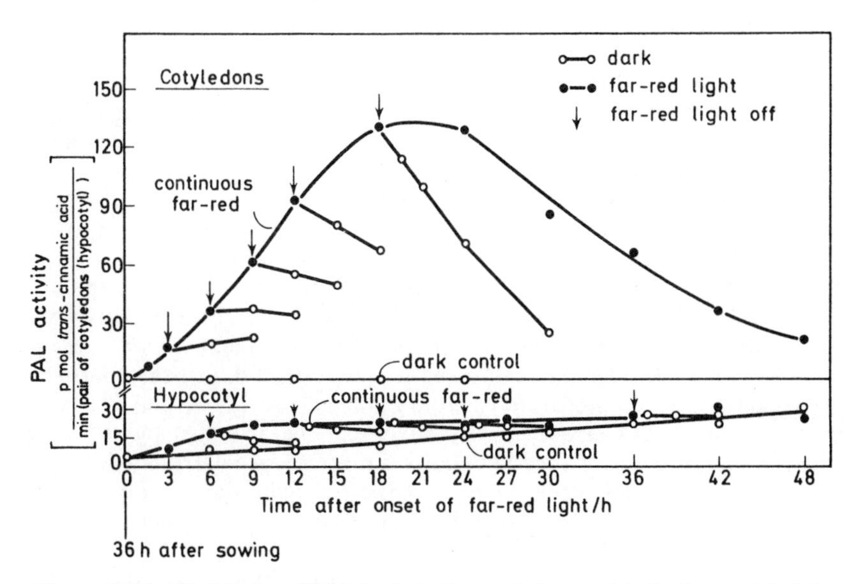

Figure 2.12 The kinetics of PAL levels in the cotyledons and in the hypocotyl of the mustard seedling in the dark and under continuous far-red light. Onset of light: 36 h after sowing (25 °C). In addition, a number of far-red → dark kinetics are indicated. This term is used to designate those kinetics of the enzyme which are observed after the standard far-red light has been turned off (at arrows). (From Dittes *et al.*[65], by courtesy of Verlag der Zeitschrift für Naturforschung.)

similar to the far-red kinetics in the cotyledons and possibly explicable in the same way (cf. Figure 2.9). Since the 'dark enzyme' and the 'P_{fr}-dependent enzyme' do not differ in gel electrophoresis[66] or in any other aspect tested so far, the conclusion is justified that the two enzymes are identical. The situation as found experimentally in the hypocotyl can then be interpreted as follows: there are tissues in the hypocotyl (e.g. in the differentiating xylem[74, 75]) which produce the enzyme in the dark. This enzyme is stable, at least during the period of experimentation. The apparent 'stability' of the enzyme can be ascribed to a lack of the 'inactivating principle' in some tissues. On the other hand, there are tissues in the hypocotyl (e.g. the anthocyanin-synthesising subepidermal layer) which produce and destroy the enzyme in precisely the same manner as do the cotyledons. These two responses of the hypocotyl are independent of one another. If this interpretation is correct, the only difference between the cotyledons and the hypocotyl with respect to PAL would be that in the one organ there is considerable dark synthesis of a stable enzyme while in the other organ dark synthesis of PAL is scarcely detectable.

While it is possible to interpret the PAL data of Figure 2.12 with more complicated models[7], it is in any case evident from the facts that the regulation of enzyme levels in organised plant tissues is complex, and that multiple mechanisms exist for controlling the level of a particular enzyme in a particular organ. Obviously there is no single or simple mechanism which controls the levels of an enzyme in all instances. Rather, as in the case of mammalian tissues[77], we are led to the conclusion that control of the level of a particular enzyme may be exerted at any point at which control can be potentially exerted, and that this will depend upon the enzyme and the tissue involved. Two further examples may further illustrate this point: (a) Recently, *repression* of PAL by phytochrome was demonstrated in tissue cultures of Jerusalem artichoke (*Helianthus tuberosus* L.)[78]. While the enzyme level in the dark is high, the level decreases dramatically under the influence of light. The light effect is clearly phytochrome-mediated (operational criteria). (b) In general enzyme induction as well as reversible enzyme activation by light in *green* tissue have been attributed to light absorption in the photosynthetic apparatus[3]. However, Tezuka and Yamamoto[79,80] have shown recently that NAD kinase activity in green cotyledons of the short day plant *Pharbitis nil* is reversibly controlled by phytochrome. It was further shown that partially purified phytochrome preparations from etiolated pea hooks and oat coleoptiles contain NAD kinase activity which can be activated by red light *in vitro*. Since the effect of a red light pulse is reversed by a corresponding far-red light pulse, there is little doubt that the red light effect is due to P_{fr}. This is the first report of the activation of an enzyme by phytochrome.

2.2.1.6 Involvement of RNA synthesis

Under the influence of continuous far-red light the amount of RNA per organ (or per unit DNA) will increase in the cotyledons of the mustard seedling. However, the lag-phase of this response is at least 6 h and the profile of the RNA (as revealed by MAK chromatography) does not change significantly. It has therefore been concluded[4] that the phytochrome-mediated increase of bulk RNA in growing tissue[33,35,81] is a non-specific secondary result rather than a cause of photomorphogenesis.

Since in short-term experiments (using a double-labelling technique of RNA with [^{14}C]-uridine and [^{3}H]-uridine and separation of the RNA on MAK columns) no significant changes in the RNA fractions could be detected in the mustard seedling cotyledons[4,82], the evidence that alterations of RNA metabolism are specifically involved in regulation of enzyme levels by phytochrome is only weak and indirect so far. A variety of phytochrome-mediated enzyme inductions in the mustard seedling (or inductions of anabolic fluxes) are *specifically* inhibited by the administration of Actinomycin D[4,8,65,83]. Characteristic of such systems, including PAL and anthocyanin induction, is the finding that whereas administration of Actinomycin D [10 μg ml^{-1}] before or at the onset of light nearly prevents apparent enzyme synthesis (or accumulation of the end product of an anabolic chain), delayed administration relative to the onset of light is much less effective. This observation suggests that RNA synthesis is necessary for initiation of specific enzyme synthesis but once RNA synthesis is accomplished the use of that RNA does not require its continued synthesis, although continued presence of P_{fr} is required for (apparent) enzyme synthesis or maintenance of anabolic fluxes[4].

So far, attempts to demonstrate changes in RNA metabolism under conditions where apparent synthesis of PAL is induced by far-red light have failed[82], however carefully they were carried out. What is urgently needed for a direct assessment of the role of RNA synthesis in enzyme induction by phytochrome is the isolation of specific messenger RNAs whose appearance can be correlated with the first appearance of P_{fr} in an etiolated seedling. The isolation of messenger RNA for haemoglobin from mouse reticulocytes[84] and the characterisation of messenger RNA from *Dictyostelium*[85] indicate the feasibility of a direct approach. However, a number of data suggest that differential synthesis of RNA is not the only point at which control over enzyme synthesis is exerted by phytochrome. Translational (or post-transcriptional) controls[86] must be envisaged as well. This aspect will be dealt with briefly in the following section.

2.2.1.7 PAL induction and phytochrome mediated anthocyanin synthesis

In the mustard seedling anthocyanin synthesis can be mediated by phytochrome without the interference of any other photochemical mechanism (cf. Figures 2.5 and 2.15). However, the action of $P_{fr(\text{ground state})}$ and of P_{fr}^{*} must be kept separate[86]. A significant finding has been that the duration of the 'initial lag-phase' is constant (3 h at 25 °C) for mustard seedlings more than 30 h old and is specific for the system, being independent of the dose, irradiance or quality of light (cf. Figures 2.5 and 2.15). The question arises of whether phytochrome is inactive in mediating anthocyanin synthesis during the 3 h of the initial lag, that is to say in operation terms, or whether an induction of anthocyanin synthesis by red light is fully reversible for a period of 3 h. The answer is given in Figure 2.13. Complete (or nearly complete) reversal of the effect of the red light is only possible for a period of the order of a few minutes after the onset of light. On the other hand, the rate of escape from reversibility is not rapid nor is the escape a total one, even after 4 h. This fact indicates that there is a continuous requirement for P_{fr} during the whole period of anthocyanin accumulation. Furthermore, P_{fr} clearly mediates anthocyanin synthesis during the lag-phase in spite of the fact that the actual synthesis of anthocyanin can proceed only after the lag-phase is overcome.

In the case of a secondary irradiation (e.g. 12 h light–18 h dark–light of the same quality and irradiance) no significant lag-phase can be detected[87]. The same is true even if the secondary irradiation differs in quality and irradiance. The only reservation is that the rate of anthocyanin synthesis mediated by secondary irradiation does not exceed the one mediated by the primary irradiation[24].

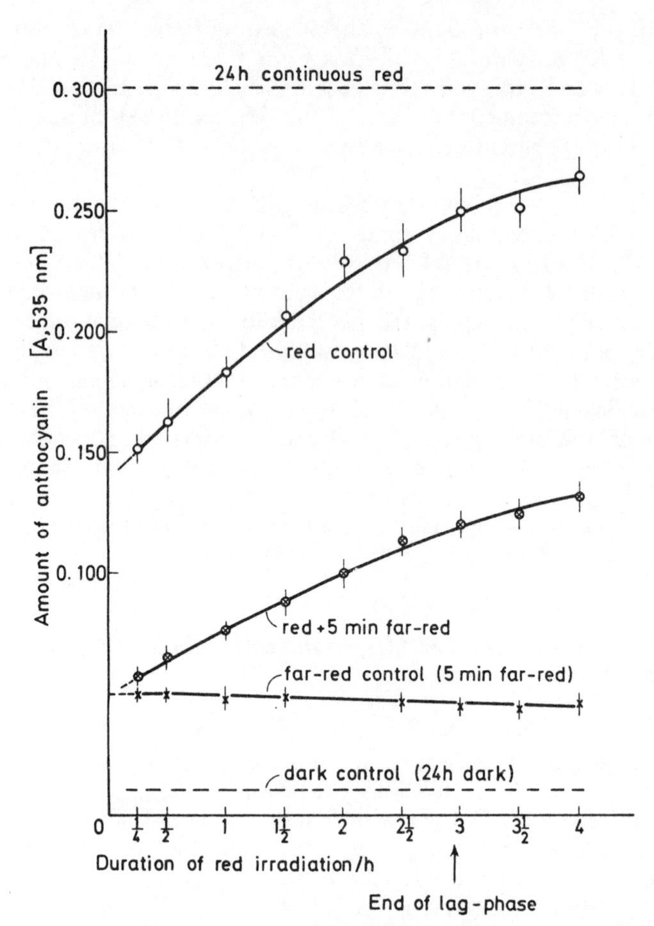

Figure 2.13 Test for reversibility in phytochrome-mediated anthocyanin induction during the initial lag-phase. After irradiation of the mustard seedlings with red light of varying duration (absicssa) the seedlings were irradiated for 5 min with long wave-length far-red (756 nm $\varphi<0.01$) and placed in the dark. Extraction of anthocyanin was performed 24 h after onset of the red light. The 'red controls' received only red light of varying durations; the 'far-red controls' received only 5 min far-red (756 nm) each at the times indicated. (From Lange *et al.*[24], by courtesy of American Society of Plant Physiologists.)

These and further data[4] have led to the conclusion that during the initial lag-phase P_{fr} exerts two functions: (a) a 'potential for biosynthesis' of anthocyanin ('capacity') is built up under the influence of P_{fr}. Since anthocyanin synthesis is very sensitive towards Actinomycin D during the lag-phase only[83, 88], the assumption seems to be justified that long-lived messenger RNA required for anthocyanin synthesis is formed during the initial lag-phase. Thus, a tentative identification of the 'capacity' with a given amount of stable messenger RNA has been made. It depends on the stability of the

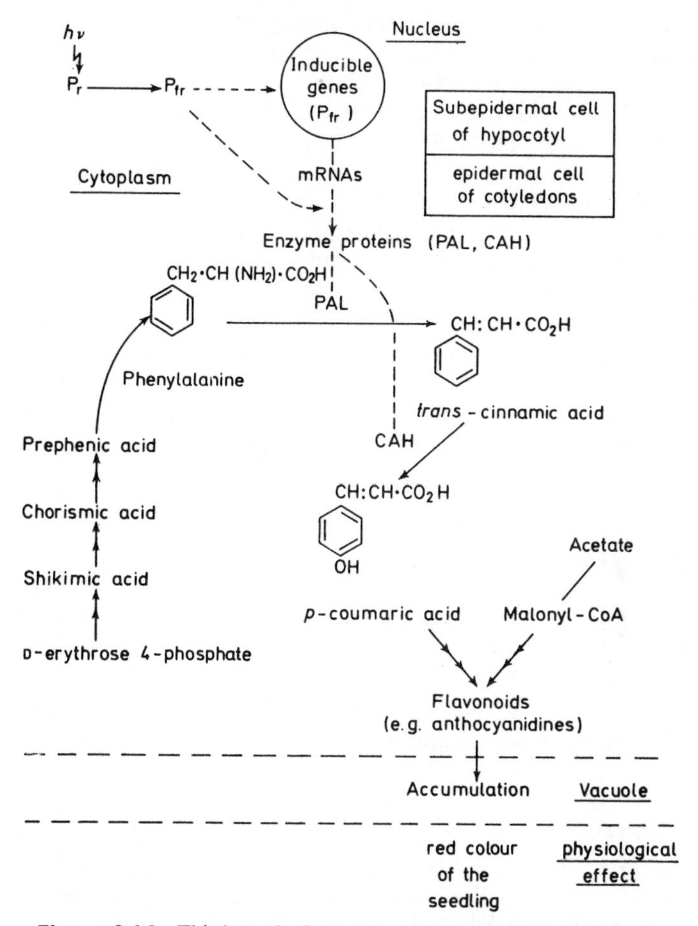

Figure 2.14 This hypothetical scheme is intended to emphasise the significance of phenylalanine ammonia-lyase (PAL) as key enzyme of the phenylpropanoid anabolism. The biosynthetic pathways which depart from *trans*-cinnamic acid eventually lead to anthocyanidins, among other secondary plant products such as lignin. The enzyme *trans*-cinnamic acid 4-hydroxylase (CAH) which catalyses the formation of *p*-coumaric acid from *trans*-cinnamic acid can also be induced by light pulses[54]. (From Mohr[4], by courtesy of Springer.)

'capacity' whether or not a secondary lag-phase can be observed[73, 87]. (b) Anthocyanin synthesis is mediated by P_{fr} ($P_{fr(ground\ state)}$ as well as P^*_{fr}) but this mediation can only become effective after the 'potential for biosynthesis' (possibly long-lived messenger RNA) has been built up. Since anthocyanin synthesis is always very sensitive towards inhibitors of protein synthesis[89], it is thought that the second effect of P_{fr} on anthocyanin synthesis might be exerted on the level of translation.

The question of the correlation of enzyme levels with anabolic pathway flux has been studied intensively in the case of PAL and anthocyanin synthesis in the mustard seedling. Although the mustard seedling forms five distinct anthocyanins, the aglycone is always cyanidin[90]. Figure 2.14 describes in principle the biochemical pathway by which cyanidin is formed, and it illustrates at the same time the hypothesis which explains the action of phytochrome on cyanidin biosynthesis in terms of 'differential enzyme induction'. The molecular skeleton of the flavonoids (including cyanidin) is derived from the cinnamic acid pathway as well as from the malonyl CoA pathway. The hypothesis assumes that the control by phytochrome is exerted along the cinnamic acid pathway through 'differential enzyme induction'. The double function of phytochrome in mediating anthocyanin synthesis in the mustard seedling (at the level of transcription as well as at the post-transcriptional

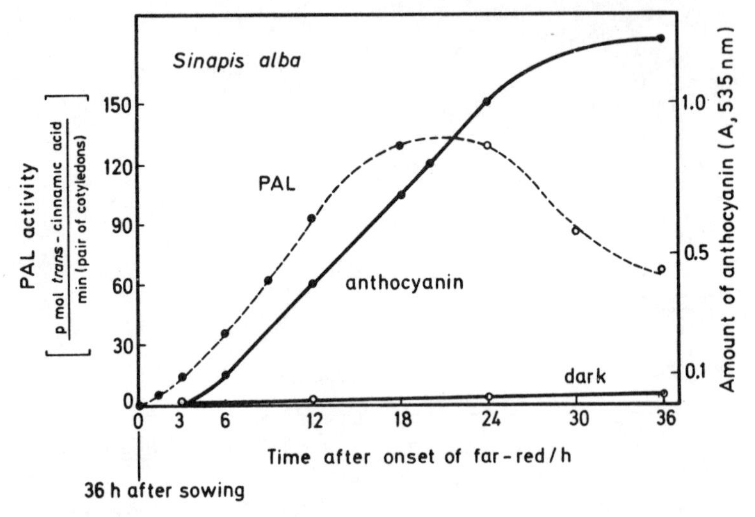

Figure 2.15 Time-courses of anthocyanin contents and PAL levels in the cotyledons of the mustard seedling. Dark values do not deviate significantly from zero in both cases. While PAL activity increases up to 18 h after the onset of light, the *rate* of anthocyanin snythesis is constant between 6 and 24 h after the onset of light. (From Mohr[4], by courtesy of Springer.)

level) is taken into account in the model. The key enzyme is PAL, which catalyses the formation of *trans*-cinnamic acid from phenylalanine and thus connects the phenylpropanoid pathway of secondary metabolism to the basic metabolism.

While the model in Figure 2.14 has been supported over the years by many experimental data[4], the problem has been whether or not PAL could be considered to be the rate-limiting enzyme for cyanidin formation. Figure 2.15 clearly shows that this is not the case. While PAL activity is a prerequisite for anthocyanin synthesis, the rate is determined by some other enzyme during the linear 'production phase'.

Formally, phytochrome-mediated anthocyanin synthesis which occurs only in epidermal cells of the cotyledons and in the subepidermal cells of the hypocotyl[91] can be used as a biochemical model system for differentiation[92]. Differential enzyme induction (including that of the non-stable enzyme PAL), mediated by phytochrome, leads to a *transient* modulation of the rate of anthocyanin synthesis (from a rate very close to zero to a very high rate, cf. Figure 2.15). Since the end product of the anabolic chain (anthocyanin) is stable (at least within the time-period of experimentation), cell differentiation occurs. The inherent problem of primary versus secondary differentiation will be touched on in a later section on phytochrome-mediated repression of lipoxygenase.

2.2.2 Ascorbate oxidase (AO) (*L-ascorbate:* O_2 *oxidoreductase*) (E.C. 1.10.3.3)

The terminal oxidase AO (soluble activity) can be induced by phytochrome in the mustard seedling[53]. The far-red——→ dark kinetics indicate that enzyme

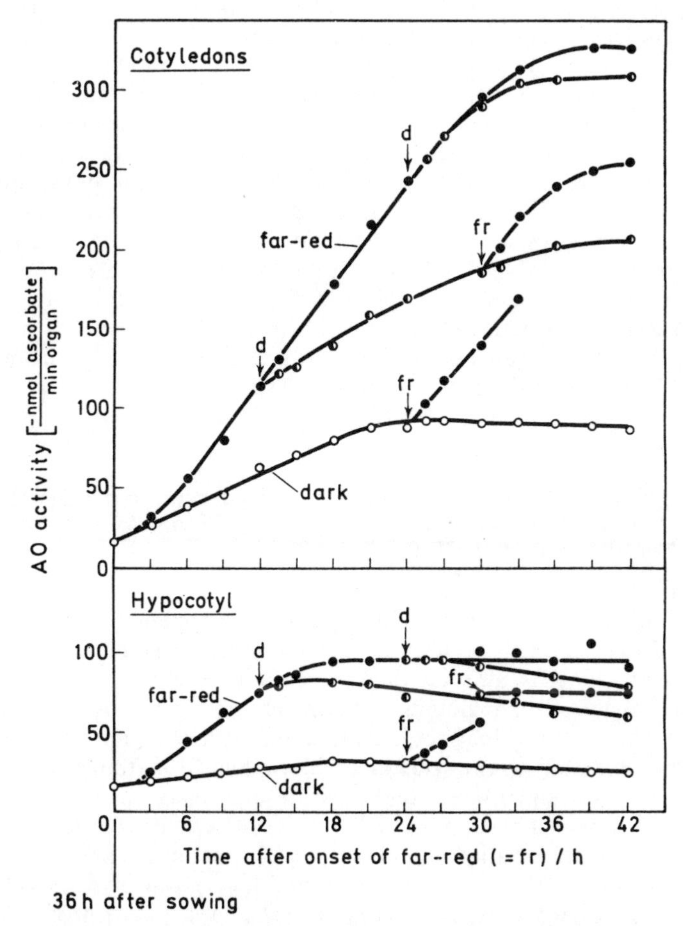

Figure 2.16 Time-courses of ascorbate oxidase (AO) in the cotyledons and in the hypocotyl of the mustard seedling in the dark and under the influence of standard far-red light (fr). In addition, far-red ——→dark kinetics are indicated (d). (From Drumm *et al.*[53], by courtesy of Springer.)

inactivation (destruction) does not come into play in the cotyledons to a significant extent (Figure 2.16, upper part). They further indicate that the action of phytochrome is continuously required to maintain a high rate of apparent enzyme synthesis. As the seedlings become older, apparent enzyme synthesis starts immediately after the onset of far-red light, i.e. without an appreciable initial lag-phase. Obviously phytochrome can induce apparent enzyme synthesis *rapidly* (even without light pretreatment) if the seedling is in the proper state of development. AO in the cotyledons seems to be representative of the majority of the enzymes which can be induced by phytochrome[4].

The behaviour of AO has been compared in cotyledons and hypocotyl of the mustard seedling (Figure 2.16). While the dark kinetics look very similar the enzyme kinetics under the influence of continuous far-red light ('basic kinetics') are conspicuously different in the two organs. The light-induced enzyme seems to be considerably less stable in the hypocotyl than in the cotyledons, at least up to 78 h after sowing. From *ca.* 18 h after the onset of light the constant AO level in the hypocotyl obviously represents a steady state (that is, the rate of far-red-mediated synthesis equals the rate of degradation). On the other hand, there are no indications of an enzyme decay in the cotyledons during the experimental period. These results and the data on PAL reported above lead to the conclusion[53] that the effect of phytochrome on enzyme induction is the same in both organs, cotyledons and hypocotyl. However, the processes of enzyme degradation are *specific* for the organ and for the enzyme. While in the case of PAL the phytochrome-induced enzyme behaves very similarly in cotyledons and hypocotyl (cf. Figure 2.12)., the far-red-mediated kinetics of AO are conspicuously different in the two organs. This difference can be attributed to a different degradation of the enzyme in the two organs during the experimental period.

2.2.3 D-glyceraldehyde-3-phosphate dehydrogenase (GPD) *(D-glyceraldehyde-3-phosphate) NADP oxidoreductase (phosphorylating)* (E.C.1.2.1.13)

Enzymes which are localised in the chloroplast compartment, such as NADP-dependent GPD, can also be induced by phytochrome (Figure 2.17). The far-red⟶dark kinetics indicate that the enzyme is stable and that P_{fr} is continuously required to maintain apparent enzyme synthesis. It must be emphasised that even a long-term irradiation with the standard far-red source[94] does not lead to a considerable chlorophyll formation[95]. However, the plastids do grow under these conditions and they reach the same size and dimensions as under normal white light[96]. The behaviour of NADP-linked GPD is possibly representative of that of other enzymes of the photosynthetic carbon reduction cycle as far as the regulation of *synthesis* by phytochrome is concerned[42], in that there is some enzyme in the dark; phytochrome exerts a strong induction; and the enzyme is relatively stable over the whole period of experimentation.

The question of where the NADP-dependent GPD is actually synthesised is still under debate[97]. The present consensus among the workers in the field seems to be that the NADP-linked DPG is encoded in the nucleus, synthesised in the cytoplasm and translocated by a specific mechanism to the plastid compartment. The mode of action of phytochrome is still obscure, and the point at which it comes into play in this process is not known. There is much evidence, however that the light-dependent resumption of plastid development as a whole is mediated by phytochrome[4]. It has long been known that the development of chloroplasts in higher plants is obligatorily dependent on light. In complete darkness only chlorophyll-free etioplasts originate, differing greatly from chloroplasts in their size, internal structure and molecular composition[98]. Chloroplast divisions have been observed very infrequently in the cells of higher plants[99, 100]. At least the majority of mature chloroplasts develop from small proplastids present in the 'undifferentiated' cells in the

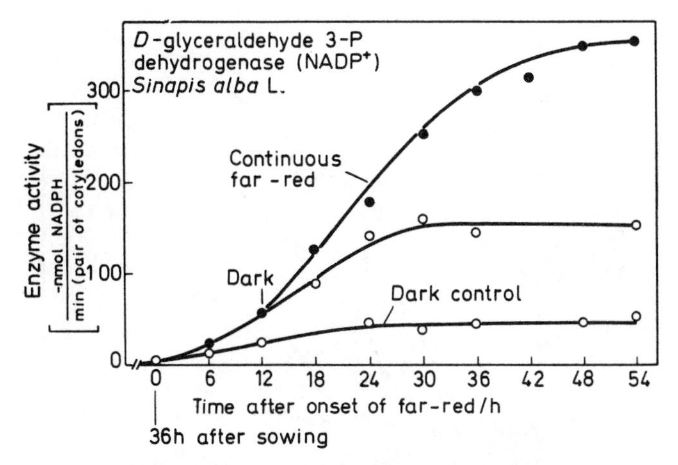

Figure 2.17 Time courses of D-glyceraldehyde -3-phosphate dehydrogenase (NADP) in the mustard seedling cotyledons in darkness and under the influence of continuous far-red light. (From Cerff,[93] by courtesy of the author.)

leaf primordia. It is not understood why plastid development in complete darkness proceeds only to a certain stage (proplastid, etioplast) and then stops. However, it is becoming clear that the light-dependent resumption of plastid development is a phytochrome-induced process.

There is ample evidence that chloroplasts contain DNA, DNA polymerase, RNA polymerase and a protein-synthesising apparatus[101]. However, there is also evidence that many genes concerned with chloroplast structure, molecular composition and function in higher plants are inherited in a Mendelian fashion and must therefore be located in the nucleus[99, 102]. In fact the formation of a fully functional chloroplast results from a complex interplay between the genetic information of the plastid, nuclear genes, hormones[103], and the external factor, light. The effect of light is twofold. The action of phytochrome has already been mentioned. The second point at which light comes into play is the phototransformation of protochlorophyllide to chlorophyllide a. There are no indications, however, that this phototransformation acts as a signal for the biosynthesis of plastidal or extraplastidal components[4,95].

2.2.4 Amylase[9] (α1,4-*glucan*) 4-*glucanohydrolase* (E.C. 3.2.1.1)

It is well-known[107,108] that the plant hormone GA_3 (gibberelic acid) will induce *de novo* synthesis of amylase in aleurone layers of embryo-free half-caryopses of some cereal grains such as barley or wheat. On the other hand, amylase can be induced in the cotyledons of the mustard seedling in the usual way (operational criteria) by the active form of phytochrome, P_{fr} (Figure 2.18). The far-red⟶ dark kinetics indicate that enzyme destruction does not

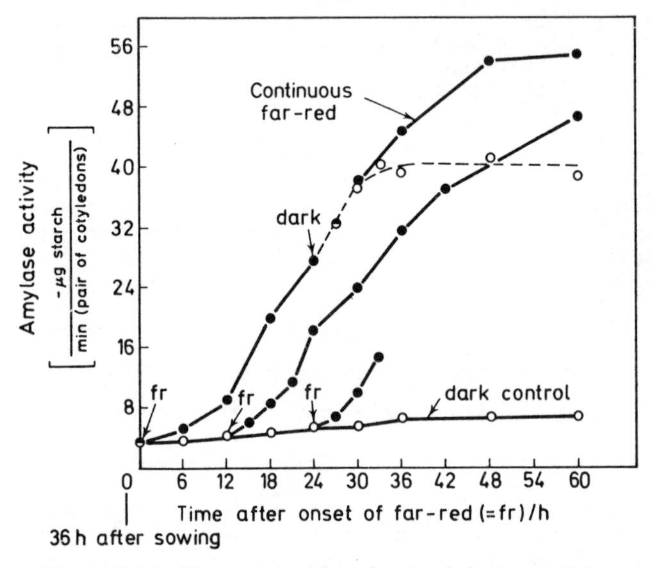

Figure 2.18 Time-courses of amylase levels in the cotyledons of the mustard seedling in darkness and under continuous far-red light (onset ↓). In addition, far-red→dark kinetics are indicated (○). (From Drumm *et al.*[9], by courtesy of Springer.)

play a significant role during the experimental period in the case of amylase either. Furthermore, it is obvious that the presence of P_{fr} is continuously required to maintain a high rate of apparent amylase synthesis.

Since the hypothesis has repeatedly been advanced[109] that the mechanism of P_{fr} action involves plant hormones at some intermediate step, an investigation was undertaken into whether or not GA_3 can induce amylase in the mustard seedling and whether or not P_{fr} can induce amylase in the embryo-free half-caryopses of barley and wheat. Table 2.5 gives the principle results. When mustard seedlings were incubated with solutions of GA_3 at concentrations which are saturating for cell elongation in the mustard hypocotyl[110], there was no induction of amylase. Rather, P_{fr}-mediated induction of amylase was significantly lower in GA_3-treated seedlings in comparison to the water controls, and this negative effect was already detectable at low concentrations. While this non-specific effect can probably be understood in terms of competition between hypocotyl and cotyledons for common pools of metabolites, the positive statement can be advanced that GA_3 does not participate in the mediation of amylase induction by P_{fr} in the cotyledons of the mustard seedling.

Table 2.5 Control of amylase increase in the mustard seedling by phytochrome (operationally, continuous far-red light = fr) and gibberellic acid (= GA_3). Submerged incubation in GA_3 or water was performed in the dark between 35 and 36 h after sowing. Onset of continuous far-red light: 36 h after sowing. (From Drumm *et al.*[53], by courtesy of Springer.)

Treatment after sowing (*dark* = d)	*Concentration of* GA_3 ($\mu g\ ml^{-1}$)	*Amylase activity* $\left[\frac{-\mu g\ starch}{min\ seedling}\right]$
35 h d	—	4.5
35 h d + 1 h water	—	4.6
35 h d + 1 h GA_3	50	4.6
35 h d + 1 h water + 24 h d	—	10.8
35 h d + 1 h GA_3 + 24 h d	50	10.0
35 h d + 1 h water + 24 h fr	—	51.5
35 h d + 1 h GA_3 + 24 h fr	50	36.2
35 h d + 1 h water + 24 h d	—	10.5
35 h d + 1 h GA_3 + 24 h d	100	7.8
35 h d + 1 h water + 24 h fr	—	48.8
35 h d + 1 h GA_3 + 24 h fr	100	32.8
35 h d + 1 h water + 24 h d	—	10.8
35 h d + 1 h GA_3 + 24 h d	0.35	10.9
35 h d + 1 h water + 24 h fr	—	50.0
35 h d + 1 h GA_3 + 24 h fr	0.35	35.8

To summarise the principal results of our investigations: synthesis of amylase in the embryo-free half-caryopsis of wheat and barley is induced by GA_3. An influence by phytochrome (operationally, continuous far-red light) cannot be detected. Synthesis of amylase in the cotyledons of the mustard seedling can be induced by P_{fr}. No positive influence of GA_3 can be detected. Obviously the amylase enzyme, which consists of several isoenzymes[9], can be induced in different systems by different effector molecules. There is no detectable interaction between phytochrome and GA_3 as far as induction of amylase is concerned.

2.2.5 Peroxidase[55, 111] (Donor: hydrogen-peroxide oxidoreductase; E.C. 1.11.1.7)

In the cotyledons of the mustard seedling continuous far-red light induces a strong increase in peroxidase activity* (Figure 2.19). The conventional

* The fact that several isoenzymes of peroxidase exist in the mustard seedling can be ignored in the present consideration.

red–far-red induction reversion experiments indicate that the operational criteria for the involvement of P_{fr} in this response are fulfilled.

The ability of the cotyledons to produce peroxidase under the influence of phytochrome is strongly dependent on the stage of development: P_{fr} is effective only when it is formed before *ca.* 96 h after sowing. However, peroxidase activity increases only after about 96 h after sowing. Thus, the formation of P_{fr} leads to apparent enzyme synthesis in a period during which the competence to respond to P_{fr} has already been lost. Obviously the induction process is clearly separated in time from the realisation of the response.

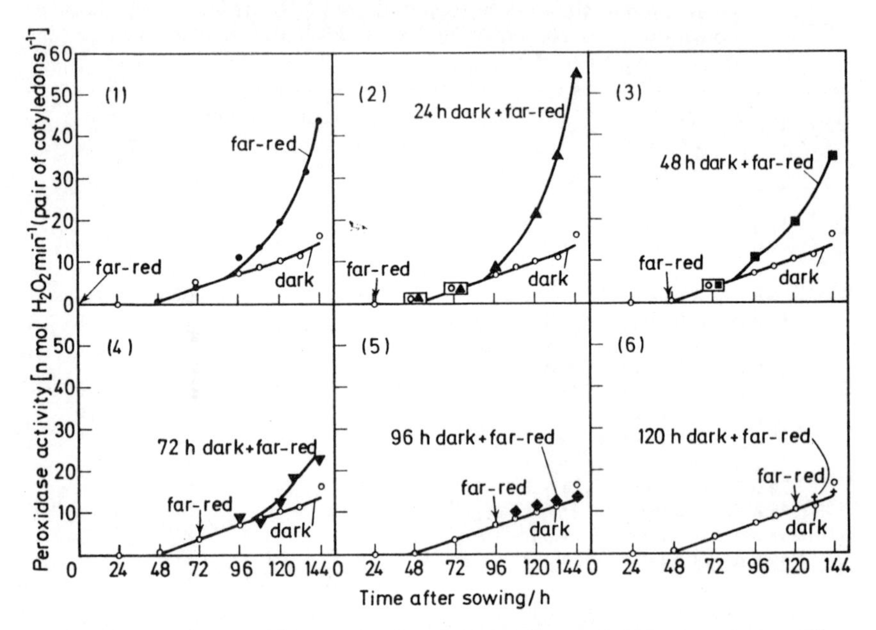

Figure 2.19 Time-courses of peroxidase activity in the cotyledons of mustard seedlings in the dark and under continuous far-red light. Onset of light: 0, 24, 48, 72, 96 or 120 h after sowing. (From Schopfer and Plachy[55], by courtesy of the authors.)

It was shown[111] that formation of P_{fr} before 48 h after sowing is also ineffective in the induction of peroxidase activity. However, if the seedlings are irradiated in the proper period of time of 'primary' differentiation (between 48 and 72 h after sowing) and then transferred to darkness, peroxidase accumulates at the same rate as under continuous far-red light for at least $2\frac{1}{2}$ days. Similar results are obtained when the irradiation lasts for 48 h longer before onset of darkness (120 h after sowing). The observed deviation of these far-red⟶dark kinetics from the far-red kinetics after *ca.* 130 h is probably an 'artifact' due to re-etiolation of the seedlings in the dark which leads to a faster depletion of reserve materials from the cotyledons.

Since P_{fr} disappears rapidly from the system in the dark, we need at least one stable intermediate in the metabolic chain between P_{fr} and apparent peroxidase synthesis which acts as a 'transmitter' of the primary effect of P_{fr}. Apparently the 'transmitter' can be formed in the presence of P_{fr} only at an early stage of 'primary' differentiation (*ca.* 48–96 h after sowing) and can act only at a later stage of 'primary' differentiation (*ca.* 96–120 h after sowing), resulting in an increased apparent synthesis of peroxidase. While the molecular nature of the 'transmitter' is still unknown at present, the formal interpretation requires that we keep two 'competences' separate: (a) the competence of the seedling for P_{fr} with respect to transmitter formation (*ca.* 48–96 h after sowing), and (b) the competence of the mustard cotyledons for transmitter with respect to apparent peroxidase synthesis (after 96 h).

2.2.6 Nitrate reductase (NAD(P)H_2: nitrate oxidoreductase; E.C. 1.9.6.1)

The extractable activity of nitrate reductase from leaves of higher plants is inducible by light and shows, under natural growth conditions, a pattern of diurnal variation[112]. In studies on the mechanism of light action green leaves were generally used as experimental material, implying that photosynthesis is involved in the induction process[113]. Recent experiments with etiolated peas have shown, however, that the light effect is mediated by phytochrome[58]. This knowledge permits a satisfactory interpretation of previous results obtained with etiolated plants or plant parts in experiments with *white* light[114,115]. Dark-grown oat and barley seedlings accumulate large amounts of nitrate in darkness but nitrate reductase activity does not increase above the low endogenous level until light is supplied .However, when dark-grown oat leaves were 'induced' in the light for 12 h and then returned to darkness, the activity continued to increase for another 24 h. The ability of corn leaves to produce an active nitrate reductase apparently depends on the presence of polyribosomes. In leaves from 10-day-old dark-grown seedlings the polyribosome level is low, and nitrate reductase above the low endogenous level cannot be induced in complete darkness, regardless of nitrate availability. Polyribosomes are formed during the initial stages of a light treatment, and after 2–4 h after the onset of light nitrate reductase activity can be induced. The specific effect of light on the formation of polyribosomes may be related to the control of messenger RNA formation and concomitant monoribosome to polyribosome transformation. Recent results[115] indicate that the ability of dark-grown shoots or leaves to form nitrate reductase in darkness decreases with increasing age. This loss of ability and the increasing lag period preceding enzyme formation in light are apparently due to a loss of active polyribosomes as the seedlings age. In conclusion, the data suggest that in the case of nitrate reductase in etiolated plant material, the process of transcription is controlled by light (phytochrome) whereas the process of translation is controlled by nitrate. However, the actual mechanism of light activation of nitrate reductase synthesis[116] is still under investigation[115].

2.3 PHYTOCHROME-MEDIATED ENZYME REPRESSION

In this section we discuss lipoxygenase (LOG) (E.C.1.99.2.1) as an example.

Lipoxygenase is a plant enzyme which catalyses the oxidation of unsaturated fatty acids containing a methylene-interrupted multiple-unsaturated system in which the double bonds are all *cis*, such as linoleic, linolenic and arachidonic acids, to the conjugated *cis-trans*, hydroperoxides. Although enzymes grouped under the general definition of lipoxygenases occur widely in the plant kingdom, and some enzymes of this type (e.g. in legume and cereal seeds) have been known and studied for years, the *physiological role* of lipoxygenases is not understood at present. Irrespective of this shortcoming, control of apparent LOG synthesis by phytochrome in the mustard seedling has been successfully used as a tool to investigate the mechanism of phytochrome-mediated enzyme repression. In this process only $P_{fr(ground\ state)}$ is involved. The HIR (Section 2.1.5) does not come into play[13].

2.3.1 Operational criteria for the involvement of phytochrome ($P_{fr(ground\ state)}$)

In the mustard seedling more than 95% of the extractable LOG activity is located in the cotyledons. Figure 2.20 shows that the apparent synthesis of LOG in the mustard seedling cotyledons is arrested immediately after the onset of far-red light, i.e. after the formation of a relatively low ($\varphi_{fr} = 0.023$) but nearly stationary concentration of the effector molecule P_{fr} in the seedling. The inhibition can be maintained for at least 12 h. Standard errors of measurement are between 0.3 and 1.5%. This precision indicates that the

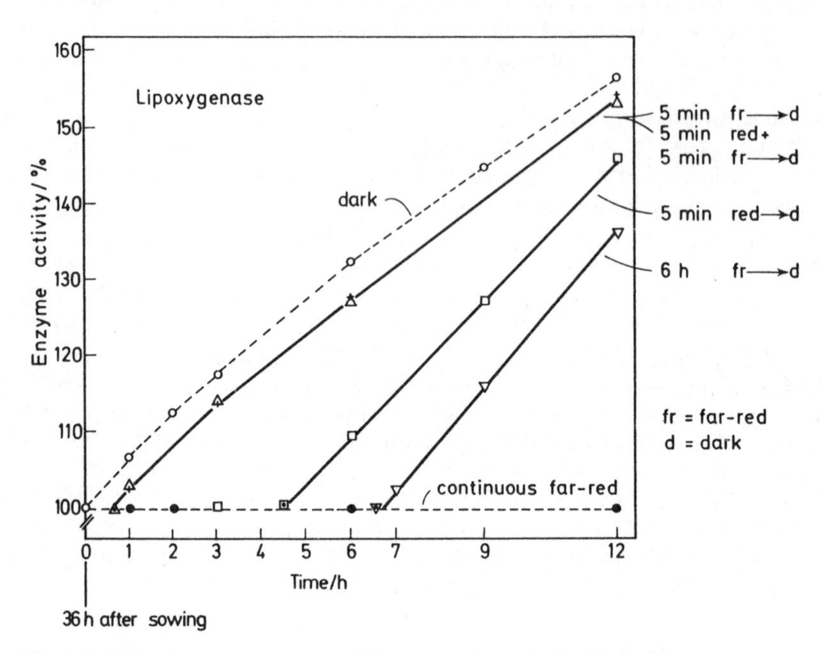

Figure 2.20 Apparent synthesis of lipoxygenase (LOG) in the dark-grown mustard seedling cotyledons is arrested by continuous standard far-red light. The time-courses of apparent LOG synthesis after a single light pulse at time zero with red, far-red, or red followed by far-red show that the operational criteria for the involvement of phytochrome ($P_{fr(ground\ state)}$) are fulfilled. (From Oelze-Karow *et al.*[13], by courtesy of Nat. Acad. Sci., Washington, D.C.)

cells of the mustard seedling which produce the enzyme form a fully-synchronised cell population. Figure 2.20 demonstrates that the usual operational criteria for the involvement of phytochrome ($P_{fr(ground\ state)}$) in arresting apparent lipoxygenase synthesis are fulfilled, and the data suggest a threshold mechanism for the action of $P_{fr(ground\ state)}$. If 80% of the total phytochrome is transferred to P_{fr} by 5 min of red light given at time zero, a delay period of 4.5 h is found in darkness before enzyme synthesis is resumed. A corresponding far-red irradiation which gives 2.3% P_{fr} at time zero ($\varphi = 0.023$) leads to a 40 min delay (duration of time between the *end* of irradiation and resumption of apparent enzyme synthesis). The delay period is also about 40 min following exposure to far-red for 6 h before returning to darkness. This is to be expected since 2.3% P_{fr} are left in a seedling when it is transferred to darkness after 6 h of far-red light. (Remember that $[P_{tot}]$ remains approximately constant in the mustard cotyledons and hypocotyl hook over at least 11 h after the onset of continuous far-red light, cf. Figure 2.6.) If 5 min of red light are followed with 5 min of far-red light, the result is that the red effect is perfectly reversed. The two curves, 5 min far-red⟶dark and 5 min red + 5 min far-red⟶dark are identical.

2.3.2 A threshold (all-or-none) mechanism

Figure 2.21 summarises the main result of the work on control of apparent lipoxygenase synthesis by phytochrome. A threshold (all-or-none) mechanism for the action of $P_{fr(ground\ state)}$ was postulated. The data are in accordance with a half-life of 45 min for P_{fr} in darkness at 36–47 h after sowing, provided a level of 1.25% P_{fr}, based on initial $[P_{tot}]$ at 36 h after sowing, is a threshold value for lipoxygenase synthesis . If the amount of P_{fr} exceeds the threshold

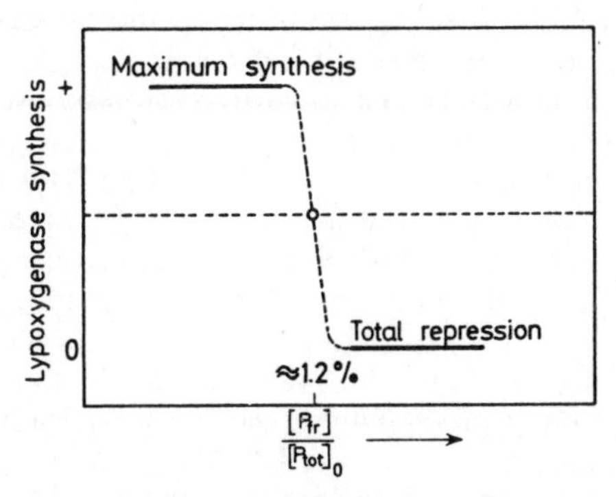

Figure 2.21 A scheme to illustrate the concept of a threshold regulation of apparent lipoxygenase synthesis by $P_{fr(ground\ state)}$. $[P_{tot}]_0$, total phytochrome at time zero (36 h after sowing). This value is a constant. For convenience, the level of P_{fr} is expressed as per cent of $[P_{tot}]_0$. Expressed in this way, the threshold value of P_{fr} is close to 1.25%. (From Mohr[4], by courtesy of Springer.)

level, apparent lipoxygenase synthesis is fully and immediately arrested. If the amount of P_{fr} decreases below the threshold level, apparent synthesis is immediately resumed at full speed.

The sharp threshold can only be understood if one assumes that the action of P_{fr} (with respect to control of apparent LOG synthesis) is strongly co-operative. It seems attractive to interpret the strong co-operativity which is indicated by the steepness of the threshold[25] in terms of co-operativity in biological membranes[117]. P_{fr} would be analogous to a ligand and the primary reactant of P_{fr} (usually designated as X) would be analogous to a pre-existing membrane, capable of performing fully-reversible conformational transitions with a high degree of co-operativity[118].

The question of whether the threshold concept can be applied to every (molecular) response mediated by $P_{fr(ground\ state)}$ must probably be answered by 'No'. In phytochrome-mediated anthocyanin synthesis of the mustard seedling, for instance, which is the prototype of a graded response, it has not been possible so far to detect any threshold for the action of $P_{fr(ground\ state)}$[24]. At present the conclusion is that any model which may be developed to explain the action of P_{fr} on apparent LOG synthesis will not explain the action of P_{fr} with respect to anthocyanin induction. There are good reasons for asking what the action mechanism of P_{fr} is, for every type of photoresponse[119].

2.3.3 Quantitative correlation between the spectrophotometrically measured phytochrome ($P_{fr(ground\ state)}$) and the physiological response

In spite of much effort the action mechanism of phytochrome ($P_{fr} + X \longrightarrow P_{fr}X \dashrightarrow$ physiological display) is still a matter of speculation and debate[1,4,16]. A major reason for this unsatisfactory situation is that it has not been possible in most cases to establish a quantitative correlation between the amount of spectrophotometrically detectable $P_{fr(ground\ state)}$ and the extent (or rate) of the physiological displays (or photoresponses)[120,121]. The only exception so far is the threshold control by $P_{fr(ground\ state)}$ of apparent LOG synthesis in the cotyledons of the mustard seedling. This response was used to determine physiologically the photostationary states φ_λ, that is, the $[P_{fr}]/[P_{tot}]$ ratios which are established by different wavelengths in the red and far-red range of the visible spectrum[25]. Under the premises (for which justification has been given) that the $[P_{fr}]/[P_{tot}]$ ratio for standard red light is 0.8 and that the decay of P_{fr} is a first-order process with a half-life of 45 min (at 25 °C), the $[P_{fr}]/[P_{tot}]$ ratios determined physiologically by means of the lipoxygenase response agree with the $[P_{fr}]/[P_{tot}]$ ratios determined spectrophotometrically by Hartmann and Spruit[22] in the hypocotyl hook of mustard seedlings but not with the φ values determined spectrophotometrically in the cotyledons. The conclusion that apparent LOG synthesis *in the cotyledons* is controlled by phytochrome located in the *hypocotyl hook* has been substantiated by further spectrophotometric[19] and physiological experiments[122]. To summarise, an explanation of the experimental facts concerning the control by P_{fr} of apparent LOG synthesis in the mustard seedling cotyledons requires a co-operative effect on the level of the reaction $P_{fr} + X \longrightarrow P_{fr}X$, a high degree of synchrony on the cellular and organismal levels, and rapid communication between the hypocotyl hook and the cotyledons[25].

2.3.4 Primary and secondary differentiation; the concept of competence

The problem has been to ascertain whether apparent LOG synthesis responds to P_{fr} throughout the whole period of the seedling's development, that is up to 84 h after sowing at 25 °C (Figure 2.22). If one irradiates with standard

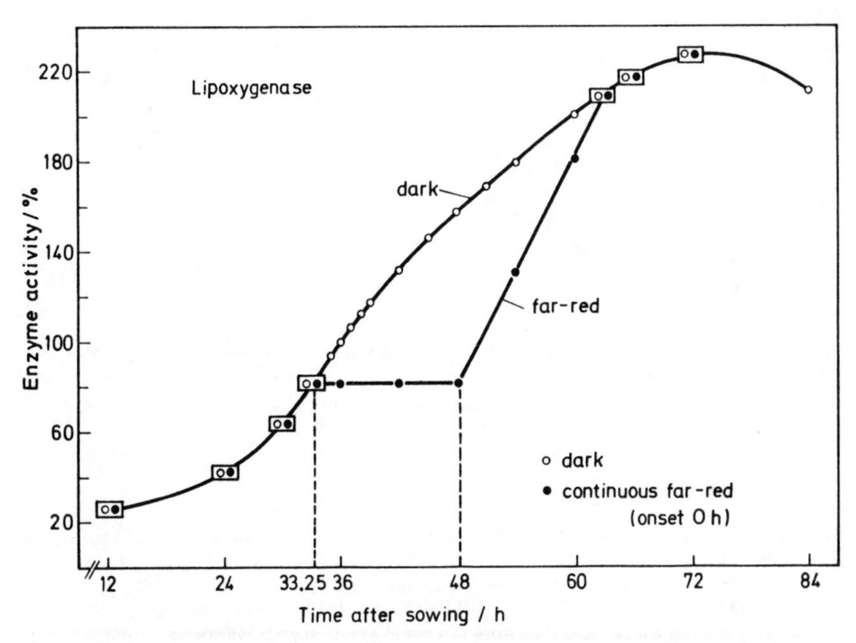

Figure 2.22 Apparent LOG synthesis in the mustard seedling cotyledons in continuous darkness and under continuous standard far-red light. The onset of far-red was at the time of sowing of the seeds (0 h). (From Oelze-Karow and Mohr[123], by courtesy of Verlag der Zeitschrift für Naturforschung.)

far-red (or red) light from the time of sowing there is no control of LOG synthesis up to *ca.* 33 h. At this point the full repression of LOG synthesis by far-red (or red)[4] light (i.e. by a P_{fr} value above the threshold) suddenly comes into play, while at 48 h after sowing, the seedling suddenly and completely escapes from control by P_{fr}. Enzyme synthesis is resumed even under continuous far-red (or red)[4] light. The kinetics after resumption return to the dark kinetics.

Exactly the same temporal pattern was observed in a number of different experiments. Under all circumstances the system escapes from control by light at 48 h after sowing (with 25 °C and standard conditions)[123]. It was concluded that the temporal pattern of response must be determined by changes in the system on which P_{fr} acts rather than by P_{fr}.

Using a terminology developed previously*, the situation (cf. Figure 2.22) can be described as follows: the repressive action of P_{fr} on apparent LOG synthesis is a function of the specific state of responsivity (or competence) of the cells and tissues. The process of differentiation which determines the changing pattern of competence is called 'primary differentiation (P_{fr})'.

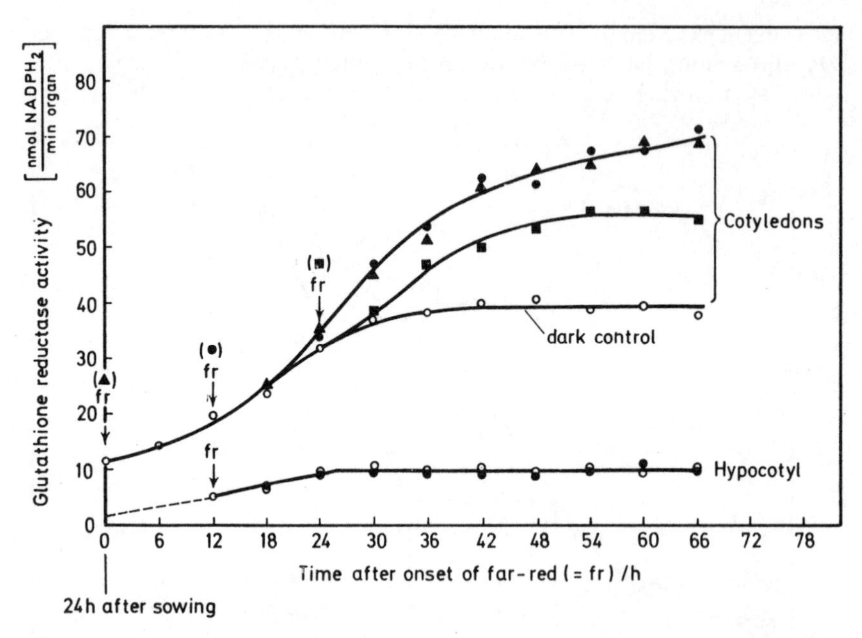

Figure 2.23 Time-courses of glutathione reductase levels in cotyledons and hypocotyl of the mustard seedling in the dark and under continuous far-red light. There is no detectable competence for phytochrome with respect to apparent glutathione reductase synthesis before *ca.* 42 h after sowing. While the actual lag-phase is not longer than 6 h (onset of light at 48 h), far-red light does not exert any influence before 48 h after sowing, irrespective of the onset of light. However, a red light pulse given at 24 h after sowing leads to a considerable increase of the enzyme level *ca.* 24 h later. Facts of this kind have led to the concept of a 'transmitter' as already mentioned in the section on peroxidase (Section 2.2.5). In addition, glutathione reductase may serve as an example of an enzyme which is induced by phytochrome in the cotyledons whereas no control by light can be detected in the hypocotyl (cf. Sections 2.2.1.5 and 2.2.2). (From Drumm[59], by courtesy of the author)

The time-course of primary differentiation (P_{fr}), whose causalities are unknown, is independent of P_{fr}. This latter statement is based on many facts[123], for example that neither the beginning nor the end of the period of control (from *ca.* 33 h to 48 h after sowing, cf. Figure 2.22) are influenced by phytochrome to any detectable extent.

* I refer to the concept of the double-action control mechanism in the course of differentiation[123]. Primary differentiation (with respect to P_{fr}) determines the response of the cell (or the system) to P_{fr}. The term secondary differentiation (with respect to P_{fr}) designates every developmental step which is triggered by P_{fr}. At the level of primary as well as at the level of secondary differentiation, fully reversible photomodulations are possible[4].

The independence of primary differentiation (P_{fr}) from phytochrome seems to be the rule rather than the exception. A thorough investigation of phytochrome-mediated anthocyanin induction[91] has led to the same conclusion[4]. Another example, briefly, is the phytochrome-mediated induction of glutathione reductase (E.C. 1.6.4.2) in the cotyledons of the mustard seedling[59] (Figure 2.23). Whether the onset of continuous far-red light is at 24 or 36 h after sowing, the time-course of the enzyme level remains the same.

2.4 ENZYME INDUCTION IN CELL SUSPENSION CULTURES

There have been attempts to find simpler systems than the intact seedling, for example, tissue and cell culture systems, to study the molecular events underlying phytochrome-mediated photoresponses. In particular, it was hoped by using cell suspension cultures to find a means of approach to the problems of cytodifferentiation and modulation of enzyme levels in plant cells. The question is whether cultured cells retain at least some regulating properties characteristic of differentiated cells in the intact plant or organ. Recent advances to be described in this section hold promise for future research.

2.4.1 Flavone glycoside and PAL synthesis in cell suspension cultures of parsley as a model system for a two-step differentiation process[124]

Since it is still difficult to obtain cell suspension cultures from mustard, Wellmann has used cell suspension cultures from parsley (*Petroselinum hortense*)[124]. These cultures were already known to form large amounts of flavone glycosides under high-irradiance white light[125]. Fresh weight was used in the experiments as a system of reference, since it was independent of the irradiation applied. However, total protein or a marker enzyme of the basic metabolism, e.g. glucose 6-phosphate dehydrogenase, could have been used instead with exactly the same result.

Preliminary 'action spectra' for the light-mediated induction of flavone glycosides in the parsley cell suspension culture were obtained with cut-off filters as well as with a monochromator. The spectra indicate that visible light is ineffective, and that the peak of activity is below 300 nm. The fluorescent white light which was used previously[125] was only effective because it contains a relatively high contamination with short-wavelength ultraviolet (u.v.). There was no indication of any damage to the cells due to the standard u.v. used. A linear increase in flavone glycoside accumulation was measured with respect to the time of u.v. irradiation up to 5 h whereas in the actual experiments only 60 min of standard u.v. were used. During the period of experimentation fresh weight increased by the same amount in either dark-treated or irradiated cultures. No differences in protein content due to the irradiation could be detected. Furthermore, the activity of glucose 6-phosphate dehydrogenase, a marker enzyme of the basic metabolism, remained unchanged.

The main results are indicated on the upper part of Table 2.6. These data show that phytochrome can become effective in the cells only after pre-irradiation with u.v. The stimulation of flavone glycoside accumulation caused by 60 min of standard u.v. was reduced by *ca.* 40% by a subsequent pulse of 10 min of far-red light. This reduction of the u.v. effect was nullified by a subsequent red irradiation. Thus, the operational criteria for the involvement of phytochrome were clearly fulfilled. However, red and far-red pulses (10 min), given without ultraviolet pretreatment, had no effect. The fact that red light given after u.v. irradiation showed no additional stimulation of flavone glycoside accumulation, was to be expected. The u.v. light by itself will lead to the formation of a large percentage of P_{fr}. Another indication for the involvement of phytochrome is the result that continuous far-red light will increase flavone glycoside accumulation to values considerably higher than those obtained after u.v. irradiation followed by darkness. It must be emphasised that there was no effect of continuous red or far-red light upon flavone glycoside formation without a preceding irradiation with u.v. light (lower part of Table 2.6).

Table 2.6 Flavone glycoside accumulation mediated by short-time irradiation with red and far-red light and with continuous far-red light in a cell suspension culture of parsley. (Fom Wellmann[124], by courtesy of Springer)

Programme	*Flavone glycosides** (A_{380})
Preirradiation with u.v. light for 60 min followed by:	
15 h dark	0.36
15 h far-red	0.41
10 min red + 15 h dark	0.355
10 min red + 10 min far-red + 15 h dark	0.255
10 min far-red + 15 h dark	0.26
10 min far-red + 10 min red + 15 h dark	0.35
Without u.v. pre-irradiation:	
16 h red	0.12
16 h far-red	0.125
10 min red + 16 h dark	0.12
10 min far-red + 16 h dark	0.125
16 h dark	0.125
Initial value before irradiation	0.12

* Standard extract = 1 g fresh weight/5 ml buffer.

The data so far obtained with the cell cultures suggest a model system for the study of a double-action control mechanism in differentiation. It can be suggested that u.v. light exerts some specific action, which changes the cells in such a way that phytochrome becomes able to act on a differentiation process as measured by the synthesis of flavone glycosides or enzyme induction (Table 2.7); in short, that u.v. can be used to make the cells competent for P_{fr}.

An alternative interpretation which is not fully ruled out at the moment is that u.v. irradiation 'activates' the phytochrome system in such a way that

Table 2.7 Control of PAL levels in cell suspension cultures of parsley by standard u.v., red and far-red (fr) light. (From Wellman[126], by courtesy of the author)

Programme	PAL *activity (relative units)*
20 min u.v. + 300 min dark	100
20 min u.v. + 5 min fr + 295 min dark	74 ± 2
20 min u.v. + 5 min fr + 5 min red + 290 min dark	99 ± 3
20 min u.v. + 5 min fr + 5 min red + 5 min fr + 285 min dark	73 ± 2
5 min fr + 315 min dark	8.5 ± 0.7
5 min fr + 5 min red + 310 min dark	8.4 ± 0.8
320 min dark	8.6 ± 0.7

red and far-red light can act in accordance with the model outlined in Figure 2.2.

2.4.2 A two-step differentiation process in isolated roots of parsley

A similar response was observed in experiments with isolated taproots of parsley plants (Table 2.8). This fact seems to indicate that the state of differentiation of the parsley cells in suspension culture (originally obtained from a

Table 2.8 Control of PAL levels in isolated taproots of parsley, kept in organ culture, by standard ultraviolet (u.v.), red and far-red (fr) light. (From Wellman[126], by courtesy of the author)

Programme	PAL *activity (relative units)*
120 min u.v. + 12 h dark	100
120 min u.v. + 5 min fr + 12 h dark	71 ± 4
120 min u.v. + 5 min fr + 5 min red + 12 h dark	102 ± 5
5 min fr + 5 min red + 14 h dark	23 + 0.3
120 min red + 12 h dark	25 ± 0.3
14 h dark	21 ± 0.4

callus at the petiole of an adult parsley plant) is similar to the state of differentiation of the cells in a parsley taproot. In any case the behaviour of the cells in a suspension culture with respect to PAL and concomitant flavonoid induction is similar to the behaviour of at least some cells in a multicellular, organised system such as a taproot.

2.5 CONCLUDING REMARKS

As was pointed out in the introduction, it is assumed that development is primarily the consequence of an orderly sequence of changes in the enzyme complement of an organism. Therefore, the investigator of photomorphogenesis will primarily try to explore those phytochrome-mediated responses in which changes in enzyme levels have a well-defined causal role in well-defined developmental steps[4]. As a consequence, recent work has concentrated on biochemical model systems of differentiation[92].

The worker in the field realises, of course, that biochemical model systems of differentiation which include only simple, water-soluble end-products (such as flavone glycosides or anthocyanins), will probably not lead to a complete solution of the real problem, which is the relationship between enzymes and form, or in other words, the relationship between enzyme specificity and structural specificity. However, it is believed that studies of simple biochemical model systems of differentiation will increase our understanding of the logical principles and molecular foundations of differentiation and thus will eventually open a path of approach to the basic problem of developmental biology, which is the relationship between modulation of enzyme synthesis and development of structural specificity, in space and time.

References

1. Shropshire, W. (1972). *Photophysiology*, **7**, 33 (A. C. Giese, editor) (New York: Academic Press)
2. Ziegler, H. and Ziegler, I. (1965). *Planta*, **65**, 369
3. Zucker, M. (1972). *Ann. Rev. Plant Physiol.*, **23**, 133 (L. Machlis, editor) (Palo Alto: Annual Reviews Inc.)
4. Mohr, H. (1972). *Lectures on Photomorphogenesis* (Heidelberg–New York: Springer)
5. Boulter, D., Ellis, R. J. and Yarwood, A. (1972). *Biol Rev.*, **47**, 113
6. Schopfer, P. and Hock, B. (1971). *Planta*, **96**, 248
7. Zucker, M. (1971). *Plant Physiol.*, **47**, 442
8. Rissland, I. and Mohr, H. (1967). *Planta*, **77**, 239
9. Drumm, H., Elchinger, I., Möller, J., Peter, K. and Mohr, H. (1971). *Planta*, **99**, 265
10. Schopfer, P. (1972). *Phytochrome*, 486 (K. Mitrakos and W. Shropshire, editors) (New York: Academic Press)
11. Andersen, R. A. and Sowers, J. A. (1968). *Phytochemistry*, **7**, 293
12. Karow, H. and Mohr, H. (1967). *Planta*, **72**, 170
13. Oelze-Karow, H., Schopfer, P. and Mohr, H. (1970). *Proc. Nat. Acad. Sci.* (*USA*), **65**, 51
14. Schopfer, P. and Mohr, H. (1972). *Plant Physiol.*, **49**, 8
15. Drumm, H., Oelze-Karow, H. and Dittes, L., unpublished data obtained in our laboratory
16. Briggs, W. R. and Rice, H. V. (1972). *Annu. Rev. Plant Physiol.*, **23**, 293 (L. Machlis, editor) (Palo Alto: Annual Reviews Inc.)
17. Schäfer, E. (1972). *Dissertation*, University of Freiburg
18. Marmé, D., Marchal, B. and Schäfer, E. (1971). *Planta*, **100**, 331
19. Schäfer, E., Schmidt, W. and Mohr, H. (1973). *Photochem. Photobiol.* (in the press)
20. Rüdiger, W. (1969). *Liebig's Ann. Chem.*, **723**, 208
21. Burke, M. J., Pratt, D. C. and Moscowitz, A. (1972). *Biochem.*, **11**, 4025
22. Hanke, J., Hartmann, K. M. and Mohr, H. (1969). *Planta*, **86**, 235
23. Borthwick, H. A., Hendricks, S. B., Toole, E. H. and Toole, V. K. (1954). *Bot. Gaz.*, **115**, 205

24. Lange, H., Shropshire, W. and Mohr, H. (1971). *Plant Physiol.*, **47,** 649
25. Oelze-Karow, H. and Mohr, H. (1973). *Photochem. Photobiol.* (in the press)
26. Hartmann, K. M. (1966). *Photochem. Photobiol.*, **5,** 349
27. Hartmann, K. M. (1967). *Naturwissenschaften*, **54,** 544
28. Hartmann, K. M. (1967). *European Photobiology Symposium, Photoreceptor Problems in Photomorphogenic Responses under High-Energy-Conditions* (U.V.–blue–far red), 29 (Book of Abstracts, Hvar)
29. Hartmann, K. M. (1967). *Z. Naturforsch.*, **22b,** 1172
30. Porter, G. (1969). *An Introduction to Photobiology*, 1 (C. P. Swanson, editor) (Englewood Cliffs: Prentice Hall)
31. Weidner, M. (1967). *Planta*, **75,** 94
32. Capesius, I. and Bopp, M. (1970). *Planta*, **94,** 220
33. Weidner, M., Jakobs, M. and Mohr, H. (1965). *Z. Naturforsch.*, **20b,** 689
34. Jakobs, M. and Mohr, H. (1966). *Planta*, **69,** 187
35. Weidner, M. and Mohr, H. (1967). *Planta*, **75,** 99
36. Schäfer, E. (1972), personal communication
37. Marcus, A. (1960). *Plant Physiol.*, **35,** 126
38. Margulies, M. M. (1965). *Plant Physiol.*, **40,** 57
39. Durst, F. and Mohr, H. (1966). *Naturwissenschaften*, **53,** 531
40. Durst, F. and Mohr, H. (1966). *Naturwissenschaften*, **53,** 707
41. Henshall, J. D. and Goodwin, T. W. (1964). *Phytochemistry*, **3,** 677
42. Feierabend, J. and Pirson, A. (1966). *Z. Pflanzenphysiol.*, **55,** 235
43. Surrey, K. (1967). *Plant Physiol.*, **42,** 421
44. Graham, D., Grieve, A. M. and Smillie, R. M. (1968). *Nature* (*London*), **218,** 89
45. Klein, A. O. (1969). *Plant Physiol.*, **44,** 897
46. van Poucke, M., Barthe, F. and Mohr, H. (1969). *Naturwissenschaften*, **56,** 417
47. Butler, L. G. and Bennet, V. (1969). *Plant Physiol.*, **44,** 1285
48. Queiroz, O. (1969). *Phytochemistry*, **8,** 1655
49. Tezuka, T. and Yamamoto, Y. (1969). *Bot. Mag. Tokyo*, **82,** 130
50. Steer, B. T. and Gibbs, M. (1969). *Plant Physiol.*, **44,** 775
51. Bottomley, W. (1970). *Plant Physiol.*, **45,** 608
52. van Poucke, M., Cerff, R., Barthe, F. and Mohr, H. (1970). *Naturwissenschaften*, **56,** 132
53. Drumm, H., Brüning, K. and Mohr, H. (1972). *Planta*, **106,** 259
54. Russell, D. W. (1971). *J. Biol. Chem.*, **246,** 3870
55. Schopfer, P. and Plachy, C. (1973). *Z. Naturforsch.*, in the press
56. Cerff, R. (1973). *Plant Physiol.*, **51,** 76
57. Acton, G. J. (1972). *Nature New Biol.*, **236,** 255
58. Jones, R. W. and Sheard, R. W. (1972). *Nature New Biol.*, **238,** 221
59. Drumm, H. (1972), personal communication
60. Hahlbrock, K., Ebel, J., Ortmann, R., Sutter, A., Wellmann, E. and Grisebach, H. (1971). *Biochim. Biophys. Acta*, **244,** 7
61. Zucker, M. (1969). *Plant Physiol.*, **44,** 912
62. Ryan, C. A. (1968). *Plant Physiol.*, **43,** 1859
63. Ryan, C. A. (1968). *Plant Physiol.*, **43,** 1880
64. Ryan, C. A. and Huisman, W. (1970). *Plant Physiol.*, **45,** 484
65. Dittes, L., Rissland, I. and Mohr, H. (1971). *Z. Naturforsch.*, **26b,** 1175
66. Schopfer, P. (1971). *Planta*, **99,** 339
67. Weidner, M., Rissland, I., Lohmann, L., Huault, C. and Mohr, H. (1969). *Planta*, **86,** 33
68. Engelsma, G. (1967). *Naturwiss.*, **54,** 319
69. Engelsma, G. (1967). *Planta*, **75,** 207
70. Siekevitz, Ph. (1972). *J. Theoret. Biol.*, **37,** 321
71. Huault, C., Larcher, G. and Malcoste, R. (1971). *Compt. Rend. Acad. Sc. Paris*, **273,** 1371
72. Engelsma, G. (1968). *Planta*, **82** 355,
73. Engelsma, G. (1970). *Acta Bot. Neerl.*, **19,** 403
74. Rubery, P. H. and Northcote, D. H. (1968). *Nature* (*London*), **219,** 1230
75. Rubery, P. H. and Fosket, D. E. (1969). *Planta*, **87,** 54
76. Drumm, H., Falk, H., Möller, J. and Mohr, H. (1970). *Cytobiologie*, **2,** 335

77. Schimke, R. T. (1969). *Current Topics in Cellular Regulation*, Vol. 1, 77 (B. L. Horecker and E. R. Stadtman, editors) (New York: Academic Press)
78. Durst, F. and Duranton, H. (1970). *Compt. Rend. Acad. Sc. Paris*, **270**, 2940
79. Tezuka, T. and Yamamoto, Y. (1969). *Bot. Mag. Tokyo*, **82**, 130
80. Tezuka, T. and Yamamoto, Y. (1972). *Plant Physiol.*, **50**, 458
81. Jaffe, M. J. (1969). *Physiol. Plant.*, **22**, 1033
82. Dittes, H. and Mohr, H. (1970). *Z. Naturforsch.*, **25b**, 708
83. Mohr, H. and Bienger, I. (1967). *Planta*, **75**, 180
84. Lockard, R. E. and Lingrel, J. B. (1969). *Biochem. Biophys. Res. Commun.* **37**, 204
85. Firtel, R. A., Jacobson, A. and Lodish, H. F. (1972). *Nature New Biol.*, **239**, 225
86. Tomkins, G. M., Gelehrter, Th.D., Granner, D., Martin, D., Samuels, H. H. and Thompson, E. B. (1969). *Science*, **166**, 1474
87. Lange, H., Bienger, I. and Mohr, H. (1967). *Planta*, **76**, 359
88. Lange, H. and Mohr, H. (1965). *Planta*, **67**, 107
89. Mohr, H. and Senf, R. (1966). *Planta*, **71**, 195
90. Havelange, A. and Schumacker, R. (1966). *Bull. Soc. Roy. Sci. Liège*, **35**, 125
91. Wagner, E. and Mohr, H. (1966). *Planta*, **71**, 204
92. Mohr, H. (1971). *Umschau*, Heft **15**, 547
93. Cerff, R. (1971). *Dissertation*, University of Freiburg
94. Mohr, H. (1966). *Z. Pflanzenphysiol.*, **54**, 63
95. Masoner, M., Unser, G. and Mohr, H. (1972). *Planta*, **105**, 267
96. Häcker, M. (1967). *Planta*, **76**, 309
97. Ellis, R. J. and Hartley, M. R. (1971). *Nature New Biol.*, **233**, 193
98. Metzner, H. (editor) (1969). *Progress in Photosynthesis Research*, Vol. 1 (Tübingen: H. Metzner)
99. Kirk, J. T. O. and Tilney-Bassett, R. A. E. (1967). *The Plastids: their Chemistry, Structure, Growth and Inheritance* (San Francisco: Freeman)
100. Wildman, S. G., Hongladarom, T. and Honda, S. J. (1962). *Science*, **138**, 434
101. Boulter, D., Ellis, R. J. and Yarwood, A. (1972). *Biol. Rev.*, Vol. 47, 130 (E. N. Willmer, editor) (London: Cambridge Philosophical Society)
102. Levine, R. P. (1969). *Ann. Rev. Plant Physiol.*, **20**, 523 (L. Machlis, editor) (Palo Alto: Annual Reviews Inc.)
103. Feierabend, J. (1969). *Planta*, **84**, 11
104. Quail, P. H. (1972). *Plant Physiol.*, in the press
105. Schäfer, E., Marchal, B. and Marmé, D. (1972). *Photochem. Photobiol.*, **15**, 457
106. Kendrick, R. E. and Frankland, B. (1968). *Planta*, **82**, 317
107. Filner, P. and Varner, J. E. (1967). *Proc. Nat. Acad. Sci.* (*USA*), **58**, 1520
108. Jacobson, J. V., Scandalios, J. G. and Varner, J. E. (1970). *Plant Physiol.*, **45**, 367
109. Galston, A. W. and Davies, P. J. (1969). *Science*, **163**, 1288
110. Mohr, H. and Appuhn, U. (1962). *Planta*, **59**, 49
111. Schopfer, P. (1972). *Proc. Symp. Tihany, Symposia biologica Hungarica*, Vol. 13, *Nucleic Acids and Proteins in Higher Plants* (G. L. Farkas, editor) (Budapest: Hungarian Academy of Sciences)
112. Beevers, L. and Hageman, R. H. (1969). *Ann. Rev. Plant Physiol.*, **20**, 495 (L. Machlis, editor) (Palo Alto: Annual Reviews Inc.)
113. Jordan, W. R. and Huffaker, R. C. (1972). *Physiol. Plant.*, **26**, 296
114. Travis, R. L., Huffaker, R. C. and Key, J. L. (1970). *Plant Physiol.*, **46**, 800
115. Travis, R. L. and Key, J. L. (1971). *Plant Physiol.*, **48**, 617
116. Zielke, H. R. and Filner, P. (1971). *J. Biol. Chem.*, **246**, 1772
117. Changeux, J. P., Thiery, J., Tung, Y. and Kittel, C. (1967). *Proc. Nat. Acad. Sci.* (*USA*), **57**, 335
118. Changeux, J. P. (1969). *Symmetry and Function of Biological Systems at the Macromolecular Level*, 235 (A. Engström and B. Strandberg, editors) (Stockholm: Almquist and Wiksell)
119. Mohr, H., Bienger, I. and Lange, H. (1971). *Nature* (*London*), **230**, 56
120. Hillman, W. S. (1967). *Ann. Rev. Plant Physiol.*, **18**, 301 (L. Machlis, editor) (Palo Alto: Annual Reviews Inc.)
121. Kendrick, R. E. and Hillman, W. S. (1972). *Physiol. Plant.*, **47**, 649
122. Oelze-Karow, H. and Mohr, H. (1973), in preparation
123. Oelze-Karow, H. and Mohr, H. (1970). *Z. Naturforsch.*, **25b**, 1282

124. Wellmann, E. (1971). *Planta*, **101**, 283
125. Hahlbrock, K. and Wellmann, E. (1970). *Planta*, **94**, 236
126. Wellmann, E. (1972). Personal communication
127. Attridge, T. H. and Smith, H. (1967). *Biochim. Biophys. Acta*, **148**, 805

Editor's Comments

In the preceding chapter, some special features of plant development were outlined. In the next two chapters, we turn to a consideration of some special features of the early development of animals. Most developmental biologists nowadays think of development in terms of the evolution of a programme or a series of co-ordinated programmes, rather like computer programmes, using a common information store. In analysing this process, three main stages can be identified. The first is the laying down of the programme, which probably occurs in the oocyte during the maturation of the egg. This stage has, to date, not proved amenable to biochemical study because of the small amounts of material available. The second phase is the evolution of a 'master programme' which selects between different subprogrammes (for different cells and tissues). Embryologists term this process 'determination'. We believe that cells become determined in the earliest stages of development following fertilisation and this is what the next two chapters consider. The third stage is the further evolution of developmental programmes in individual cells, often called 'maturation'; it is the subject of later chapters.

The biochemistry of the development of the early embryo has been studied in rather more detail in animals than in plants but it seems not unlikely that similar phenomena are to be encountered in both.

It is important to grasp some of the biological features of early development in order to see these studies in perspective. In animals, two different kinds of development are recognised. The first, the *mosaic* type of development, is characteristic of insects and has interesting implications. In many insect eggs the pronucleus is situated at one pole. Following fertilisation, the nucleus undergoes repeated divisions, without any cellular division, and a cluster of nuclei forms at one end of the egg. These nuclei then migrate to different parts of the egg and, only after this has occurred, do cell walls form. Many years ago the important observation was made that an injury to a particular region of an unfertilised egg could give rise to a specific and roughly predictible defect in the insect which developed from it. The implication clearly is that some of the information for the development of the insect organ is localised in the cytoplasm of the egg.

A rather different course of events is observed in *regulative* development which occurs, for example, in vertebrates. Following fertilisation of the regulative egg, the nucleus undergoes a series of very rapid doublings and divisions but, in this instance, each division is accompanied by segregation of the daughter nuclei and cytoplasmic cleavage. These cleavages result in the formation of a clump of cells called a morula. In the early stages of morula formation, the cells seem to have a high degree of equivalence in that if some of them are removed, the remaining cells can from a perfectly normal, if smaller, embryo. Organisms which show regulative development (e.g. the mouse) therefore have a more flexible regulatory machinery in the early stage of development than do organisms, such as insects, which exhibit mosaic development.

It cannot be concluded, however, that there are no cytoplasmic determinants in regulative eggs. The sea urchin egg exhibits many of the characteristics of regulative development but, as is well-known, centrifugation of the sea urchin egg can completely disrupt subsequent development, leading to a bizarre distribution of tissues. Hence, there is no really clear-cut difference between regulative and mosaic development although the two do represent extreme alternatives.

In the following two essays, the early development of the sea urchin and the frog are described. These animals have proved particularly useful for biochemical study and, as a result, we have a rather clear picture of some of the phenomena which occur both before and immediately after fertilisation. One of the most intriguing phenomena to occur before fertilisation is the amplification of the ribosomal genes; this is not confined to *Xenopus* but has been studied in some detail in this frog and is discussed by Dr Denis. The reason for this amplification becomes clear when we consider the events which occur immediately after fertilisation of many embryos. In the first few hours, there is an enormous synthesis of DNA and of the histones necessary to complex with it to form the basis of chromatin. The demands for protein synthesis are such that an enormous reservoir of ribosomes has to be provided in the egg. Moreover, there is no striking synthesis of RNA during much of this early development although there must be a great requirement for some messenger RNA species as well as ribosomal RNA. So far as we can determine, this is achieved partly by making RNA from some genes which are present as multiple copies (e.g. the histone genes) and partly by using a pool of preformed 'masked' messengers which accumulates in the egg. This latter phenomenon has been particularly studied in the sea urchin and is described in the next chapter.

3
Macromolecule Synthesis in Sea Urchin Development

J. PAUL
The Beatson Institute for Cancer Research, Glasgow

3.1 INTRODUCTION

Many of the most important biochemical events in differentiation occur within a few hours of fertilisation. The timing and sequence of events vary from species to species but the same general pattern applies to insects, many other invertebrates and vertebrates and probably to plants. Biochemical studies have been conducted in all these but the most detailed studies have been done in amphibia (as described in Chapter 4) and in sea urchins[1-5].

Both sea urchins and amphibia are examples of holoblastic embryos, i.e. embryos which arise by complete cleavage of the zygote. The outlines of early development of the sea urchin are shown in Figure 3.1. Cell differentiation becomes recognisable in the blastula; at the late blastula state (the mesenchymal blastula) mesenchymal cells emerge.

During cleavage, intense DNA and protein synthesis occur but the total volume, mass and protein content of stages up to morula do not change greatly and are essentially the same as in the unfertilised egg. With the formation of the blastocoel, the embryo enlarges but the total protein content does

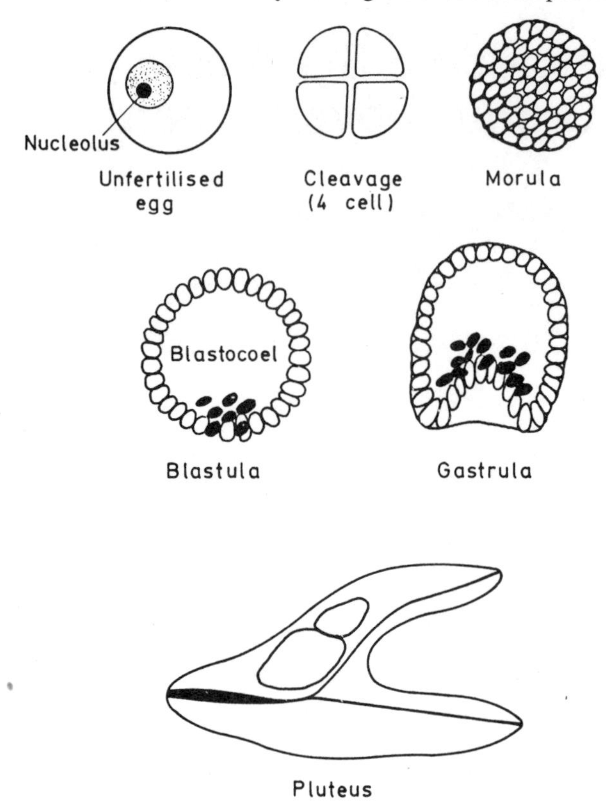

Figure 3.1 The main stages of early development of the sea urchin embryo. The early mesenchymal blastula is formed in many species about 12 h after fertilisation and the early gastrula about 8 h later

not increase markedly until late gastrulation by which time cell differentiation is quite far advanced and tissue structures are beginning to appear.

Although there is no net increase in the protein content of the embryo during the earliest (cleavage) stages of development, a wide range of proteins is made. Moreover, protein synthesis is essential for early development; in its absence, one cleavage may occur but no more.

It is now clear that the polysomes which synthesise these proteins are at first composed of ribosomes and messenger RNA which were made during oogenesis, sometimes a year or more earlier. Moreover, the mRNA stored in the egg is not attached to ribosomes but is segregated from them in ribonucleoprotein particles. The maternal mRNA sequestered in these particles is sometimes referred to as 'masked messenger'.

Soon after fertilisation, new mRNA is made and new ribosomal components are synthesised. In many species, however, development can proceed to the gastrula stage in the absence of RNA synthesis; protein synthesis during this time is therefore mainly dependent on mRNA and ribosomes made in the egg. Some species depend on new RNA synthesis earlier than this; on the other hand, there are no records of development proceeding beyond gastrula in the absence of RNA synthesis. Synthesis of mRNA usually starts immediately after fertilisation but new ribosomal species, including 5S RNA and tRNA, commonly do not appear until blastula. Some of the experimental evidence for these statements, especially as they relate to the sea urchin, will be discussed in this review.

3.2 NEW PROTEIN AND RNA SYNTHESIS IN SEA URCHIN EMBRYOS

The egg membrane is rather impermeable to most molecules and special techniques (such as injecting labelled amino acids into the egg) have to be used for the introduction of isotopes to study the synthesis of new protein. In properly conducted experiments it would appear that there is a low level of protein synthesis, even in the unfertilised egg. Following fertilisation, the incorporation of precursors into proteins increases dramatically and continues at a high level. However, there is no net increase in protein content of the embryo up to the blastula stage although a wide spectrum of proteins is made[6,7]. The nature of all the proteins formed at this stage is not known. Indeed, Davidson argues, from the complexity of mRNA in the egg, that many thousands may be made[1]. Early experiments by Gross and Cousineau[8] and Stafford and Iverson[9] indicated that much of this new protein is associated with the nuclear apparatus and autoradiographic studies suggested that it might be mitotic spindle protein. More recent work has shown that an important component of the newly synthesised proteins in early embryogenesis is histones[3].

It is not certain whether RNA synthesis goes on in the mature egg but some nuclear RNA synthesis can already be shown by autoradiography in the uncleaved zygote. The synthesis of DNA-like RNA was demonstrated at the four-cell stage many years ago[11-13], and there is abundant evidence for new mRNA synthesis quite soon after fertilisation[14-19].

In contrast, Nemer and Infante[20] have shown very conclusively that, in the sea urchin, ribosomal synthesis does not start until the blastula stage. In these experiments, a variant of the sea urchin *Strongylocentrotus purpuratus* was used which synthesises a labile ribosomal RNA in which the 18S component can be split into two 13S fragments by brief heat treatment. This characteristic can be exploited to follow the fate of maternal ribosomes in the developing embryo by studying the emergence of normal 18S RNA in heterozygotes and its replacement of the abnormal RNA. Nemer and Infante[20] found that the concentration of maternal ribosomal RNA did not change until the mesenchymal blastula stage. Only thereafter did newly synthesised rRNA appear in significant amounts. Moreover, sedimentation analysis of RNA newly made (as determined by incorporation of labelled precursors) during early cleavage shows no evidence of ribosomal species[13, 21, 22]. Comb[23] has found that newly incorporated methyl groups cannot be demonstrated in ribosomal RNA until the pluteus stage and it has also been shown by Cowden and Lehman[24] that nucleoli first appear in the sea urchin embryo at the very early gastrula stage. The evidence is very strong, therefore, that the main ribosomal species are not synthesised until the late blastula or early gastrula stage in the sea urchin. Comb and his associates[25] have studied the appearance of 5S ribosomal RNA and report that it is synthesised at gastrulation also.

The evidence concerning mRNA contrasts strikingly with that for rRNA. It is apparently made in very early cleavage and there is quite good experimental evidence that some of the new RNA species are different from those present in the egg immediately after fertilisation. This evidence has been adduced from DNA/RNA hybridisation studies by Glisin *et al.*[26] and by Whiteley *et al.*[27]. In the first set of experiments, unlabelled RNA from different developmental stages was hybridised in competition with labelled RNA from the blastula stage. The findings indicated near identity of RNAs from the unfertilised egg, cleavage embryos and the blastula but clear differences between RNAs from the gastrula or pluteus stages and RNA from the blastula. The experiments by Whiteley *et al.*[27] were, in a sense, the converse of these in that labelled RNA from the pluteus stage was hybridised in competition with unlabelled RNA from earlier stages. They found evidence for complete homology of labelled pluteus RNA only with RNA from the pluteus stage. There was reduced homology with RNA from the gastrula stage and much less effective competition by RNA from the blastula and unfertilised egg. These experiments can be broadly interpreted as implying that there is little change in the RNA populations in the stages from the unfertilised egg to the blastula but that thereafter there is a progressive accumulation of new RNA species and the disappearance of some of those originally present.

Among the types of RNA made during the early stages, one is of particular interest. This is a 9S RNA presumed to be histone mRNA. During the rapid synthesis of histones which occurs in early development, a class of small polyribosomes becomes very prominent especially during the peak of DNA synthesis[28]. A rapidly-labelled 9S DNA can be isolated from this class of polyribosomes[29, 30]. This RNA is presumed to be histone messenger because it disappears if DNA synthesis is blocked, and, as will be discussed later, RNA of this size class has been extracted from sea urchin embryos by Gross *et al.*[15] and shown to direct histone synthesis in a cell-free system derived from Krebs ascites cells. Weinberg *et al.*[31] have been able to separate the messenger into components similar to the components of histone mRNA, corresponding to the five histones. Moreover, Skoultchi and Gross[19] have shown that the rapidly-labelled 9S RNA which appears early in sea urchin development competes in DNA hybridisation with the 9S RNA which can be shown to direct histone synthesis. On the other hand, 18 and 26S ribosomal RNA, globin mRNA and 9S histone mRNA from a different sea urchin do not compete out this 9S RNA.

3.3 EFFECTS OF INHIBITORS ON EARLY EMBRYONIC DEVELOPMENT

Early embryological studies[1, 32], which involved studying hybrid embryos formed between species, showed that some parental component was essential for normal development after gastrulation but, equally, did not show any such requirement before gastrulation. Moreover, Harvey[33], in his classical experiments, demonstrated that non-nucleated fragments of sea urchin eggs could undergo the characteristic cleavages exhibited by normal embryos. With the availability of specific inhibitors of protein and RNA synthesis in the early 60s, new and intriguing facts began to emerge. First, Hultin[34] showed that puromycin completely prevents cleavage. This provided presumptive evidence that protein synthesis was essential for this process; the possibility that the phenomenon might be due to a side effect of puromycin was rendered unlikely by similar observations using cycloheximide[35], which also completely inhibits cleavage. These observations, combined with Harvey's earlier findings, led to the intriguing suggestion that protein synthesis might be able to occur in the absence of a nucleus and, presumably therefore, of RNA synthesis, in early embryogenesis. This impression was strengthened by experiments by Gross and Cousineau[8, 36] and others which showed that cleavage could occur in embryos in which RNA synthesis had been almost completely inactivated by actinomycin D.

Indeed active protein synthesis was directly demonstrated by Denny and Tyler[37] and Brachet *et al.*[38] in parthogenetically-activated enucleated halves of eggs and it occurred in them at the same rate as in controls. Moreover, in the early cleavage stages, protein synthesis is apparently qualitatively perfectly normal although no RNA synthesis is taking place[39-41]. Finally, Malkin *et al.*[42] isolated polysomes from actinomycin D-treated cleavage embryos and showed that these are, to all intents and purposes, normal.

Taken together, the clear implication of these findings is that a complete protein synthesising apparatus is necessary for early development and that this is already present before fertilisation. Practically all the early protein synthesis in the embryo appears to be due to this pre-existing machinery while informational RNA which is made during early cleavage is probably required for protein synthesis which occurs, not in early cleavage but from the blastula stage onwards. Very little, if any, RNA synthesis occurs in the mature egg and it seems likely that informational RNA is synthesised during the lampbrush stage (diplotene) in the oocyte, which may occur months or even years before the egg is eventually fertilised. Ribosomal RNA also accumulates during oogenesis and there is a special mechanism for its synthesis, as discussed in the next chapter.

The question which has therefore intrigued biochemists and embryologists is: if all the components necessary for protein synthesis are present in the egg, why are proteins not made? The two main hypotheses which have been considered are, either that mRNA is associated with ribosomes to form polysomes which are, for an unknown reason, inactive or, alternatively, that mRNA is sequestered from the ribosomes (the so-called 'masked messenger' hypothesis). There is some evidence for both of these. This seems to show, on the one hand, that maternal messenger in the egg is in a masked form, and on the other, that in early embryogenesis, some of the polysomes which are formed are not immediately active in protein synthesis.

3.4 INACTIVE POLYSOMES AND 'INFORMOSOMES'

Most of the evidence for inactive polysomes comes from the work of Spirin and Nemer[43] and Infante and Nemer[44]. They were able to demonstrate that labelled DNA-like RNA, heterogeneous in size, accumulates in early embryogenesis. However, it was observed that the spectrum of proteins formed in the early embryo is virtually the same whether they are treated with actinomycin D or not[41] and, moreover, there is little change in the quantitative synthesis of proteins, at least up to the blastula stage, as a result of an actinomycin D block[36, 42].

Spirin and Nemer[43] also obtained more direct experimental evidence to show that polysomes, containing RNA newly made during early embryogenesis, are not engaged in protein synthesis. In these experiments, they labelled newly-made RNA with one isotope and nascent protein with another. Polysomes from embryos labelled in this way were then sedimented in a sucrose gradient. The RNA label was found almost entirely in small classes of polysomes but nearly all the peptide synthesis was identified in larger classes which contained negligible amounts of newly-synthesised RNA. This very direct demonstration of a dichotomy between newly-made polysomes and polysomes making proteins in the early stages of embryogenesis complements studies by Barros *et al.*[45] and Giudice *et al.*[46] which showed that when RNA synthesis was blocked at different stages during cleavage, the effects were not seen until much later, implying that the transcription of RNAs essential for development of the gastrula, occurs during cleavage some hours earlier. The above studies therefore argue for the presence of a class of polysomes, formed with RNA synthesised during early cleavage which are either completely inactive in protein synthesis or, more probably, synthesise proteins very slowly until early gastrula.

Spirin and his colleagues[5, 43] have described yet another class of particles containing newly-made RNA which appears in the cytoplasm of cleavage stage embryos. These particles sediment in sucrose gradients more slowly than ribosomes and are, therefore, clearly distinct from the class of small ribosomes described above. They are heterogeneous in size and the RNA extracted from them hybridises well to DNA[43] and therefore has the characteristics of mRNA rather than of any of the ribosomal species. Spirin has called these particles 'informosomes'.

3.5 'MASKED mRNA' IN THE EGG

The evidence for inactive polysomes and informosomes is all derived from studies in the early embryo. There is a considerable accumulation of evidence to indicate that yet a different mechanism is involved in the storage of maternal mRNA in the egg. Perhaps the most crucial experiments demonstrating this were the classical study of Hultin[34] extended and substantiated by other workers[47-52]. In these experiments, it was found that when polysomes were isolated from unfertilised sea urchin eggs and incubated with isotopically labelled amino acids in the presence of necessary factors, no detectable polypeptide synthesis occurred but when polyuridylic acid was added as a template to such an incubation mixture, the polysomes proved to be extremely efficient in synthesising polyphenylalaline. In contrast, the polysomes from cleavage embryos exhibited a high level of endogenous polypeptide synthesis and were less readily stimulated with polyU. Although these expriments were interpreted in different ways, it is now faily clear that the egg contains large numbers of uncommitted ribosomes and that almost immediately after fertilisation, these become programmed to form polysomes. Direct demonstrations of this were provided in the work of Monroy and Tyler[53], Stafford and Iverson[9], Malkin *et al.*[42], Infante and Nemer[44] and Cohen and Iverson[54]. These workers showed by sucrose gradient sedimentation and other means that ribosomes extracted from unfertilised eggs were almost all free whereas in the cleavage embryo, they were present as polysomes.

The question therefore arose: Where is the mRNA in the egg? That messenger is present in the unfertilised egg is, of course, strongly inferred from the studies which have already been described; direct evidence for this has now accumulated. Several groups have extracted RNA from unfertilised eggs and shown that some of this is a very effective template for peptide synthesis in a cell-free system. It has also been demonstrated that some of this RNA hybridises effectively with DNA and has other general characteristics of mRNA[48,55-57]. A clue to the actual location of the messenger came from the work of Afzelius[58] who discovered a class of heavy RNA particles in the sea urchin egg. Subsequently Stavy and Gross[59] claimed that most of the template active RNA was associated with these particles.

3.6 HISTONE mRNA

Quite recently, much more precise evidence has been obtained about one specific class of mRNAs, the messengers for histones. In an earlier section, evidence was presented for the new synthesis of histone mRNA very soon after fertilisation. It is clear, however, that not all histone mRNA is made after fertilisation since histones are made in the presence of actinomycin D.

Jacobs-Lorena *et al.*[60] demonstrated that a class of RNA molecules sedimenting at 7–9S could be isolated from HeLa cells and that these acted as a template for the synthesis of histones in a cell-free system derived from Krebs ascites cells. Gross *et al.*[14] have found that RNA of this size can be isolated from 20S ribonucleoprotein particles which occur in the unfertilised sea urchin egg and that, in a cell-free system from Krebs ascites cells, it can also direct the synthesis of histones. The ribonucleoprotein particles were isolated from a post-ribosomal supernatant. RNA isolated from components which ran faster than 40S incorporated more aspartic acid than lysine into newly-synthesised polypeptides but a component sedimenting to the light side of 40S ribosomal subunits, incorporated lysine at a considerably higher rate than aspartic acid. The labelled protein newly-synthesised in the *in vitro* system co-electrophoresed with the histone components of a pure histone preparation run as markers. The identity of the translation products with histones was further investigated by preparing tryptic peptides using protein which had been synthesised in the presence of labelled phenylalaline as the only labelled amino acid. This procedure yields a typical peptide fingerprint for histones to which the translation product in these experiments was found to conform. This work seems to provide rather conclusive evidence that, in the unfertilised egg, histone mRNA is already present.

The histone mRNA isolated in these experiments sediments at about 9S, like the histone messenger described in other systems. Skoultchi and Gross[19] have further investigated the hybridisation characteristics of this RNA. It competes with 9S RNA synthesised in early embryonic development but not with any of the ribosomal RNA species. Moreover, by competitive hybridisation experiments, it has been shown that histone mRNA can be identified only in the ribosomal supernatant component sedimenting at about 20S.

3.7 MATERNAL AND NEWLY-SYNTHESISED mRNA IN EARLY EMBRYOGENESIS IN OTHER SPECIES

Some of the evidence relating to the amphibians will be described in the next chapter. The situation there seems to be similar in many ways to that in the sea urchin. The requirement for protein synthesis has been demonstrated by the fact that inhibitors such as puromycin block development of the embryo[61]. However, as in the sea urchin, cleavage occurs and protein synthesis goes on at a high level in the presence of actinomycin D[38, 61, 62, 78], and also in enucleated eggs[63]. Davidson *et al.*[64] have also obtained evidence for template mRNA in the amphibian egg.

There is ample evidence for similar phenomena in many other animal species. For example, cleavage and gastrulation occur in the presence of actinomycin D in the snail *Ilyanasa obsoleta*[65, 66]. In insect eggs too, development proceeds up to the gastrulation after injection of actinomycin D into the egg[67]. Moreover, a complete polysomal apparatus has been demonstrated in the cleavage stages of amphibians[68], teleosts[69] and nematodes[70]. Even when RNA synthesis is blocked, all protein synthesis occurs on maternal ribosomes in teleosts[69, 71] and insects (*Oncopeltus*)[72] since ribosomal synthesis does not start in the embryo till the gastrula stage. On the other hand, in mammals[73, 74] and in birds[75, 76] ribosomal synthesis starts earlier, during cleavage. Indeed, in the mouse, there is evidence that ribosomal synthesis starts as early as the four to eight cell stage[73, 74]. These latter observations indicate that the timing of the switch from synthesis of proteins on maternal polysomes to synthesis of proteins on embryonic polysomes, may occur at different stages depending on species. However, the universal phenomenon seems to be that probably in all animal species, so far studied, protein synthesis immediately after fertilisation occurs with the aid of maternal ribosomes and is directed by maternal messenger RNA.

Moreover, although the great majority of studies on this phenomenon have been carried out with a variety of animal systems, it seems rather clear from the work of Weeks and Marcus[77] that masked mRNA co-exists with uncommitted ribosomes in the wheat germ and that the formation of polysomes quickly follows activation. Hence the storage of maternal RNA in the gamete and its use for synthesis of proteins in the earliest stages of embryogenesis is quite likely to prove a universal phenomenon in multicellular eukaryotes.

References

1. Davidson, E. H. (1968). *Gene Activity in Early Development*. (New York and London: Academic Press)
2. Gross, P. R. (1972). *Molecular Genetics and Developmental Biology*. (Englewood Cliffs, N.J.: Prentice Hall Inc.)
3. Gross, P. R., Gross, K. W., Skoultchi, A. I. and Ruderman, J. V. (1973). *6th Karolinska Symposium in Research Methods in Reproductive Endocrinology*, 244
4. Nemer, M. (1967). *Prog. Nucleic Acid Res. Mol. Biol.*, **7,** 243
5. Spirin, A. S. (1966). *Current Topics Devel. Biol.*, **1,** 1
6. Baker, R. F. (1966). Ph.D. Thesis, Brown University
7. Monroy, A., Vittorelli, M. L. and Guarneri, R. (1961). *Acta Embryol. Morphol. Exp.*, **4,** 77
8. Gross, P. R. and Cousineau, G. H. (1963). *J. Cell. Biol.*, **19,** 260
9. Stafford, D. W. and Iverson, R. M. (1964). *Science*, **143,** 580
10. Selvig, S. E., Greenhouse, G. A. and Gross, P. R. (1972). *Cell Differentiation*, **1,** 5
11. Glisin, V. R. and Glisin, M. V. (1964). *Proc. Nat. Acad. Sci. U.S.*, **52,** 1548
12. Gross, P. R., Malkin, L. I. and Moyer, W. A. (1964). *Proc. Nat. Acad. Sci. U.S.*, **51,** 407
13. Wilt, F. H. (1964). *Develop. Biol.*, **9,** 299
14. Gross, K. W., Jacobs-Lorena, M., Baglioni, C. and Gross, P. R. (1973). *Proc. Nat. Acad. Sci. U.S.*, **70,** 2614
15. Gross, K. W., Ruderman, J., Jacobs-Lorena, M., Baglioni, C. and Gross, P. R. (1973). *Nature New Biology*, **241,** 272
16. Hogan, B. and Gross, P. R. (1971). *J. Cell Biol.*, **48,** 692
17. Raff, R. A., Colot, H. V., Selvig, S. R. and Gross, P. R. (1972). *Nature* (*London*), **235,** 211
18. Ruderman, J. V. and Gross, P. R. (1973). *Develop. Biol.* Submitted for publication
19. Skoultchi, A. I. and Gross, P. R. (1973). *Proc. Nat. Acad. Sci. U.S.*, **70,** 2840
20. Nemer, M. and Infante, A. A. (1965). *Science*, **150,** 217
21. Nemer, M. (1963). *Proc. Nat. Acad. Sci. U.S.*, **50,** 230
22. Siekevitz, P., Maggio, R. and Catalano, C. (1966). *Biochim. Biophys. Acta*, **129,** 145
23. Comb, D. G. (1965). *J. Mol. Biol.*, **11,** 851
24. Cowden, R. R. and Lehman, H. E. (1963). *Growth*, **27,** 185
25. Comb, D. G., Katz, S., Branda, R. and Pinzino, C. J. (1965). *J. Mol. Biol.*, **14,** 195
26. Glisin, V. R., Glisin, M. V. and Doty, P. (1966). *Proc. Nat. Acad. Sci. U.S.*, **56,** 285
27. Whiteley, A. H., McCarthy, B. J. and Whiteley, H. R. (1966). *Proc. Nat. Acad. Sci. U.S.*, **55,** 519
28. Kedes, L. H. and Gross, P. R. (1969). *Nature*, **223,** 1335
29. Kedes, L. H. and Birnstiel, M. L. (1971). *Nature New Biol.*, **230,** 165
30. Kedes, L. H., Hogan, B., Cognetti, G., Selvig, S., Yanover, P. and Gross, P. R. (1969). *Cold Spring Harbor Symp. Quant. Biol.*, **34,** 717
31. Weinberg, E. S., Birnstiel, M., Purdom, I. F. and Williamson, R. (1972). *Nature*, **240,** 225
32. Hyman, L. H. (1955). The Invertebrates; Echinodermata, New York: McGraw-Hill
33. Harvey, E. B. (1936). *Biol. Bull.*, **71,** 101
34. Hultin, T. (1961). *Experentia*, **17,** 410
35. Karnofsky, D. A. and Simmel, E. B. (1963). *Prog. Exp. Tumour Res.*, **3,** 254
36. Gross, P. R. and Cousineau, G. H. (1964). *Exp. Cell Res.*, **33,** 368
37. Denny, P. C. and Tyler, A. (1964). *Biochem. Biophys. Res. Commun.*, **14,** 245
38. Brachet, J., Ficq, A. and Tencer, R. (1963). *Exp. Cell Res.*, **32,** 168

39. Gross, P. R. (1967). *Current Topics Develop. Biol.*, **2,** 1
40. Spiegel, M., Ozaki, H. and Tyler, A. (1965). *Biochem. Biophys. Res. Commun.*, **21,** 135
41. Terman, S. A. and Gross, P. R. (1965). *Biochem. Biophys. Res. Commun.*, **21,** 595
42. Malkin, L. I., Gross, P. R. and Romanoff, P. (1964). *Develop. Biol.*, **10,** 378
43. Spirin, A. S. and Nemer, M. (1965). *Science*, **150,** 214
44. Infante, A. A. and Nemer, M. (1967). *Proc. Nat. Acad. Sci. U.S.*, **58,** 681
45. Barros, C., Hand, G. S. Jr. and Monroy, A. (1966). *Exp. Cell Res.*, **43,** 167
46. Giudice, G., Mutolo, V. and Donatuti, G. (1968). *Arch. Entwicklungsmech., Organ*
47. Brachet, J., Decroly, M., Ficq, A. and Quertier, J. (1963). *Biochim. Biophys. Acta*, **72,** 660
48. Maggio, R., Vittorelli, M. L., Rinaldi, A. M. and Monroy, A. (1964). *Biochem. Biophys. Res. Commun.*, **15,** 436
49. Nemer, M. (1962). *Biochem. Biophys, Res. Commun.*, **8,** 511
50. Nemer, M. and Bard, S. G. (1963). *Science*, **140,** 664
51. Tyler, A. (1963). *Amer. Zoologist*, **3,** 109
52. Wilt, F. H. and Hultin, T. (1962). *Biochem. Biophys. Res. Commun.*, **9,** 313
53. Monroy, A. and Tyler, A. (1963). *Arch. Biochem. Biophys.*, **103,** 431
54. Cohen, G. H. and Iverson, R. M. (1967). *Biochem. Biophys. Res. Commun.*, **29,** 349
55. Crippa, M., Davidson, E. H. and Mirsky, A. E. (1967). *Proc. Nat. Acad. Sci. U.S.*, **57,** 885
56. Davidson, E. H., Allfrey, V. G. and Mirsky, A. E. (1964). *Proc. Nat. Acad. Sci. U.S.*, **52,** 501
57. Slater, D. W. and Spiegelman, S. (1966). *Proc. Nat. Acad. Sci. U.S.*, **56,** 164
58. Afzelius, B. A. (1957). *Z. Zellforsch.*, **45,** 660
59. Stavy, L. and Gross, P. R. (1967). *Proc. Nat. Acad. Sci. U.S.*, **57,** 735
60. Jacobs-Lorena, M., Baglionii, C. and Borun, T. W. (1972). *Proc. Nat. Acad. Sci. U.S.*, **69,** 2095
61. Brachet, J., Denis, H. and de Vitry, F. (1964). *Develop. Biol.*, **9,** 398
62. Wallace, H. and Elsdale, T. R. (1963). *Acta Embryol. Morphol. Exp.*, **6,** 275
63. Smith, L. D. and Ecker, R. E. (1965). *Science*, **150,** 777
64. Davidson, E. H., Crippa, M., Kramer, F. R. and Mirsky, A. E. (1966). *Proc. Nat. Acad. Sci. U.S.*, **56,** 856
65. Collier, J. R. (1966). *Current Topics Develop. Biol.*, **1,** 39
66. Feigenbaum, L. and Goldberg, E. (1965). *Amer. Zoologist.*, **5,** 198
67. Lockshin, R. A. (1966). *Science*, **154,** 775
68. Gansen, P. van (1967). *Exp. Cell Res.*, **47,** 157
69. Belitsina, N. V., Aikhozhin, M. A., Gavrilova, P. L. and Spirin, A. S. (1964). *Biochem. (USSR)*, **29,** 315
70. Kaulenas, M. S. and Fairbairn, D. (1966). *Develop. Biol.*, **14,** 481
71. Aithozhin, M. A., Belitsina, N. V. and Spirin, A. S. (1964). *Biochemistry (USSR)*, **29,** 145
72. Harris, S. E. and Forrest, H. S. (1967). *Science*, **156,** 1613
73. Mintz, B. (1964). *J. Exp. Zool.*, **157,** 85
74. Mintz, B. (1964). *J. Exp. Zool.*, **157,** 273
75. Lerner, A. M., Bell, E. and Darnell, J. E. Jr. (1963). *Science*, **141,** 1187
76. Solomon, J. (1957). *Biochim. Biophys. Acta*, **24,** 584
77. Weeks, D. P. and Marcus, A. (1971). *Biochim. Biophys. Acta*, **232,** 671
78. Brachet, J. and Denis, H. (1963). *Nature (London)*, **198,** 205

4
Nucleic Acid Synthesis during Oogenesis and Early Embryonic Development of the Amphibians

H. DENIS
Centre de Génétique moléculaire C.N.R.S., France

4.1 INTRODUCTION

Growing embryos contain an increasing variety of protein and RNA molecules[1-3]. It is therefore very likely that the release of genetic information in embryonic cells is primarily controlled at the level of transcription. In other words, the nucleus of each type of differentiating cell synthesises and sends to the cytoplasm a particular set of messenger molecules which are later translated into organ-specific proteins. A certain amount of control is also exercised at the level of translation. The mere appearance of a messenger molecule in the cytoplasm does not automatically cause the synthesis of the corresponding protein. Translation of a given message can be delayed for some time or even subjected to variations in time according to the physiological state of the cell[4]. The mechanism which enables the differentiating cell to select a definite part of its genetic information for transcription and translation is far from clear at the present time. There will be little hope of elucidating this mechanism as long as the control of gene expression in differentiated tissues is not fully understood.

In recent years, many embryologists have devoted their attention to the process of nucleic acid accumulation in oocytes and embryos. The eggs of all animals contain much more DNA and RNA than any somatic cell of the same species. If the nature and the distribution of all macromolecules stored in the egg were known with accuracy, it is felt that many problems related to early differentiation would be clarified to a much greater extent. Attention has been mainly focused in the past few years on ribosomal and transfer-RNA which are easier to study than informational RNA (messenger-RNA and heterogeneous nuclear RNA). Since there is much more ribosomal and transfer-RNA in the egg than in other cells the following question has been asked: how does the oocyte, which contains one single nucleus, manage to accumulate as much RNA as several thousand somatic cells? The main purpose of the present review is to answer this question. This article will be exclusively devoted to the Amphibians. The most important experiments in the field under review have been performed on the eggs of frogs and newts and especially on those of *Xenopus laevis*, the South African clawed toad (which is in fact a primitive frog belonging to the Pipidae family).

4.2 MAIN MORPHOLOGICAL FEATURES OF OOGENESIS AND EMBRYONIC DEVELOPMENT IN AMPHIBIANS

In Amphibians, as in other Vertebrates, the ovary differentiates as a mesodermal fold attached to the roof of the body cavity. This fold is later invaded by germ cells coming from the endoderm. During the larval life of the female, the germ cells undergo several rounds of division without increasing in size. At the time of metamorphosis, some of the germ cells transform into oocytes and begin to grow. Their chromosomes condense and become visible. In a few weeks, the oocyte goes through the leptotene, zygotene and pachytene stages of the first meiotic prophase (Figure 4.1) and stops in diplotene[5]. The oocyte remains at this stage during the whole period of its growth which will last for several months.

Oogenesis in *Xenopus* as in other Amphibians can be divided into two periods. During the first period, the diameter of the oocyte increases from 10 μm to 200–300 μm. The nucleus (also called the germinal vesicle) enlarges in a proportionate fashion. The cytoplasm remains clear and almost entirely free from the microstructures that are usually seen with the electron microscope in somatic cells, i.e. endoplasmic reticulum, Golgi body and ribosomes[6-8]. Mitochondria are the only type of organelles which are abundant in small oocytes.

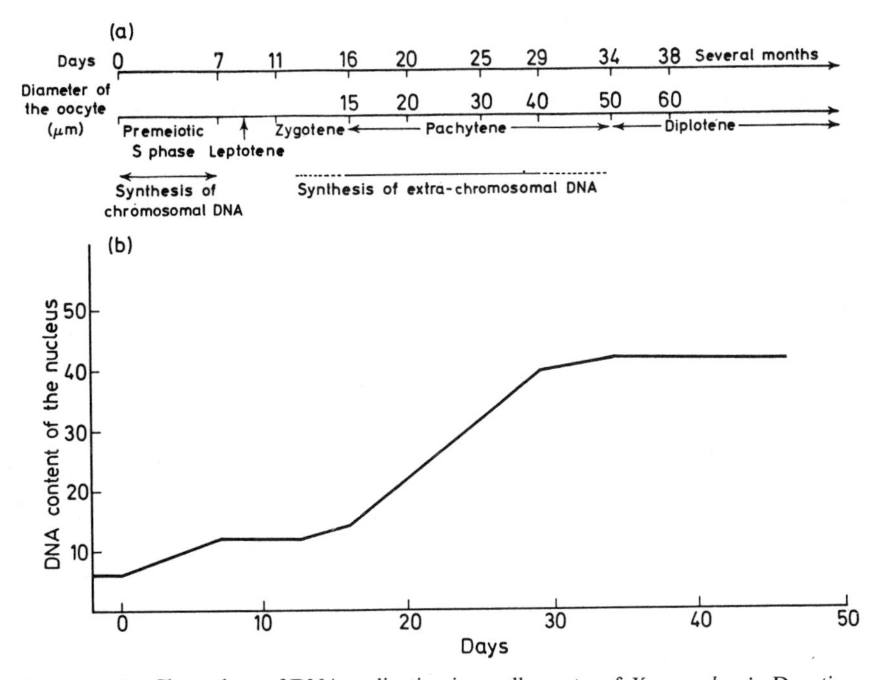

Figure 4.1 Chronology of DNA replication in small oocytes of *Xenopus laevis*. Duration of the stages of the first meiotic prophase is shown in (a) together with the size reached by the oocyte during the early stages of meiosis. The increase in DNA content of the germinal vesicle is represented schematically in (b). The slope of curve (b) is arbitrary since the rate of DNA synthesis is not necessarily constant during both periods of replication. (Data collected from Refs. 5 and 18.)

During the second period of its growth (vitellogenesis), the oocyte becomes progressively loaded with ribosomes and various kinds of inclusions: yolk platelets, pigment grains, oil droplets and glycogen granules. Mitochondria continue to accumulate in large oocytes as in small ones.

The germinal vesicle also undergoes important morphological changes during oogenesis. The diplotene chromosomes become progressively longer and thinner as the nucleus swells and take up a characteristic configuration usually referred to as 'lampbrush'. A few weeks after the beginning of the diplotene stage, a large number of nucleoli appear in the germinal vesicle. These nucleoli remain present until the end of oogenesis. They are not attached to the chromosomes as in somatic cells. The number of nucleoli per germinal vesicle varies from one Amphibian species to another. According to Buongiorno-Nardelli, Amaldi and Lava-Sanchez[9], this number is not random since the oocytes of various species of Anurans and Urodeles have been found to contain 250, 500 or 1000 nucleoli but never more than 1000 or less than 250.

When the oocyte reaches its maximum size (about 1 mm in *Xenopus*), it becomes dormant. The chromosomes contract and gather in the centre of the germinal vesicle. The nucleoli also move to the centre of the nucleus. The oocyte remains quiescent until it receives the hormonal stimulus that causes ovulation and maturation. The germinal vesicle bursts and the oocyte completes its first meiotic division by sending out its first polar body. The second meiotic division starts immediately and stops in metaphase. Meanwhile, the oocyte is expelled from the ovary and travels through the oviduct where it is covered with several layers of albuminous material (jelly coat). Completion of the second meiotic division and emission of the second polar body occurs only after fertilisation.

Embryonic development starts with a period of rapid cell division (cleavage). No important displacement of cells takes place during cleavage. Morphogenesis begins with gastrulation. The three primitive layers of the embryo (ectoderm, mesoderm and endoderm) separate and arrange themselves in concentric fashion. Gastrulation is immediately followed by neurulation which leads to organisation of the central nervous system in the outer layer (ectoderm). At the end of neurulation, all the rudimentary organs have reached their definitive positions and begin to differentiate. The embryo soon starts contracting spasmodically. A few days later, it swims actively and reacts first to mechanical and later to visual stimuli. The last organ which differentiates is the gut. When all intestinal yolk has disappeared, the tadpole is ready to feed. The time needed for the egg to transform into a tadpole varies from one species to another. In *Xenopus*, embryonic development is particularly rapid: only 5 days elapse between fertilisation and the uptake of food.

4.3 DNA SYNTHESIS IN OOCYTES AND EMBRYOS

4.3.1 Nuclear DNA

The nucleus of the oocyte contains DNA not only inside but also outside the chromosomes. The existence of extra-chromosomal DNA in the germinal vesicle of the Amphibians has been known for many years[10], but it is only recently that the nature and function of this type of DNA has been given a correct interpretation. Chromosomal and extra-chromosomal DNAs differ both in their role during oogenesis and in the amount of genetic information that they contain. We will therefore discuss them separately.

4.3.1.1 Chromosomal DNA

The germ cell contains the same number of chromosomes and the same amount of chromosomal DNA as a somatic cell. One week before the onset of the first meiotic division, replication of chromosomal DNA begins (Figure 4.1). No synthesis of chromosomal DNA takes place in the oocyte during the whole length of its growth in the ovary[11,12]. Until completion of the first meiotic division and expulsion of the first polar body, there is twice as much chromosomal DNA in the oocyte as in a somatic cell. The ovarian oocyte is therefore tetraploid. The unfertilised egg is diploid and remains so after fertilisation because the amount of DNA which is expelled in the second polar body is replaced by an equal amount of DNA brought in by the spermatozoon.

Synthesis of chromosomal DNA begins immediately after fertilisation. In fact, the male and the female pro-nuclei start replicating their DNA even before fusing[13]. During early cleavage, the cells divide very rapidly. There is no interval between the end of each mitosis (M phase) and the next period of DNA replication (S phase)[14]. At the end of the first 10 h of cleavage, the blastula of *Xenopus* already contains 15 000 cells. Accordingly, the chromosomal DNA content of the embryo increases 15 000-fold during this period of development[15]. After the beginning of gastrulation, the frequency of mitoses decreases considerably. A short G_1 phase appears between the M and S phases and the G_2 phase also lengthens[14]. When the embryo of *Xenopus* begins to feed, it contains 10^6 times as much chromosomal DNA (6 μg) as the uncleaved egg (6 pg)[15]. The relative length of each phase (M, G_1, S and G_2) of the cell cycle is now approximately the same as in adult tissues: the S phase represents one-fifth to one-third of the interval between two successive mitoses[13].

4.3.1.2 Extra-chromosomal DNA

(a) *Amplification of ribosomal genes in oocytes*—During the pachytene stage, the DNA content of the germinal vesicle increases more than 3-fold (Figure 4.1). A conspicuous mass of extra-chromosomal DNA appears in the nucleus and condenses into a crescent-shaped cap[16-18]. The extra-chromosomal cap of *Xenopus* contains approximately 30 pg of DNA, i.e. 2.5 times as much as the whole set of chromosomes[18]. In the late pachytene and early diplotene stage, a few nucleoli begin to form inside the DNA cap. Soon after, the DNA cap disperses, so that the nucleoli become free in the nuclear sap[19]. Microscopical and biochemical evidence strongly suggests that the DNA which makes up the pachytene cap is the same as the DNA which is found in the nucleoli of larger oocytes[17,19,20]

The function of extra-chromosomal DNA was elucidated in 1968 by Brown and Dawid[21], Gall[17] and Evans and Birnstiel[22]. This DNA codes for the two main species (28S and 18S) of ribosomal RNA. Instead of containing 1000 28S and 18S genes as each somatic cell of *Xenopus* does[23], the oocyte contains about 2×10^6 of these genes. In other words, the oocyte *amplifies* its ribosomal genes at the beginning of its growth period. As pointed out as early as 1942 by Painter and Taylor[24], the Amphibian oocyte is highly polyploid with respect to the nucleolar organiser and tetraploid with respect to the remainder of the genome.

(b) *Properties of amplified DNA*—The amplified DNA of *Xenopus* has been studied in great detail by Brown and Dawid[21] and by Dawid, Brown and Reeder[25]. Ribosomal DNA from somatic cells and amplified ribosomal DNA from oocytes were found to be identical in all respects save in their buoyant density. Amplified ribosomal DNA has a slightly greater density (1.729 g cm^{-3}) than somatic ribosomal DNA (1.724 g cm^{-3}). According to Dawid, Brown and Reeder[25], this difference can be explained by the following observation. Somatic ribosomal DNA contains a substantial amount of 5-methyl deoxycytidylic acid (13% of all cytidylic residues), whereas amplified DNA contains no detectable amount of this nucleotide. The higher degree of methylation observed in somatic ribosomal DNA is sufficient to account for the lower buoyant density (5 mg cm^{-3}) of this DNA. Both types of ribosomal DNA are formed of repeating units comprising a transcribed stretch and a non-transcribed stretch (spacer) of approximately equal length (see Figure 4.2)[26, 27]. Each transcribed stretch contains a 28S sequence, a 18S sequence and an extra sequence of about 900 nucleotide pairs that is not represented in 28S- and 18S-RNA[25]. The primary product of the ribosomal gene in both somatic cells and oocytes is a 40S molecule[28, 29] which is the common precursor of 28S- and 18S-RNA. Soon after transcription, the 40S precursor is split into an 18S and a 30S molecule which in turn gives rise to 28S-RNA[29]. The final products of the ribosomal genes are apparently identical in oocytes and in somatic cells.

(c) *Mechanism of DNA amplification*—The mechanism of DNA amplification in early oocytes is not fully elucidated at the present time. The template used for producing the extra copies of ribosomal DNA is the nucleolar organiser itself[30]. An alternative mechanism proposed by Wallace, Morray and Langridge[31], which suggests that the amplified genes would be copied on an episome transmitted from mother to daughter through the female germ cell line, has proved to be wrong[30].

As far as the molecular mechanism of amplification is concerned, two different models can be envisaged (Figure 4.2). Either each replica of amplified DNA is directly copied on the chromosome (direct amplification), or each primary copy is itself used as a template for the next round of replication (cascade amplification). The second model is probably correct for the following reasons. First, amplification is a rather fast process since the oocyte needs less than 2 weeks to synthesise about 500 copies of each nucleolar organiser, whereas one week is necessary to replicate each molecule of chromosomal DNA (Figure 4.1). The cascade model most easily explains the rapidity of the amplification process. Second, the oocytes of *Xenopus* females heterozygous for the O nu mutation[32] contain as many nucleoli and as much amplified

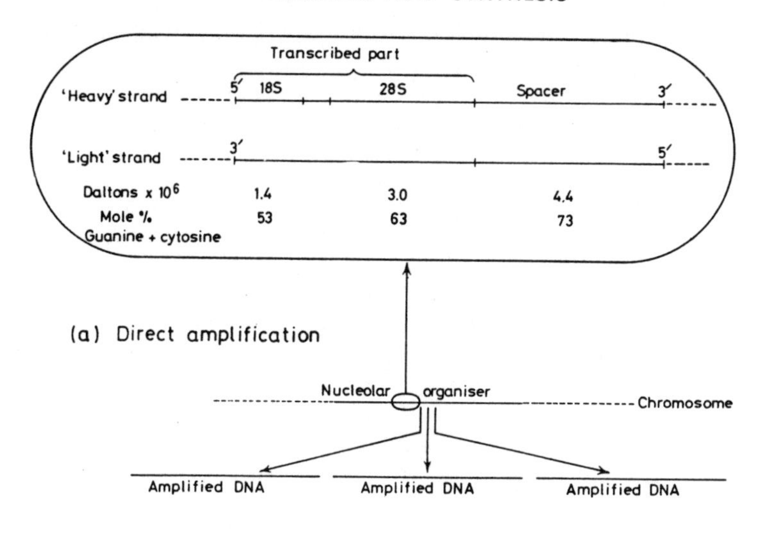

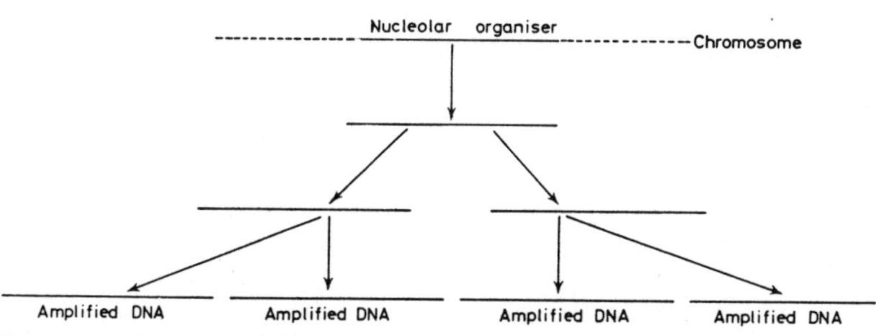

Figure 4.2 Two models of DNA amplification in the oocyte. According to model (a), each replica of amplified DNA is directly copied from chromosomal DNA. According to model (b), each primary copy of amplified DNA is used as a template for a further round of replication. Details of the structure of ribosomal DNA in *Xenopus laevis* are given in the upper part of the figure. Each nucleolar organiser contains about 500 repeating units identical to that shown here. The molecules of amplified DNA are represented in linear form although many and perhaps all of them are circular[40,41,140]. (Structure of ribosomal genes slightly modified from Ref. 25.)

DNA as normal oocytes[33], although the mutant oocytes derive from germ cells possessing only one nucleolar organiser (instead of two) and 50% less ribosomal genes than normal diploid cells[34]. This shows that the amount of amplified DNA finally present in the oocyte is not determined by the amount of ribosomal DNA that the cell contained before the amplification started. In mutant oocytes, the amplified DNA would simply undergo one more round of replication than in normal oocytes (Figure 4.2(b)). A third argument in favour of the cascade model is provided by the observation of Buongiorno-Nardelli, Amaldi and Lava-Sanchez[9] on the number of nucleoli in the oocytes of various Amphibian species. The striking coincidence of the number found with powers of two can be best understood by assuming that each nucleolus contains one copy of the nucleolar organiser and that the nucleolar organisers are duplicated at least six times in cascade (Figure 4.2).

The enzyme which produces the extra copies of ribosomal DNA in oocytes was first thought to be DNA polymerase. However, it has been recently proposed that reverse transcriptase instead of DNA polymerase is involved in this process[35, 36]. The evidence for the involvement of reverse transcriptase in amplification is not very strong. It first rested on inhibition experiments by means of a rifampicin derivative (2′,5′-dimethyl-*N*(4′)-benzyl-*N*(4′)-(desmethyl) rifampicin) which is known to affect reverse transcriptase and not DNA polymerase[37]. This antibiotic was found to reduce the synthesis of extra-chromosomal DNA more than that of chromosomal DNA in pachytene oocytes[35, 36]. More recently, attempts have been made to isolate an RNA intermediate from amplifying oocytes that would be used by reverse transcriptase to generate new copies of ribosomal DNA[38, 39]. This RNA intermediate must be a transcript not only of the 28S and 18S sequences but also of the spacer sequences (Figure 4.2), since the amplified DNA also contains non-transcribed spacers[25, 40, 41]. Pachytene cells indeed contain 47S-RNA molecules which are longer than the usual 40S precursor[35], but rigorous proof that this RNA is used as a substrate by reverse transcriptase is still lacking.

Very recently, Bird, Rogers and Birnstiel[141] tried to repeat the experiments of Crippa and Tocchini-Valentini[35,38,39] but they could not reproduce any of the results obtained by these authors.

4.3.2 Cytoplasmic DNA

The egg of all Amphibians contains far more DNA than a somatic cell of the same species[42-46]. The estimated amount of DNA per egg varies from one author to another according to the method used for estimating the DNA. The latest measurements indicate that the egg of *Xenopus* contains from 300 to 500 times as much DNA as a diploid cell[15]. Most of the DNA in the egg is located in the cytoplasm since the DNA content of the egg nucleus is only 5 to 10 times as high as that of a somatic cell. DNA from *Xenopus laevis* and *Rana pipiens* eggs has the same buoyant density and hence the same base composition as nuclear DNA of the same species[15]. However, there is no detectable sequence homology between egg DNA and erythrocyte DNA. Egg DNA has been convincingly shown by Dawid[15, 47] to be of mitochondrial origin. The egg of *Xenopus* contains about 100 000 times as many mitochondria as a liver cell[48] and mitochondrial DNA accounts for at least 65% of the DNA present in the egg[47]. The remainder of egg DNA may be located in the yolk platelets[49-51].

There is apparently no difference between mitochondrial DNA from eggs and oocytes and mitochondrial DNA from adult liver[54]. Both types of DNA consist of double-stranded circular molecules with a molecular weight of 10.6×10^6 and a contour length of 5.7 μm in *Xenopus laevis* and *Rana pipiens*[52-54]. Synthesis of mitochondrial DNA is thought to occur continuously in growing oocytes since mitochondria multiply during the whole of oogenesis[6].

Mitochondria stored in the egg become distributed among the blastomeres when cleavage begins. No net synthesis of mitochondrial DNA occurs in the embryo until the swimming stage[48]. This observation suggests that the number of mitochondria per embryo does not increase during early development.

Mitochondria of somatic tissues are endowed with partial autonomy since they can synthesise DNA, RNA and protein[55,56]. The same is true for the mitochondria of oocytes and embryos. Synthesis of mitochondrial DNA in oocytes is certainly not coupled with that of chromosomal DNA since mitochondria multiply from the beginning to the end of oogenesis, whereas no DNA synthesis occurs in the chromosomes[11,12]. The same absence of coupling is observed during embryonic development. A consequence of the autonomous multiplication of the mitochondria is that all mitochondria of the adult derive from those of the oocyte. Direct proof of maternal and cytoplasmic inheritance for mitochondrial DNA has recently been given by Dawid and Blackler[57].

4.4 RNA SYNTHESIS IN OOCYTES AND EMBRYOS

The Amphibian egg contains far more RNA than DNA. There is in a single egg of *Xenopus laevis* as much DNA as in 300–500 somatic cells[15], but as much RNA as in 300 000 somatic cells[58]. About 95% of the RNA stored in the egg is ribosomal in nature; 2% is transfer-RNA and 2–5% is DNA-like RNA.

4.4.1 Synthesis and accumulation of ribosomal RNA in oocytes and embryos

The ribosome of the Amphibian, like that of all Eukaryotes, contains three molecules of RNA sedimenting at 28S, 18S and 5S. Two of these molecules (28S and 5S) are located in the larger (60S) sub-unit of the ribosome whereas 18S-RNA is a component of the smaller sub-unit (40S).

The mitochondrial ribosome differs from the cytoplasmic ribosome in two main respects. First, it lacks 5S-RNA. Second, its two RNA components are much smaller than the corresponding ones of the cytoplasmic ribosome. For the sake of clarity, the two ribosomal RNA species of the Amphibian mitochondrion will here be called 21S and 13S[59], although these figures give a misleading impression of the size of these molecules[60-62].

Synthesis of 28S- and 18S-RNA and the major steps of ribosome assembly occur in the nucleolus. The process begins with transcription of 40S-RNA which is the common precursor of 28S- and 18S-RNA[28,29]. Soon after transcription, the 40S precursor becomes associated with proteins. It is then methylated and cleaved into two main pieces that later give rise to 28S- and 18S-RNA. In contrast to 28S- and 18S-RNA, 5S-RNA is not methylated. It is synthesised in the nuclear sap, then moves into the nucleolus and becomes incorporated into the larger ribosome sub-unit at the end of the assembly process. Complete ribosomal sub-units are released from the nucleolus into the nucleoplasm and thence into the cytoplasm, where protein synthesis takes place.

4.4.1.1 *Synthesis of ribosomal RNA in oocytes*

(a) *28S- and 18S-RNA*—When observed with the electron microscope, pre-vitellogenic oocytes of *Xenopus laevis* are seen to contain very few, if any, ribosomes[6-8]. Ribosomes are not totally absent in very small oocytes (less than 100 μm in diameter) but disappear completely in larger oocytes (100–300 μm in diameter)[19, 63]. Biochemical analysis of the RNA extracted from small oocytes confirms the electron-microscopical observations. Only 20–25% of this RNA has a high enough molecular weight to be eluted in the non-retarded peak when filtered through a column of Sephadex G-100 (Figure 4.3). High-molecular weight RNA from small oocytes has an unusual

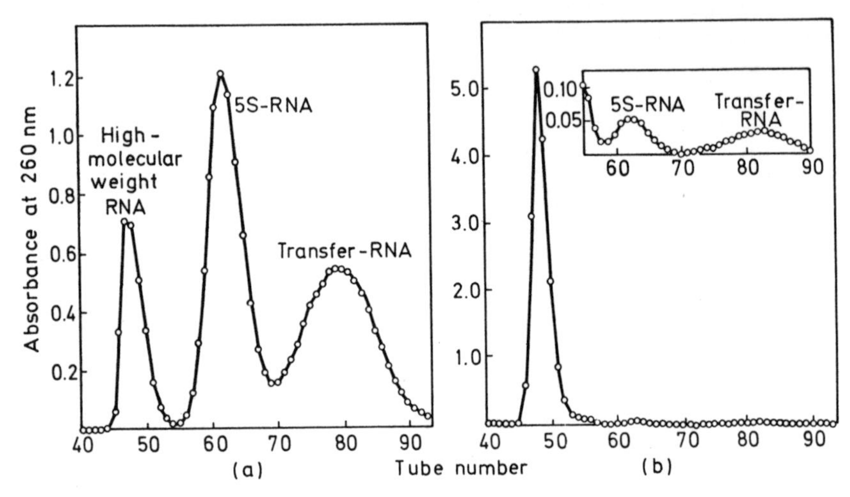

Figure 4.3 Analysis of the RNA present in (a) small and (b) large oocytes of *Xenopus laevis*. RNA was extracted from whole ovaries from (a) immature females and (b) from vitellogenic oocytes freed from somatic cells. The average diameter of the oocytes was 200 μm in (a) and 1.000 μm in (b). RNA was analysed by filtration through a column of Sephadex G-100. High-molecular weight RNA in (a) is a mixture of D-RNA and (28S + 18S)-RNA (see text), whereas high-molecular weight RNA in (b) is mostly (28S + 18S)-RNA. (Original data of the author.)

base composition: it contains only 50% guanine + cytosine[64], whereas (28S + 18S)-RNA contains 60% guanine + cytosine[26]. High-molecular weight RNA is therefore a mixture of DNA-like RNA (40% guanine + cytosine) and ribosomal RNA (60% guanine + cytosine). It follows that as little as 12% of the RNA content of the small oocyte is (28S + 18S)-RNA (see Table 4.1)[64]. This percentage is probably an over-estimate since RNA of pre-vitellogenic oocytes has been extracted from the whole ovaries of immature females. The somatic cells of the ovaries are likely to contribute most of the 28S- and 18S-RNA present in homogenates of the gonad because these cells contain far more ribosomes per unit volume than the oocytes[6-8]. Evidence has recently been obtained showing that all ribosomes present in homogenates of immature ovaries indeed derive from somatic cells[65].

Soon after the beginning of vitellogenesis, 28S- and 18S-RNA become the

Table 4.1 RNA content of oocytes and embryos of Xenopus laevis*

Stage	*Total RNA in each RNA class (%)*					*Relative number of molecules*				
	28S	18S	5S	*Transfer-RNA*	*D-RNA*	28S	18S	5S	*Transfer RNA*	*D-RNA*
Pre-vitellogenic oocyte (150 μm)	8.5	4.0	39.8	35.2	12.5	1	1	175†	250†	—
Unfertilised egg	65.0	31.0	1.8	2.2	not measured	1	1	1	2	—
Stage 35 embryo	60.0	28.4	2.3	9.3	not measured	1	1	1.4	9.3	—

* From Ref. 64.

† These figures are minimum values because the amount of (28S + 18S)-RNA in small oocytes was overestimated. Many somatic cells adhere to these oocytes and probably contain more ribosomes than the oocytes themselves.

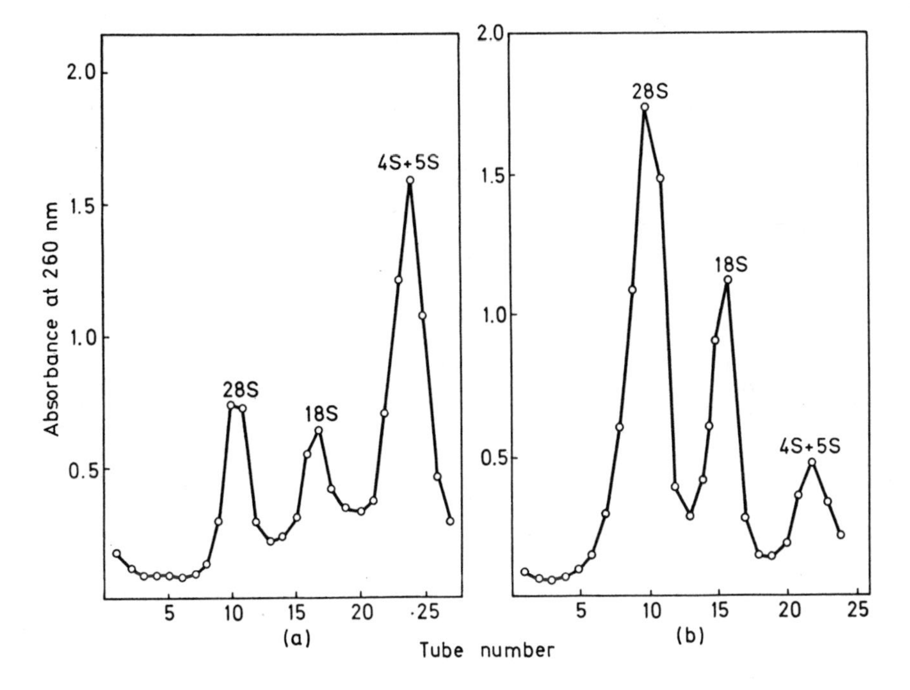

Figure 4.4 Analysis of the RNA present (a) in small and (b) in large oocytes of *Xenopus laevis*. The average diameter of the oocytes was 90 μm in (a) and 375 μm in (b). RNA was analysed by centrifugation in a sucrose density gradient. The left-hand part of the gradient contains not only 28S- and 18S-RNA but also a substantial amount of D-RNA (see text). (From Mairy and Denis[64], by courtesy of Academic Press)

most abundant RNA species (Figures 4.3 and 4.4) and remain so until the end of oogenesis. 28S- and 18S-RNA accumulate far more rapidly than all other RNA species and make up 95% of the RNA content of the unfertilised egg[64,66,67]. Large amounts of ribosomes can actually be observed with the electron microscope in the cytoplasm of vitellogenic oocytes[6-8]. These ribosomes are free in the cell sap and not attached to a membrane system as in many somatic cells. The egg of *Xenopus* contains 1.1×10^{12} ribosomes, i.e. 300 000 times as many as a liver cell[33].

It may be concluded from the data summarised above that ribosome accumulation is not a continuous process during oogenesis. The oocyte does not accumulate any important amounts of ribosomes or of (28S + 18S)-RNA during the first period of its growth. The 28S and 18S genes which are amplified at the pachytene stage (Figure 4.1) are therefore not fully utilised before the onset of vitellogenesis. It is hard to decide whether the 28S and 18S genes are completely repressed in small oocytes. The nucleoli of these oocytes are strongly basophilic and incorporate tritiated nucleosides into RNA[12]. Nucleolar activity is generally interpreted as an indication of ribosomal RNA synthesis and ribosome assembly in somatic cells[68]. Other types of RNA could, however, be synthesised in the nucleolus or move rapidly into it after their synthesis. The nascent RNA which is seen by autoradiography to appear in the nucleolus might also be ribosomal RNA that has failed to be incorporated into the ribosomes and to be processed into 28S- and 18S-RNA. This possibility is rather unlikely since no important synthesis of precursor 40S-RNA was observed in small oocytes[64,69]. In order to find out whether or not synthesis of (28S + 18S)-RNA is completely inactive in small oocytes, it will be necessary either to obtain these oocytes free from follicle cells or to isolate nucleoli from oocytes and to study their synthetic activity *in vitro*. Whatever the outcome of these experiments, the available data clearly show that a considerable activation of the 28S and 18S genes occurs at the beginning of vitellogenesis. The mechanism of this activation remains to be elucidated. Again isolation of nucleoli from small and large oocytes might help to solve this problem.

(b) *5S-RNA*—Synthesis of 5S-RNA in oocytes is not at all co-ordinate with that of 28S- and 18S-RNA[64,69]. The oocyte first accumulates large amounts of 5S-RNA which makes up 40–50% of the RNA content of the cell at the end of the pre-vitellogenic period (Figure 4.3). At this point in its growth, the oocyte contains at least 175 molecules of 5S-RNA for every molecule of 28S- and 18S-RNA (Table 4.1). In later oogenesis, 5S-RNA makes up a declining proportion of the total RNA. In the unfertilised egg, 28S-, 18S- and 5S-RNA are present in roughly equimolar amounts[64]. Accumulation of 5S-RNA, which is much faster than that of 28S- and 18S-RNA in small oocytes, becomes much slower after the beginning of vitellogenesis. This is not due to a decrease in the rate of production of 5S-RNA in large oocytes but rather to an increase in the rate of production of 28S- and 18S-RNA (see above). 5S-RNA therefore accumulates during the whole period of the oocyte growth whereas 28S- and 18S-RNA accumulate only during half of this period.

The absence of coordination between the synthesis of 5S-RNA and that of 28S- and 18S-RNA raises a number of questions. First, is the excess 5S-RNA accumulated in small oocytes conserved in the cells until enough ribosomes are being assembled to accommodate this RNA? Second, in what state and in what cell compartment is 5S-RNA stored during the initial period of oogenesis? Third, how does the oocyte manage to synthesise as many 5S-RNA molecules as 28S- and 18S- molecules although much smaller numbers of 5S genes are available?

The following observations throw light on the fate of 5S-RNA accumulated in small oocytes. When small oocytes of *Xenopus* are exposed for several weeks to tritiated uridine *in vivo*, all types of RNA become labelled (Figure 4.5(a)). 5S-RNA contains almost one-half of the total radioactivity present in RNA. If small oocytes labelled as in (Figure 4.5(a)) are allowed to grow in the female for one year, so that they have now entered vitellogenesis, these cells still contain appreciable amounts of radioactivity (Figure 4.5(b)). At the end of the experiment, the distribution of radioactivity among the three main classes of RNA (high-molecular weight RNA, 5S-RNA and transfer-RNA) is roughly the same as one year before. But high-molecular weight RNA which amounts to only 25% of the total RNA in small oocytes (Figure 4.5(a)) is now the dominant class of RNA (Figure 4.5(b)). The three main species of ribosomal RNA that were equally labelled at the beginning of the experiment (Figure 4.5(a)) have widely different specific activities 10 months later (Figure 4.5(b)). The specific activity of 5S-RNA is *ca.* 30 times as high as that of 28S- and 18S-RNA.

This experiment may be explained by assuming that all or at least a large part of the 5S-RNA accumulated in small oocytes is conserved and remains present in the labelled form after the beginning of vitellogenesis. During this period, the oocytes synthesise far more 28S- and 18S-RNA than 5S-RNA, so that the latter is far less diluted than the former by the unlabelled molecules which appear when the nucleotide pool of the cells is no longer radioactive. The experiments described in Figure 4.5 and several others not reported here[71] do not prove that *all* 5S-RNA synthesised in pre-vitellogenic oocytes is conserved until the end of oogenesis. But it is clear that at least *some* 5S molecules formed in early oogenesis have a long enough half-life to be still present in mature oocytes[71].

5S-RNA appears therefore to be far more stable in oocytes than in somatic cells. Since the half-life of ribosomal RNA is about 3.5 days in embryos[48], it follows that the stability of 5S-RNA is at least one order of magnitude lower in embryos than in oocytes. This differential stability may be due to several causes. First, the concentration of nucleases is probably much lower in oocytes than in somatic tissues. Second, a large part of oocyte 5S-RNA is stored in nucleoprotein particles which protect it against the degradative enzymes of the cell (see below). Finally, oocyte and somatic 5S-RNA differ in both their primary and secondary structures[72, 73]. Oocyte 5S-RNA might have a better resistance to hydrolysis than somatic 5S-RNA.

The intracellular distribution of 5S-RNA in growing oocytes has also been studied. Homogenates of small and large oocytes give very different sedimentation profiles when centrifuged in sucrose density gradients (Figure 4.6). Vitellogenic oocytes yield a prominent peak of ribosomes and a superficial peak containing all the soluble material of the cell (Figure 4.6(b)). Pre-vitellogenic oocytes give a more complicated sedimentation profile: two

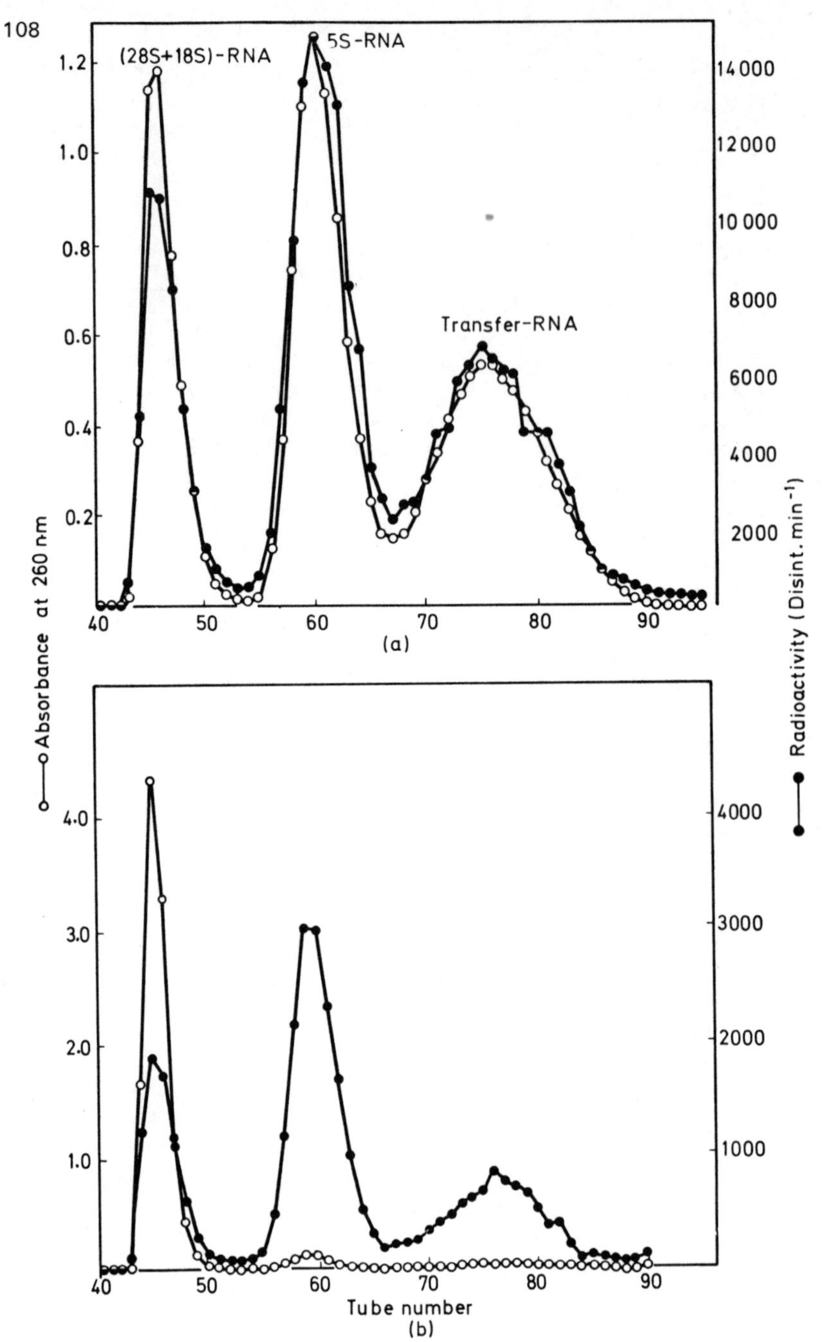

Figure 4.5 Fate of 5S-RNA synthesised in pre-vitellogenic oocytes of *Xenopus laevis*. RNA was labelled in small oocytes by injecting 100 μCi of tritiated uridine into several immature females. After (a) two months and (b) one year, RNA was extracted from the ovaries of the injected females and analysed on a Sephadex G-100 column. The average diameter of the oocytes was 100 μm in experiment (a) and 375 μm in experiment (b). (From Mairy and Denis[70], by courtesy of Springer Verlag)

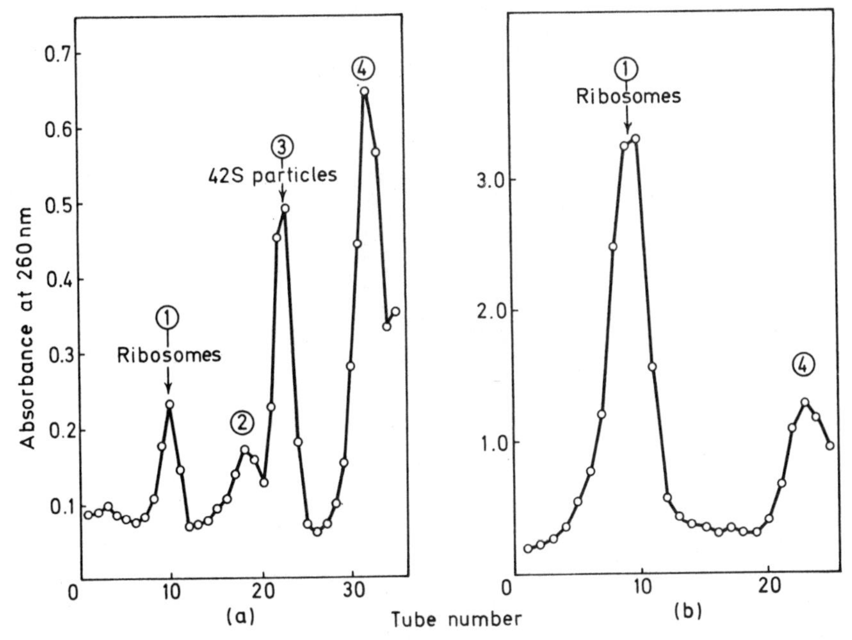

Figure 4.6 Sedimentation profiles of homogenates of (a) small and (b) large oocytes of *Xenopus laevis*. Whole ovaries from (a) immature females and (b) vitellogenic oocytes freed from somatic cells were homogenised in buffered saline and centrifuged at low speed in order to remove yolk and cell debris. The supernatant was layered on top of a 15–30% sucrose solution and centrifuged at 35 000 r.p.m. for 3h. The average diameter of the oocytes was 90 μm in experiment (a) and 375 μm in experiment (b). (From Denis and Mairy[71] by courtesy of Springer Verlag)

extra peaks with sedimentation coefficients of 42S and 58S can be observed (Figure 4.6(a)). All four peaks of Figure 4.6(a) differ from each other in their RNA and protein content (Table 4.2 and Figure 4.7). High-molecular weight RNA is present in peaks 1 and 2 and 5S-RNA is present in peaks 3 and 4. Most of transfer-RNA is present in peak 3. The molar ratio of transfer-RNA to 5S-RNA in peak 3 is 3:1. About 15 different proteins can be detected in peak 3 by electrophoresis in polyacrylamide gels (Figure 4.7). None of these proteins is present in large amounts either in the ribosomes (peak 1) or in the cell sap (peak 4).

In order to interpret the sedimentation profiles shown in Figure 4.6, it is essential to check that the nucleoprotein peaks which are observed in homogenates of small oocytes are not artefacts arising during homogenisation of the cells. Various controls suggest that the 42S and 58S peaks really do correspond to nucleoprotein particles that exist in the cytoplasm of the oocytes[71]. When the RNA and protein components of the 42S peak were separated and brought together under the conditions used for homogenisation, no aggregation was found to occur[71]. As shown in Table 4.2, about 40% of the 5S-RNA stored in small oocytes is located in the 42S particles while the remainder is apparently free in the cell sap. Particle-bound 5S-RNA differs from free 5S-RNA in at least one respect; 5S-RNA from the 42S particles carries two

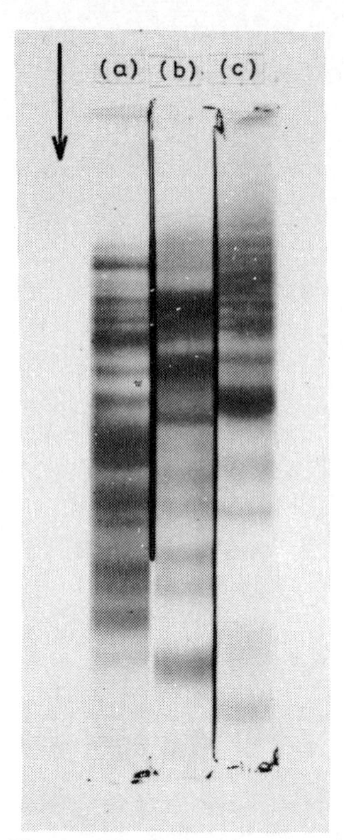

Figure 4.7 Polyacrylamide gel electrophoresis of the proteins from (a) peak 1, (b) peak 3 and (c) peak 4 of Figure 4.6(a). Proteins were solubilised in urea (0.5 mol l^{-1}) and sodium dodecylsulphate (0.1 %) and analysed in 10 % gels. Gel (a) corresponds to the ribosomes, gel (b) to the 42S particles (see text) and gel (c) to the cell sap. The arrow indicates the direction of electrophoresis. (From Mairy[74])

Table 4.2 RNA and protein content of the four peaks of Figure 4.6(a)*

Component	*Total RNA and protein in various peaks (%)*			
	Peak 1 (ribosomes)	*Peak 2 (58S)*	*Peak 3 (42S)*	*Peak 4 (cell sap)*
D-RNA + (28S + 18S)-RNA	58.3	26.3	9.3	6.1
5S-RNA	0.9	4.3	38.7	56.1
Transfer-RNA	1.8	10.2	79.7	8.3
Protein	14	11	26	49

* From Ref. 71

or three phosphate groups on its 5′-terminal nucleotide whereas 5S-RNA from the cell sap carries only one phosphate group on this nucleotide[65]. By all other criteria, the 5S-RNA contents of the two cell compartments are indistinguishable[65].

These observations argue strongly against the 42S peak being due to an artificial aggregation arising during homogenisation of the oocytes. It is most unlikely that after lysing the cells a segregation occurs among the 5S molecules on the sole basis of the degree of 5′-terminal phosphorylation, the triphosphorylated molecules being preferentially included in nucleoprotein complexes while the others remain free in the homogenate. Dephosphorylation of unbound 5S-RNA during homogenisation is also improbable since newly-made 5S-RNA was found to be triphosphorylated both in the 42S particles and in the cell sap (see below). It may be concluded that a substantial part of 5S-RNA becomes sequestered in nucleoprotein particles soon after being synthesised and stays in this state until the beginning of vitellogenesis. At this point, the 42S particles disaggregate and release the RNA that they contain[71]. A large amount of 5S RNA and 5S-RNA thus becomes available for protein synthesis and utilisation of the ribosomes. Conclusions similar to these have also been reached by Ford[69].

Little information is available on the role and structure of the 42S particles observed in pre-vitellogenic oocytes[69, 71]. These particles contain *ca.* 50% of the RNA and 25% of the protein present in the homogenates (Table 4.2). The RNA components of the 42S particles are quite easy to identify (Table 4.2), but identification of the protein components will certainly be much more difficult. Since no protein is apparently shared by the ribosomes and the 42S particles (Figure 4.7), these cannot act as a storage place for ribosomal proteins. Other functions could be imagined for the protein moiety of the 42S particles. When vitellogenesis starts, the ribosome content of the oocyte rises rapidly. In order to make its own proteins, the oocyte needs not only ribosomes and transfer-RNA but also a variety of enzymes and factors which play a role in protein synthesis. The 42S particle may contain at least some of these factors and release them when the ribosomes become abundant in the cell. Some preliminary evidence has been obtained which points to the presence of aminoacyl t-RNA synthetases in the 42S particles[76] but this is all that is known about the protein components of these particles at present.

Whatever the nature of the proteins contained in the 42S particles they certainly protect transfer-RNA and 5S-RNA against the degradative enzymes of the cell[65]. It has already been mentioned that 5S-RNA from the cell sap carries a smaller number of phosphate groups on its 5′-terminal nucleotide than 5S-RNA from 42S particles. All newly-formed molecules of 5S-RNA are triphosphorylated at the 5′ end in all cell compartments[65]. Dephosphorylation thus proceeds much more rapidly in the cell sap than in the 42S particles. This suggests that the RNA inside these particles is protected in some way against the phosphatases of the cell.

A similar protection against the endonucleases may be demonstrated more indirectly[65]. After a short period of labelling, the specific activity of 5S-RNA was followed in the main cell compartments of the oocyte[65]. After 20 h had elapsed following the end of the pulse, 5S-RNA was found to be equally labelled in all cell compartments. With time, the specific activity of 5S-RNA decreased more rapidly in the cell sap than in the 42S particles. Particle-bound 5S-RNA therefore seems to turn over less rapidly than the 5S-RNA free in the cell sap. Further evidence in favour of the protective role of the 42-S particles has been obtained from a comparison of the ribonuclease sensitivity of purified RNA with the sensitivity of RNA included in the particles. Free RNA or RNA released from the particles by deoxycholate treatment was found to be much less resistant to ribonuclease than particle-bound RNA[65].

The third question relating to the expression of 5S genes in oocytes and in somatic cells has been at least partially answered by the following studies. The oocytes of *Xenopus* contain about 2×10^6 28S and 18S genes, but only 100 000 5S genes[21, 22, 65, 77]. If all these genes were transcribed with equal efficiency during the whole length of oogenesis, the egg would contain many more molecules of 28S and 18S-RNA than of 5S-RNA. To avoid this, the oocyte switches on its 5S genes from the beginning to the end of oogenesis but keeps its 28S and 18S genes in a repressed state during the first half of its growth period. This causes the observed accumulation of 5S-RNA in small oocytes[64, 69] and the progressive incorporation of the excess 5S-RNA into the ribosomes after the onset of vitellogenesis[70].

Since the amplified 28S and 18S genes are most probably eliminated before fertilisation[78], all the cells of the embryo and of the adult contain about 50 times less 28S and 18S genes than 5S genes[23]. This is exactly the opposite of what is observed in oocytes. If all ribosomal genes were transcribed an equal number of times during each mitotic cycle, somatic cells would produce a large excess of 5S-RNA that could not be incorporated into the ribosomes. In fact, many of the 5S genes are expressed in the oocyte but are repressed in somatic cells. This may be demonstrated by comparing the nucleotide sequence of 5S-RNA from oocytes with that of 5S-RNA from somatic cells[73, 79].

The sequence studies were prompted by previous observations on the chromatographic and electrophoretic properties of 5S-RNA[72]. 5S-RNA from cultured cells and from embryos is retained more tightly by columns of methylated albumin–Kieselguhr than 5S-RNA from oocytes (Figure 4.8). Furthermore, somatic 5S-RNA has a higher electrophoretic mobility in polyacrylamide gels than oocyte 5S-RNA[72]. Treatment with denaturing agents does not suppress these differences. This suggests that somatic and oocyte 5S-RNAs differ both in their primary structure and in their secondary structure[72]. The nucleotide sequences of the two types of 5S-RNAs were indeed found to be different (Figure 4.9). Somatic 5S-RNA consists of a homogeneous population of molecules. Only one type of 5S-RNA with the sequence shown in Figure 4.9 was detected in kidney cells, in spleen and in testis. Oocyte 5S-RNA is far more heterogeneous, and several types of 5S-RNA exist in oocytes. One of them has the same sequence as somatic 5S-RNA and accounts for between 10 and 20% of oocyte 5S-RNA. The remainder of oocyte 5S-RNA differs from somatic 5S-RNA by at least six residues (Figure 4.8). A seventh substitution was detected around residue 90 but occurred only in about one-half of the molecules. It follows that there are at least three types of 5S-RNA in oocytes; one type has the same sequence as somatic 5S-RNA and the other two types differ from each other by one nucleotide around residue 90. The results outlined above show that the 5S genes expressed in somatic cells are not the same as those expressed in oocytes. It is very likely, although not yet

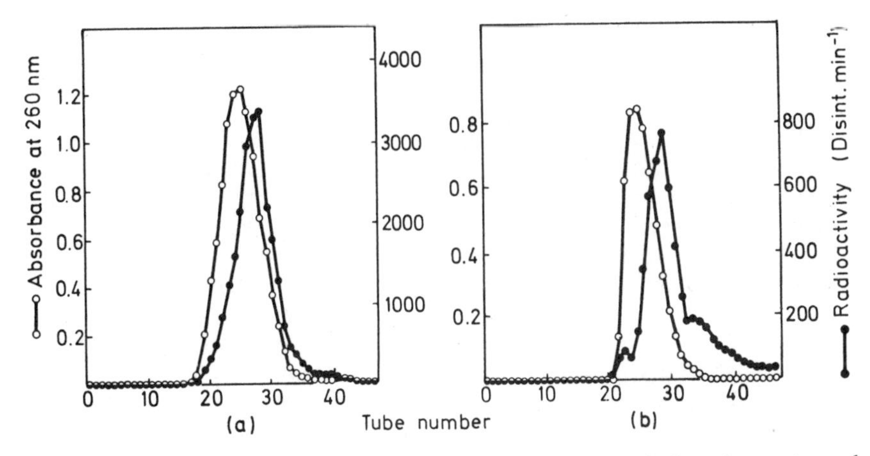

Figure 4.8 Chromatographic behaviour of 5S-RNA from pre-vitellogenic oocytes and from somatic cells of *Xenopus laevis*. Unlabelled ovaries of immature females were mixed (a) with cultured cells labelled *in vitro* or (b) with stage 35 embryos (swimming tadpoles) labelled from the gastrula stage onwards. 5S-RNA was extracted, purified (see Figure 4.3) and chromatographed on a column of methylated albumin–Kieselguhr. In both experiments, the radioactivity profile corresponds to somatic 5S-RNA whereas the absorbance profile corresponds to oocyte 5S-RNA. (From Denis *et al.*[72], by courtesy of the Editor, *Biochimie*)

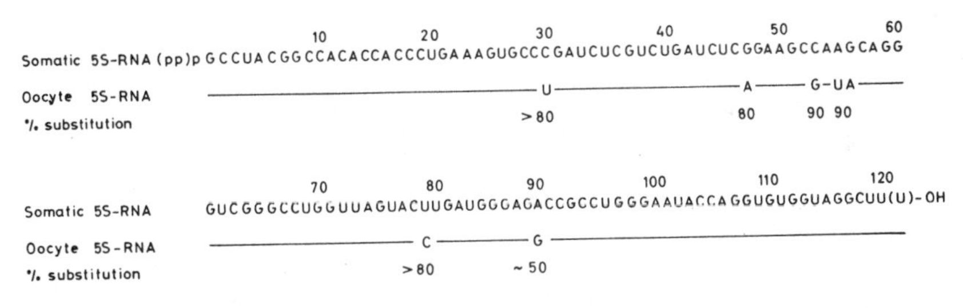

Figure 4.9 Nucleotide sequence of somatic and oocyte 5S-RNAs of *Xenopus laevis*. Somatic 5S-RNA was purified from cultured kidney cells. Oocyte 5S-RNA was purified from the 42-S particles of immature ovaries. It was therefore not contaminated by somatic RNA. A percentage substitution of 80% at a given residue means that 80% of oocyte 5S molecules have the nucleotide indicated on the lower line whereas the other 20% have the nucleotide indicated on the top line. The sequence of somatic 5S-RNA was elucidated by Wegnez, Monier and Denis[73] and Brownlee, Cartwright, McShane and Williamson[80]. The sequence of oocyte 5S-RNA was determined by Wegnez, Monier and Denis[73] and Ford and Southern[79]. The sequence shown here for oocyte 5S-RNA is only tentative. Substitution 6 (U→C) could be at nucleotide 79 or 80. Substitution 7 could not be localised with accuracy. It could occur at nucleotide 90 as shown here or at nucleotide 91. (Slightly modified from Refs. 73, 79 and 80)

proved, that *all* 5S genes are active in oocytes whereas only *part* of these genes are active in somatic cells.

The sequence heterogeneity observed in oocyte 5S-RNA implies that all 5S genes are not the same. Such a diversity is not surprising, since the 5S genes are distributed on several chromosomes[81] and not clustered in one single locus, like the 28S and 18S genes (Figure 4.2). It is conceivable that each cluster of 5S genes evolved independently in the ancestors of the present day frogs. Only one cluster of 5S genes (out of 10) would be expressed in somatic cells and this could explain the observed homogeneity of somatic 5S-RNA. It should be noted in this respect that the 5S genes active in somatic cells of *Xenopus* are closer in structure to the corresponding genes of Mammals than the 5S genes active in oocytes[73, 79].

The large number of 5S genes present in all cells of *Xenopus* may be considered as an adaptation for producing large eggs containing many ribosomes. In Mammals whose eggs are much smaller than those of the Amphibians, the 5S genes are only five times as numerous as the 28S and 18S genes[82]. The oocyte of *Xenopus* can accumulate about 300 000 times as much 5S-RNA as a somatic cell without amplifying the corresponding genes because of this high gene dosage. It might appear simpler for the frog to multiply the number of 28S and 18S genes in all cells, instead of amplifying them in the oocyte. If this system had been adopted, about one-half of the DNA content of each cell would have to be ribosomal DNA for the oocyte to produce the huge stock of ribosomes that the egg needs. Most of the 28S and 18S genes would not be needed in somatic cells but would nevertheless have to be replicated at each cell division and somehow protected against the mutations that tend to introduce diversity in clusters of identical genes. Devoting one-half of the cell's genetic information to producing ribosomal RNA is probably not a good solution for the organism. As we have seen, the Amphibian keeps a low number of 28S and 18S genes in somatic cells but a high number of 5S genes[23]. The highly redundant 5S genes are not too heavy a burden for the cell since these genes make up less than 1 % of the genome length[27].

4.4.1.2 Synthesis of ribosomal RNA in embryos

(a) *Cytoplasmic ribosomal RNA*—Synthesis of ribosomal RNA stops almost completely when the oocyte reaches its maximum size and becomes dormant[83]. During cleavage, the egg does not synthesise any detectable amount of ribosomal RNA[58]. At the onset of gastrulation, nucleoli appear in all embryonic cells. Production of ribosomal RNA starts first in ecto-mesodermal tissues and later in endodermal tissues[84]. The amount of ribosomal RNA per embryo remains approximately constant during the whole of gastrulation and neurulation[58]. After the end of neurulation, the RNA content of the embryo begins to rise slowly. The rate of increase does not vary until the time of feeding. At this stage, the tadpole contains approximately twice as many ribosomes and twice as much ribosomal RNA as the unfertilised egg[58]. Activation of the ribosomal genes is therefore not immediately followed by a change in the RNA content of the embryo. This lag is due to the large amount of ribosomes stored in the egg. Many hours of synthesis are necessary to increase this stock in a detectable way.

All types of ribosomal genes (28S, 18S and 5S) become active at the beginning of gastrulation[85]. Synthesis of 5S-RNA in *Xenopus* is not coordinate with that of 28S- and 18S-RNA since the embryo synthesises more molecules of 5S-RNA than molecules of 28S- and 18S-RNA[85]. Only a part of the 5S molecules is taken up by the ribosomes. The remainder is probably degraded. A similar situation has been observed in the cultured cells of Mammals by Leibowitz, Weinberg and Penman[86]. The over-production of 5S-RNA in embryos is not too surprising. All somatic cells contain far more 5S genes than 28S and 18S[23]. Although many 5S genes are permanently repressed in somatic cells[73], it is likely that the number of 5S genes that remain active in embryos and adult tissues is still higher than the number of 28S and 18S genes. This could explain the excess production of 5S-RNA observed in embryos of *Xenopus*.

The amount of ribosomes present in the egg is sufficient to support protein synthesis during the early stages of embryonic development. This point has been demonstrated by Brown and Gurdon in their elegant study of the anucleolate mutant of *Xenopus laevis*[68]. As in many other Amphibian species, there are in *Xenopus* two nucleolar organisers per diploid genome. One mutant has, however, been discovered which has only one nucleolar organiser and one nucleolus per cell[32]. One-half of the eggs layed by females heterozygous for the anucleolate mutation are devoid of nucleolar organiser. If these eggs are fertilised by a male carrying the same mutation in the heterozygous state, one-half of the resulting zygotes will possess one single nucleolar organiser and will develop into viable embryos. One-quarter of the offspring contains two nucleolar organisers, the remaining quarter being anucleolate. The embryos derived from the latter type of eggs begin to develop normally but die after hatching[87]. The anucleolate embryos do not synthesise any detectable amount of 28S- and 18S-RNA[68]. Accordingly, these embryos contain much less ribosomes and ribosomal RNA than their normal siblings at the stage of developmental arrest. There are enough ribosomes in the egg to maintain a sufficient level of protein synthesis during the first part of embryonic development. As time goes by, the stock of ribosomes of maternal origin becomes insufficient and the anucleolate embryo dies, presumably of protein deficiency. The molecular basis of the anucleolate mutation was found by Wallace and Birnstiel[34] to consist of an almost complete loss (deletion) of the 28S and 18S genes. The studies described above convincingly demonstrate that the early steps of embryonic development (gastrulation, neurulation, beginning of differentiation) can occur without any endogenous synthesis of ribosomes.

Until recently, there was no evidence in the literature for the existence of any heterogeneity in the ribosome population of the egg and of the embryo. Since 5S-RNA from somatic cells and 5S-RNA from oocytes are different (Figure 4.9), it follows that there are probably two classes of ribosomes in the developing embryo. One class is present in the egg and contains 5S-RNA of oocyte type. The other class is assembled after fertilisation. It contains 5S-RNA of somatic type (Figure 4.8(b)). As development proceeds, the proportion of ribosomes deriving from the egg diminishes whereas the proportion of newly-made ribosomes increases. Could the observed differences between somatic and oocyte 5S-RNAs influence the properties of the corresponding ribosomes? It is difficult to answer this question because a good test system is not available. Judging by the ability of the oocyte ribosomes to translate a somatic message that has been injected into the cell[88-90], it seems very likely that both classes of ribosomes present in developing embryos have the same properties. Since the ribosomes of the oocyte are ready to translate any messenger-RNA made available to them by injection, it is most probable that the translation machinery of the egg has the same properties as that of differentiated cells.

(b) *Mitochondrial ribosomal RNA*—Since the egg contains a very large number of mitochondria, it contains much more 21S- and 13S-RNA than any somatic cell. Ribosomal RNA from mitochondria makes up about 0.5% of the total RNA in the egg of *Xenopus*[48]. The amount of mitochondrial RNA per embryo increases from the gastrula stage onwards, and both types of ribosomal RNA (21S and 13S) accumulate coordinately. Synthesis of ribosomal RNA therefore starts at the same time both inside and outside the mitochondria[48,58], but these phenomena are not coupled obligatorily since the anucleolate mutant of *Xenopus* synthesises normal amounts of mitochondrial RNA[48].

4.4.2 Synthesis and accumulation of transfer-RNA in oocytes and embryos

There are at least 40 different species of transfer-RNA in all somatic cells of the Eukaryotes. Transfer-RNA makes up from 10–15% of the total RNA content in differentiated tissues. This corresponds to a ratio of about 10 molecules of transfer-RNA per ribosome. This ratio varies considerably during oogenesis and embryonic development.

4.4.2.1 Synthesis of transfer-RNA in oocytes

Transfer-RNA behaves in oocytes very much in the same way as 5S-RNA[64,69]. Pre-vitellogenic oocytes of *Xenopus* accumulate large amounts of transfer-RNA which make up about 35% of the RNA content of the cell (Figure 4.3). More than 90% of the transfer-RNA of small oocytes is included in 42S particles[71]. Almost all transfer-RNA molecules are thus sequestered and apparently not available for protein synthesis. It must be recalled that small oocytes contain very few ribosomes and cannot synthesise much protein. Transfer-RNA must therefore be regarded as a storage product that will be active in protein synthesis only after the beginning of vitellogenesis.

During the final period of oogenesis, transfer-RNA makes up a decreasing proportion of the total RNA (Figure 4.3). This proportion falls to 2% in mature oocytes and unfertilised eggs[64]. This corresponds roughly to two molecules of transfer-RNA per ribosome (Table 4.1). The transfer-RNA content of the egg (80 ng) is about 60 000 times as high as that of a somatic cell.

The properties of transfer-RNA from somatic cells and oocytes have not been compared as carefully as those of 5S-RNA. Several experimental data suggest that somatic transfer-RNA might be different from oocyte transfer RNA.

First, the base composition of both types of RNAs is not the same[64]. Second, transfer-RNA from somatic cells is eluted from columns of Sephadex G-100 slightly behind transfer-RNA from oocytes[64]. Third, both types of transfer-RNA have different electrophoretic properties in polyacrylamide gels and different chromatographic behaviours on columns of methylated albumin–Kieselguhr[91]. These observations cannot be interpreted unambiguously at the present time. Transfer-RNA might be transcribed on different genes in somatic cells and in oocytes. To prove this, it would be necessary to purify at least one isoacceptor species from both types of cells and to determine its nucleotide sequence.

4.4.2.2 Synthesis of transfer RNA in embryos

Synthesis of transfer-RNA does not start in the embryo before the onset of gastrulation, although a certain amount of terminal labelling occurs during cleavage[92]. From gastrulation onwards, the *Xenopus* embryo doubles its content of transfer-RNA with each doubling of the cell number[92]. The feeding tadpole contains about 10 times as much transfer-RNA as the unfertilised egg[64,92]. Transfer-RNA therefore accumulates much more rapidly than ribosomal RNA. There are about 10 molecules of transfer-RNA per ribosome in the tadpole, instead of two as in the egg (Table 4.1).

Transfer-RNA could play an important regulatory role during embryonic development not only by controlling the overall level of protein synthesis but also by determining which proteins are to be synthesised in each embryonic tissue. A given protein might fail to be synthesised in a given cell because this cell lacks the transfer-RNA molecules that could recognise one or several codons in the messenger-RNA corresponding to this protein. If this type of regulation is operating in embryonic development, it should be possible to detect changes in the transfer-RNA population of the growing embryo and differences from one differentiating tissue to another.

Such changes have been looked for in the embryo of *Xenopus*[93]. The properties of transfer-RNA were found to be the same in embryos and in adult tissues. All species of transfer-RNA are probably present in the embryo from fertilisation until the time of feeding. Even in tissues specialised in the production of one or few proteins, such as the silk gland of *Bombyx mori*, all species of transfer-RNA are present, although some of them are found in larger amounts than others[94-97]. The most abundant transfer-RNA species in the silk gland correspond to those amino acids which occur most frequently in silk (glycine, alanine, serine and tyrosine). A similar adaptation of the transfer-RNA population to protein synthesis has been observed in the lens[98]. This phenomenon must be regarded as a way of increasing the efficiency of protein synthesis in highly-specialised cells, rather than as a way of selecting a particular message for preferential translation.

4.4.3 Synthesis of messenger-RNA in oocytes and embryos

4.4.3.1 Ways of studying messenger-RNA

For several reasons, messenger-RNA is far more difficult to study than ribosomal RNA and transfer-RNA. First, messenger-RNA makes up a small proportion of the RNA content of the cell. Second, there are in principle as many species of messenger-RNA in a cell as species of protein. Since the length of the proteins varies considerably, it follows that the messenger population of most types of cells is very heterogeneous, not only in sequence but also in size. Third, the properties of messenger-RNA are similar in many respects to those of a class of RNA usually referred to as heterogeneous nuclear RNA whose metabolism is mainly restricted to the nucleus. Heterogeneous nuclear RNA has a base composition close to that of messenger-RNA and DNA* (about 40% guanine + cytosine in all Vertebrates). It is very heterogeneous in size and its average length far exceeds that of messenger-RNA[99-102]. An increasing amount of evidence suggests that heterogeneous nuclear RNA might be the precursor of messenger-RNA. According to the current view, messenger-RNA appears in the nucleus as a giant molecule containing 10 000 or more nucleotides[103]. Soon after synthesis, the precursor is modified at the 3′ end by the addition of a large number (several dozen to several hundred) of adenylic residues[104-107]. The precursor is then progressively shortened at the 5′ end[108-110]. The major part of the precursor molecule is therefore eliminated inside the nucleus; only the 3′ end containing the added poly(A) sequence will finally be conserved and sent to the cytoplasm.

The techniques used for the study of ribosomal RNA and transfer-RNA cannot usually be utilised for the study of messenger-RNA. Methods must be developed that can physically separate messenger-RNA from bulk RNA or at least offer a way of detecting it. Three methods are available for the study of messenger-RNA: (a) molecular hybridisation with DNA, (b) annealing with polyuridylic acid and (c) artificial translation.

Method (a) does not permit discriminations between messenger-RNA and heterogeneous nuclear RNA, but allows a comparison between several populations of messenger-RNAs with different nucleotide sequences. The efficiency of the discrimination based on sequence differences is, however, greatly restricted by the internal redundancy that is observed in the genome of all Eukaryotes[111]. RNA transcribed on the repetitive portion of the genome most often forms hybrids of poor quality when annealed with single-stranded DNA. These hybrids contain many mismatched sequences, for RNA can form duplexes not only with the DNA stretch on which it has been transcribed but also with a wide variety of similar, but not identical, stretches. It follows that several closely similar populations of RNA may hybridise with the same sequences in DNA although they are not identical. Little unambiguous information can therefore be obtained by means of the hybridisation technique unless special precautions are taken.

Method (b) is suitable for separating messenger-RNA from ribosomal and

* Heterogeneous nuclear RNA and messenger-RNA are given here the common name of DNA-like RNA (D-RNA) when they cannot be distinguished. D-RNA might also contain other species of RNA which are not yet characterised.

transfer-RNAs which do not contain poly(A) sequences. This method, however, neither separates messenger-RNA from heterogeneous nuclear RNA nor several species of messenger-RNA from one another. A further drawback of the poly(U) procedure is that the messenger molecules which are devoid of poly(A) sequences (the histone messengers, for instance[106, 112]) behave like bulk RNA and are not separated from ribosoml and transfer-RNA.

Method (c) can be used for detecting one particular messenger species in a complex mixture of molecules. A sensitive test is needed to follow the appearance of the protein coded by the messenger under study. Two different protein-synthesising systems are available for the demonstration of messenger activity. The first system functions *in vitro*. It consists of the translation machinery of normal or tumoral cells supplemented with ATP[113-115]. The second system consists of a living cell into which the RNA is injected with a microneedle. Mature oocytes or unfertilised eggs of *Xenopus* are used for this purpose[88]. As we have seen earlier, these cells contain large amounts of ribosomes and of transfer-RNA which are not fully utilised during oogenesis. Introducing a foreign species of messenger-RNA into the egg causes the appearance of the protein coded by this messenger.

4.4.3.2 Synthesis of messenger-RNA and heterogeneous nuclear RNA in oocytes

Pre-vitellogenic oocytes of *Xenopus* accumulate large amounts not only of 5S-RNA and transfer-RNA but also of DNA-like RNA (D-RNA). As mentioned earlier, high-molecular weight RNA from small oocytes (Figure 4.3(a)) has an unusual base composition; it contains 50% guanine + cytosine[64]. This RNA is therefore probably a mixture of D-RNA (40% guanine + cytosine) and (28S + 18S)-RNA (60% guanine + cytosine). When centrifuged in a sucrose density gradient, RNA from small oocytes separates into three peaks; a small peak of 28S-RNA, a small peak of 18S-RNA and a large peak of 5S-RNA + transfer-RNA (Figure 4.4(a)). No discrete peak of D-RNA is observed. A partial separation between ribosomal RNA and D-RNA can be achieved by centrifuging homogenates of pre-vitellogenic oocytes in sucrose density gradients[71]. Part, and perhaps all, of the D-RNA of the cell sediments as a 58S nucleoprotein peak whereas 28S- and 18S-RNA sediment with the ribosome peak (Figure 4.6(a)). When RNA is extracted from the 58S peak and analysed in a sucrose gradient, it gives rise to a 19S and a 28S peak[71]. The latter is thought to be a ribosomal contaminant. The 19S component is the only discrete species of D-RNA that can be observed in small oocytes. This component does not appear in sucrose density analyses of bulk RNA (Figure 4.4(a)) because it sediments at the same level as 18S-RNA.

D-RNA of small oocytes is apparently as stable as 5S-RNA and transfer-RNA[64]. It makes up more than 10% of the RNA content of the oocyte (Table 4.1). No other cell has so far been reported to contain such a high proportion of D-RNA. Like transfer-RNA and 5S-RNA, D-RNA is included in nucleoprotein particles which could protect it against the degradative enzymes of the cell.

D-RNA present in small oocytes is probably stored for use in later oogenesis, as was shown to be the case for 5S-RNA[70]. After the beginning of vitellogenesis, D-RNA continues to be synthesised at a high rate. Evidence for this is given by the properties of nascent high-molecular weight RNA. This RNA has a base composition of approximately 50% guanine + cytosine. About one-half of the newly-made RNA is therefore D-RNA, the other half being 28S- and 18S-RNA[64]. D-RNA does not, however, accumulate in large oocytes as it apparently did in small ones, since D-RNA constitutes from 2 to 5% of the total RNA in large oocytes and in eggs[116-119].

Autoradiographic observations suggest that D-RNA is synthesised in the oocyte on the lateral loops of the lampbrush chromosomes[120,121]. Upon exposing the oocyte to tritiated uridine, some loops can be seen to incorporate a large amount of radioactive material into the RNA. If the oocytes are left in the presence of the precursor for a long enough period of time, all loops eventually become labelled[83]. Since the loops contain about 5% of the DNA present in the lampbrush chromosomes, it follows that about one-twentieth of the DNA is transcribed during the final period of oogenesis[83]. This conclusion is directly confirmed by DNA–RNA hybridisation experiments. If a fixed amount of denatured DNA of *Xenopus* is annealed with increasing amounts of labelled RNA from large oocytes, it is found that 1.7% of the DNA hybridises with RNA[122]. In other words, about 3.5% of the total number of genes are expressed in vitellogenic oocytes. Since the conditions used in the experiment just mentioned did not allow the non-repeated sequences of DNA to hybridise with RNA, the observed saturation level is probably underestimated.

Some of the D-RNA molecules synthesised during vitellogenesis are still present in the oocytes at the end of its growth[122]. The other molecules are degraded soon after being synthesised. Part of the D-RNA population of the oocyte is therefore stabilised and stored for several months.

When the oocyte reaches its maximum size, synthesis of D-RNA stops almost completely. It resumes soon after the oocyte has received the hormonal stimulus that elicits ovulation and maturation. The oocyte synthesises 1–6 pg of D-RNA during ovulation[123]. This amount is surprisingly high given the small length of time (12 h) necessary for maturation and ovulation to be completed. The nature of the RNA synthesised during maturation remains unknown. This RNA certainly plays an important role in maturation since inhibition of RNA synthesis with actinomycin D prevents the germinal vesicle from breaking down[124-126].

D-RNA present in the unfertilised egg has been studied with great care[118-119]. This RNA constitutes from 2 to 5% of the RNA content of the egg and is complementary to more than 3% of the DNA. It is produced both by the repetitive and by the non-repetitive portion of the genome. The non-repetitive sequences that detectably hybridise with egg RNA represent 1.2% of the total sequences of DNA[118,127]. This means that the egg contains many different molecules of RNA. If all the D-RNA of the egg had messenger properties, it could code for 40 000 proteins of the same length as one globin chain[118]. This, however, does not mean that the egg is indeed able to synthesise so many proteins. No precise role can be ascribed to the D-RNA stored in the egg. At least part of this RNA must be messenger in nature because RNA extracted from unfertilised eggs stimulates protein synthesis in a cell-free system[116], but the proteins synthesised under these conditions have not been characterised.

4.4.3.3 *Synthesis of messenger-RNA and heterogeneous nuclear RNA in embryos*

Protein synthesis in the cleaving egg is not affected by doses of actinomycin D that completely suppress RNA synthesis[128, 129]. Enucleation of the egg does not reduce the level of amino-acid incorporation into protein[130]. Protein synthesis in the egg is therefore supported by messenger molecules that were present in the oocyte before fertilisation. In agreement with this, RNA extracted from blastulae and early gastrulae of *Xenopus* has been found to be partially homologous with RNA synthesised by vitellogenic oocytes[122]. It may be concluded that all proteins made by the embryo during the first hours of its development are organised by messenger molecules deriving from the oocyte. Nothing is known about the nature of the proteins synthesised by the embryo on maternal templates. RNA present in sea urchin eggs is known to contain the information required for the synthesis of several proteins and enzymes involved in the process of DNA replication and cell division[131-136]. No evidence of this kind is available for the Amphibians.

Cleaving eggs mostly translate messenger molecules of maternal origin but they already synthesise small amounts of D-RNA[58]. For many hours, this type of RNA is actually the only one that is made by the embryo. The rate of synthesis of D-RNA increases considerably a little before the onset of gastrulation[58, 123]. D-RNA accumulates rapidly during gastrulation and more slowly thereafter[123].

D-RNA synthesised at the gastrula and neurula stages rapidly turns over[2]. Protein synthesis, which is largely resistant to actinomycin D until the end of cleavage, becomes more sensitive to this antibiotic after the appearance of the blastoporal lip[129]. This probably means that the gastrula does not rely any longer upon maternal templates for protein synthesis, but now translates newly-made messengers. When differentiation sets in, an increasing proportion of the D-RNA population of the embryo becomes stabilised and remains present for many hours after being synthesised[2].

It is crucial to know if the RNA population of the embryo changes during development. If release of the genetic information in differentiating cells is controlled at the level of transcription, the D-RNA content of the embryo may be expected to vary both qualitatively and quantitatively during the course of embryogenesis. Many experiments have been undertaken to test this hypothesis by means of DNA–RNA hybridisation[2, 137, 138]. D-RNA present in growing embryos has been found to be complementary to an increasing portion of the genome[2, 139]. Embryonic development is therefore accompanied by a progressive release of the genetic information enclosed in DNA. The D-RNA population of blastulae and early gastrulae changes very rapidly[137] and differs from one region of the embryo to another[138]. However, the experiments just mentioned suffer from several ambiguities. First, they were carried out under conditions that allow only the repetitive sequences of DNA to hybridise with RNA[111]. This results in a poor matching of the DNA–RNA duplexes. Second, by the hybridisation tests used, it was not possible to distinguish between heterogeneous nuclear RNA and messenger-RNA. The RNA that binds to DNA under the usual annealing conditions is likely to be mostly heterogeneous nuclear RNA. It is therefore desirable that the experiments referred to above be repeated with the aid of the most recent improvements of the hybridisation technique. Messenger-RNA can be separated from heterogeneous nuclear RNA by isolating cellular sub-fractions. Purified nuclei predominantly contain heterogeneous nuclear RNA whereas polysomes yield upon extraction most of the messenger-RNA of the cell. The stringency of the annealing tests can be enhanced by using only the non-repetitive portion of the genome for the DNA–RNA hybridisation experiments. Finally, the poly-(U) technique should be used in order to accurately determine the amount of poly(A)-containing RNA in growing embryos and differentiating tissues. The only measurements of the D-RNA content of eggs and embryos have been obtained in indirect ways[2,123].

References

1. Denis, H. (1961). *J. Embryol. Exp Morph.*, **9,** 422
2. Denis, H. (1966). *J. Mol. Biol.*, **22,** 285
3. Denis, H. (1966). *Activité des gènes au cours du développement embryonnaire* (Liège: Desoer)
4. Tomkins, G. M., Gelehrter, T. D. Granner, D., Martin, D., Samuels, H. H. and Thompson, E. B. (1969). *Science*, **166,** 1474
5. Coggins, L. W. and Gall, J. G. (1972). *J. Cell Biol.*, **52,** 569
6. Balinsky, B. I. and Devis, R. J. (1963). *Acta Embryol. Mroph. Exp.*, **6,** 55
7. Thomas, C. (1967). *Arch. Biol.*, **78,** 347
8. Thomas, C. (1969). *J. Embryol. Exp. Morph.*, **21,** 165
9. Buongiorno-Nardelli, M., Amaldi, F. and Lava-Sanchez, (1972). *Nature* (*New Biol.*), **238,** 134
10. King, H. D. (1908). *J. Morphol.*, **19,** 369
11. Brachet, J (1960). *The Biochemistry of Development* (New York: Pergamon Press)
12. Ficq, A. (1961). *Symposium on Germ Cells and Earliest Stages of Development*, 121 (Milano: Ist Lombardo, Fondatione A. Baselli)
13. Gross, P. R. (1968). *Ann. Rev. Biochem.*, **37,** 631
14. Graham, C. F. and Morgan, R. W. (1966). *Develop, Biol.*, **14,** 439
15. Dawid, I. B. (1965). *J. Mol. Biol.*, **12,** 581
16. Ficq, A. (1968). *Exp. Cell Res.*, **53,** 691
17. Gall, J. G. (1968). *Proc. Nat. Acad. Sci. USA*, **60,** 553
18. MacGregor, H. C. (1968). *J. Cell Sci.*, **3,** 437
19. Van Gansen, P. and Schram, A. (1972). *J. Cell Sci.*, **10,** 339
20. Gall, J. G. and Pardue, M. L. (1969). *Proc. Nat. Acad. Sci. USA*, **63,** 378
21. Brown, D. D. and Dawid, I. B. (1968). *Science*, **160,** 272
22. Evans, D. and Birnstiel, M. L. (1968). *Biochim. Biophys. Acta*, **166,** 274
23. Brown, D. D. and Weber, C. S. (1968). *J. Mol. Biol.*, **34,** 661
24. Painter, T. S. and Taylor, A. N. (1942). *Proc. Nat. Acad. Sci. USA*, **28,** 311
25. Dawid, I. B., Brown, D. D. and Reeder, R. H. (1970). *J. Mol. Biol.*, **51,** 341
26. Birnstiel, M. L., Speirs, J., Purdom, I., Jones, K. and Loening, U. E. (1968). *Nature* (*London*), **219,** 454
27. Brown, D. D. and Weber, C. S. (1968). *J. Mol. Biol.*, **34,** 681
28. Landesman, R. and Gross, P. (1968). *Develop. Biol.*, **18,** 571
29. Loening, V. E., Jones, K. W. and Birnstiel, M. L. (1969). *J. Mol. Biol.*, **45,** 353
30. Brown, D. D. and Blackler, A. W. (1972). *J. Mol. Biol.*, **63,** 75
31. Wallace, H., Morray, J. and Langridge, W. H. R. (1971). *Nature* (*London*), **230,** 201
32. Elsdale, T. R., Fischberg, M. and Smith, S. (1958). *Exp. Cell Res.*, **14,** 642

33. Perkowska, E., MacGregor, H. C. and Birnstiel, M. L. (1968). *Nature* (*London*), **217,** 649
34. Wallace, H. and Birnstiel, M. L. (1966). *Biochim. Biophys. Acta*, **114,** 296
35. Crippa. M. and Tocchini-Valentini, G. P. (1971). *Proc. Nat. Acad. Sci. USA*, **68,** 2769
36. Ficq, A. and Brachet, J. (1971). *Proc. Nat. Acad. Sci. USA*, **68,** 2774
37. Gurgo, C., Thiry, K. K. and Green, M. (1971). *Nature* (*New Biol.*), **229,** 111
38. Brown, R. D. and Tocchini-Valentini, G. P. (1972). *Proc. Nat. Acad. Sci. USA*, **69** 1746
39. Mahdavi, V. and Crippa, M. (1972). *Proc. Nat. Acad. Sci. USA*, **69,** 1749
40. Miller, O. L. and Beatty, B. R. (1969). *Science*, **164,** 955
41. Miller, O. L. and Beatty, B. R. (1969). *J. Cell Physiol.*, **74,** Suppl. 1, 225
42. Baltus, E. and Brachet, J. (1962). *Biochim. Biophys. Acta*, **61,** 157
43. Gregg, J. R. and Løvtrup, S. (1955). *Biol. Bull.*, **108,** 29
44. Hoff-Jørgensen, E. and Zeuthen, E. (1952). *Nature* (*London*), **169,** 245
45. Kuriki, Y. and Okasaki, R. (1959). *Embryologia* (*Nagoya*), **4,** 337
46. Sze, L. C. (1953). *Physiol. Zool.*, **26,** 212
47. Dawid, I. B. (1966). *Proc. Nat. Acad. Sci. USA*, **56,** 269
48. Chase, J. W. and Dawid, I. B. (1972). *Develop. Biol.*, **27,** 504
49. Baltus, E., Hanocq-Quertier, J. and Brachet, J. (1968). *Proc. Nat. Acad. Sci. USA*, **61,** 469
50. Hanocq-Quertier, J., Baltus, E., Ficq, A. and Brachet, J. (1968). *J. Embryol. Exp. Morph.*, **19,** 273
51. Hanocq, F., Kirsch-Volders, M., Hanocq-Quertier, J., Baltus, E. and Steinert, G. (1972). *Proc. Nat. Acad. Sci. USA*, **69,** 1322
52. Dawid, I. B. and Wolstenholme, D. R. (1967). *J. Mol. Biol.*, **28,** 233
53. Wolstenholme, D. R. and Dawid, I. B. (1967). *Chromosoma*, **20,** 445
54. Wolstenholme, D. R. and Dawid, I. B. (1968). *J. Cell. Biol.*, **39,** 222
55. Borst, P. and Kroon, A. M. (1969). *Int. Rev. Cytol.*, **26,** 107
56. Roodyn, D. B. and Wilkie, D. (1968). *The Biogenesis of Mitochondria* (London: Methuen and Co).
57. Dawid, I. B. and Blackler, A. W. (1972). *Develop. Biol.*, **29,** 152
58. Brown, D. D. and Littna, E. (1964). *J. Mol. Biol.*, **8,** 669
59. Swanson, R. F. and Dawid, I. B. (1970). *Proc. Nat. Acad. Sci. USA*, **66,** 117
60. Dawid, I. B. (1972). *J. Mol. Biol.*, **63,** 201
61. Dawid, I. B. and Chase, J. W. (1972). *J. Mol. Biol.*, **63,** 217
62. Grivell, L. A., Reijnders, L. and Borst, P. (1971). *Eur. J. Biochem.*, **19,** 64
63. Thomas, C. (1972). *Exp. Cell Res.*, **74,** 547
64. Mairy, M. and Denis, H. (1971). *Develop. Biol.*, **24,** 143
65. Wegnez, M. and Denis, H. (1973). *Biochimie*, in press
66. Brown, D. D. and Littna, E. (1964). *J. Mol. Biol.*, **8,** 688
67. Davidson, E. H., Allfrey, V. G. and Mirsky, A. E. (1964). *Proc. Nat. Acad. Sci. USA*, **52,** 501
68. Brown, D. D. and Gurdon, J. B. (1964). *Proc. Nat. Acad. Sci. USA*, **51,** 139
69. Ford, P. J. (1971). *Nature* (*London*), **233,** 561
70. Mairy, M. and Denis, H. (1972). *Eur. J. Biochem.*, **24,** 535
71. Denis, H. and Mairy, M. (1972). *Eur. J. Biochem.*, **25,** 524
72. Denis, H., Wegnez, M. and Willem, R. (1972). *Biochimie*, **54,** 1189
73. Wegnez, M., Monier, R. and Denis, H. (1972). *FEBS Lett.*, **25,** 13
74. Mairy, M. (1973). *Ph.D. Thesis*, University of Liège
75. Wegnez, M. and Denis, H. (1972). *Biochimie*, **54,** 1069
76. Denis, H. and Hentzen, D. (1973). Unpublished results
77. Gall, J. G. (1969). *Genetics*, **61,** Suppl. 1, 123
78. Brachet, J., Hanocq, F. and Van Gansen, P. (1970). *Develop. Biol.*, **21,** 157
79. Ford, P. J. and Southern, E. M. (1973). *Nature* (*New Biol.*), **241,** 7
80. Brownlee, G. G., Cartwright, E., McShane, T. and Williamson, R. (1972). *FEBS Lett.*, **25,** 8
81. Pardue, M. L. (1972), Brown, D. D. and Birnstiel, M. L. (1973). *Chromosoma*, **42,** 191
82. Quincey, R. V. and Wilson, S. H. (1969). *Proc. Nat. Acad. Sci. USA*, **64,** 981
83. Davidson, E. H. (1968). *Gene Activity in Early Development* (New York: Academic Press)

84. Woodland, H. R. and Gurdon, J. B. (1968). *J. Embryol. Exp. Morph.*, **19,** 363
85. Abe, H. and Yamana, K. (1971). *Biochim. Biophys. Acta*, **240,** 392
86. Leibowitz, R. D., Weinberg, R. A. and Penman, S. (1973). *J. Mol. Biol.*, **73,** 139
87. Wallace, H. (1960). *J. Embryol. Exp. Morph.*, **8,** 405
88. Gurdon, J. B., Lane, C. D., Woodland, H. R. and Marbaix, G. (1971). *Nature (London)*, **233,** 177
89. Lane, C. D., Marbaix, G. and Gurdon, J. B. (1971). *J. Mol. Biol.*, **61,** 73
90. Moar, V. A., Gurdon, J. B., Lane, C. D. and Marbaix, G. (1971). *J. Mol. Biol.*, **61,** 93
91. Denis, H. (1973). Unpublished results
92. Brown, D. D. and Littna, E. (1966). *J. Mol. Biol.*, **20,** 95
93. Marshall, R. and Nirenberg, M. (1969). *Develop. Biol.*, **19,** 1
94. Chavancy, G., Daillie, J. and Garel, J. P. (1971). *Biochimie*, **53,** 1187
95. Garel, J. P., Mandel, P., Chavancy, G. and Daillie, J. (1970). *FEBS Lett.*, **7,** 327
96. Garel, J. P., Mandel, P., Chavancy, G. and Daillie, J. (1971). *FEBS Lett.*, **12,** 249
97. Garel, J. P., Mandel, P., Chavancy, G. and Daillie, J. (1971). *Biochimie*, **53,** 1195
98. Garel, J. P., Virmaux, N. and Mandel, P. (1970). *Bull. Soc. Chim. Biol.*, **52,** 987
99. Darnell, J. E., Philipson, L., Wall, R. and Adesnik, M. (1971). *Science*, **174,** 507
100. Houssais, J. F. and Attardi, G. (1966). *Proc. Nat. Acad. Sci. USA*, **56,** 616
101. Scherrer, K., Marcaud, L., Zajdela, F., London, I. M. and Gross, F. (1966). *Proc. Nat. Acad. Sci. USA*, **56,** 1571
102. Soeiro, R., Birnboim, H. C. and Darnell, J. E. (1966). *J. Mol. Biol.*, **19,** 362
103. Granboulan, N. and Scherrer, K. (1969). *Eur. J. Biochem.*, **9,** 1
104. Edmonds, M. and Caramela, M. G. (1969). *J. Biol. Chem.*, **244,** 1314
105. Edmonds, M., Vaughan, M. H. and Nakazato, H. (1971). *Proc. Nat. Acad. Sci. USA*, **68,** 1336
106. Greenberg, J. R. and Perry, R. P. (1972). *J. Mol. Biol.*, **72,** 91
107. Mendecki, J., Lee, Y. and Brawerman, G. (1972). *Biochemistry*, **11,** 792
108. Darnell, J. E., Wall, R. and Tushinski, R. J. (1971). *Proc. Nat. Acad. Sci. USA*, **68,** 1321
109. Kates, J. (1970). *Cold Spring Harbor Symp. Quant. Biol.*, **35,** 743
110. Lee, Y., Mendecki, J. and Brawerman, G. (1971). *Proc. Nat. Acad. Sci. USA*, **68,** 133
111. Britten, R. J. and Kohne, D. E. (1968). *Science*, **161,** 529
112. Adesnik, M. and Darnell, J. E. (1972). *J. Mol. Biol.*, **67,** 397
113. Aviv, H., Boime, I. and Leder, P. (1971). *Proc. Nat. Acad. Sci. USA*, **68,** 2303
114. Lim, L. and Canellakis, E. S. (1970). *Nature (London)*, **227,** 710
115. Lockard, R. E. and Lingrel, J. B. (1969). *Biochem. Biophys. Res. Commun.*, **37,** 204
116. Cape, M. and Decroly, M. (1969). *Biochim. Biophys. Acta*, **174,** 99
117. Davidson, E. H., Crippa, M., Kramer, F. R. and Mirsky, A. E. (1966). *Proc. Nat. Acad. Sci. USA*, **56,** 856
118. Davidson, E. H. and Hough, B. R. (1971). *J. Mol. Biol.*, **56,** 491
119. Hough, B. R. and Davidson, E. H. (1972). *J. Mol. Biol.*, **70,** 491
120. Gall, J. G. and Callan, H. G. (1962). *Proc. Nat. Acad. Sci. USA*, **48,** 562
121. Izawa, M., Allfrey, V. G. and Mirsky, A. E. (1963). *Proc. Nat. Acad. Sci. USA*, **49,** 544
122. Crippa, M., Davidson, E. H. and Mirsky, A. E. (1967). *Proc. Nat. Acad. Sci. USA*, **57,** 885
123. Brown, D. D. and Littna, E. (1966). *J. Mol. Biol.*, **20,** 81
124. Brachet, J. (1967). *Exp. Cell Res.*, **48,** 233
125. Dettlaff, T. A. (1966). *J. Embryol. Exp. Morph.*, **16,** 183
126. Schuetz, A. W. (1967). *J. Exp. Zool.*, **166,** 347
127. Davidson, E. H. and Hough, B. R. (1969). *Proc. Nat. Acad. Sci. USA*, **63,** 342
128. Denis, H. (1964). *Develop. Biol.*, **9,** 458
129. Denis, H. (1964). *Develop. Biol.*, **9,** 473
130. Smith, L. D. and Ecker, R. E. (1965). *Science*, **150,** 777
131. Gross, P. R. and Cousineau, G. H. (1963). *J. Cell Biol.*, **19,** 260
132. Gross, P. R. and Cousineau, G. H. (1964). *Exp. Cell Res.*, **33,** 368
133. Gross, P. R., Malkin, L. I. and Moyer, W. A. (1964). *Proc. Nat. Acad. Sci USA*, **51,** 407
134. Noronha, J. M., Sheys, G. H. and Buchanan, J. M. (1972). *Proc. Nat. Acad. Sci. USA*, **69,** 2006
135. Raff, R. A., Colot, H. V., Selvig, S. E. and Gross, P. R. (1972). *Nature (London)*, **235,** 211

136. Stafford, D. W. and Iverson, R. M. (1964). *Science*, **143,** 580
137. Davidson, E. H., Crippa, M. and Mirsky, A. E. (1968). *Proc. Nat. Acad. Sci. USA*, **60,** 152
138. Flickinger, R. A., Greene, R., Kohl, D. M. and Miyagi, M. (1966). *Proc. Nat. Acad. Sci. USA*, **56,** 1712
139. Denis, H. (1968). *Adv. Morphogenesis*, **7,** 115
140. Hourcade, D., Dressler, D. and Wolfson, J. (1973). *Proc. Nat. Acad. Sci. USA.*, **70,** 2926
141. Bird, A., Rogers, E. and Birnstiel, M. (1973). *Nature New Biology*, **242,** 226

Editor's Comments

After differentiation has been initiated during early development and the process of determination has occurred, most cells in animals are committed to a restricted pathway of development. Some cells are very rapidly committed to terminal differentiation. For example, in most vertebrates, most of the neurones present at birth represent the final stage in development; they never undergo subsequent division and may survive essentially unchanged for 100 years or more. In other tissues, most cells do not go on to terminal differentiation immediately but form a pool of 'stem cells' which provides a self-renewing population of partially differentiated cells. A good example of cells of this kind is provided by the imaginal discs in insects which were referred to in the Introduction. These cells exhibit none of the overt characteristics of the tissues into which they will eventually differentiate but are nonetheless, committed to differentiate into one kind of tissue only. While they are in this state, they can propagate very extensively as was shown by the work of Hadorn and his colleagues. Given the correct environment, they then undergo terminal differentiation.

There are many examples of this kind of differentiation and, since these lend themselves rather readily to biochemical investigation, some of them have been studied in great detail. The maturation of erythroid cells is an excellent example. In the adults, there is a small pool of stem cells which are capable of developing into erythrocytes or leukocytes. These stem cells have not been positively identified by morphological criteria but their presence can be shown quite unequivocally by inoculating bone marrow into mice which have received a dose of irradiation such as to eliminate their own blood forming capacity. When a very small amount of bone marrow is injected, individual colonies develop in the spleen. Some of these form erythrocytes, some leukocytes and some both. Stem cells from a donor animal can eventually re-populate the recipient; hence the stem cells can multiply. They seem to have two alternative fates. They can reproduce themselves like the cells of the imaginal disc but they can also go on to form three other kinds of cells, one committed to form red cells, one to form polymorphonuclear white cells and one to form platelets. The eventual maturation of the erythrocyte and leukocyte precursors seems to depend on humoral factors which have been designated erythropoietin and granulopoietin. The 'erythropoietin responsive cell' responds to erythropoietin by an increase in cell number and entry into the final maturation pathways which, of course, involves the synthesis of haemoglobin and the formation of the mature erythrocyte.

The development of many of the secondary sex organs is dependent on the steroid sex hormones and it has been suspected that a situation rather similar to that obtaining in erythropoiesis exists. Although the biology of the stem cells in this system has not been so well established, the later events have been very clearly studied and one of the best systems has turned out to be the chick oviduct, the development of which is exclusively dependent on the presence of oestrogens and progesterone.

5
Mammalian Erythroid Cell Differentiation

P. A. MARKS, R. A. RIFKIND and A. BANK
Columbia University, New York

5.1 INTRODUCTION

Understanding the mechanisms of mammalian cell differentiation is important with respect to normal cellular function and will probably provide insights into several abnormal biological states, such as cancer, genetically-determined defects in cell function and virus-induced cell dysfunction. There is no single biological system which is ideal for investigating all aspects of the regulation of differentiation of eukaryotic cells. Erythroid cell differentiation has provided a useful model for studying a number of aspects of this problem[1,2]. A great deal is known about the structural and biochemical aspects of erythroid cell differentiation. Erythroid cells are relatively easily obtained in quantity. Erythroid cells synthesise a predominant class of proteins, globins, the structure and function of which are among the most extensively elucidated of any proteins[3]. Experimentally, erythroid cells can be analysed, both *in vivo* and *in vitro* as whole cells and as cell-free systems, for synthesis of differentiated proteins. Among the difficulties in working with erythroid cells to study mechanisms of cell differentiation are the limited genetic variants available for *in vitro* manipulation[4] and the lack of a differentiating erythroid cell line which can be sustained in culture. There are promising beginnings in both of these areas with the description of a variety of genetically-determined disorders of erythropoiesis in mice[4] and the recent discovery that erythroid cell differentiation occurs in cultures of leukaemic cells infected with Friend virus[5].

This review summarises our present understanding of mammalian erythroid cell differentiation, drawing most heavily on foetal mouse erythropoiesis, the best studied mammalian erythroid system. This system has provided insight into a number of questions related to the regulation of erythroid cell differentiation including: (1) the cellular basis for changing patterns of haemoglobin synthesis during foetal development; (2) the relationship between primitive and definitive erythroid cell lines; (3) mitotic activity of differentiating erythroblasts in relation to the synthesis of differentiated proteins, the haemoglobins; (4) the number of structural genes coding for globins; (5) the characteristics of mRNA for globin and relation to the synthesis of proteins during erythroid cell differentiation; (6) the cell site of action of the hormone erythropoietin; (7) the nature of erythropoietin effect on cell differentiation and haemoglobin synthesis, and (8) the nature of the erythroid cell differentiation which proceeds in cultures of viral infected leukaemic cells.

5.2 ORGAN SITES OF ERYTHROPOIESIS IN DEVELOPING FOETUSES

The mouse has a gestation period of 21 days. The first morphologically identifiable site of erythropoiesis in the developing foetal mouse occurs in the blood islands of the yolk sac at approximately the eighth day of gestation[6-8]. Precursor cells of circulating nucleated erythroid cells proliferate in these yolk-sac blood islands from about the eighth to the tenth day of gestation. Immature nucleated erythroblasts enter the foetal circulation on the ninth day[8,9] (Figure 5.1). The-yolk sac erythroid cells proliferate and continue to differentiate in the foetal circulation from the tenth day through, at least, the 13th

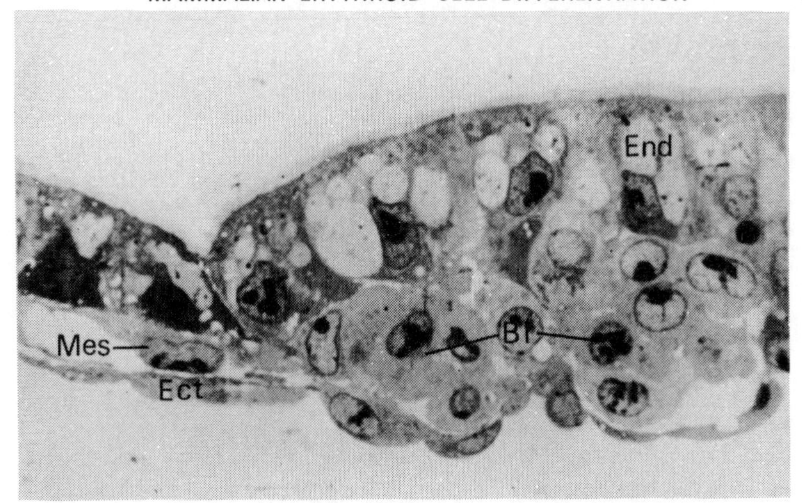

Figure 5.1 Light micrography (×520) (reduced ⅔rds on reproduction) of foetal mouse yolk sac. Day 8 of gestation. Endodermal layer (End). Ectodermal layer (Ect). Early blood islands (BI) are present as small clusters of cells in the mesenchymal (Mes) layer. These cells, neither yet free floating nor fully formed, are precursors of yolk-sac erythroid cells

day of gestation[8,10] (Figure 5.2). Mitosis in these yolk-sac derived foetal erythroblasts may be observed through day 13 of gestation[11] (Figure 5.2). In the early stages of differentiation of these proerythroblasts, polyribosomes and mitochondria are abundant in the cytoplasm. Cytoplasmic organelles decrease progressively in concentration as these cells differentiate and as haemoglobin accumulates. These cytoplasmic changes occur concomitantly with progressive condensation of nuclear chromatin, disappearance of a nucleolus and development of overall nuclear pycnosis and shrinkage. The mature circulating erythrocyte of yolk-sac blood-island origin is nucleated but the cytoplasm contains no mitochondria or ribosomes (Figure 5.2).

A second site of erythropoiesis in the foetal mouse becomes morphologically detectable in the liver during the tenth day of gestation[12] (Figure 5.3). The erythroid cell precursors appear to be derived from mesenchymal cells adjacent to the cords of hepatic epithelial cells of ectodermal origin. These erythroid cell precursors, referred to as haemocytoblasts, give rise to proerythroblasts which differentiate through a series of morphologically identifiable stages to non-nucleated reticulocytes. Unlike the terminal differentiation of yolk-sac blood-island derived erythroid cells, the differentiation of liver erythroid cells is characterised by nuclear expulsion prior to the loss of cytoplasmic ribosomes and mitochondria. It is of interest to note that the structural changes associated with the differentiation of the yolk-sac erythroid cells are analogous to those characteristic of avian and amphibian erythropoiesis[13]. Liver erythroid cell differentiation appears to involve structural changes characteristic of erythropoiesis in adult mammalian bone marrow (Figure 5.4).

The development of the yolk sac erythroid cells from precursors in the blood islands appears to proceed as a cohort. Morphologically, at each

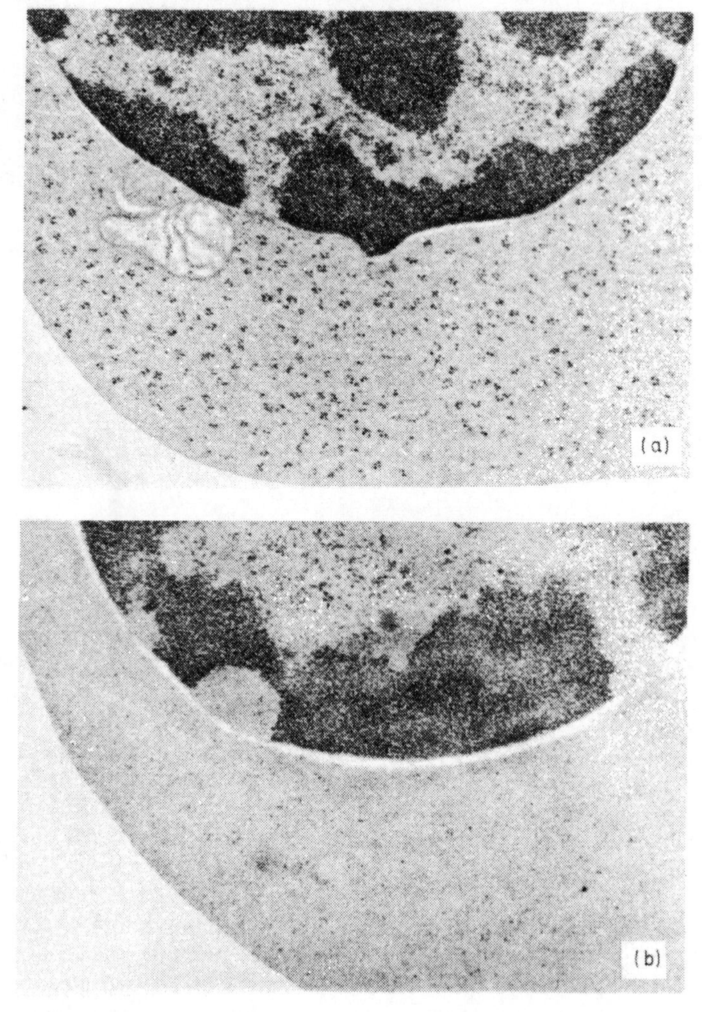

Figure 5.2 (a) Foetal mouse nucleated erythroblast taken from peripheral blood. Day 11 of gestation. Cytoplasm contains many polyribosomes and mitochondria. Electron micrograph $\times 21\,250$ (reduced $\frac{8}{10}$ths on reproduction)

(b) Foetal mouse nucleated erythroblast taken from peripheral blood Day 15 of gestation. Cytoplasm contains no ribosomes or polyribosomes. Electron micrograph $\times 21\,250$ (reduced $\frac{8}{10}$ths on reproduction)

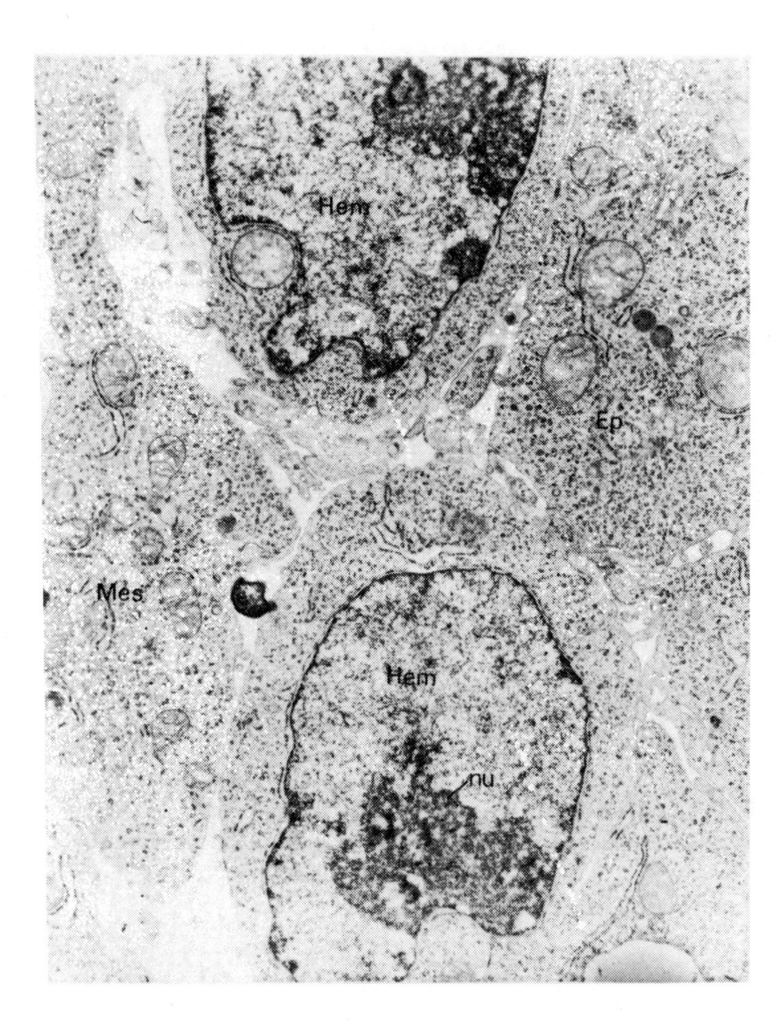

Figure 5.3 Interface between mesenchyme (Mes) and hepatic endodermal epithelium (Ep) in transverse septum of foetal mouse. Day 10–11 of gestation. The nucleoli (nu) of the two mesenchymal cells in contact with the hepatic epithelium are enlarged, exhibiting characteristics of haematopoietic cell precursors, haemocytoblasts (Hem). The mesenchymal cells themselves are free floating and rounded. Electron micrograph $\times$10 625 (reduced $\frac{8}{10}$ths on reproduction)

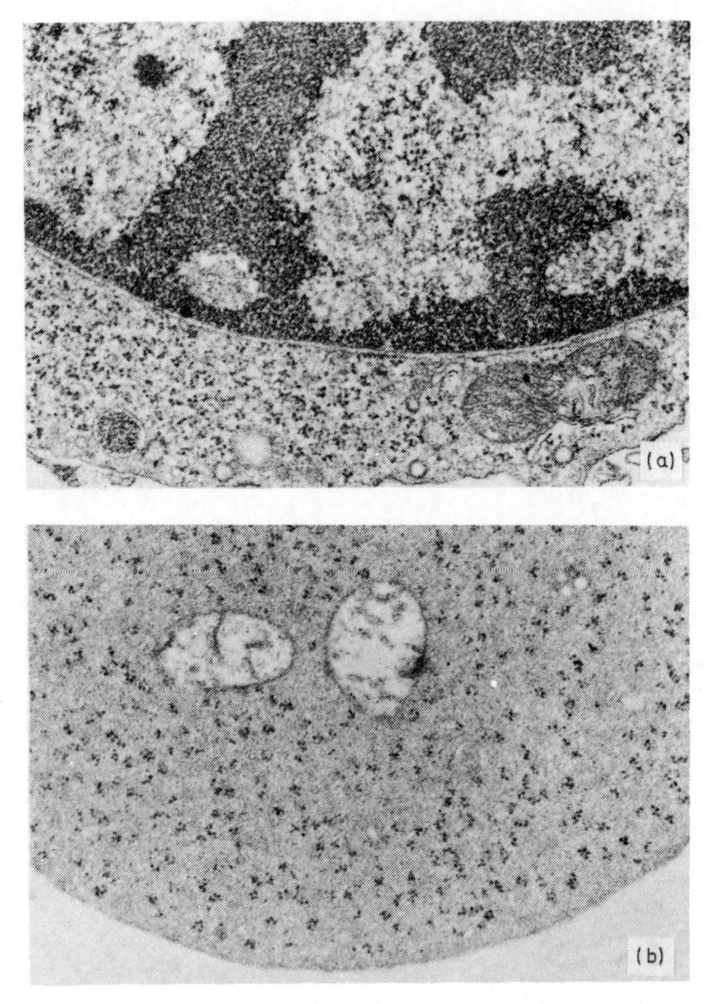

Figure 5.4 (a) Foetal mouse erythroblast taken from the liver. Day 15 of gestation. Nuclear chromatin pattern moderately condensed. Cytoplasm contains polyribosomes and mitochondria. Electron micrograph $\times 10\ 625$ (reduced $\frac{8}{10}$ths on reproduction)

(b) Foetal mouse non-nucleated reticulocyte taken from peripheral blood. Day 15 of gestation. Cytoplasm contains a few degenerating mitochondria and many polyribosomes. Electron micrograph $\times 10\ 625$ (reduced $\frac{8}{10}$ths on reproduction)

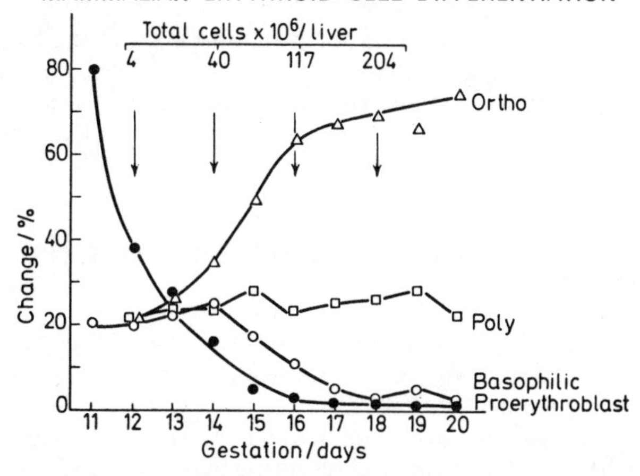

Figure 5.5 Liver erythroid cell population of foetal mice on days 11–18 of gestation. Cells classified morphologically, (△) orthochromic erythroblasts, (□) polychromatophilic erythroblasts, (○) basophilic erythroblasts, (●) proerythroblasts. Total erythroid cells per liver noted at top of figure are taken from studies by Barker *et al.*[9]. (From Marks and Rifkind[42], by courtesy of American Association for the Advancement of Science.)

stage of gestation between days 9 and 15, yolk-sac erythroid cells differentiate in a relatively homogeneous fashion[11]. In contrast to yolk-sac erythroid cells, erythropoiesis in the foetal liver proceeds as a heterogeneous population with regard to cell stage. Liver erythropoiesis, at least for a transient period, involves a self-perpetuating precursor cell which yields differentiating erythroblast over a period of time. From day 11 of gestation, the population of liver erythroid cells becomes increasingly more differentiated[7,9,12,14-16]. On day 11, *ca.* 80% of the erythroid cells present in the liver are at a very immature stage of differentiation, but by day 14, the proportion of these immature cells decreases sharply, to $< 5\%$. Concomitantly, there is an increase in the proportion of orthochromic erythroblasts which have synthesised haemoglobin (Figure 5.5). During this period, there is generally a greater than tenfold increase in the total number of erythroid cells in the liver.

The data summarised above indicate that there are two distinct populations of erythroid cells appearing at different times during mouse foetal development, namely, the primitive cell line developing in yolk-sac blood islands and the more definitive cell line which appears initially in the liver.

Direct evidence with respect to the morphological characteristics of yolk-sac and liver-erythroid cell lines in the developing foetal mouse is perhaps better than that available for any other species. These findings of shifting sites of erythropoiesis with foetal development are analogous to observations in other species, including man[17,18] where the first erythroid cells of the human embryo are detectable morphologically in yolk-sac blood sinuses at *ca.* 18 days of gestation and the appearance of immature definitive erythrocytes has been described in the liver of 12 mm human embryos (5–6 weeks of gestation). Evidence for shifts from a primitive to a definitive erythroid cell line have been described in the chicken[19, 20] and in the metamorphosis of the tadpole to the adult frog[21-25]. It is likely that similar changes occur in other species, but less definitive studies are available documenting the specific aspects of erythropoiesis during foetal development in the rat[26], monkey[27], cattle, pigs, sheep[28-30] and goats[31].

5.3 PATTERNS OF GLOBIN SYNTHESIS DURING FOETAL DEVELOPMENT

Changes in the types of haemoglobin synthesised during foetal development have been described not only for the mouse but for several other mammalian and lower animal species[3, 20, 32-39].

Studies in the mouse have provided the most direct evidence on the relationship between alterations in erythropoietic cell line and the types of haemoglobin formed. In the erythroid cells of the yolk-sac blood islands of strain C57B1/6J mice, three haemoglobins are formed[40]. These have been characterised as embryonic haemoglobin E_I, composed of x- and y-chains, embryonic haemoglobin E_{II}, composed of α- and y-chains and embryonic haemoglobin E_{III}, composed of α- and z-chains (Figures 5.6 and 5.7). There is no detectable synthesis of β-globin in yolk-sac erythroid cells.

Structural studies of the globin chains to date suggest that α-, β-, x-, y- and

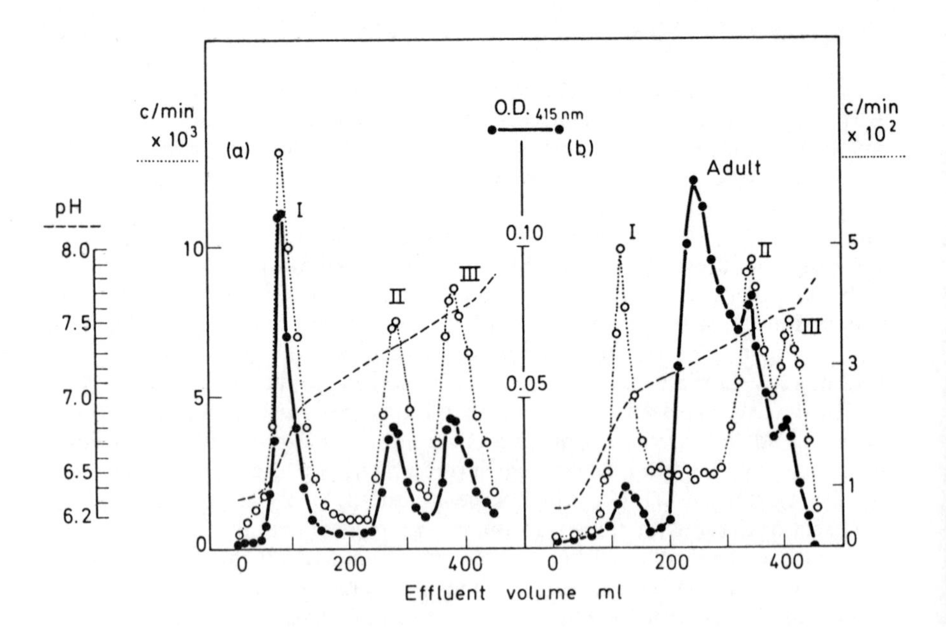

Figure 5.6 Pattern of elution of haemoglobins synthesised in erythroid cells of yolk sacs of foetal mice incubated with [^{14}C]valine. Chromatography of haemolysates of cells taken at day 11 of gestation on carboxymethyl cellulose. (a) No added carrier. (b) Added unlabelled haemolysate of adult mouse erythrocytes. (●——●) Optical density. (○ . . . ○) Radioactivity elution pattern, (- - -) pH.; (I) Hb E_I; (II) Hb E_{II}; (III) Hb E_{III}. (From Fantoni *et al.*[40], by courtesy of American Association for the Advancement of Science.)

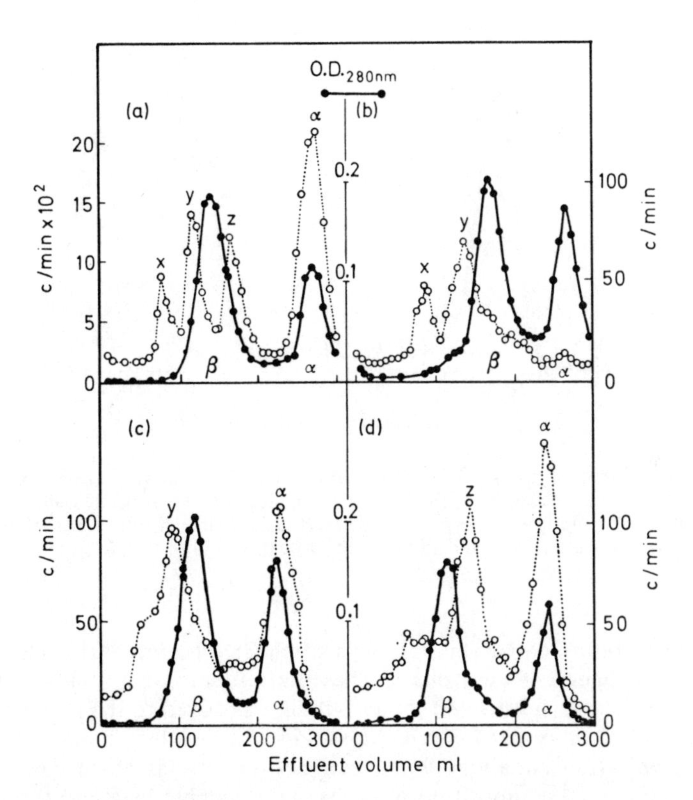

Figure 5.7 Pattern of elution of globin synthesised in erythroid cells of yolk sacs of foetal mice incubated with [^{14}C]valine. Chromatography of haemolysates of cells taken at day 11 of gestation. Added, unlabelled, carrier of adult mouse *α*- and *β*-globin. (●——●) Optical density. (○ . . . ○) Radioactivity. *α*- and *β*-adult mouse globin chains indicated respectively at bottom of each frame; (*α*), (x, y and z) embryonic globin chains, indicated at upper part of each frame. (a) Sample from total haemolysate. (b) Sample from Hb E_I, isolated by chromatographic method (see Figure 5.6). (c) Sample from Hb E_{II}, isolated by chromatographic method (see Figure 5.6). (d) Sample from Hb E_{III}, isolated by chromatographic method (see Figure 5.6). (From Fantoni *et al.*[40], by courtesy of the American Association for the Advancement of Science.)

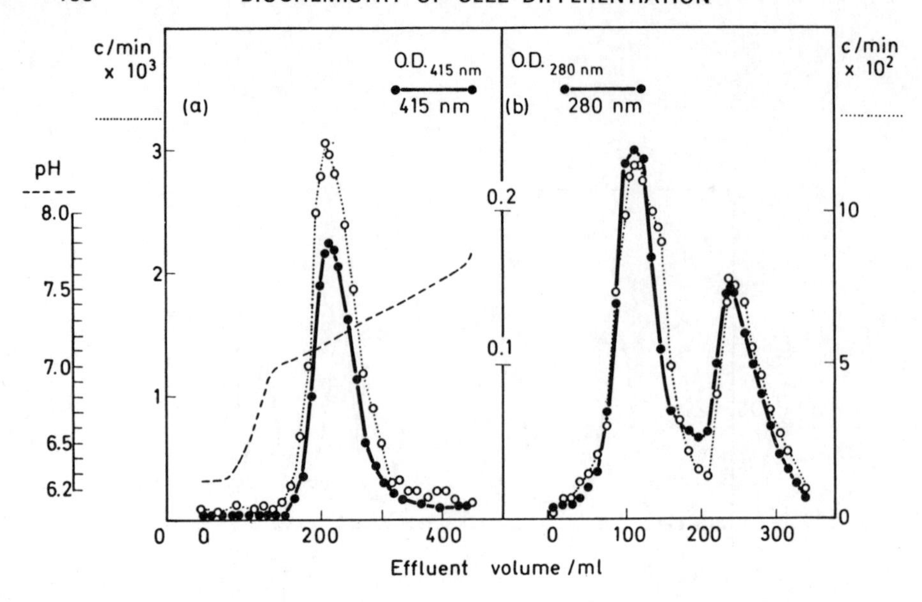

Figure 5.8 Pattern of elution of haemoglobin (a) and globin (b) taken from foetal mouse erythroid cells incubated with [^{14}C]valine. Added marker of adult mouse haemoglobin. (●——●) Optical density. (○ . . . ○) Radioactivity. (From Fantoni *et al.*[40], by courtesy of the American Association for the Advancement of Science.)

z-chains are controlled by separate genes[34, 41]. Steinheider *et al.*[34] have analysed peptide digests of the z-chain. These data suggest that about 20 of the 146 amino acids are substituted in the z-chain compared to the β-chain. This difference is less than that between human α- and β-globin chains. The same authors have partially analysed the structure of the x-globin chain. They found that the amino acid composition of the x-chain matches better to the mouse α-chain than mouse β-chain. They estimate a minimum of 50 amino acid differences between the mouse α- and mouse x-chain, yielding a maximum of 64% correspondence. It is of further interest that tryptic digest of mouse x-chain yields two peptides which are identical to human Hb ε-peptides[34]. The human ε-chain appears to be more like mouse x-chain than mouse α-chain. This similarity suggests an interesting common evolutionary ancestry for mouse x- and human ε- embryonic globin chains.

Mouse foetal liver erythroid cells synthesise a single type of haemoglobin with a globin composition indistinguishable from that of the haemoglobin ($\alpha_2\beta_2$) present in the adult of this species[40] (Figure 5.8). The change in pattern of haemoglobin synthesis from embryonic to adult type of haemoglobin during mouse foetal development is associated with the substitution of liver erythropoiesis for yolk-sac erythropoiesis. This cellular basis for changing patterns of haemoglobin synthesis during foetal development is a more general phenomenon in the animal kingdom (Table 5.1)[42]. Evidence indicates that an analogous shift from a primitive to a definitive erythroid cell line is associated with the change in the types of haemoglobin synthesis in man[43-45],

Table 5.1 Patterns of haemoglobin synthesis related to changes in erythropoietic cell lines during foetal development

Species	*Cell line*			
	Embryonic		*Definitive*	
	Site	*Haemoglobin*	*Site*	*Haemoglobin*
Mouse[40]	Yolk sac	E_{I} (xy)	Liver and bone marrow	A ($\alpha\beta$)
		E_{II} (αy)		
		E_{III} (αz)		
Man[43-45]	? Yolk sac	Gower I (ε)	Liver and bone marrow	F ($\alpha\gamma$)
		Gower II ($\alpha\varepsilon$)		A ($\alpha\beta$)
		Portland I ($\xi\gamma$)		A ($\alpha\beta$)
Rabbit[34]	? Yolk sac	E_{I} ($x\varepsilon$)	Liver and bone marrow	A ($\alpha\beta$)
		E_{II} ($\alpha\varepsilon$)		
Chicken[46]	Yolk sac	E	Yolk sac and bone marrow	A (α?)
		P (αA ?)		D (α?)
Frog[47]	Liver	Type I	Liver and bone marrow	Frog I
		Type II		Frog II

chick[46] and tadpole[45, 47]. An analogous progression from embryonic to adult haemoglobin has also been reported for rabbits[34], sheep[35] and goats[37].

In addition to this pattern of changes in erythropoietic site and cell lines, which appear to determine changes in haemoglobins formed, a further suggestion of pattern may be discerned in the structural alterations in the haemoglobins synthesised in primitive and definitive cell lines. In mouse, man, chicken, goat and sheep, conversion from embryonic to adult-type haemoglobins includes the substitution of one globin chain. In certain species, such as man, in addition to embryonic and adult haemoglobin so-called foetal haemoglobins occur. These foetal haemoglobins are synthesised in cells in later stages of foetal development than are embryonic haemoglobins. In addition, the foetal haemoglobins are formed in adult cells and, specifically in man, both HgA, adult haemoglobin and HgF, foetal haemoglobin can be formed in the same cell[28]. In the mouse and man, and probably in rabbit, goat, sheep and chick, the substitution is in a β-like chain. The α-chain is constant in structure in the embryonic and adult haemoglobins, presumably reflecting continued activity for the structural gene for the α-chain and shift in activity for the structural gene for the β-type chain. This does not appear to be the case with the tadpole and frog haemoglobins which have no common peptide chain[46, 47]. It is of further ontogenetic interest that an embryonic haemoglobin has been characterised in mouse, haemoglobin E_{I}[40], in man, haemoglobin Portland[43] and in rabbit, haemoglobin E_{I}[34] which is composed of two different globin chains neither of which are structurally identical to the α- or β-globin of the adult of the species.

5.4 RATES OF SYNTHESIS OF GLOBIN CHAINS DURING FOETAL DEVELOPMENT

The embryonic globins are synthesised in the yolk-sac blood islands. The earliest reported experiments are with day 9 yolk sacs in which the synthesis of α-, x-, y- and z-globin chains are demonstrated[2]. The syntheses of embryonic haemoglobin E_I and E_{III} cease by about day 11 of gestation. Embryonic haemoglobin E_{II} continues to be synthesised at a linear rate through at least day 13[48]. These data suggest that syntheses of embryonic globin chains x and z are terminated by day 11 while the syntheses of α-chain and embryonic globin chain y continue for at least 1 to 2 subsequent days of yolk-sac erythroid

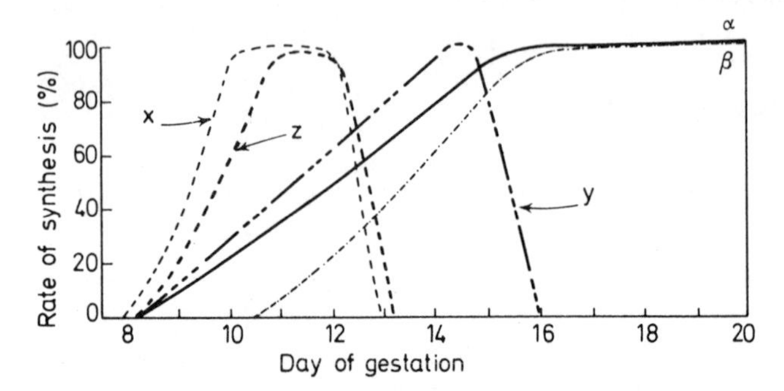

Figure 5.9 Schematic representation of relative rates of synthesis of globins during foetal development. For representational purposes 100% is set as the maximum rate of synthesis for each globin chain. This does not indicate rates of synthesis of a given globin chain relative to any other

cell development. Similar but not identical data have been reported by Steinheider *et al.*[34]. The main difference between these authors' data and those from our laboratory is that the rate of synthesis of z-chain proceeds at about the same level as that of the y-chain, while the x-chain synthesis is terminated on day 10–11. By day 10–11 the onset of α- and β-globin chain synthesis in liver erythroid cells is first detectable. Thereafter, the synthesis of α and β continues at relatively equal rates (Figure 5.9).

5.5 PRIMITIVE AND DEFINITIVE ERYTHROID CELL LINES

The relationship between the primitive and definitive erythroid cell lines has not been elucidated. Two possibilities can be considered: (1) that the primitive erythroid cell line, the yolk-sac erythroid cells, is precursor of the definitive haemopoietic cell lines; (2) alternatively, the yolk-sac erythroid cell lines and the definitive haemopoietic cell lines might diverge from a common pluripotential precursor cell at a stage of foetal development prior to the development of yolk-sac blood islands (Figure 5.10). The yolk-sac blood island erythropoiesis in foetal mice proceeds as a cohort, while foetal liver erythropoiesis, and subsequent sites of definitive haemopoiesis have a capacity for

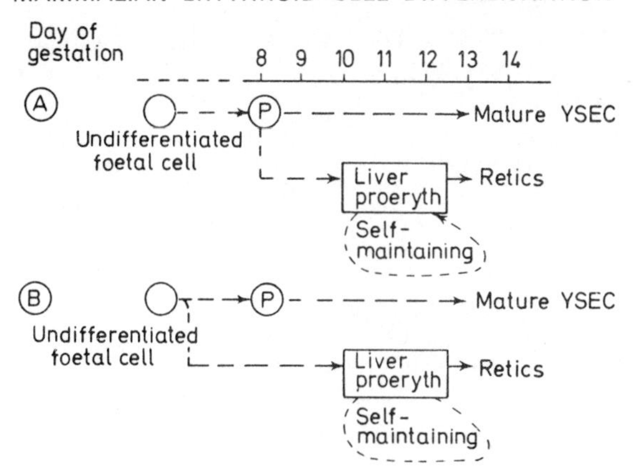

Figure 5.10 Schematic representation of alternative relationship between primitive (yolk-sac erythroid cell) line and definitive (foetal liver and adult bone marrow) erythroid cell lines

self-maintenance, as well as the production of differentiated progeny. There is no definitive evidence in any species to indicate that the primitive erythroid cells are direct precursors of definitive erythroid cells. Moore and Metcalf[49] have demonstrated that foetal mouse yolk sacs contain pluripotential cells capable of producing erythroid, granulocytic and megakaryocytic colonies, assayed by the spleen colony method of Till and McCulloch[50] as well as *in vitro* colony-forming cells, granulocyte and macrophage precursors. Their observations are compatible with the hypothesis that the yolk sac serves as the primary source for colonisation of subsequent sites of haematopoiesis. The data, however, fall short of direct evidence for this hypothesis. Further, since the *in vitro* colony-forming assay employed to evaluate the presence of precursor cells in explanted yolk sacs yields only granulocytes and macrophages and no erythroid cells, there are no data bearing on the question of the source of the erythroid precursor cell or the pluripotentional haematopoietic cell. It is of interest that although pluripotential haemopoietic cells are detectable by the spleen colony-forming assay[50] of *in vivo* yolk sacs, no demonstrable megakaryocytic or granulocytic cell formation occurs in foetal mouse yolk-sac blood islands *in situ*. These findings suggest that in the yolk-sac environment, differentiation is restricted to the erythropoietic cell lineage. The relationship between primitive and definitive erythroid cell lines is a fundamental area requiring intensive investigation.

5.6 DNA SYNTHESIS, MITOSIS AND HAEMOGLOBIN FORMATION DURING ERYTHROID CELL DIFFERENTIATION

As indicated above, yolk-sac erythroid cells differentiate as a cohort of cells *in vivo*[11]. Globin synthesis occurs in cells still capable of DNA synthesis and of cell division. This was demonstrated in studies with yolk-sac erythroid cells of 11 day foetuses which were incubated with [^{3}H]thymidine to evaluate DNA synthesis. Approximately 80% of these cells were labelled with thymidine after 4 h of incubation. By day 11, yolk-sac erythroid cells had synthesised *ca.* 45% of all the haemoglobin to be formed by these cells. Between days 10 and 14 of gestation, yolk-sac erythroid cells undergo approximately two cell divisions. With each succeeding division, there is a smaller increment in haemoglobin content per cell (Table 5.2). Between days 10 and 11, the increase

Table 5.2 Number of yolk-sac erythroid cells, content of embryonic haemoglobins, and mitotic index on days 10 to 14 of gestation. These data are derived from Ref. 11

Day	*No. of cells* (× 10^6/embryo)	*Content of embryonic haemoglobins** μg/embryo	μg/10^6 cells	*Increment in haemoglobin content per cell per day* (μg/10^6 cells)	*Mitotic index*†
10	0.31	3.5	11.3		3.2
				42.3	
11	0.70	37.5	53.6		3.8
				10.4	
12	1.22	78.0	64.0		3.4
				7.3	
13	1.41	102.0	72.3		0.4
				7.3	
14	1.50	121.0	80.6		0.0

* These values represent the average of at least three separate experiments on each day. The values for μg of haemoglobin per 10^6 cells were calculated as the quotient of the average value for μg of haemoglobin per embryo divided by the average number of cells

† Based on counts of 2000 cells from blood pooled from at least 100 foetuses on each day

in content of haemoglobin per cell is six times greater than that which occurs in cells between days 12 and 13. While there may be haemoglobin synthesis following terminal mitosis, it is at a considerably lower rate than that occurring in earlier stages of yolk-sac erythroid cell differentiation[48].

It has been suggested that the synthesis of specialised proteins does not occur in cells which are actively proliferating[51]. The data cited above indicate that this is not the case for differentiation of erythroid cells derived from yolk-sac blood islands. Analogous data have been reported for differentiating cells of the tadpole–frog erythroid line[23], galea of the silk moth[52], fibroblasts capable of collagen synthesis[53] and in cells producing antibodies[54].

Erythroid cell differentiation may be a variation of the theme of obligatory separation of mitotic activity and synthesis of differentiated proteins, which does appear valid for differentiation of skeletal muscle[55]. This hypothesis states that in differentiation there is a critical mitosis, following which the definitive commitment to the synthesis of differentiated proteins occurs. In the case of the erythroid cell, the definitive mitosis need not be the terminal division of a cell line. It differs from divisions for self-renewal of the precursor cells by virtue of initiation of expression of the programme of differentiation.

This programme determines the transition from precursor cell to erythroblast including the number of divisions and the amount of haemoglobin to be synthesised by the cell and its progeny[11,55]. This hypothesis would further suggest that the definitive mitosis results in a process which, *in vivo*, is irreversible. The study summarised above in foetal mice from our laboratory[11] and those reported by Holtzer and his co-workers with chick erythropoiesis[55-57] are compatible with this hypothesis which would predict that the number of cell division and the amount of haemoglobin synthesis by these cells is programmed at the point of definitive mitosis and is basically independent of specific external influences thereafter.

Fantoni *et al.*[58] have reported additional data consistent with this hypothesis and also suggesting that the mitotic history and haemoglobin synthesis of yolk-sac erythroid cells are not linked in a mandatory fashion. Exposure of foetal mice to a suitable dose of x-radiation resulted in a decrease in the number of mitoses which the yolk-sac erythroid cells undergo, by one, on the average. This treatment did not affect the content of haemoglobin per embryo which reached a concentration per cell of twice normal in the irradiated foetus. These data are consistent with the concept that yolk-sac erythroid cells are programmed to synthesise a specific amount of haemoglobin at a relatively early stage in differentiation and that this programme is independent of that which determines the number of cell divisions that the erythroblasts undergo.

5.7 mRNA FOR GLOBIN

5.7.1 Control at the level of DNA

A crucial question with respect to the regulation of gene expression during erythroid cell differentiation is the number of copies of the globin genes that exist in the genome. Approximately 95% of the total proteins synthesised during the course of erythroid cell differentiation are globins[59]. Mechanisms regulating the rate of synthesis of differentiated protein operate at the level of DNA or at the level of protein synthesis. At the level of DNA there are at least two models which could lead to synthesis of large amounts of a single protein. The first would involve multiple copies of the globin structural gene in the DNA of erythroid cells. This is referred to as gene reiteration. A second possibility is that in erythroid cells, there occurs specific replication of the globin gene, referred to as gene amplification. In the synthesis of ribosomal RNA both gene reiteration and gene amplification seem to occur and serve as a precedent for suggesting these two models[60,61]. Alternatively, to these control models, the genetic locus determining the structure of globin may be unique and the large amount of globins synthesised during erythroid cell differentiation reflect selective rates of transcription, processing and/or translation of globin mRNA.

Earlier experimental approaches to determining the number of copies of globin genes in the genome employed techniques of DNA–RNA hybridisation where partially-purified globin mRNA was added in excess[62,63]. In these studies with the mouse and chicken, it was estimated that globin mRNA hybridised with an amount of DNA corresponding to 30 000 or 60 000 copies of globin, respectively. More recently, in experiments performed under conditions where duck 9S RNA was hybridised with excess duck DNA, the results suggested that approximately five globin genes existed per genome[64].

Another approach involves the use of copy DNA. Reverse transcriptase from avian myeloblastosis virus has been used with purified globin mRNA to prepare partial DNA copies (cDNA) of the globin mRNA. This cDNA provides a specific approach to measuring the number of copies of globin genes in the genome by cDNA–DNA hybridisation[65-67]. Using DNA prepared with globin mRNA of duck[68], and of mouse[69], data have been obtained indicating that the globin gene frequency is less than five times that of the non-reiterated portion of the genome. In addition, Packman *et al.*[68] demonstrated that the number of globin genes in duck reticulocytes does not differ substantially from those in duck liver. These data indicate that the globin gene in the mouse and duck genome is unique and that neither gene reiteration nor amplification accounts for the large amount of globin made as erythroid cells differentiate. The much higher values for the number of globin genes in the genome estimated from studies employing DNA–RNA hybridisation in the presence of RNA excess are misleading and are possibly accounted for by part of the globin mRNA being hybridised with portions of many other genes. In other words, the globin mRNA is transcribed from a single unique locus in the genome, the nucleotide sequences of which are unique through a predominant stretch that may have a portion which is related to many other gene loci. Further, these data suggest that the globin gene is present in non-erythroid cells, but not expressed.

5.7.2 Characteristics of mRNA for globins

mRNA for globin had initially been identified tentatively on the basis of kinetics of labelling and size[70,71], but it was not until the demonstration of the capacity of mRNA preparations to direct α- and β-globin chain synthesis in heterologous cell-free systems, that definitive evidence with respect to identification of globin mRNA was available[72,73]. Earlier studies had provided evidence that RNA fractions, corresponding to sedimentation coefficient of *ca.* 9S, stimulated amino acid incorporation in cell-free systems into globin or globin-like peptide fragments[74-76]. Subsequent to the studies of Lingrel and his co-workers[72,73,76], mRNA for globin active in heterologous cell-free systems and in *Xenopus* oocytes have been purified from human[77-79], duck[80], mouse[72,79,81], guinea-pig[72], and rabbit[72,79,82-84]. The isolation and characterisation of mRNA globin by assay for its biological activity has permitted analysis of the appearance of biologically active mRNA during the course of erythroid cell differentiation induced by erythropoietin[85] and occurring in cultures of Friend virus-infected cells[86]. These are discussed more fully later.

mRNA for globin has been characterised not only in terms of its biological activity, but also chemically and physically. It is known that globins have, in general, a mol. wt. of *ca.* 16 000, which implies that the mRNA necessary for coding the structure of globin should have a mol. wt. of *ca.* 160 000 (calculated on the basis of 3 nucleotides coding for each amino acid). Studies on biological activity of mRNAs have indicated that globin mRNA is found in a fraction of RNA corresponding to a sedimentation coefficient of 5 S–18 S and generally at a peak of 9S, which approximates to a mol. wt. of 200 000.

On the basis of direct measurements, the estimate of molecular weight for globin mRNA ranges from 2.1×10^5 to 2.35×10^5 [87-89]. If these estimates of the molecular weight of globin mRNA are correct, it suggest that the isolated globin mRNA exceeds by *ca.* 50 000–75 000 that necessary to code for the structure of globin. It has been suggested that at least some of this non-structural coding portion of globin mRNA reflects the presence of adenylic acid sequences at the 3′ end of the molecule[90-92]. Lingrel *et al.*[72] have reported that the 3′ end of the globin mRNA isolated from rabbits consists of an adenylic acid sequence of 5 or 6 bases long and that there is an adenylic acid-rich region elsewhere in the molecule. Both Lingrel[72] and Williamson *et al.*[89] provided evidence for considerable secondary structure in the mRNA for globin.

The observation that mRNA for α- and β-globin are active in cell-free systems developed from heterologous tissues[72,79,80,84] and, even more elegantly, in intact *Xenopus* oocytes[82] suggests that the specificity for translation of mRNA resides primarily is the structure of the mRNA. In each of these heterologous systems, globin mRNA in translated without the addition of homologous tissue specific initiation factors[93-95]. With more fractionated cell-free systems in which separated ribosomes and supernatant factors are reconstituted, the requirement for mRNA specific factors has been reported[96-98]. Thus, while the issue of the requirement for tissue specific factors for translation of globin mRNA remains incompletely resolved, it is clear that its translation occurs in heterologous systems without the requirement for specific homologous translation factors. It is possible that specific factors are required for the optimal translation of globin mRNA.

5.7.3 Relationship of HnRNA to mRNA

Recent studies on the relationship between heterogeneous nuclear RNA (HnRNA) and mRNA have provided some insights into the synthesis and processing of mRNA. In the course of erythroid cell differentiation, there is evidence that the bulk of the RNA, i.e. rRNA and tRNA, are synthesised at earlier stages in the process of cell differentiation, while mRNA is synthesised in later stages[88,99].

There is considerable evidence to suggest that heterogeneous nuclear RNA is a precursor to cytoplasmic mRNA. Studies with competition hybridisation of RNA to DNA have shown that there is a degree of similarity between HnRNA and mRNA[100-103]. These studies are not definitive, owing to the possibility that there are competing sequences in the two types of RNA molecules which are similar but not identical and, in any event, represent only a small fraction of the total sequences. Scherrer *et al.*[104] provided evidence to suggest that there is a precursor–product relationship between giant Hn-RNA and cytoplasmic mRNA on the basis of studies designed to optimise hybridisation of the unique sequences of DNA. Darnell and his co-workers[105] have shown that polyadenylic acid segments containing 150–250 nucleotides appear to be covalently linked to heterogeneous nuclear RNA and to mRNA. There is also evidence that the poly A in mRNA and HnRNA is of similar size and composition[106] and that the poly A in HnRNA becomes labelled before that in mRNA[105-107]. Further evidence that HnRNA is a precursor of mRNA is provided by Melli and Pemberton[108] who demonstrated homology between duck erythrocyte nuclear HnRNA and the antimessenger of duck erythroid cell 9S RNA (globin mRNA). The antimessenger was synthesised using micrococcus lysodeikticus RNA polymerase. On the basis of their data, they calculated that *ca.* 10% of the heavier HnRNA and 20% of the lighter fraction of HnRNA are complementary to the antimessenger RNA. These authors suggest that approximately two molecules of mRNA exist in each molecule of the heavier fraction of HnRNA. Most HnRNA is not a precursor of mRNA but is synthesised and degraded within the cell nucleus[109, 110]. In the processing of HnRNA to mRNA it has been suggested[105] that the attachment of poly A at a terminus of the heterogeneous RNA molecule may serve as a recognition site for the processing of mRNA so that these molecules may reach the cytoplasm. While it is still uncertain whether adenine-rich stretches are involved in either the processing or the translation of globin mRNA, there is a correlation between the extent of the adenine-rich content of 9S RNA and its activity as globin messenger in a cell-free system[111]. The fraction of 9S RNA which is low in poly A sequences has a much reduced level of biological activity for mouse α- and β-globin sequences compared to a fraction which is high in poly A sequences.

Recently, Williamson *et al.*[112] reported that HnRNA from mouse foetal liver erythroblasts contain mRNA sequences for globin, demonstrated by injection HnRNA preparations into Xenopus oocytes. The HnRNA preparation is translated in the Xenopus oocytes to yield globin.

5.7.4 Stability of messenger RNA

There are several lines of evidence to indicate that the mRNA for globin is relatively stable and may have a lifetime of several days. Perhaps the most convincing evidence is that globin synthesis proceeds in reticulocytes, non-, nucleated cells which do not synthesise RNA[70, 113-115]. The question of when, during the course of erythroid cell differentiation, mRNA becomes stable has been more difficult to answer. One approach has been to correlate the capacity for new RNA synthesis with globin and non-globin protein synthesis. Such studies have been done with foetal mouse yolk-sac erythroid cells[116]. In yolk-sac erythroid cells, between days 11 and 15 of gestation there is a progressive decrease in the content of RNA. The capacity for RNA synthesis falls sharply between days 11 and 13 of gestation. Paralleling this decrease in the rate of RNA synthesis is a decrease in the formation of non-haemoglobin protein, while the capacity for haemoglobin synthesis remains relatively unchanged during this period (Figure 5.11). Over 90% of the non-haemoglobin proteins synthesised in yolk-sac erythroid cells are nuclear proteins[117]. Of this nuclear protein, *ca.* 50% is insoluble in acid. These data suggest that globin mRNA is more stable than non-globin mRNAs by day 11 of yolk-sac erythroid cell differentiation. This interpretation assumes that the availability

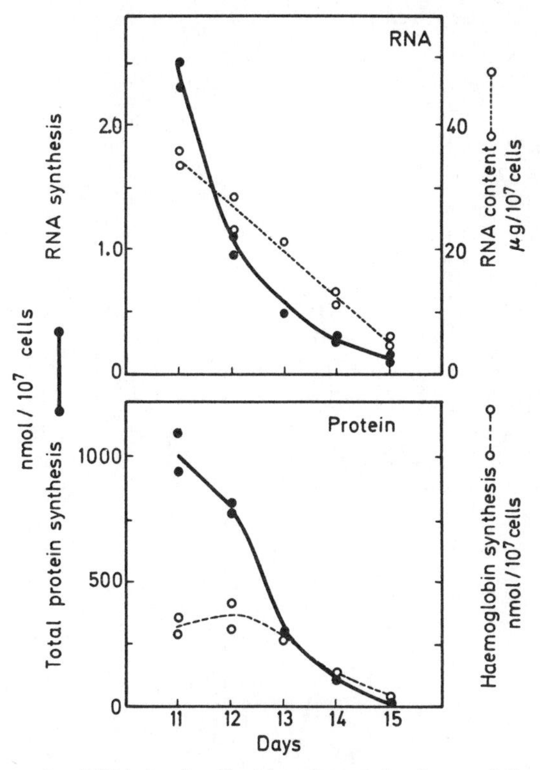

Figure 5.11 Haemoglobin, total protein and RNA synthesis and RNA content in erythroid cells of foetal mouse yolk-sac. Cells taken from peripheral blood at days 11, 12, 13, 14 and 15 of gestation. RNA content is expressed as μg/10^7 erythroid cells during a 5 min incubation at 37°C. (For details of methodology see Fantoni *et al.*[116]). (a) Content of RNA (○ . . . ○); rate of RNA synthesis (●——●). (b) Rate of haemoglobin synthesis (○ . . . ○); rate of total protein synthesis (●——●)

of mRNA is the limiting factor in determining the rate of globin and non-globin protein synthesis as yolk-sac erythroid cells differentiate. An alternative explanation of these data is that mRNA for globin is formed in excess of non-globin mRNA, relative to the level of mRNA which is rate limiting in the synthesis of these proteins.

Another approach to estimating the stability of globin mRNA during erythroid cell differentiation is to inhibit new RNA synthesis with actinomycin and measure the effect on protein formation[116]. Incubation of 11-day yolk-sac erythroid cells with actinomycin D has little or no effect on the rate of synthesis of haemoglobins, but inhibits non-globin protein synthesis by at least 90%[116]. This suggests that globin formation is directed by relatively stable mRNA formed at some time prior to day 10 of gestation. On the other hand, in 11-day yolk-sac erythroid cells, the formation of non-globin proteins may proceed on relatively short-lived molecules of mRNA.

In contrast with yolk-sac erythropoiesis, it has been possible to demonstrate in liver erythroid cells a transition from haemoglobin synthesis which is sensitive to inhibition by actinomycin D on days 11 and 12 to insensitivity to the effects of this antibiotic by day 13 of gestation[116]. This development of resistance to actinomycin D appears to reflect an alteration in the environment in which erythropoiesis is proceeding. On both days 12 and 15, actinomycin D inhibits RNA synthesis by more than 90% at all stages of differentiation between proerythroblasts and orthochromic erythroblasts[118]. On day 12, the antibiotic inhibits the uptake of labelled iron or leucine employed as

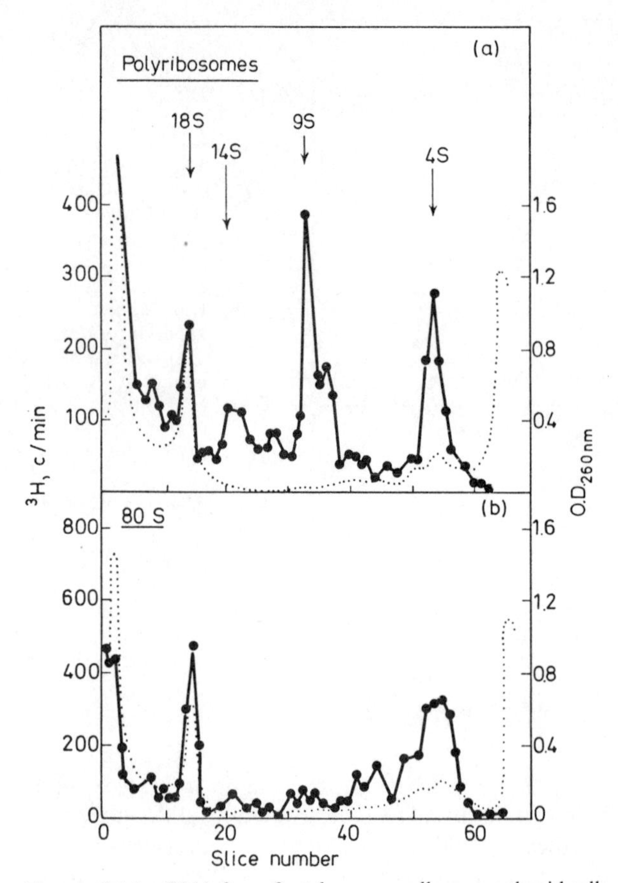

Figure 5.12 RNA from foetal mouse yolk-sac erythroid cells incubated with [^{3}H]uridine for 60 min. Day 11 of gestation. (---) Optical density. (●——●) Radioactivity CPM. (a) Polyribosome RNA showing two major peaks of radioactivity between 4S and 18S RNA corresponding to 9S and 14S. (b) 80S ribosome RNA showing no major peaks of radioactivity between 4S and 18S. (From Terada *et al.*[117], by courtesy of *J. Molec. Biol.*)

precursors of haemoglobin synthesis, at all stages of erythroblasts. By day 15, the antibiotic has little or no effect on the uptake of these isotopes into polychromatophilic and orthochromic erythroblasts which synthesise haemoglobin. These observations suggest that the stabilisation of the haemoglobin synthetic capacity in liver erythroid cells occurs at a specific stage of foetal development.

It must be emphasised that these interpretations of the actinomycin studies are subject to verification by direct assay of mRNA content. Singer and Penman[119] have shown that actinomycin inhibits RNA and, subsequently, protein synthesis in HeLa cells, with little degradation of mRNA as measured by hybridisation.

There is evidence that the degradation or inactivation of polyribosomal mRNA is related to translation. Marks *et al.*[120] demonstrated that there was essentially complete preservation of the biological activity of mRNA during NaF induced inhibition of translation in whole reticulocytes. Following treatment with NaF, the mRNA activity is fully recoverable either by purification and assay in the cell-free system[121] or by release of inhibition of protein synthesis in whole cells upon washing out of the NaF[120]. By contrast, in cells incubated for comparable periods of time but without inhibition of protein synthesis, the rate of synthesis of globin decreases progressively. A relation between mRNA degradation and translation has been suggested on theoretical grounds by Sussman *et al.*[122].

5.7.5 Patterns of mRNA synthesis during yolk-sac erythroid cell differentiation

At 11 days, only one-third of the total protein formed in mouse yolk-sac erythroid cells are haemoglobins, but by day 13 essentially all of the proteins synthesised in the cells are haemoglobins. The formation of nuclear and other non-haemoglobin proteins decreases to zero by day 13[116].

These alterations in the pattern of proteins formed are associated with changes in the types of RNA synthesised at different stages of differentiation. RNA purified from polyribosomes of day 11 erythroid cells contain a major peak of rapidly-synthesised RNA, corresponding to 9S[117]. This 9S RNA is not recovered from 80S ribosomes and is not a degradation product of ribosomal RNA (Figure 5.12). Between days 11 and 13 of gestation, while there is a marked decrease in the overall rate of synthesis of RNA, the rate of synthesis of 9S RNA falls almost to zero, in parallel with the decrease in non-haemoglobin protein formation. It was further shown that the 9S RNA synthesised in 11-day cells is not identical, by the criteria of electrophoresis in agarose acrylamide gel, with the messenger RNA for haemoglobin isolated from polyribosomes of adult mouse reticulocytes[117]. Because mouse yolk-sac erythroid cells synthesise three types of embryonic haemoglobins, two of which contain chains indistinguishable from the α-chains of adult haemoglobin, one would anticipate that if the 9S RNA synthesised in these cells is mRNA for globin, it should be identical to the mRNA for globin found in adult mouse reticulocyte polyribosomes. It has also been found that agents such as actinomycin D or hydroxyurea, which block the synthesis of 9S RNA in yolk-sac erythroid cells, inhibit nuclear protein formation but not haemoglobin synthesis[117]. The characteristics of this 9S RNA, namely, that it turns over rapidly, that it is electrophoretically distinct from adult reticulocyte mRNA for globin, and that it is not synthesised in the presence of actinomycin D or hydroxyurea, agents which inhibit the synthesis of nuclear proteins but not haemoglobins, suggest it may be mRNA for one or more of the nuclear proteins. mRNA for globin in yolk-sac erythroid cells appears to be synthesised prior to the tenth day of gestation and remains stable while these erythroblasts proliferate fourfold. The stable mRNAs for haemoglobins may be distributed to daughter cells through, on the average, two cell divisions.

5.8 ERYTHROPOIETIN-INDUCED ERYTHROID CELL DIFFERENTIATION

5.8.1 Target cell of action

One of the most challenging problems in molecular biology of eukaryote cell differentiation is the mechanism of action of hormones. These substances comprise a spectrum of chemical entities and exhibit remarkable cellular selectivity in inducing specific responses. These responses may involve a sequence of events that effect genetic expression and require synthesis of new RNAs and proteins. Erythropoietin is the erythropoietic hormone. It has not been prepared in pure biologically-active form. Analysis of partially-purified preparations[123] suggest that it has a mol. wt. of 46 000 and is a glycoprotein.

Central to our understanding of the mechanism of erythroid cell differentiation and the action of erythropoietin is identification and isolation of the immediate precursor of the haemoglobin-forming erythroid cell (erythropoietin-responsive cell). Among tissues which have been shown to be responsive to erythropoietin are adult rat bone marrow[124] and foetal mouse liver[14, 16]. Erythropoietin may stimulate haemoglobin synthesis in yolk-sac blood islands of 8-day foetuses, but subsequent stages of differentiation of these cells are minimally or not at all responsive to the hormones[125].

Foetal mouse liver has proved to be a suitable tissue to approach the problem of identification of the erythropoietin-responsive cell. On day 11 of gestation, when the foetal liver becomes a site of erythropoiesis, *ca.* 80% of the erythroid cells present are at a very immature stage and are classified morphologically as proerythroblasts[12, 126]. These cells are very active in incorporation of uridine into RNA and leucine into protein[42]. This is indicated by a rate of uptake of [^{3}H]uridine and [^{3}H]leucine by these cells several-fold higher than that in morphologically more differentiated cells in the same foetal livers and in morphologically comparable cells on subsequent days of gestation. Erythropoiesis in the foetal liver proceeds as a heterogeneous population with regard to cell stage, unlike yolk-sac erythroid cells which differentiate in a relatively homogeneous cohort of cells. Foetal liver is a transitional site of erythropoiesis (by birth, the liver is no longer a site of erythropoiesis) depending on a self-perpetuating precursor cell which yields differentiating erythroblasts, at least over a short period of time. The disappearance of metabolitically-active proerythroblasts may be the cellular basis for the loss in capacity for sustained erythropoiesis in the foetal liver, as these cells may be the erythropoietin responsive cells.

The identification and purification of the erythropoietin-responsive cell from foetal liver was facilitated with the elucidation of certain aspects of erythropoietin action. In foetal liver erythroid cells, the hormone-induced increase in haemoglobin formation does not result from a direct effect on the rate of haemoglobin synthesis per cell[16, 126]. This is indicated by the observation that erythropoietin does not increase the uptake of [^{3}H]leucine per polychromatophilic or per orthochromatic erythroblast in cultures in which the hormone causes an approximately twofold increase in haemoglobin formation. The erythropoietin-stimulated increase in haemoglobin synthesis is caused by an increase in the number of cells synthesising haemoglobin. The hormone

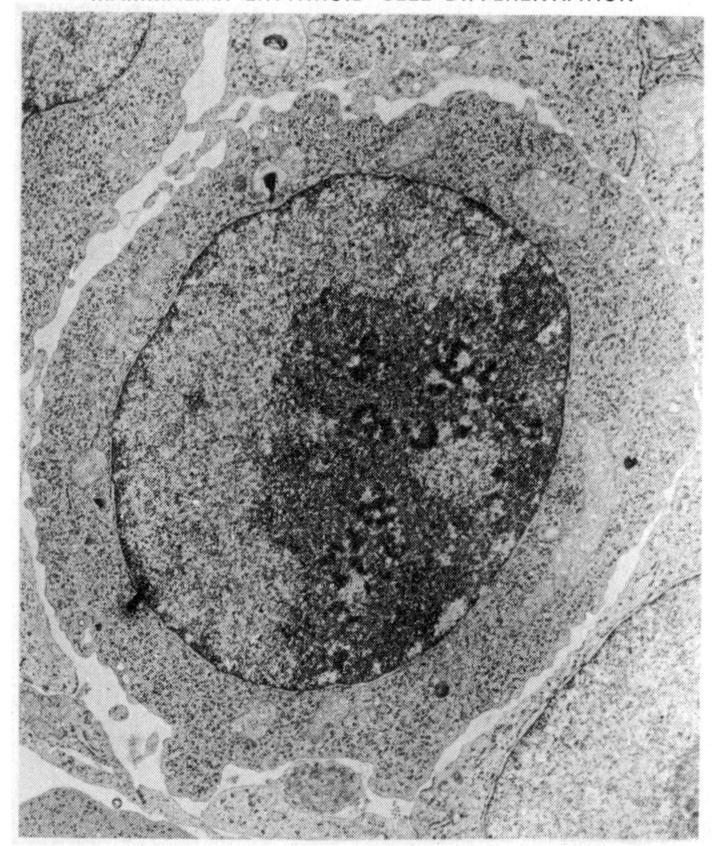

Figure 5.13 Typical cell from population of erythropoietin-sensitive precursor cells taken from foetal mouse liver. Day 11 of gestation. In cytoplasm are free polyribosomes, mitochondria and relatively few cisternae of endoplasmic reticulum. Characteristic are a large nucleus and nucleolus, with almost exclusively extended chromatin in the nucleoplasm. ×9 000 (reduced $\frac{8}{10}$ on reproduction)

acts to maintain the number of immature proerythroblasts and basophilic erythroblasts in the population and to increase the total number of haemoglobin-forming cells. Erythropoietin is required for renewal of the immature population of erythroid cell precursors under these conditions *in vitro*. Studies from several laboratories[127] are consistent with this concept that erythropoietin stimulates erythroid cell differentiation.

In erythroid cells of foetal liver, the first detectable effect of erythropoietin is a selective stimulation of RNA synthesis in the most immature cells, the proerythroblasts[16, 126]. The hormone had no effect on RNA formation in more differentiated erythroid cells or in non-erythroid cells in foetal livers. The stimulation of RNA synthesis by erythropoietin was used as a criterion for the identification of the erythropoietin-responsive cell. These erythropoietin-responsive cells are included in a class of precursor cells designated proerythroblasts on the basis of cytological criteria. (It must be recognised that

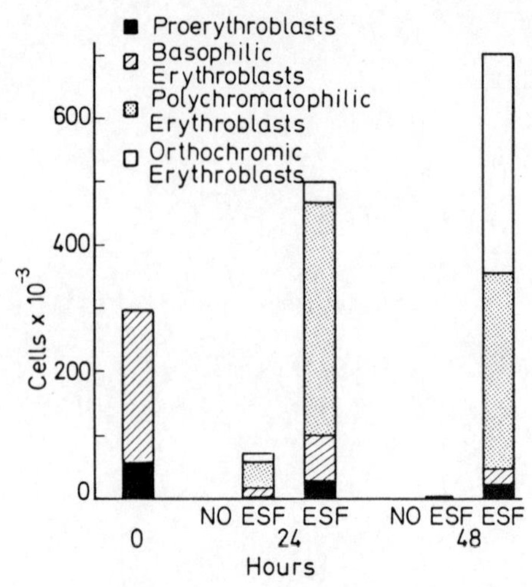

Figure 5.14 Effect of erythropoietin on the preparation of erythroid cell precursors purified from 13-day foetal mouse livers, by antibody treatment as described in Ref. 132. Cells were maintained in culture with and without erythropoietin for 48 h

cytological criteria are inadequate to distinguish whether, or not, an erythroblast has initiated the process of haemoglobin synthesis.) The cells in this population are large and by light and electron microscopy display a high ratio of nuclear to cytoplasmic material and an extended chromatin pattern within the large nucleus (Figure 5.13). The cytoplasm contains abundant polyribosomes, mitochondria and sparse elements of the endoplasmic reticulum.

There is considerable evidence that the precursor cell response to erythropoietin is distinct from the pluripotential haematopoietic stem cell[128]. Stephenson and Axelrod[129], for example, employing velocity sedimentation of mouse foetal liver cells, obtained a partial separation of erythropoietin-responsive cells from haematopoietic spleen colony-forming cells. Exploitation of antigen differences between erythroid cell precursors and more differentiated erythroid cells[130, 131] provided an effective basis for isolating a population of immature erythroid precursors. In the presence of complement and antiserum to mature mouse erythroid cells, polychromatophilic erythroblasts and more differentiated cells are haemolysed. Addition of erythropoietin to cultures of the unlysed, purified erythroid cell precursors stimulated proliferation and differentiation to erythroblasts with the initiation of haemoglobin synthesis[132] (Figure 5.14).

5.8.2 Erythropoietic effect on macromolecule synthesis

Erythropoietic stimulation of RNA synthesis occurs within 1 h of culture of foetal liver erythroid cells with the hormone[126]. The erythropoietin stimulation of RNA synthesis is not dependent on any hormone-mediated effect on DNA synthesis[133, 134]. Inhibition of DNA synthesis by hydroxyurea, cytosine arabinoside or 5-fluorodeoxyuridine does not prevent the erythropoietin-stimulated RNA synthesis in proerythroblasts[133-135]. While the effects of inhibition of protein synthesis may depend on the nature of the inhibitor, cycloheximide inhibition of protein formation does not prevent the early hormone-stimulated RNA synthesis[136]. Erythropoietin-stimulated synthesis

Table 5.3 Activity of mRNA for globin in precursor cells culture with erythropoietin

Source of RNA	*Time of culture*	*Total proteins*	*Globin*
	hr	*cpm*	*cpm*
Total cell	0	51 540	980
	10	35 360	6940
Cytoplasmic fraction	0	32 040	500
	10	39 920	7420

RNA was extracted from precursor cells before and after culture with erythropoietin for 10 h. 2/5 of the cells were used for extraction of total cell RNA and the remainder were used for extraction of RNA from the cytoplasmic fraction. The numbers of cells used were 7.5×10^7 and 6.2×10^7 for 0 and 10 h preparation, respectively The 6–16S fractions were prepared by sucrose-gradient centrifugation and assayed in the Krebs ascites tumour cell-free system as described in Ref. 85.

of RNA in foetal liver proerythroblasts in culture precedes detectable increase in cell number, the appearance of biologically-active mRNA for globin or haemoglobin synthesis[85, 134]. A hormone effect on DNA synthesis has been reported as an early phenomenon, i.e. within 1 h of culture of foetal liver erythroid cells[137]. Such an early effect on DNA synthesis was not confirmed in studies by Chui *et al.*[126]. An effect of erythropoietin on DNA formation was observed after 8–10 h of culture. Inhibition of RNA or of protein synthesis eliminates the erythropoietin effect on DNA formation[133, 138]. The early erythropoietin-stimulated RNA in rat marrow involves a variety of species including 150S, 55 – 65S, 45S, 28S, 18S, 9S, 6S and 4S[139, 140]. Similarly, after 3 h of culture with foetal liver erythroblasts, erythropoietin stimulates several species of RNA, including ribosomal precursor RNA, HnRNA and RNA sedimenting at > 60S[135].

With the preparation of purified populations of erythropoietin-responsive precursor cells, the effect of erythropoietin on RNA synthesis and, more specifically, mRNA for globin has been examined[85, 137]. Erythropoietin stimulates RNA synthesis in cultured erythroid precursor cells within 1 h. The increase in RNA synthesis observed early involves HnRNA 45S, 32S and 4S and subsequently, 28S and 18S[138]. No biologically active mRNA for globin was recoverable from these erythroid precursor cells prior to incubation with erythropoietin[85]. After 10 h of culture with the hormone, there was a marked increase in globin mRNA activity. The appearance of globin mRNA activity correlated with the stimulation of globin synthesis in these cells, which occurred between 5 and 10 h of incubation (Table 5.3).

5.8.3 Hypothesis for erythropoietin action

The findings summarised above suggest the following:

That the erythropoietin-responsive precursor cell contains little or no globin mRNA in biologically-active form; that erythropoietin is necessary to induce the transition from precursor stage to erythroblast stages capable of haemoglobin synthesis; that this transition involves the appearance of active globin mRNA, which could reflect either initiation of transcription of globin genes or an increase in the rate of processing of transcribed globin mRNA present in the precursor cells in an inactive form, such as a component of nuclear HnRNA; that the erythropoietin-responsive cell is itself differentiated from a progenitor pluripotential stem cell.

This hormone responsive cell may have a receptor recognition site for erythropoietin, possibly on the cell membrane. The initial effect of the hormone is to stimulate nuclear RNA synthesis. The increase in RNA synthesis leads to an increase in protein formation—not specifically globin—and, as a consequence, increased DNA synthesis and mitosis. Mitosis affords the opportunity for reprogramming the genes with consequent initiation of globin mRNA synthesis and subsequent globin formation[141].

This reasoning is highly speculative, but is consistent with the following known facts: erythropoietin acts selectively on proerythroblasts, stimulation of a variety of RNAs being the earliest-detected effect on macromolecular synthesis. Erythropoietin-stimulated DNA synthesis and mitosis is blocked by inhibitors of RNA formation. Globin mRNA is not present in an active form in the erythropoietin-responsive cell and appears only after several hours of incubation with the hormone. Inhibition of RNA or of DNA synthesis prevents erythropoietin stimulation of globin formation. This hypothesis can be tested specifically with the isolated erythropoietin-responsive cell. The critical issues of the site of erythropoietin action and the mechanism of initiation of globin mRNA transcription and its relation to mitosis remain to be examined.

5.9 ERYTHROID CELL DIFFERENTIATION IN FRIEND VIRUS-INFECTED CULTURES

Friend[142] discovered a virus-induced murine leukaemia which is characterised by a rapid proliferation of reticulum cells of spleen and liver, lymphorytosis, erythroblastosis, anaemia and thrombocytopenia[143]. This so-called Friend virus appears to be a mixture of several viruses, at least one of which produces a hypervolemic polycythaemia within 2–3 weeks of infection[144]. This virus is capable of re-establishing erythopoiesis in a hypertransfused polycythaemic mouse. The effect of the virus is not inhibited by anti-erythropoietin antibodies[144]. Administration of erythropoietin to Friend virus-infected mice accelerated the appearance of polycythaemia. These *in vivo* studies suggested that Friend virus-induced polycythaemia is independent of erythropoietin production and the effect of the virus and erythropoietin may be additive in leading to increased erythropoiesis.

Cells from the Friend virus-induced leukaemic mice can be propagated in continuous culture[5]. Throughout years of serial transfer, a small fraction of these cells, generally $< 10\%$ and frequently $< 1\%$ were observed to differentiate to erythroid cells. Recently, Friend and her co-workers[145] reported that addition of dimethyl sulphoxide (DMSO) to these culture media may induce 80% or more of the cells to differentiate to erythroid cells. These erythroid cells have been characterised morphologically[146] and shown to synthesise haeme and α- and β-globin[145, 147-149]. The study of globin chain composition and tryptic peptide analysis demonstrated that the globins formed in these cells, whether or not induced to differentiate with DMSO, were identical to α- and β-chains of adult haemoglobin of DBA/2J mice from which the cells originated.

The induction of erythroid cell differentiation by addition of DMSO to virus-infected cells in culture is associated with the appearance of mRNA for globin[86]. Globin mRNA is virtually undetectable in the non-DMSO treated Friend virus-infected leukaemic cell cultured under the condition of Ross and his co-workers[86]. Addition of DMSO to these cultures is associated with the appearance of mRNA for globin within 2 days, which reaches a maximum within 4 days. In these studies[86], globin mRNA was measured by using cDNA ([^{3}H]DNA synthesised with reverse transcriptase employing purified mouse globin mRNA as a template) as a probe for hybridisation with globin mRNA. The hybridisable cytoplasmic RNA induced in these cells had a coefficient of sedimentation of *ca.* 9S. These results are comparable to those summarised above[132] with respect to the erythropoietin induction of erythroid cell differentiation and the appearance of biologically-active globin mRNA. The appearance of globin mRNA following DMSO induction may reflect the onset of transcription of the globin gene.

The nature of the relationship between Friend virus and the propagation in culture of erythroid precursor cells is not known. Tambourin and Wendling[150] have provided indirect evidence that a Friend virus may have as its target cell the erythropoietin-responsive cell. The effect of the virus may be to transform erythroid cell precursors, i.e. cells programmed for differentiation to erythroid cells and haemoglobin synthesis, so that they develop the capacity to propagate in culture. As with other differentiated cells, uninfected erythroid cell precursors cannot be propagated in continuous culture, as can the virus-infected cells[132]. The mechanism by which DMSO induces erythroid cell differentiation in the virus-infected cell cultures is not elucidated.

5.10 SUMMARY

Erythroid cell differentiation provides an important model to investigate several aspects of the regulation of mammalian cell differentiation.

(1) Foetal development in many animal species is associated with a change in the types of globins formed as gestation proceeds. In mice, alteration in the type of haemoglobin synthesised during foetal development is associated with the substitution of liver erythropoiesis for yolk-sac blood-island erythropoiesis. There appears to be a greater constancy in transcription of the α- than the β-type structural gene as sequential erythroid cell lines appear during gestation.

(2) Mouse yolk-sac erythroid cell differentiation proceeds as a cohort. There are at least two classes of proteins distinguishable with respect to relative dependence on continued mRNA synthesis. The major portion of nuclear proteins appear to be dependent on relatively short-lived mRNAs while synthesis of the 'differentiated' proteins, the globins, are dependent on relatively-stable mRNA molecules.

(3) During differentiation of yolk-sac erythroid cells, DNA synthesis and cell division proceed in cells synthesising haemoglobin.

(4) Erythropoietin acts selectively on the most immature morphologically-identifiable erythroid cell precursor. This population of erythropoietin responsive cells can be purified from foetal mouse livers and requires the hormone to proliferate and differentiate in culture.

(5) The erythropoietin responsive precursor cell appears to contain no globin mRNA recoverable in a biologically-active form. The initial effects of erythropoietin on macromolecular formation in these cells are to stimulate synthesis of RNAs of a variety of classes, but not specifically, the appearance of biologically-active globins mRNA. Only after 5–10 h in culture with the hormone do these cells contain globin mRNA in a biologically active form. A hypothesis for the action of erythropoietin is presented.

(6) Friend virus-infected cell cultures contain erythroid cell precursors which can be propagated continuously. DMSO induces erythroid cell differentiation in these virus-infected cell cultures.

Acknowledgement

Studies reviewed in this article which are from the laboratories of the authors were supported in part by grants from the National Institute of General Medical Sciences (GM-14552, and GM-18153) and National Science Foundation (GB-4631, GB-27388). Arthur Bank is a scholar of the American Cancer Society.

References

1. Gordon, A. S. (1970). *Regulation of Hematopoiesis*, Vol. 1 (New York: Appleton-Century-Crofts)
2. Marks, P. A. (1972). *Harvey Lectures, Series 66*, 43 (New York: Academic Press)
3. Ingram, V. M. (1963). *The Hemoglobins in Genetics and Evolution* (New York: Columbia University Press)
4. Russell, E. S. (1970). *Regulation of Hematopoiesis*, Vol. 1, 649 (A. S. Gordon, editor) (New York: Appleton-Century-Crofts)
5. Friend, C., Patuleia, M. C. and DeHarven, E. (1966). *National Cancer Institute Monograph*, **22,** 505
6. Attfield, M. (1951). *J. Genet.*, **50,** 250
7. Russell, E. S. and Bernstein, S. E. (1966). *Biology of the Laboratory Mouse*, 351 (E. L. Green, editor) (New York: McGraw-Hill)
8. Bank, A., Rifkind, R. A. and Marks, P. A. (1970). *Regulation of Hematopoiesis*, Vol. 1, 701 (A. S. Gordon, editor) (New York: Appleton-Century-Crofts)
9. Barker, J. E., Keenan, M. A. and Raphals, L. (1969). *J. Cell. Physiol.*, **74,** 51
10. Fantoni, A., de la Chapelle, A., Chui, D., Rifkind, R. A. and Marks, P. A. (1969). *Ann. N.Y. Acad. Sci.*, **165,** 194

11. de la Chapelle, A., Fantoni, A. and Marks, P. A. (1969). *Proc. Nat. Acad. Sci. USA*, **63,** 812
12. Rifkind, R. A., Chui, D. H. K. and Epler, H. (1969). *J. Cell. Biol.*, **40,** 343
13. Andrew, W. (1965). *Comparative Hematology* (New York: Grune and Stratton)
14. Cole, R. J. and Paul, J. (1966). *J. Embryol. Exp. Morphol.* **15,** 245
15. Paul, J., Conkie, D. and Freshney, R. I. (1969). *Cell Tissue Kinet.*, **2,** 283
16. Rifkind, R. A., Chui, D., Djaldetti, M. and Marks, P. A. (1969). *Trans. Amer. Assoc. Phys.*, **82,** 380
17. Bloom, W. and Bartelmez, G. W. (1940). *Amer. J. Anat.*, **67,** 21
18. Knoll, W. and Pigel, E. (1949). *Acta Haematol.*, **2,** 369
19. Lemez, L. (1964). *Advances in Morphology*, **3,** 197
20. Wilt, F. H. (1967). *Advances in Morphology*, **6,** 89
21. Moss, B. and Ingram, V. M. (1968). *J. Molec. Biol.*, **32,** 481
22. Moss, B. and Ingram, V. M. (1968). *J. Molec. Biol.*, **32,** 493
23. Maniatis, G. M. and Ingram, V. M. (1971). *J. Cell. Biol.*, **49,** 373
24. Maniatis, G. M. and Ingram, V. M. (1971). *J. Cell. Biol.*, **49,** 380
25. Maniatis, G. M. and Ingram, V. M. (1971). *J. Cell. Biol.*, **49,** 390
26. Hunter, J. A. and Paul, J. (1969). *J. Embryol. Exp. Morphol.*, **21,** 361
27. Kitchen, H., Eaton, J. W. and Stenger, V. G. (1968). *Arch. Biochem. Biophys.*, **123,** 227
28. Kleihauer, E. and Stoffler, G. (1968). *Molec. Gen. Genet.*, **101,** 59
29. Kleihauer, E., Brauchle, E., and Brandt, G. (1966). *Nature* (*London*), **212,** 1272
30. Grimes, R. M. and Duncan, C. W. (1959). *Arch. Biochem. Biophys.*, **84,** 393
31. Adams, H. R., Wrightstone, R. N., Miller, A. and Huisman, T. H. J. (1969). *Arch. Biochem. Biophys.*, **132,** 223
32. Kovach, J. S., Marks, P. A., Russell, E. S. and Epler, H. J. (1967). *J. Molec. Biol.*, **25,** 131
33. Baglioni, C. and Sparks, C. E. (1963). *Develop. Biol.*, **8,** 272
34. Steinheider, G., Medleris, H. and Ostertag, W. (1971). *Syntheses, Struktur und Funktion des Hamoglobins* 225 (Martin and Nowicki, editors) (Munchen: J. F. Lehmanns, Verlag)
35. Tucker, E. M. (1971). *Biol. Rev.*, **46,** 341
36. Nienhuis, A. and Anderson, W. F. (1972). *Proc. Nat. Acad. Sci., USA*, **69,** 2184
37. Wrightstone, R. N., Wilson, J. B. and Huisman, T. H. J. (1970). *Arch. Biochem. Biophys.*, **138,** 451
38. Huisman, T. H. J., Lewis, J. P., Blunt, M. H., Adams, H. R., Miller, A., Dozy, A. M., and Boyd, E. M. (1969). *Ped. Res.*, **3,** 189
39. Kleihauer, E. and Tautz, Ch. (1972). *Res. Exp. Med.*, **158,** 219
40. Fantoni, A., Bank, A. and Marks, P. A. (1967). *Science*, **157,** 1327
41. Vulpis, G. and Bank, A. (1970). *Biochim. Biophys. Acta*, **207,** 390
42. Marks, P. A. and Rifkind, R. A. (1972). *Science*, **175,** 955
43. Capp, G. L., Rigas, D. A. and Jones, R. T. (1970). *Nature* (*London*), **228,** 278
44. Huehns, E. R. Flynn, F. V., Butler, E. A. and Beaven, G. H. (1961). *Nature* (*London*), **189,** 496
45. Huehns, E. R., Dance, N., Beaven, G. H., Keil, J. V., Hecht, F. and Motulsky, A. G. (1964). *Nature* (*London*), **201,** 1095
46. Ingram, V. M. (1972). *Nature* (*London*), **235,** 338
47. Maniatis, G. M. and Ingram, V. M. (1970). *Advances in the Bio-Sciences*, **6,** 529
48. Fantoni, A., de la Chapelle, A. and Marks, P. A. (1969) *J. Biol. Chem.*, **244,** 675
49. Moore, M. A. S. and Metcalf, D. (1970). *Brit. J. Haematol.*, **18,** 279
50. Till, J. E. and McCulloch, E. A. (1961). *Rad. Res.*, **14,** 213
51. Ebert, J. D. and Kaighn, M. E. (1966). *Major Problems in Developmental Biology*, 29 (M. Locke, editor) (New York: Academic Press)
52. Kafatos, F. X. and Feder, N. (1968). *Science*, **161,** 470
53. Davies, L. M., Priest, Z. H. and Priest, R. E. (1968). *Science*, **169,** 91
54. Szienberg, A. and Cunningham, A. J. (1968). *Nature* (*London*), **217,** 747
55. Holtzer, H. (1970). *Symp. Internatl. Soc. Cell. Biology, Gene Expression in Somatic Cells*, **9,** 69 (H. Padykula editor) (London: Academic Press)
56. Campbell, G. LeM., Weintraub, H., Mayall, B. H. and Holtzer, H. (1971). *J. Cell Biol.*, **50,** 669
57. Weintraub, G., Campbell, G. LeM. and Holtzer, H. (1971). *J. Cell. Biol.*, **50,** 652

58. Fantoni, A., Ghiara, L. and Pozzi, L. V. (1972). *Biochim. Biophys. Acta*, **269**, 141
59. London, I. (1961). *Harvey Lectures, Series, 56*, 151 (New York: Academic Press)
60. Birnstiel, M. L., Grunstein, M., Speirs, J. and Hennig, W. (1969). *Nature* (*London*), **223**, 1265
61. Attardi, J. and Amaldi, F. (1970). *Ann. Rev. Biochem.*, **39**, 183
62. Williamson, R., Morrison, M. and Paul, J. (1970). *Biochem. Biophys. Res. Commun.*, **40**, 740
63. Fanches, Z., De Jimenez, E., Dominquez, J. L., Weed, F. H. and Bock, R. M. (1971). *J. Molec. Biol.*, **61**, 59
64. Bishop, J. O., Pemberton, R. and Baglioni, C. (1972). *Nature New Biol.*, **235**, 231
65. Kacian, D. L., Spiegelman, S., Bank, A., Terada, M., Metafora, S., Dow, L. and Marks, P. A. (1972). *Nature New Biol.*, **235**, 167
66. Verma, I. M., Temple, G. H., Mann, H. and Baltimore, D. (1972). *Nature New Biol.*, **235**, 163
67. Ross, J., Aviv, H. Scolnick, E. and Leder, P. (1972). *Proc. Nat. Acad. Sci. USA*, **69**, 264
68. Packman, S., Aviv, H., Rose, J. and Leder, P. (1972). *Biochem. Biophys. Res. Commun.* **49**, 813
69. Harrison, P. R., Hell, A., Birnie, G. D. and Paul, J. (1972). *Nature*, **239**, 219
70. Marks, P. A., Willson, C., Kruh, J. and Gros, F. (1962). *Biochem. Biophys. Res. Commun.*, **8**, 9
71. Marbaix, G., Burny, A., Huez, G. and Chantrenne, H. (1966). *Biochim. Biophys. Acta.*, **114**, 404
72. Lingrel, J. B., Lockard, R. E., Jones, R. F., Burr, H. E. and Holder, J. W. (1971). *Series Haematol.*, IV, 3, 37
73. Lockard, R. E. and Lingrel, J. B. (1971). *Nature New Biol.*, **233**, 204
74. Schapira, G., Dreyfus, J. C. and Maleknia, N. (1968). *Biochem. Biophys. Res. Comm.*, **32**, 558
75. Grossbard, L., Banks, J. and Marks, P. A. (1968). *Arch. Biochem. Biophys.*, **125**, 580
76. Lockard, R. E. and Lingrel, J. B. (1969). *Biochem. Biophys. Res. Commun.*, **37**, 204
77. Benz, E. J., Jr. and Forget, B. G. (1971). *J. Clin. Invest.*, **50**, 2755
78. Nienhuis, A. and Anderson, W. F. (1971). *J. Clin. Invest.*, **50**, 2458
79. Metafora, S., Terada, M., Dow, L. W., Marks, P. A. and Bank, A. (1972). *Proc. Nat. Acad. Sci. USA*, **69**, 1299
80. Pemberton, R. E., Housman, D., Lodish, H. E. and Baglioni, C. (1972). *Nature New Biol.*, **235**, 99
81. Lockard, R. E. and Lingrel, J. B. (1972). *J. Biol. Chem.*, **247**, 4174
82. Gurdon, J. B., Lane, C. D., Woodland, H. R. and Marbaix, G. (1971). *Nature* (*London*) **233**, 177
83. Marbaix, G. and Lane, C. D. (1972). *J. Molec. Biol.*, **67**, 517
84. Aviv, H., Boime, I. and Leder, P. (1971). *Proc. Nat. Acad. Sci. USA*, **68**, 2303
85. Terada, M., Cantor, L., Metafora, S., Rifkind, R. A., Bank, A. and Marks, P. A. (1972). *Proc Nat. Acad. Sci. USA*, **69**, 3575
86. Ross, J., Ikawa, Y. and Leder, P. (1972). *Proc. Nat. Acad. Sci. USA*, **69**, 3620
87. Blobel, G. (1971). *Proc. Nat. Acad. Sci. USA*, **68**, 832
88. Gaskill, P. and Kabat. D. (1971). *Proc. Nat. Acad. Sci. USA*, **68**, 72
89. Williamson, R., Morrison, M., Lanyon, G., Eason, R. and Paul, J. (1971). *Biochemistry*, **10**, 3014
90. Lim, L. and Canellakis, E. S. (1970). *Nature* (*London*), **227**, 710
91. Burr, H. and Lingrel, J. B. (1971). *Nature New Biol.*, **233**, 41
92. Molloy, G. R., Sporn, M. B., Kelly, D. E. and Perry, R. P. (1972). *Biochem.*, **11**, 3256
93. Mathews, M. B., Osborn, M. and Lingrel, J. B. (1971). *Nature* (*London*), **233**, 206
94. Stavnezer, J. and Huang, R. C. C. (1971). *Nature* (*London*), **230**, 172
95. Sampson, J. and Borghetti, A. F. (1972). *Nature New Biol.*, **238**, 200
96. Heywood, S. M. (1970). *Nature* (*London*), **225**, 696
97. Heywood, S. M. (1970). *Proc. Nat. Acad. Sci. USA*, **67**, 1782
98. Cohen, B. B., (1971). *Biochim. Biophys. Acta*, **247**, 133
99. DeBellis, R. H., Gluck, N. and Marks, P. A. (1964). *J. Clin. Invest.*, **43**, 1329
100. Birnbaim, H., Pene, J. and Darnell, J. (1967). *Proc. Nat. Acad. Sci. USA*, **158**, 320
101. Arion, B. Y. A. and Georbrev, G. P. (1967). *Proc. Nat. Acad. Sci. USSR*, **172**, 716
102. Shaeren, R. W. and McCarthy, B. J. (1967). *Biochem.*, **6**, 283

103. Soeiro, R. and Darnell, J. E. (1970). *J. Cell. Biol.*, **44,** 467
104. Scherrer, K., Spohr, G., Granboulan, N., Morel, C., Grosclaude, J. and Dhezzi, C. (1970). *Cold Spring Harbor Symp. Quant. Biol.*, **35,** 539
105. Darnell, J. E., Philipson, L., Wall, R. and Adesnik, M. (1971). *Science*, **174,** 507
106. Benjamin, J. (1966). *J. Molec. Biol.*, **16,** 539
107. Darnell, J. E., Wall, R. and Tushinski, R. J. (1971). *Proc. Nat. Acad. Sci. USA*, **68,** 1321
108. Melli, M. and Pemberton, R. E. (1972). *Nature New Biol.*, **236,** 172
109. Penman, S., Scherrer, K., Becker, Y. and Darnell, J. E. (1963). *Proc. Nat. Acad. Sci. USA*, **49,** 654
110. Soeiro, R., Vaughan, M. H., Warner, J. R. and Darnell, J. E. (1968). *J. Cell Biol.*, **39,** 112
111. Morrison, M. R., Gorski, J. and Lingrel, J. E. (1972). *Biochem. Biophys. Res. Commun*, **49,** 775
112. Williamson, R., Drewienkiewicz, C. and Paul, J. (1973). *Nature New Biol.*, **241,** 66
113. Kruh, J., Ross, J., Dreyfus, J. C. and Schapira, G. (1961). *Biochim. Biophys. Acta*, **49,** 509
114. Bishop, J., Favelukes, G., Schweet, R. and Russell, E. (1961). *Nature* (*London*), **191,** 1365
115. Borsook, H (1958). *Conf. on Hemoglobin, Nat. Acad. of Sci.*, **557,** 111 (1958 Washington, D.C. *National Research Council*)
116. Fantoni, A., de la Chapelle, A. and Marks, P. A. (1968). *J. Molec. Biol.*, **33,** 79
117. Terada, M., Banks, J. and Marks, P. A. (1971). *J. Molec. Biol.*, **62,** 347
118. Djaldetti, M., Chui, D., Marks, P. A. and Rifkind, R. A. (1970). *J. Molec. Biol.*, **50,** 345
119. Singer, R. H. and Penman, S., (1972). *Nature* (*London*), **240,** 100
120. Marks, P. A., Burka, E. R., Conconi, F., Perl, W. and Rifkind, R. A. (1965). *Proc. Nat. Acad. Sci. USA*, **53,** 1437
121. Terada, M., Metafora, S., Banks, J., Dow, L. W., Bank, A. and Marks, P. A. (1973). *Biochem. Biophys. Res. Commun.*, **47,** 766
122. Sussman, M. (1970). *Nature* (*London*), **225,** 1245
123. Goldwasser, E. and Kung, C. K. H. (1972). *J. Biol. Chem.*, **247,** 5159
124. Goldwasser, E. (1966). *Current Topics in Developmental Biology*, 73 (A. Monroy and A. A. Moscona, editors) (New York: Academic Press)
125. Bateman, A. E. and Cole, R. J. (1971). *J. Embryol. Exp. Morphology*, **26,** 475
126. Chui, D., Djaldetti, M., Marks, P. A. and Rifkind, R. A. (1971). *J. Cell. Biol.*, **51,** 585
127. Krantz, S. B., and Jacobson, L. O. (1970). *Erythropoietin and Regulation of Erythropoiesis*, 118 (Chicago: University of Chicago Press)
128. McCulloch, E. A. (1970). *Regulation of Hematopoiesis*, Vol. 1, 133 (A. S. Gordon, editor) (New York: Appleton-Century-Crofts)
129. Stephenson, J. R. and Axelrod, A. A. (1971). *Blood*, **37,** 417
130. Borsook, H., Ratner, K. and Tattrie, B. (1969). *Nature* (*London*), **221,** 1261
131. Minio, F., Howe, C., Hsu, K. C. and Rifkind, R. A. (1972). *Nature New Biol.*, **237,** 187
132. Cantor, L. N., Morris, A. J., Marks, P. A. and Rifkind, R. A. (1972). *Proc. Nat. Acad. Sci. USA*, **69,** 1337
133. Gross, M. and Goldwasser, E. (1970). *J. Biol. Chem.*, **245,** 1632
134. Djaldetti, M., Preisler, H., Marks, P. A. and Rifkind, R. A. (1972). *J. Biol. Chem.*, **247,** 731
135. Nicol, A. G., Conkie, D., Lanyon, W. G., Drewienkiewicz, C. E., Williamson, R. and Paul, J. (1972). *Biochim. Biophys. Acta*, **277,** 342
136. Gross, M. and Goldwasser, E. (1972). *Biochim. Biophys. Acta*, **287,** 514
137. Paul, J. and Hunter, J. A. (1969). *J. Molec. Biol.*, **42,** 31
138. Maniatis, G. M., Rifkind, R. A., Bank, A. and Marks, P. A. (1973). *Proc. Nat. Acad. Sci., USA* (in press)
139. Gross, M. and Goldwasser, E. (1969). *Biochemistry*, **8,** 1795
140. Gross, M. and Goldwasser, E. (1971). *J. Biol. Chem.*, **246,** 2480
141. Gurdon, J. B. (1969). *Proc. XII Intern. Congr. Genetics, Nucleo-cytoplasmic Interactions During Cell Differentiation*, **3,** 191
142. Friend, C. (1957). *J. Exp. Med.*, **105,** 307

143. Metcalf, D., Furth, J. and Buffett, R. (1959). *Cancer Res.*, **19,** 52
144. Mirand, E. A. (1970). *Regulation of Hematopoiesis*, Vol. 1, 635 (A. S. Gordon, editor) (New York: Appleton-Century-Crofts)
145. Friend, C., Scher, W., Holland, J. G. and Sato, T. (1971). *Proc. Nat. Acad. Sci. USA*, **68,** 378
146. Sato, T., Friend, C. and de Harven, E. (1971). *Cancer Res.*, **31,** 1402
147. Sassa, S., Takaku, F., Nako, K., Ikawa, Y. and Sugano, H. (1968). *Proc. Soc. Exp. Biol. Med.*, **127,** 527
148. Ostertag, W., Melderis, H., Steinheider, G., Kluge, N. and Dube, S. (1972). *Nature New Biol.*, **239,** 231
149. Scher, W. J., Holland, G. and Friend, C. (1971). *Blood*, **37,** 428
150. Tambourin, P. and Wendling, F. (1971). *Nature New Biol.*, **234,** 230

6
Oestrogen-induced Differentiation of Target Tissues

A. R. MEANS and B.W. O'MALLEY
Baylor College of Medicine, Houston

6.1 INTRODUCTION

Growth and differentiation is a complex process that must require an integrated synthesis of DNA along with synthesis of a full complement of RNA and protein for each new cell type. Differentiation can refer either to morphological changes or to the appearance of new biochemical function. In certain animal models, tissue growth and differentiation can be induced with specific chemical effectors. Initiation of active metabolism and cell division in these tissues allows a precise temporal analysis of the biochemical events that result in the differentiation response.

Several steroid hormones acting upon their target tissues provide examples of hormone-mediated growth and differentiation[1-5,58,59]. Hormone-mediated biochemical differentiation refers to the initiation of a new cell capacity that normally appears during the maturation process. Thus, these responses are specific for embryonic or immature tissues. The effect of glucocorticoids on the differentiation of chick embryo retinal cells provides one example of hormone-dependent biochemical differentiation[1,3]. Action of the insect steroid ecdysone in the initiation of the maturation or moulting process of insect larvae provides another example[4]. Dihydrotestosterone is the hormone responsible for induction of differentiation of the male phenotype in normal embryos[58,59]. Likewise, the decidual cell reaction in the uterus in response to progesterone is manifest by a proliferation of cells similar to that occurring upon implantation of a blastocyst[5]. Thus, steroid hormones are capable of stimulating prematurely the appearance of new proteins and target-cell functions under well-defined conditions. These events are part of the normal adult response during maturation and presumably are under the direction of endogenous hormones.

The manner by which hormones regulate growth and differentiation of target tissues has been the topic of numerous investigations. One model system which has been particularly useful in this regard is the oviduct of the immature chick[6-8]. Administration of oestrogen into these animals gives a rapid and pronounced cytodifferentiation[6,9,10]. Moreover, the new cell types which appear synthesise large quantities of specific proteins which are easily quantified by chemical and immunochemical techniques[6,11]. Thus, a single steroid hormone, oestrogen, regulates both biochemical and morphological differentiation of the oviduct in a highly ordered fashion. The remainder of this chapter will deal with oestrogen regulation of oviduct growth and differentiation and will contain a review of the complex series of molecular events which accompany this process.

6.2 OESTROGEN RECEPTORS

The oviduct contains an oestrogen-binding macromolecule with similar properties to those of other target tissues such as the uterus[12]. The binder is apparently protein and is tissue-specific and heat labile. Oestradiol-17β and diethylstilboestrol are bound with high affinity (K_d *ca.* 10^{-10} l mol^{-1}). Moreover, this affinity is higher than for other naturally-occurring oestrogenic steroids such as oestrone and oestriol. Finally, this binding protein has little or no affinity for androgens, progestins or glucocorticoids. Thus, it appears that oestrogen initiates its action in oviduct by first combining with a cytoplasmic receptor protein as has been shown to be the case for nearly all of the steroid hormones[13,14]. It is probable that this complex is transferred subsequently into the nucleus where it attaches, in some specific manner, to the nuclear chromatin[15-17]. This interaction is presumably required in order for oestrogen to initiate, in a precise temporal sequence, the well-known effects on morphological and biochemical differentiation.

6.3 MORPHOLOGICAL CORRELATES OF OESTROGEN-INDUCED DIFFERENTIATION

6.3.1 Histological changes

The oviduct of the chick may be compared to the mammalian uterus. The major segment is called the 'magnum', and it is in this segment that the egg-white substance is added to the yolk. Consequently, this is the area of the oviduct where the major egg-white proteins, such as ovalbumin, are synthesised. The magnum of the oviduct is under direct control by oestrogen and in response to this steroid undergoes marked morphological cytodifferentiation. Eventually, three distinct types of epithelial cells differentiate from the primitive cells of the oviduct mucosa. Two of these cell types synthesise cell-specific proteins which can be measured readily by biochemical and immunochemical techniques and therefore can be used as markers for the differentiation process. Thus, tubular gland cells synthesise the major egg-white proteins such as ovalbumin, conalbumin, ovomucoid and lysozyme[6,18], whereas the goblet cells synthesise avidin in response to administration of another steroid hormone, progesterone[6,9,10]. The third type of epithelial cell is ciliated columnar cells and appears to be concerned with motility.

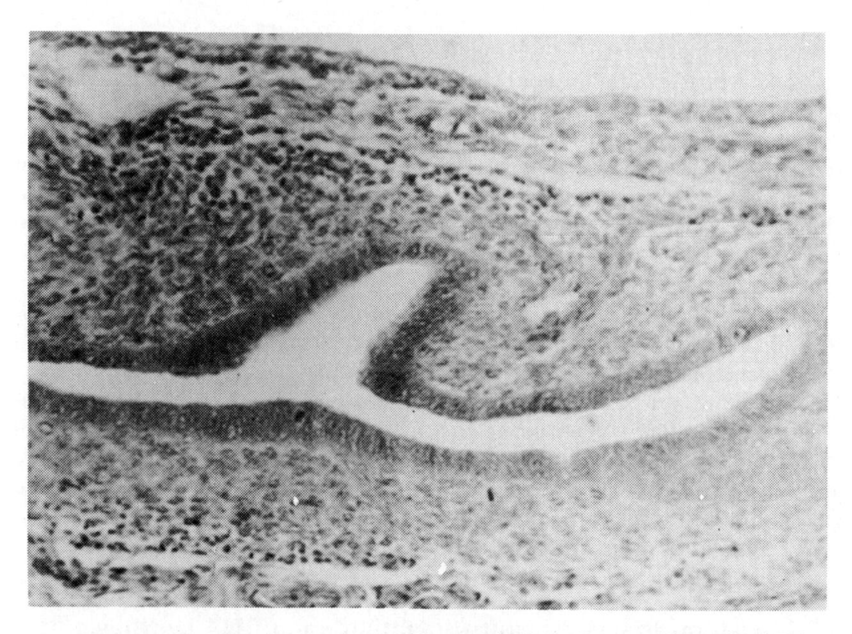

Figure 6.1(a) Light micrograph of oviduct epithelium of an undifferentiated 7-day-old chick. Columnar epithelial cells with prominent nucleoli are seated upon dense stroma. Stromal nuclei are rounded and contain nucleolar clumps ($\times 250$). (Reduced $\frac{2}{3}$rds on reproduction)

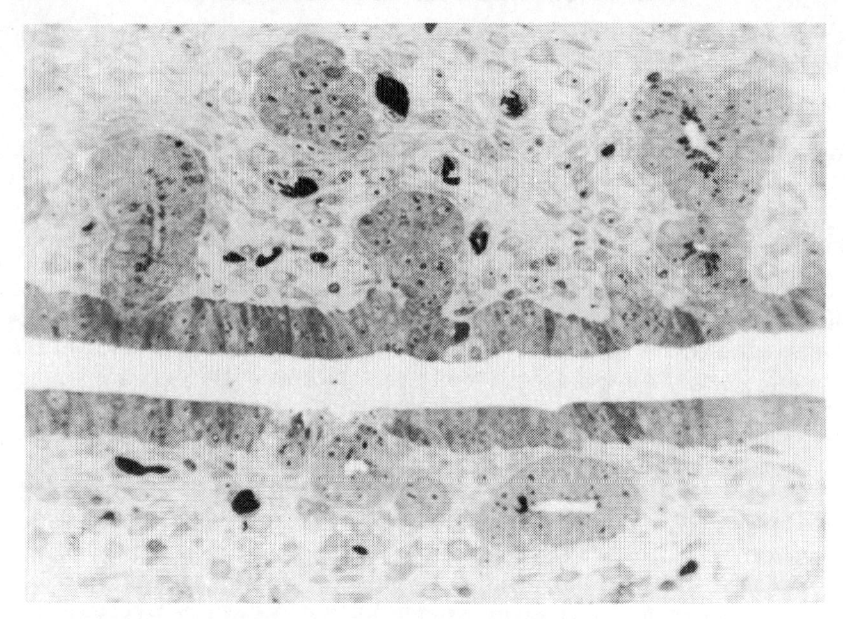

Figure 6.1(b) Early tubular gland formation from surface epithelium of chick oviduct. Immature chicks (7 days old) were stimulated for 4 days with oestrogen. Stromal oedema has advanced to involve subepithelial region (×250). (Reduced ⅔rds on reproduction)

In the immature chick the oviduct mucosa consists of a thin layer of pseudostratified columnar epithelium (Figure 6.1a) which rests on a compact stroma of polygonal cells[6]. Only sparse endoplasmic reticulum exists in the primitive epithelial cells and ribosomes, although abundant, are dispersed in the hyaloplasm. When oestrogen is administered to these chicks, some of the earliest histological changes involve stromal oedema, aggregation of ribosomes into clusters and progressive development of the rough endoplasmic reticulum[6]. These changes are first demonstrable during the first day of oestrogen treatment.

Continued oestrogen treatment results within 36 h, in the formation of bud-like invaginations of the epithelium. These cell clusters invade the stroma and subsequently become the tubular gland cells which secrete ovalbumin (Figure 6.1b). Ciliated cells begin to appear in the surface layer of the original epithelium by six days of hormone treatment. Finally, goblet cells which will eventually synthesise avidin in response to progesterone begin to differentiate after the ninth day of oestrogen. Thus, the differentiation of the primitive epithelium into three distinct cell types, each with a distinctly different function, requires approximately 10 days of continued stimulation by oestrogen[6].

6.3.2 Cell proliferation: mitotic activity and DNA synthesis

Cell proliferation plays an important role during the cytodifferentiation of the oviduct. Administration of oestrogen gives a marked increase in cell number, concomitant with increases in the synthesis[19] and content of DNA[6]. Changes in DNA synthesis have been observed within 24 h after hormone administration and DNA content increases for the first 10 days of oestrogen treatment[6,19,20]. In the oviduct of the immature chick mitoses are very infrequent as evidenced by a low rate of DNA synthesis and a mean mitotic index (MI) of 0.43[20]. These low values reflect the slow natural growth of the oviduct that normally continues until sexual maturation at 5–6 months. Inhibitors of DNA synthesis such as 5-fluorodeoxyuridine and hydroxyurea have been used to estimate the duration of G_2 in the unstimulated oviduct[20]. The only cells that enter mitosis after treatment with these inhibitors should be G_2 cells since the antimetabolites block cells in the S or synthetic phase of the cell cycle. Both inhibitors produce a drop in MI and the time interval required to produce a 50% decrease in MI provides an estimate of the mean duration of

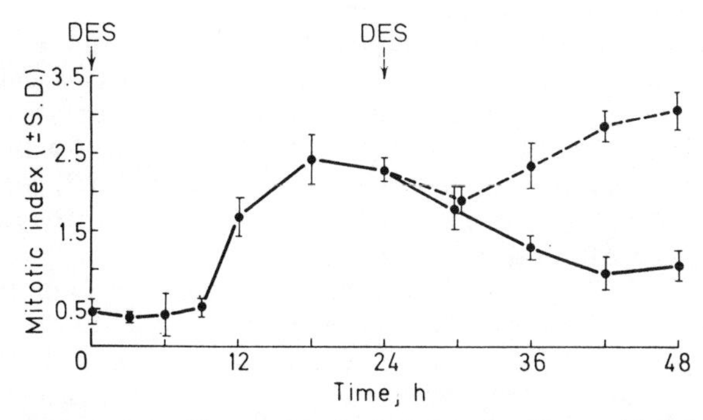

Figure 6.2 Changes in mitotic index of oviduct epithelial cells following treatment of immature chicks with oestrogen. Hormone was administered at 0 h (●——●) or at 0 and 24 h (●- - -●). (From Socher and O'Malley[20] by courtesy of Academic Press.)

G_2 plus mitosis over 2. In the unstimulated chick oviduct this value is 2.25 h. Therefore, if it is assumed that the duration of mitosis is 1 h, G_2 is equivalent to 1.75 h. Appearance of labelled mitoses after the injection of tritiated thymidine has also revealed the duration of G_2 to be between 1.5 and 2 h. Studies with tritiated thymidine have also revealed the mean labelling index of the surface epithelium to be 3.16, indicating that only a small fraction of cells in the immature oviduct are synthesising DNA. These data have been used to estimate the duration of the S-phase as 7.3 h.

A single injection of oestrogen stimulates cells to divide in the unstimulated oviduct (Figure 6.2). Within 9 to 12 h there is a fourfold increase in the frequency of cells in mitosis[20]. The mitotic index reaches a peak at 18 h and begins to fall by 24 h. If a second injection of oestrogen is given on the descending portion of the MI curve, an elevation of mitotic frequency is again observed. Analysis of data derived from these experiments led to the suggestion that oestrogen stimulates a single population of cells to divide in a partially synchronous manner. Maintenance of this synchrony requires repeated injections of oestrogen and this steroid hormone apparently stimulates

G_1 cells to enter the DNA synthetic phase. Indeed oestrogen treatment gives a similar stimulation of the rate of DNA synthesis in oviduct which leads to a dramatic increase in the total content of DNA. All of these data are consistent with the induction of cell proliferation in the oviduct following oestrogen administration.

6.4 ALTERATION IN NUCLEAR CHROMATIN

The newly-differentiated tubular gland cells synthesise specific protein such as ovalbumin and lysozyme in response to oestrogen administration. Based upon the hypothesis that alterations in gene transcription during differentiation reflect, in part, changes in the tissue specific pattern of gene restriction, it seems logical to investigate the properties of oviduct chromatin during oestrogen-mediated tissue differentiation.

As an initial approach, the qualitative chemical composition of chromatin was studied[21]. Chromatin was isolated from purified nuclei of chick oviduct at various stages of oestrogen-mediated development. Histones were removed from the isolated chromatin with high salt-urea and four fractions of non-histone chromosomal proteins termed AP_1, AP_2, AP_3 and AP_4, were removed from the dehistonised chromatin by differential solubility extractions. This fractionation procedure, although crude, seems to be effective in that it can be used to obtain large bulk preparations of each fraction. It has been demonstrated to be reproducible with regard to the quantitative analysis of protein. No changes in the species of histones could be found at any stage of oviduct development[21]. Moreover, the quantitative level of each of the major histone fractions remained constant. On the other hand, changes were observed in the quantitative levels of non-histone chromosomal proteins (NHP) with respect to the concentration of DNA in the chromatin (Figure 6.3). During the first 4 days of oestrogen treatment NHP levels increased. This was followed by a gradual decrease in the milligrams NHP per milligram of DNA, until by 14 days of oestrogen treatment the level was below that of the unstimulated chick. Concentrations of chromatin-associated RNA followed a pattern similar to that of the acidic proteins, with peak amounts again being present in chromatin from 4-day stimulated chicks. Analysis of the four subfractions of NHP revealed that the major changes which result in response to oestrogen occurred in AP_2[22]. The qualitative changes in AP_2 were further studied by submitting this fraction to amino acid composition studies and electrophoresis on SDS polyacrylamide gels[22]. Changes were noted in the amino acid composition of this subfraction of NHP with the most striking change being in the number of cysteine residues. Five cysteine molecules per 1000 amino acid residues were noted in AP_2 from unstimulated chicks. By day 19 this value had increased to 45. Less dramatic changes were also demonstrable in several other amino acid species. Examination of the NHP subfractions by gel electrophoresis revealed fraction AP_2 to be the most heterogeneous fraction. Again, although each fraction undergoes some alteration in banding patterns during oviduct development, fraction AP_2 shows the greatest changes. Thus, the electrophoretic patterns and the amino acid composition data support the suggestion that qualitative changes occur in the non-histone chromosomal protein fractions during oviduct development.

The structural relationships of the role of NHP in the tissue-specific restriction of DNA was further investigated by examining the antigenic properties of the NHP–DNA complexes during development by the immunochemical method of micro-complement fixation[23]. Antibodies were prepared against

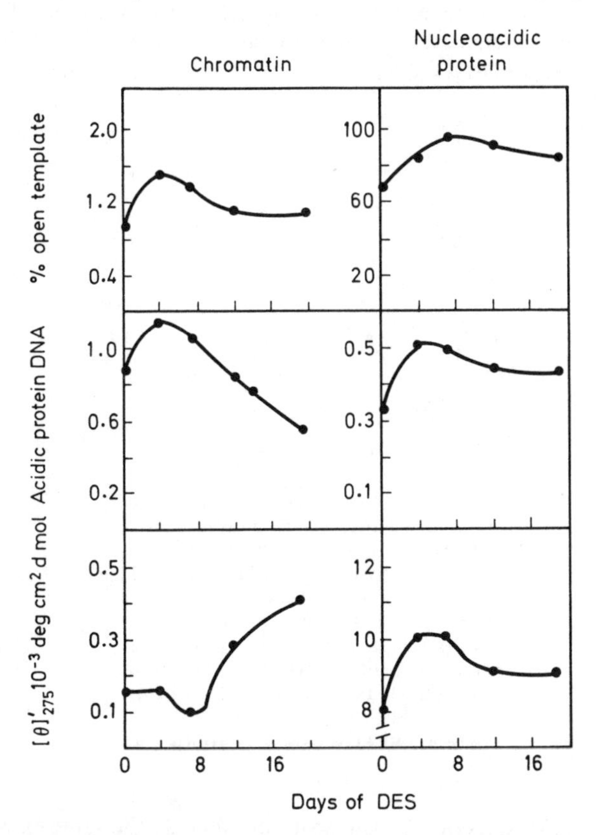

Figure 6.3 Alterations in chromatin during oestrogen-induced differentiation of the oviduct. Percentage template available for transcription, quantitative levels of non-histone chromosomal proteins and circular dichroic spectra of chick oviduct chromatin and non-histone protein are plotted as a function of time following oestrogen administration. (From Spelsberg, Mitchell, Chytil, Wilson and O'Malley[22], by courtesy of Elsevier.)

NHP–DNA complexes isolated from oviducts of chicks treated with oestrogen for 15 days. To assure specificity, nucleo-acidic protein was isolated from the chromatins of chick heart, liver and spleen and tested for antigenicity, using the same antiserum prepared against oviduct NHP, as described above. The NHP from other organs showed a very limited affinity for the antibody prepared against oviduct NHP. This limited affinity suggests that a large number of antigenic sites present in oviduct preparations were tissue-specific and indicated a considerable dissimilarity in the antigenic sites of NHP–DNA complexes present in the chromatins of different organs. Moreover, preparations of NHP from undifferentiated oviducts showed very little antigenicity using antiserum prepared against NHP from oestrogen stimulated and differentiated oviduct. On the other hand, fixation of complement by NHP which were isolated from oviducts of chicks injected with oestrogen for various periods of time showed a gradual appearance of antigenicity with the duration of oestrogen treatment (Figure 6.4) until NHP from oviducts of 12-day

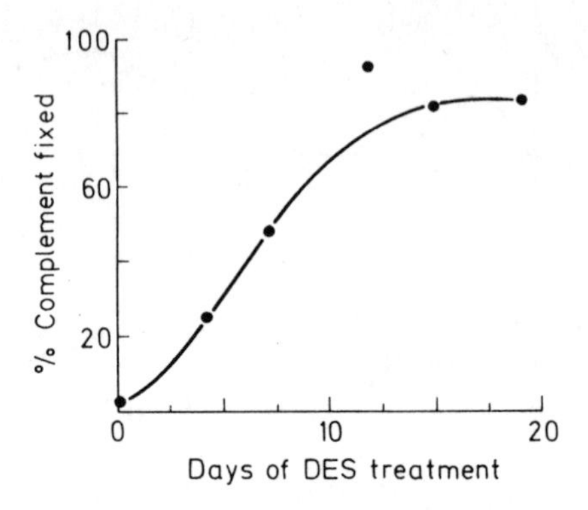

Figure 6.4 Changes in antigenicity of non-histone chromosomal protein-DNA complexes during hormone-mediated tissue differentiation. NHP (10 mg DNA) from each stage of oviduct development were added to complement fixation assays. The antisera in these assays was prepared against NHP from oviducts of chicks treated with oestrogen for 15 days. (From Spelsberg, Mitchell, Chytil, Wilson and O'Malley[22], by courtesy of Elsevier.)

stimulated chicks fixed complement to approximately the same extent as those from 15- or 19-day treated animals[22]. Thus, development of antigenicity of the NHP showed a developmental change which coincided with the oestrogen-induced morphologic development of this organ.

These data show that the NHP–DNA complexes are good immunogens which give complement-fixing antibody. These antibodies react strongly with preparations from homologous organ, whereas the affinity for preparations of NHP from heterologous sources is very low. This indicates that the arrangement of the antigenic sites inherent to the NHP in chromatin is organ-specific. Furthermore, during development of chick oviduct the antigenic sites for NHP undergo marked alterations which probably involve changes in the species of NHP, in addition to possible structural alterations of already-existing proteins. These data add credence to the hypothesis that NHP may be important in the tissue-specific restriction of the DNA.

Values for template efficiency of oviduct chromatin during differentiation show similar changes to those reported above in the quantitative amounts of NHP[22]. Increases in template efficiency are noted for the first four days, followed by a gradual decline until 19 days of oestrogen treatment (Figure 6.3). Changes in the template efficiency of chromatin were also accompanied by changes in the circular dichroism (c.d.) spectrum obtained under highly standardised conditions[22]. The magnitude of the ellipticity at 275 nm for chromatin is much less than that for pure DNA. This decreased ellipticity has been suggested to be caused by an altered geometry of the DNA when complexed with proteins[24]. As differentiation of the oviduct progresses in response to oestrogen, the chromatin DNA displays a gradual increase in the magnitude of ellipticity (Figure 6.3). Other studies suggest that this increase may represent an opening of the DNA, which might be expected to be reflected in the template efficiency of the chromatin[24]. This does appear to be the case in the chick oviduct system in response to oestrogenic steroids. Analysis of c.d. spectra of chromatin at 210 nm also suggests that significant conformational changes occur during oviduct development. Spectral data obtained in this far u.v. region largely represent conformational changes of chromatin associated proteins[25]. Thus, the c.d. analysis of chromatin during oestrogen-mediated differentiation leads to the concept that a major alteration of the steric conformation of target cell chromatin occurs during this hormone-induced process[22]. Coupled with the data showing major quantitative and qualitative changes in the NHP of chromatin, these studies provide strong evidence that differentiation represents progressive alterations in chromatin which may result in changes in cell structure and function.

6.5 CHANGES IN RNA SYNTHESIS AND RNA POLYMERASE

During oestrogen-mediated differentiation of chick oviduct all classes of RNA are eventually increased. This is true both for the rate of synthesis of each species and for accumulation in oviduct tissue. Within 1 day after stimulation by oestrogen of the undifferentiated oviduct, changes occur in the pattern of ribosomal precursor RNA as shown by polyacrylamide gel electrophoresis[26]. By 2–4 days, ribosomal 28S, 18S and 5S RNA species are increased[26,27]; marked effects on the synthesis of transfer RNA are also seen by this time[28]. These increased rates of synthesis of ribosome associated RNA species are accompanied for the first 7 days after oestrogen treatment by a continuous increase in the cytoplasmic accumulation of newly-formed ribosomes[27,29,30]. Oestrogen has also been shown to increase the activity of RNA polymerase in isolated nuclei during hormone-induced oviduct differentiation. In this tissue there seems to occur a concomitant increase in the activity of polymerase I and polymerase II[6,31]. Recent studies by Cox *et al.*[32] have reported stimulation of both polymerase I and II activities by 2 h after administration of oestrogen (Figure 6.5). Continued oestrogen administration also apparently increases the number of polymerase molecules extractable from oviduct nuclei. It seems likely, therefore, that oestrogen-induced increase in polymerase must occur both by enhancing the activity of existing enzyme molecules as well as by increasing the number of initiation sites on the DNA due to an increased number of active polymerase molecules.

The available data suggest that oestrogen significantly affects gene transcription during hormonal stimulation of the undifferentiated chick oviduct. Nearest-neighbour frequency analysis of the RNA products synthesised by nuclei or chromatin *in vitro* and competition-hybridisation experiments using repeating sequences in total-cell DNA have indicated a qualitative change in the nuclear RNA species synthesised during oestrogen-mediated oviduct differentiation[6,33,34]. Moreover, the ability of oviduct nuclear RNA to direct the synthesis of [^{14}C]phenylalanine into polyphenylalanine using a translation

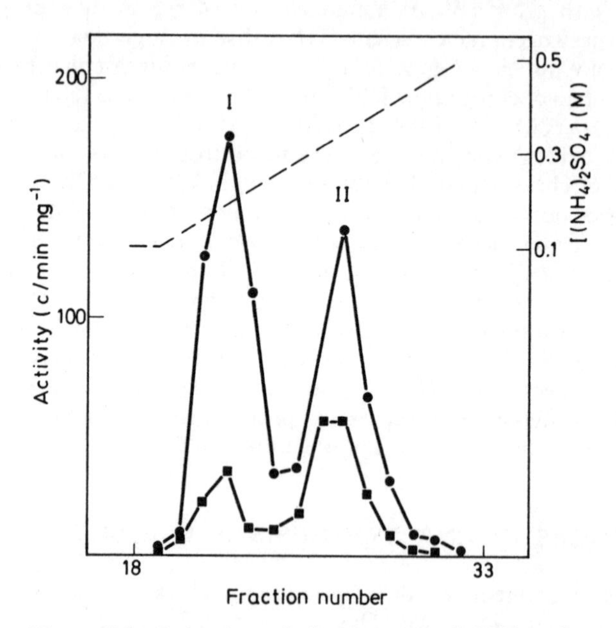

Figure 6.5 Estimation of the amounts of RNA polymerase I (nucleolar) and II (nucleoplasmic) extractable from oviduct nuclei. Activity profiles obtained from unstimulated (■——■) or 24 h oestrogen treated chicks (●——●) are expressed as c/min mg^{-1} DNA. (From Cox, Haines and Carey[32], by courtesy of Springer.)

system derived from *E. coli* is also stimulated by oestrogen[27]. In all of these studies, maximal increases were noted between 3 and 6 days of oestrogen treatment. These data offer strong evidence that oestrogen must promote the synthesis of new species of RNA which precede the appearance of cell-specific proteins in the cytoplasm.

One way by which to obtain evidence to support the above hypothesis is the use of nucleic acid hybridisation. Our initial studies on DNA–RNA hybridisation were performed under the standard conditions of low nucleic acid concentration and short incubation times[33,34]. These conditions have been shown to involve mainly the rapidly-reassociating highly-repetitive sequences to the exclusion of the more slowly-reassociating sequences which have been defined operationally as unique sequences[35]. Furthermore, the presence of closely-related sequences in eukaryotic DNA may have resulted in a lack of gene locus specificity such as the mismatching of base sequences in these early experiments[36]. We have, therefore, utilised the technique of RNA excess hybridisation to unique species of eukaryotic deoxyribonucleic acid[37-39] in an attempt to answer the following questions: (a) are unique sequences transcribed in the chick oviduct and, if so, to what extent? (b) how much of this transcribed nuclear RNA is processed and translated as mRNA in the cytoplasm? and (c) what effect does oestrogen have on the transcription and processing of oviduct RNA synthesised from unique DNA during hormone-mediated growth and differentiation of the oviduct?

Initially, we examined the complexities of the chick oviduct genome and

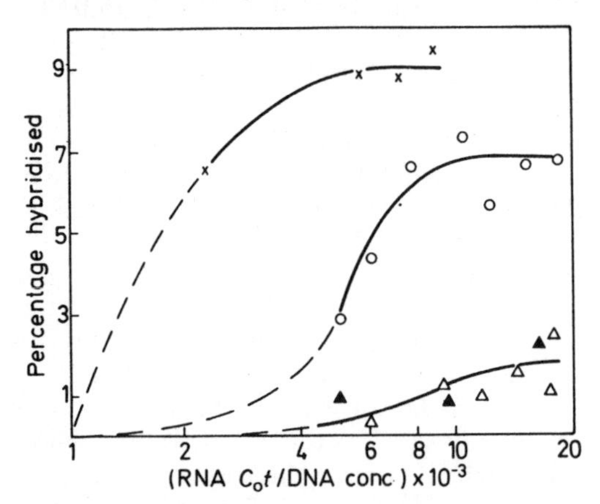

Figure 6.6 Saturation of chick [^{3}H]unique DNA with nuclear or cytoplasmic RNA extracted from oviducts of unstimulated or 18-day oestrogen-treated chicks. Nuclear RNA from unstimulated (○) or stimulated (×). Cytoplasmic RNA from unstimulated (▲) or stimulated (△). Hybridisation reactions were performed in 0.4 M phosphate buffer at 70 °C. RNA concentrations were greater than 15 mg ml^{-1}. (From Liarakos, Rosen and O'Malley[41], by courtesy of the American Chemical Society.)

the effect of oestrogen on the kinetics of chick DNA renaturation[40]. The methods utilised in these DNA renaturation experiments involved the use of purified native DNA, sheared to *ca.* 400 nucleotide lengths, and renaturation under reaction conditions which maximise specific base pair interactions and minimise mismatching. These studies demonstrate that the chick oviduct genome is composed of *ca.* 30% repetitive sequences and 70% unique sequences. Although the oviduct genome does not contain the highly repeated sequences found in mouse satellite DNA, 5% of the oviduct genome may represent a family of related sequences of 1 000 nucleotide pairs reiterated some 70 000 times. Moreover, the identical renaturation kinetics of unstimulated, 3 or 20 day oestrogen-treated and adult hen oviduct DNA suggest that major gene duplication, deletion or alteration probably does not account for the marked oestrogen-induced growth and differentiation. This type of phenomenon (specifically, ribosomal gene amplification) has been reported to occur during amphibian oogenesis. Thus, it appears that the new proteins required for growth must arise via differential gene transcription.

In order to investigate the various RNA populations present during different stages of growth and differentiation, the chick oviduct genome was subdivided on the basis of nucleotide sequence homology. Chick DNA was allowed to renature to a C_0t value of 420. The slowly-renaturing unique sequences were separated on hydroxylapatite columns and then recycled to assure the absence of contaminating repeated sequences. The effect of oestrogen on the extent of unique-sequence DNA transcription was then studied using both nuclear and cytoplasmic RNA extracted from immature and fully-differentiated chick oviducts[40]. Saturation hybridisation experiments were conducted at 70% in 0.4 M phosphate buffer, using vast excess of RNA. There appears to be no significant difference in the amount of unique DNA hybridised to cytoplasmic RNA from the immature or oestrogen-differentiated oviduct (Figure 6.6). However, unique DNA does appear to react to a greater extent with oviduct nuclear RNA from oestrogen stimulated chicks than with oviduct nuclear RNA from unstimulated animals. This difference in apparent saturation values with nuclear RNA implies an increase in the amount of unique DNA transcribed after oestrogen treatment. Our data also suggest that only 20–25% of the total unique DNA sequences transcribed in the nucleus may be represented in the cytoplasmic RNA sequences, and that this value is not appreciably altered by oestrogen treatment.

The saturation values obtained for chick unique DNA with nuclear RNA from the immature animal was 7%, whereas that from chicks treated for 18 days with oestrogen was 9%[41]. These values are at best minimum estimates of unique DNA transcription. True saturation values are most difficult to determine, since RNA preparations are heterogeneous and therefore contain a distribution of base sequences at different frequencies. Whereas RNA species present at high to moderate concentrations will react with DNA over the course of the hybridisation reaction, those species present at low frequencies may not react prior to breakdown, making it difficult to obtain a true saturation value. Therefore, the saturation values should not be interpreted as an absolute measure of unique DNA transcription. On the other hand, the difference in saturation values for nuclear RNAs from unstimulated and oestrogen-treated oviduct are reproducible. Similar caution must be expressed in the interpretation of hybridisation-saturation experiments obtained with cytoplasmic RNA. Observed reaction values may be influenced by nuclear RNA contamination of the cytoplasmic RNA preparations. This could result from leakage of RNA from nuclei or from nuclei breakage during the tissue fractionation procedure. It is clear, however, that a specific increase in the extent of unique sequence DNA gene transcription does occur in chick oviduct during oestrogen-stimulated differentiation.

6.6 HORMONAL REGULATION OF mRNA

Oestrogen brings about an increase in the synthesis of oviduct ribosomes within 1 day of administration to the immature chick[27]. For at least 7 days of hormone treatment the oviduct content of ribosomes continues to increase, but by 10 days has begun to decline. Concomitant with the increased synthesis of ribosomes are oestrogen-induced changes in the distribution of ribosomes and polysomes, analysed by sucrose-gradient centrifugation[27,30]. A large proportion of particles exist as monomers in the unstimulated oviduct. On the other hand, after 4 days of treatment with oestrogen more than 90% of the cytoplasmic RNA particles exist as aggregates of two or more ribosomes.

Ribonucleoprotein preparations were tested for their ability to synthesise protein in the cell-free system[27,29]. Oestrogen administration results in a doubling of incorporation activity within 24 h. By 4 days of hormone treatment polysome protein synthesis assayed *in vitro* reaches a maximum before beginning to decline at 7 days of oestrogen treatment. The marked stimulation of incorporation activity at 4 days is in keeping with the striking increases in ribosome synthesis and conversion of monomers to polysomes noted at the same time. Again, the decline in protein synthesis *in vitro* occurs in coordination with a decreased synthesis of ribosomes and a further shift in the polysome pattern.

Since major changes occur in the population of soluble proteins in the oviduct after oestrogen administration, and since all proteins are made on cell polysomes, experiments were undertaken to determine whether qualitatively different peptides were synthesised by isolated polyribosomes in the cell-free system. This was accomplished by analysing the products translated *in vitro* by polyacrylamide gel electrophoresis[27]. Polyribosomes were incubated in the cell-free protein synthesising system and following incubation these preparations were pooled and carried through the remaining procedures together. Marked differences were apparent between 0 and 4 days of oestrogen when corrected counts per minute was plotted. The ratio of ^{3}H to ^{14}C at various points on the gel varied considerably, thus demonstrating that striking changes occur in the population of peptides synthesised *in vitro* before and after oestrogen treatment. Taking appropriate controls into account, we attributed these changes in the peptides synthesised *in vitro* to reflect hormone-induced changes in the new population of target tissue mRNA.

Only *ca.* 25–35% of the acid-insoluble radioactivity was released from the ribosomes following incubation in the cell-free system[29]. In order to demonstrate that the soluble radioactivity was present as completed tissue proteins, it was necessary to determine the nature of these peptides. Ovalbumin comprises *ca.* 60% of the total oviduct protein in chicks treated for 15 days with oestrogen. A specific antibody for ovalbumin was prepared and characterised in our laboratories. This preparation was reacted with dialysed supernatant fluid obtained from the cell-free system. A time dependent increase in antibody precipitable radioactivity accounted for 25% of the total acid insoluble counts present in the ribosome-free supernatant fluid[29]. On the other hand, no radioactive material precipitated with anti-ovalbumin if the cell-free reaction included polysomes from unstimulated oviducts. These data suggest that it is possible to demonstrate the synthesis of an oviduct-specific protein in a polysomal cell-free system. Although immunological competence cannot be taken as conclusive evidence for identity, it is strongly suggestive. Furthermore, the fact that the polysomes from unstimulated oviduct failed to synthesise antibody-reacting material *in vitro* strengthens the argument since the oviduct of the unstimulated chick does not produce ovalbumin *in vivo*. The implication of these studies is that mRNA for ovalbumin is associated with ribosomes only during the period in which oestrogen is being administered to the animals.

Further evidence for this assumption was obtained by incubation of oestrogen-stimulated chick or hen oviduct minces in the presence of tritiated cytidine and adenosine. This results in a labelling pattern of polyribosomes in the region of 6–15 ribosome aggregates[42]. Ovalbumin, which has a molecular weight of 45 000, would be translated on the average by a ribosome to messenger ratio of *ca.* 13:1. Since ovalbumin appears to be the major protein synthesised under these conditions, we had reason to believe that the label would serve as a marker for its mRNA. Extraction of these polysomes by detergent treatment, followed by sucrose gradient centrifugation, resulted in a typical RNA profile with the bulk of radioactivity found as a broad peak with a sedimentation value of 16–18 S, which corresponds closely to the expected sedimentation value for the ovalbumin mRNA[42].

Direct assessment of mRNA can only be obtained by demonstrating the ability of an RNA fraction to support the *de novo* synthesis of a specific protein in an *in vitro* translation system. We chose to use a modification of the reticulocyte lysate system as first described by Stavnezer and Huang[43], for our studies involving *in vitro* translation of oviduct mRNA. Prior to addition of the chick RNA, the radio-labelled [^{14}C]valine is incorporated almost entirely into globin chains. Addition of chick RNA results in the appearance of a [^{14}C]-labelled protein peak which is coincident with authentic ovalbumin upon analysis by polyacrylamide gel electrophoresis. Verification of the product of the heterologous translation system as ovalbumin has been confirmed by four separate methods. These methods have included precipitation with a specific antibody raised against purified ovalbumin, the coelectrophoresis of the solubilised antigen on SDS-acrylamide gels and cochromatography on carboxymethylcellulose[42]. In addition, Rhoads, McKnight and Schimke[44] have shown that peptide maps of ovalbumin synthesised *in vitro* in response to total hen oviduct RNA, were similar to those prepared from authentic ovalbumin.

Synthesis of [^{14}C] ovalbumin shows linear dependence on exogenous oviduct mRNA[42,45]. Oviduct ribosomal RNA or transfer RNA, or total RNA from liver or spleen demonstrated no capacity to direct *in vitro* synthesis of [^{14}C]-labelled ovalbumin. Hen oviduct contained large quantities of ovalbumin mRNA, whereas RNA from immature chicks contained no mRNA activity. However, ovalbumin mRNA could be induced in immature chicks by treatment with oestrogen. Withdrawal of the hormone resulted in a disappearance of ovalbumin mRNA, but readministration of oestrogen once again induced marked accumulation of ovalbumin mRNA[42,45].

When ovalbumin and ovalbumin mRNA are measured in the same tissue samples (Figure 6.7), a striking correlation over a period of 0–17 days can be demonstrated between ovalbumin accumulation in the stimulated oviduct, and its mRNA activity[46]. This relationship requires clarification since during this same time of differentiation and growth there are other dramatic changes occurring, particularly in the cellular content of nucleic acid. In order to better study the control of ovalbumin synthesis, we used oviduct minces from chicks which had been withdrawn from oestrogen for 16 days and then killed at various times following readministration of a single dose of this steroid. Rate of ovalbumin synthesis was assayed by incubating 2 h in the presence of ^{14}[C]lysine[47]. Synthesis of ovalbumin was again quantified using the specific-antibody procedure. The rate of ovalbumin synthesis was found to be time dependent, peaking at 18 h after steroid induction, at which time the rate of synthesis begins to decline. An approximate half-life for the ovalbumin mRNA under these hormonal conditions can be calculated from the

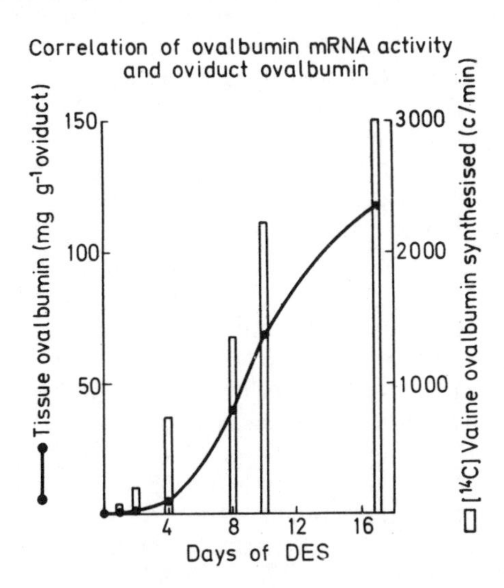

Figure 6.7 Effects of oestrogen on ovalbumin mRNA activity and tissue concentrations of ovalbumin. Immature chicks received daily injections of oestrogen. Tissue levels of ovalbumin were determined immunochemically. Messenger RNA was partially purified on Millipore filters and assayed for its ability to promote the synthesis of ovalbumin in a heterologous protein synthesis system. (From Comstock, Rosenfeld, O'Malley and Means[46], by courtesy of the National Academy of Sciences.)

decline in the rate of synthesis, to be 8–10 h. Measurement of the rate of total protein synthesis in oviduct after oestrogen demonstrates that peak synthesis occurs some 6 h before the peak synthesis of ovalbumin. Ovalbumin mRNA was then extracted from oviduct and quantified during this same period, following a single injection of oestrogen. A remarkable parallelism exists between the rate of ovalbumin synthesis and the available mRNA[47]. Prior to injection of oestrogen at zero time very little ovalbumin mRNA was detected. Maximum induction occurred at 18 h and mRNA content returned to barely detectable levels by 72 h. From the decline in the activity of translatable RNA it can be calculated that again the half-life of ovalbumin mRNA appears to be 8–10 h. These studies indicate that oestrogen acts at the level of gene transcription leading to the accumulation of a specific mRNA during differentiation of the oviduct. The appearance of the message seems to be a rate-limiting factor in determining the rate and extent of synthesis of the species specific protein, ovalbumin. Additional experiments in our laboratory have also revealed oestrogen-induced increases in the quantity of mRNA coding for another specific protein, lysozyme.

In order to continue our studies on the kinetics of synthesis and properties of the ovalbumin mRNA, a more highly-purified preparation was required. Brawerman and co-workers have demonstrated that many mRNA molecules contain a sequence of polyadenylate residues at their 3′ terminal end[48]. This characteristic allows these molecules to be adsorbed selectively on nitrocellulose filters[49]. We were able to demonstrate that ovalbumin mRNA in fact had a sequence of polyA[42] and were then able to utilise this fact to effect a partial purification by the Millipore technique[50]. This procedure results in a simple one-step 50-fold purification of ovalbumin mRNA activity. When this partially purified mRNA fraction was centrifuged on a 5–20% sucrose gradient, the bulk of the mRNA activity was localised at the 16–18S and 33–35S fractions of the gradient. Since the mRNA molecule for ovalbumin would be expected to have a sedimentation coefficient of *ca.* 16–18S, the activity associated with the 16–18S RNA was expected. More interesting, however, was the ovalbumin mRNA activity associated with the 33–35S fraction of the sucrose gradient. This peak of activity was not present when mRNA was isolated from polyribosomes. Moreover, when run in dimethyl sulphoxide gradients the activity remains at 33–35S. These results suggest that this material is not merely an aggregate of the monomer 16–18S ovalbumin mRNA. It is possible that this fraction represents the nuclear precursor molecule to the cytoplasmic ovalbumin mRNA. If this proves to be the case, then such studies could be utilised to gain information on the nuclear synthesis, processing and transport of a specific mRNA molecule during hormone-mediated tissue differentiation.

Further purification of the ovalbumin mRNA was achieved by centrifuging filter-bound RNA on 5–20% sucrose gradients and collecting the 16–18S peak which contained significant activity of ovalbumin mRNA when tested in the heterologous translation system. This 16–18S fraction was again placed on Millipore filters. After elution from the filters and reprecipitation from ethanol, this RNA fraction was tested in the reticulocyte translation system, and shown to contain a high concentration of ovalbumin mRNA activity; in fact, this RNA fraction was enriched *ca.* 100-fold for ovalbumin mRNA, compared to total nucleic acid. Thus, it could be calculated that this RNA preparation should now be *ca.* 90% pure ovalbumin mRNA. Recently Palacios *et al.*[51] have also reported purification of ovalbumin mRNA. At any rate, the purity was considered sufficient to begin studies designed to show that this steroid hormone-induced mRNA is transcribed from single-copy DNA. First, nascent ovalbumin chains on oviduct polysomes were adsorbed to a solid matrix containing antiovalbumin. RNA was then extracted and placed on Millipore filters. This procedure also gave a preparation of RNA *ca.* 90% pure ovalbumin mRNA.

Partially purified ovalbumin mRNA was incubated in an *in vitro* reaction system with RNA-directed DNA polymerase (RT), isolated from avian myeloblastosis virus (AMV). The reaction system contained mRNA, tritium labelled deoxynucleotides and actinomycin D. Production of complementary DNA copy to ovalbumin mRNA ($cDNA_{ov}$) was shown to be dependent on the addition of oligo dT primer and was inhibited completely by the addition of ribonuclease. The tritiated DNA procedure was analysed on alkaline sucrose gradients and ranged in size from 150 to 1 600 nucleotides. The average size of the product was 250–300 nucleotides. When the tritiated $cDNA_{ov}$ was reacted with excess ovalbumin mRNA, 90% of the tritiated DNA formed a stable hybrid with the ovalbumin mRNA, indicating that the tritiated DNA was indeed a complementary copy of the mRNA. A fraction of this tritiated $cDNA_{ov}$ isolated from alkaline sucrose gradients and containing *ca.* 400 nucleotides, was then used in a DNA excess hybridisation experiment. Whole chick DNA sheared to 400 nucleotides was mixed with a small amount of the tritiated DNA (unlabelled DNA to tritiated $cDNA_{ov}$ ratio 10^7:1). Hybridisation experiments were then carried out at 60 °C. Complementary tritiated DNA_{ov} hybridised with chick DNA with a $C_0t_{\frac{1}{2}}$ of 560 and the reaction exhibited second-order kinetics. Under these conditions single copy or unique sequence DNA hybridises with a C_0t of *ca.* 600. Thus, this complementary tritiated DNA_{ov} has hybridisation properties suggesting that the ovalbumin gene may only be represented a single time in the chick genome. These results, although quite preliminary in nature, have important regulatory implications in that under conditions of oestrogen stimulation *ca.* 65% of the total protein synthesised by oviduct is ovalbumin. These data suggest then that during oviduct differentiation oestrogen may act at the level of transcription to stimulate production of numerous copies of a single gene. This would lead ultimately to a high intracellular concentration of ovalbumin mRNA, and subsequently of the ovalbumin protein.

6.7 GENE ACTIVATION AND AMPLIFICATION DURING OESTROGEN-INDUCED DIFFERENTIATION

The appearance of cytological differentiation in the chick oviduct must follow prior biochemical differentiation and result in marked changes in the patterns of protein synthesis. Indeed, oestrogen stimulates synthesis of nuclear rapidly-labelled RNA and nuclear RNA polymerase activity within a few minutes after the initial administration to immature chicks. This activation of gene transcription results in the appearance of new species of nuclear RNAs as evidenced by nearest-neighbour dinucleotide analysis and DNA–RNA hybridisation studies. Analysis of the mRNA directed synthesis of oviduct proteins on polysomes *in vitro* has revealed an oestrogen-induced change in polysomal bound mRNA which is both qualitative and quantitative. Adult tissue-specific proteins, such as ovalbumin and lysozyme, are then synthesised at the translation level in response to the appearance of the specific mRNAs which code for these proteins. This results in cytological differentiation and the subsequent production of new tissue-specific proteins.

Differentiation of the oviduct in response to oestrogen must require an increase in the intracellular concentrations of many selected proteins. This necessitates a change in the entire population of intracellular proteins leading to the integrated appearance of completely new morphological cell types, tissue-specific proteins and metabolic functions. One of the earliest effects of oestrogen on the activity of a specific protein is the rapid stimulation of oviduct ornithine decarboxylase, first demonstrable 2 h after injection of oestrogen to immature chicks[52]. This response is completely blocked by cycloheximide and has been observed both *in vivo* or *in vitro*. Stimulation of this enzyme and polyamine synthesis have also been linked with the chemical induction of growth in other cell types and it is possible that polyamines may trigger, or more probably simply support, the necessary cell processes that lead to rapid nucleic acid and protein synthesis and subsequent cell growth.

In addition to this very early stimulation of ornithine decarboxylase activity, oestrogen induces synthesis of several secretory proteins that eventually are found in egg white and causes a marked stimulation in overall protein synthesis[27]. Ovalbumin, which constitutes *ca.* 65% of the total protein in the oviduct is undetectable in undifferentiated oviduct and first appears *ca.* 24 h after the single injection of oestrogen to the chick[6,47,53]. Synthesis of other egg-white proteins that exist in lower concentrations have also been quantified during differentiation of the oviduct. Thus, synthesis of lysozyme, ovomucoid and conalbumin are also stimulated by oestrogen[6,42,18,54]. In addition to the induction of proteins such as ovalbumin and lysozyme, oestrogen also stimulates the synthesis and accumulation of protein translation factors which are required for the initiation of protein synthesis on cytoplasmic ribosomes[46]. Another unique protein whose synthesis is regulated by oestrogen is the receptor for progesterone[45] that is localised primarily within the goblet cells of the differentiated epithelium[11]. Increasing amounts of this protein magnifies the response to progesterone which ultimately is manifest by the specific induction of the egg-white protein, avidin[6,56,57]. Oestrogen also apparently induces synthesis and accumulation of its own cytoplasmic receptor. Indeed, oestrogen gives a 2–3% increase in unique gene transcription. This may be a reflection of the induced heterogeneity of cell types in the differentiated oviduct. We may form the hypothesis that in the differentiated gland each of the three cell types may have some unique RNA sequences in common and some peculiar to each cell type, which may reflect its function. Increases in transcription for the oviduct as a whole could then be the result of the appearance of those unique RNA transcripts peculiar to each cell type and function.

6.8 SUMMARY

Definition of the manner by which oestrogen induces differentiation in chemically-precise terms will require the continued efforts of many investigators. It seems certain, however, that the primary response must be initiated by the steroid hormone entering the cell and interacting with a specific cytoplasmic receptor molecule (Figure 6.8). This complex is then transferred to the nucleus where it combines in some yet unknown way with the nuclear chromatin. This interaction must activate certain nuclear genes which results in the coordinated synthesis of new species of mRNA, as well as increased amounts of ribosomal and transfer RNA. These nucleic acids are then processed and transported into the cytoplasm where they form the

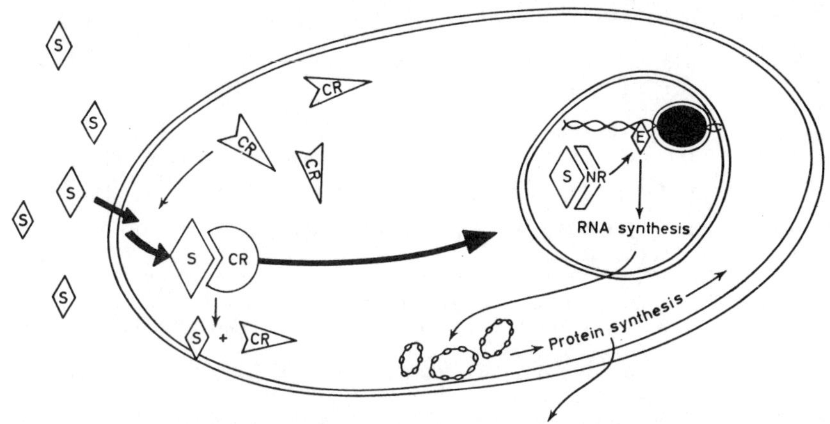

Figure 6.8 Proposed temporal sequence of biochemical events initiated by a steroid hormone in its early action on cells of a target tissue S-steriod; CR-cytoplasmic receptor; NR-nuclear receptor

protein-synthesising machinery necessary to support the synthesis of large numbers of cell-specific proteins. Operation of this temporal sequence of events, although still speculative in certain respects, would permit an amplification response for the hormone effector molecule which would give in the integrated synthesis of a host of nucleic acid molecules, structural proteins and metabolic enzymes needed for growth and differentiation of the target organ.

References

1. Moscona, A. A. and Piddington, R. (1966). *Biochim. Biophys. Acta*, **121,** 409
2. Reif-Lehrer, L. and Amos, H. (1968). *Biochem. J.*, **106,** 425
3. Schwartz, R. J. (1972). *Nature New Biol.*, **237,** 121
4. Karlson, P. and Sekeris, C. E. (1966). *Rec. Progr. Horm. Res.*, **22,** 473
5. DeFeo, V. J. (1967). *In: Cell Biol. of the Uterus*, p. 191 (R. M. Wynn, editor) (New York: Appleton-Century-Crofts) p. 191
6. O'Malley, B. W., McGuire, W. L., Kohler, P. O. and Korenman, S. G. (1969). *Rec. Progr. Horm. Res.* **25,** 105
7. Means, A. R. and O'Malley, B. W. (1972). *Metabolism*, **21,** 357
8. O'Malley, B. W. and Means, A. R. (1972). In: *Estrogen Target Tissues and Neoplasia* (T. L. Dao, editor) p. 3 (Chicago: University of Chicago Press)
9. Kohler, P. O., Grimley, P. M. and O'Malley, B. W. (1969). *J. Cell Biol.*, **40,** 8
10. Oka, T. and Schimke, R. T. (1969). *J. Cell Biol.*, **41,** 816
11. Kohler, P. O., Grimley, P. M. and O'Malley, B. W. (1968). *Science*, **160,** 86
12. Cox, R. F., Catlin, G. H. and Carey, N. H. (1971). *Europ. J. Biochem.*, **22,** 46
13. O'Malley, B. W. (1971). *Metabolism*, **20,** 981
14. Jensen, E. V. and DeSombre, E. R. (1972). *Ann. Review Biochem.*, **41,** 203
15. Spelsberg, T. C., Steggles, A. W. and O'Malley, B. W. (1971). *J. Biol. Chem.*, **246,** 4188
16. Steggles, A. W., Spelsberg, T. C., Glasser, S. R. and O'Malley, B. W. (1971). *Proc. Nat. Acad. Sci. U.S.A.*, **68,** 1479
17. Cox, R. F. and Carey, N. H. (1971). *Biochem. J.*, **124,** 71P
18. Palmiter, R. D. (1972). *J. Biol. Chem.*, **247,** 6450
19. Oka, T. and Schimke, R. T. (1969). *J. Cell Biol.*, **43,** 123
20. Socher, S. H. and O'Malley, B. W. (1973). *Develop. Biol.*, **30,** 2
21. Spelsberg, T. C., Steggles, A. W. and O'Malley, B. W. (1971). *Biochim. Biophys. Acta*, **254,** 129

22. Spelsberg, T. C., Mitchell, W. M., Chytil, F., Wilson, E. M. and O'Malley, B. W. (1973). *Biochim. Biophys. Acta* (In the press)
23. Chytil, F. and Spelsberg, T. C. (1971). *Nature New Biol.*, **233,** 215
24. Wagner, T. and Spelsberg, T. C. (1971). *Biochemistry*, **10,** 2599
25. Henson, P. and Walker, I. O. (1970). *J. Biochem.*, **16,** 524
26. Kapadia, G., Means, A. R. and O'Malley, B. W. (1971). *Cytobios.*, **3,** 33
27. Means, A. R., Abrass, I. B. and O'Malley, B. W. (1971). *Biochemistry*, **10,** 1561
28. O'Malley, B. W., Aronow, A., Peacock, A. C. and Dingman, C. W. (1968). *Science*, **162,** 567
29. Means, A. R. and O'Malley, B. W. (1971). *Acta Endocr. Suppl.*, **153,** 318
30. Palmiter, R. D., Christensen, A. K. and Schimke, R. T. (1970). *J. Biol. Chem.*, **245,** 833
31. McGuire, W. L. and O'Malley, B. W. (1968). *Biochim. Biophys. Acta*, **157,** 187
32. Cox, R. F., Haines, M. E. and Carey, N. H. (1973). *Europ. J. Biochem.*, **32,** 513
33. O'Malley, B. W., McGuire, W. L. and Middleton, P. A. (1968). *Nature* (*London*), **218** 1249
34. O'Malley, B. W. and McGuire, W. L. (1968). *Proc. Nat. Acad. Sci. U.S.A.*, **60,** 1527
35. Britten, R. J. and Kohne, D. E. (1968). *Science*, **161,** 529
36. McCarthy, B. J. and Church, R. B. (1970). *Annu. Rev. Biochem.*, **39,** 131
37. Gelderman, A. H., Rake, A. V. and Britten, R. J. (1971). *Proc. Nat. Acad. Sci. U.S.A.*, **68,** 172
38. Melli, M. and Bishop, J. O. (1969). *J. Molec. Biol.*, **40,** 117
39. Grouse, L., Chilton, M. D. and McCarthy, B. J. (1972). *Biochemistry*, **11,** 798
40. Rosen, J. M., Liarakos, C. D. and O'Malley, B. W. (1973). *Biochemistry*, **12,** 2803
41. Liarakos, C. D., Rosen, J. M. and O'Malley, B. W. (1973). *Biochemistry*, **12,** 2809
42. Means, A. R., Comstock, J. P., Rosenfeld, G. C. and O'Malley, B. W. (1972). *Proc. Nat. Acad. Sci. U.S.A.*, **69,** 1146
43. Stavnezer, J. and Huang, R. C. (1971). *Nature New Biol.*, **230,** 172
44. Rhoads, R. E., McKnight, G. S. and Schimke, R. T. (1971). *J. Biol. Chem.*, **246,** 7407
45. Rosenfeld, G. C., Comstock, J. P., Means, A. R. and O'Malley, B. W. (1972). *Biochem. Biophys. Res. Commun.*, **46,** 1695
46. Comstock, J. P., Rosenfeld, G. C., O'Malley, B. W. and Means, A. R. (1972). *Proc. Nat. Acad. Sci. U.S.A.*, **69,** 1146
47. Chan, L., Means, A. R. and O'Malley, B. W. (1973). *Proc. Nat. Acad. Sci. U.S.A.*, **70,** 6
48. Lee, S. Y., Mendecki, J. and Brawerman, G. (1971). *Proc. Nat. Acad. Sci. U.S.A.*, **68,** 1331
49. Brawerman, G., Mendecki, J. and Lee, S. Y. (1972). *Biochemistry*, **11,** 637
50. Rosenfeld, G. C., Comstock, J. P., Means, A. R. and O'Malley, B. W. (1972). *Biochem. Biophys. Res. Commun.*, **47,** 387
51. Palacios, R., Sullivan, D., Summers, M. N., Kiely, M. L. and Schimke, R. T. (1973). *J. Biol. Chem.*, **248,** 540
52. Cohen, S., O'Malley, B. W. and Stastny, M. (1970). *Science*, **170,** 336
53. Palmiter, R. D. and Wrenn, J. T. (1971). *J. Cell Biol.*, **50,** 598
54. Palmiter, R. D. and Schimke, R. T. (1973). *J. Biol. Chem.*, **248,** 1502
55. Toft, D. O. and O'Malley, B. W. (1972). *Endocrinology*, **90,** 1041
56. O'Malley, B. W. and McGuire, W. L. (1968). *J. Clin. Invest.*, **47,** 654
57. O'Malley, B. W., Means, A. R. and Sherman, M. R. (1971). *The Sex Steroids*, p. 315 (K. W. McKerns, editor) (New York: Appleton-Century-Crofts)
58. Wilson, J. D. and Lasnitzki, I. (1971). *Endocrinology*, **89,** 659
59. Wilson, J. D. (1972). *N. Engl. J. Med.*, **287,** 1284

Editor's Comments

In studying the differentiation of tissues in which a specific polypeptide is synthesised, it is natural that much attention should have focused on transcription of the gene. Hence, a good deal of recent work has been concerned with the accumulation of messenger RNA in these tissues. However, even before the current interest in messenger RNA had developed its full momentum, investigators studying enzyme induction in eukaryotic, especially mammalian, systems, had drawn attention to the fact that many induced enzymes are apparently unstable. This was almost predictable. In *E. coli* a reduction in the cellular content of an enzyme can easily be achieved by interrupting its synthesis, because cells continue to multiply rapidly and existing enzyme is soon diluted out by other proteins. However, in many eukaryotic tissues, for example the mammalian liver, such a mechanism is not possible because the rate of cell division is extremely low. Hence, if all enzymes, once induced, were stable, elevated enzyme levels might persist for months. In organisms of this kind, therefore, degradation of macromolecules might be expected to play almost as important a part in control as synthesis. That this is indeed the case is discussed extensively in the next two essays, in which it will be shown that regulation probably occurs not only through synthesis of messenger RNA and protein but also by degradation.

7
Protein Synthesis and Degradation in Animal Tissue

R. T. SCHIMKE
Stanford University

7.1 INTRODUCTION

During the past 20 years there has been a remarkable increase in understanding the regulation of enzyme levels and protein synthesis as studied in microbial systems. Equally remarkable has been the accumulation of observations that enzyme (and other protein) levels in intact animals and in a variety of animal cells in culture can be altered by physiological, developmental, nutritional and hormonal changes, by the administration of pharmacological agents, or as a consequence of mutational events[1-3]. The list of enzymes (proteins) so affected is large, increases yearly, and includes examples from essentially every major metabolic pathway in one or more tissues.

This chapter will describe the present status of the field, with sufficient examples to provide the reader with an indication of the systems and methodology employed, as well as current concepts of underlying molecular mechanisms. Current knowledge of cellular regulatory mechanisms has been obtained largely from studies with *E. coli*[4,5]. This highly opportunistic unicellular organism, selected for rapid growth under a variety of nutritional conditions, is vastly different from the individual cell in a multicellular organism. Such a cell is often not growing, may carry out a highly specialised function, and is associated with similar and dissimilar cells in tissues and organs. In such an environment, so different from that of a unicellular organism, new regulatory problems have arisen, and new 'solutions' have been found or superimposed on regulatory mechanisms common to all organisms. One such 'solution' has been the elaboration of a variety of hormones necessary for the integrated metabolic functioning and development of various tissues and organs. It is not surprising, therefore, to find that various hormones have profound effects on levels of a number of proteins. Another 'solution' concerns the problem of how to effect changes in metabolic machinery and structural components in response to environmental and developmental changes, a process that includes removal of unneeded enzymes and other macromolecular complexes, as well as the synthesis of those newly required. In bacteria the removal process can result from dilution during phases of rapid growth. On the other hand, in animal tissues, where no growth and little cellular division takes place, the process of protein degradation becomes increasingly significant as a means of removing unneeded metabolic machinery and hence as a means of regulating specific enzyme levels.

The flux, i.e. both synthesis and degradation, is a continual process in essentially all eukaryotes[6,7,170], and involves essentially all elements of the individual cells, except for the DNA. Thus the changes in enzyme levels must be considered in the context of a complex series of continual and extensive changes in various macromolecular structures. Thus the only constancy is, in fact, that of continual change. It is this continual flux that will be emphasised in this chapter.

7.2 EXPERIMENTAL SYSTEMS EMPLOYED

Clearly the ultimate goal for understanding the regulation of enzyme levels in higher organisms involves an analysis of the fundamental biochemistry and molecular biology of the series of processes whereby a protein is synthesised and degraded, and how they are controlled. Ideally one wishes a single experimental system in which to study this entire process, and which will allow for the isolation and characterisation of specific genes and their controlling elements, specific mRNAs and potential regulatory factors concerned with their translation and degradation. Clearly the degree of sophistication of the question and success in obtaining definitive answers are dictated, to a large extent, by the nature of the system under study. It is unfortunate that investigators often utilise systems that are basically unable to provide meaningful answers to the questions posed.

A variety of systems have been employed, including intact animals, perfused organs, organ cultures, tissue explants, and isolated cells in continuous culture. Each system provides certain advantages as well as limitations. An 'ideal' system might have the following properties: (a) uniformity of cell population, (b) a sufficient quantity of material to allow for the purification and characterisation of a specific protein and the macromolecules (DNA and mRNA) involved in its synthesis; (c) wide fluctuations of protein content as affected by experimental variables; (d) well-defined agents or simple procedures for perturbing enzyme content; (e) availability of mutations affecting both structure and content of the protein; (f) easy manipulation of isotopic tracers.

There is essentially no single system that satisfies all of these criteria. A system which potentially comes closest is cells in continuous culture. However, even this approach has previously been limited because of lack of suitable mutants and because few cells when cultured continue to synthesise large amounts of a specific differentiated protein or retain differentiated regulatory properties characteristic of those of the intact animal. Recent advances in the isolation and growth of functional cell types with characteristics similar to the intact tissue give promise for the future [8-10]. However, even with this advantage the amounts of cells necessary for obtaining sufficient amounts of mRNA (for instance) may well limit the use of cultured cells.

7.3 A MATHEMATICAL MODEL FOR DESCRIBING CHANGES IN ENZYME LEVELS

In view of the fact that there is a continual synthesis and degradation of essentially all proteins of liver, any description of changing enzyme levels must consider both synthesis and degradation. Similar formulations have been made by Segal and Kim[11], Price *et al.*[157], and Berlin and Schimke[12]. Thus a change of an enzyme level can be expressed by:

$$\frac{\mathrm{d}E}{\mathrm{d}t} = K_s - K_d E \tag{7.1}$$

where E is the content of enzyme (units $\times$ mass^{-1}), K_s is a zero-order rate constant of enzyme synthesis* (units $\times$ time^{-1} $\times$ mass^{-1}), and K_d is a first-order rate constant of degradation† (time^{-1}). In general there is little if any change in total mass of tissue, for example, liver, during an experimental time period and consequently an expression for a change in total tissue mass is not included.

In the steady state when $\mathrm{d}E/\mathrm{d}t = 0$, then

$$K_s = K_d E \tag{7.2}$$

and

$$E = K_s/K_d \tag{7.3}$$

Thus, in the steady state, the amount of enzyme is a function of both the rate of synthesis and the rate of degradation. An alteration in either rate can affect the level of the enzyme.

The time course describing the approach of E to a new steady state following manipulation of the experimental system, in which K_s is changed to K_s' and K_d is changed to K_d', where E_t is the activity at any time t, and where E_0 the enzyme activity under the steady-state conditions defined by K_s and K_d, is given by:

$$\frac{E_t}{E_0} = \frac{K_s'}{K_d'} - \left(\frac{K_s'}{K_d' E_0} - 1\right) e^{-k'dt} \tag{7.4}$$

If E_0 is taken as equation (7.1), equation (7.4) would represent the 'fold' increase in enzyme activity, an expression commonly used in studies of mammalian enzyme regulation. This equation indicates that, although the new steady state is determined by the values of K_s' and K_d, the time course of change from one steady state to another, in other words, the time required for any change in enzyme level, is determined only by the rate constant of degradation, K_d'.

* The rate of synthesis of a specific protein is determined by a number of factors, including the number of ribosomes, amount of messenger ribonucleic acid (mRNA), levels of amino acids and transfer ribonucleic acid (tRNA), availability of initiation and transfer factors, etc. In this simplified model, the separate roles of such variables have been included under a general notation of a rate of enzyme synthesis.

† The rate of degradation of a protein is expressed in terms of a first-order rate constant because in all cases studied in liver, the degradation of specific intracellular proteins has followed first-order kinetics. Rate constants of degradation are often expressed in terms of a half-life, where half-life $= \ln 2/k_d$.

Equation (7.4) shows then, that enzymes with more rapid turnover rates will have more rapid changes in enzyme activity, even if the rates of synthesis of all proteins are affected to the same extent. One might propose that for enzymes in which the absolute level of enzyme controls the rate of flux through the reaction catalysed by that enzyme (as opposed to activation–inhibition

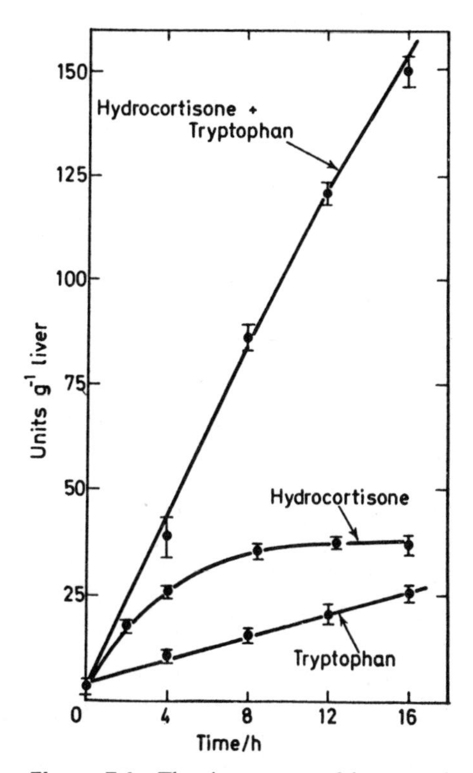

Figure 7.1 The time course of increases in tryptophan pyrrolase produced by repeated administrations of hydrocortisone 21-phosphate and tryptophan. Adrenalectomised rats weighing 150–170 g each were given injections every 4 h as follows: 150 mg of L-tryptophan in 12 ml of 0.85% NaCl intra-peritoneally and 5 mg of hydrocortisone 21-phosphate subcutaneously. At the onset of the experiment an additional 5 mg of hydrocortisone 21-phosphate were given intraperitoneally. Every 4 h, the livers of four animals were assayed for tryptophan pyrrolase activity. See Ref. 16 for details

of existing enzyme, or substrate availability), and in which fluctuations in that rate are required for appropriate metabolic control, a rapid rate of turnover is desirable. Thus only with enzymes that turn over rapidly can rapid increases and decreases occur by altering either the rate of synthesis or the rate of degradation of the enzyme.

This type of simplified model has been applied to a number of studies on the time course of changes in enzyme levels in an attempt to define the mechanism of nutritional[13, 14] and hormonal[12, 15] effects on enzyme levels. One such analysis is given in Figure 7.1 for the effects of hydrocortisone and of L-tryptophan on the activity of rat liver tryptophan oxygenase (pyrrolase)[16]. Repeated administrations of hydrocortisone at 4 h intervals result in a rapid increase of activity reaching a plateau after 8 h. Repeated administrations of L-tryptophan result in a slow virtually linear increase of activity. The combination of hormone and substrate results in a virtually linear increase over a 12 h period with levels 25-fold greater than basal levels. These results are consistent with equation (7.4) if hydrocortisone increases the rate of synthesis approximately four- to five-fold, and L-tryptophan decreases the rate of enzyme degradation essentially to zero. This general pattern, in which various experimental conditions can affect either K_s or K_d, is a recurring theme in enzyme regulation in animal tissues.

Examples of how enzyme levels are changed by altering K_s or K_d will be the subject of a subsequent section.

7.4 GENERAL COMMENTS ON THE METHODOLOGY OF STUDY OF ENZYME REGULATION

There is a logical sequence of experimental questions in a study of mechanisms for altered enzyme levels in higher organisms. The first question is whether the altered enzyme activity results from changes in the catalytic activity of an existing protein or the actual content of enzyme protein. The use of inhibitors of protein and/or RNA synthesis are often employed as a first approximation to answering this question. Conclusions based on the effects of inhibitors of RNA or protein synthesis on enzyme activity are tentative at best. Firstly, it is never clear whether such inhibitors may inhibit the synthesis of RNA or protein(s) that are necessary for the activation of an existing protein or for translation of mRNA, as opposed to an effect of specific mRNA or protein synthesis, as is often assumed. Secondly, these agents are profoundly toxic, and may have effects on cells not understood at present[17-21]. Thirdly, their use is limited to intact animal systems in which enzyme changes occur rapidly, since doses that effectively inhibit macromolecular synthesis, lead to death of the animals within 12–18 h.

More definitive is the use of immunological techniques for determining the content of immunologically-reactive protein, as employed for analysis of tryptophane oxygenase[16], tyrosine aminotransferase[22, 23], arginase[24], serine dehydratase[25], glutamic–alanine transaminase[171] of rat liver, catalase[27] and δ-aminolevulinate dehydratase[28] of mouse liver, folate reductase[29], tyrosine aminotransferase[30] and arginase[31] in cultured cells, malic enzyme in chicken liver[32] and carbamyl phosphate synthetase of frog liver[33]. Despite the ambiguities of use of metabolic inhibitors, it is interesting to find how consistent has been the finding that altered enzyme levels as detected by activity measurements correspond to alterations in immunologically-reactive protein.

Once it is established that a difference in enzyme activity results from altered enzyme content, the next queston is whether there is an alteration in the rate of enzyme synthesis, or enzyme degradation (or both). Ideally this question should be answered by measuring independently K_s, K_s', K_d and K_d' (from equation (7.4)). Theoretically this can be done by measurements of time courses of changes in enzyme activities, but the techniques involve numerous assumptions. For example K_d could be obtained theoretically using agents that inhibit protein synthesis, whereupon the decay of activity (in the basal state) should reflect the inherent degradation rate. However, there are instances in which inhibition of protein degradation by drugs results in inhibition of enzyme degradation (inactivation)[23, 34-36], although such inhibition is by no means general for all enzymes. Thus this method must be questioned in each instance. If K_d is known, K_s can be calculated from equation (7.3). K_s' and K_d' can be estimated from the time course of change in enzyme levels by equation (7.4) during and after withdrawal of the stimulus, and thereafter remain constant during the time of analysis of the changing time course. These assumptions have not been determined in most instances.

The combined use of isotopes and isolation of the specific enzyme, generally by use of specific precipitating antibodies, allows for more definitive answers. The rate of synthesis is routinely determined by a short-term incorporation of isotope into specific protein[37]. Correction for differences in free amino acid pools between two experimental conditions is generally made by comparing specific incorporation with incorporation into total protein. The time of the pulse of incorporation is critical, since it must be short relative to the half-life of the protein. Thus if an enzyme is turning over rapidly (half-lives of several hours), a long period of incorporation will not only reflect altered rates of synthesis, but will also encompass sufficient time that isotope incorporation may also reflect altered rates of the enzyme.

The use of combined isotope and immunological procedures to establish half-lives of enzymes is not without problems, particularly for studies involving the decay of pulse labelled enzymes. A major problem is the extensive re-utilisation of isotope, both in intact animals, and in cells in culture[38-40], a re-utilisation that can be as much as 50–80% of the label. Thus half-lives determined by the isotope-decay method must be considered as 'apparent'. As recently emphasised by Poole[41] the half-lives of enzymes with a rapid turnover will be markedly underestimated by the isotope-decay technique. In addition the problem of re-utilisation becomes particularly important if the rates of degradation are compared in two physiological states of the experimental system, e.g. different nutritional states of an animal, or different phases of growth of cultured cells, since variations in the degree of re-utilisation of previously-administered isotope can markedly affect the apparent degradation rates of the protein.

In addition, it is necessary to ensure the specificity of the immunoprecipitation, since a major problem is the extent of non-specific co-precipitation of radioactivity. One technique commonly used involves a second precipitation of protein from the labelled mixture of proteins with additional unlabelled enzyme and antibody, i.e. after the specific precipitate has been collected[16]. This is not an adequate demonstration of specificity. It is also necessary to show that the labelled protein specifically immunoprecipitated migrates on SDS-acrylamide gels as a unique band(s), with the migratory properties similar to that of the known subunits of that protein. More detailed descriptions and comments of various techniques for studying enzyme and synthesis degradation are given by Schimke[42].

7.5 PROPERTIES OF PROTEIN TURNOVER

In order to understand the dynamic flux of proteins in cells in relation to enzyme regulation, certain general properties of protein turnover are here listed. Although these properties are described in terms of studies with rat liver, the same general properties are common to all tissues and cells.

7.5.1 Turnover is extensive

Studies of Swick[14], Buchanan[43] and Schimke[24] have indicated that essentially all proteins of rat liver take part in the continual replacement process. These studies have used the general technique of continuous administration of isotope of known specific activity and subsequent comparison of the specific activity of isotope isolated from protein with that of the administered isotope. For example, in studies using an algal diet of constant [^{14}C] specific activity, Buchanan[43] estimated that *ca.* 70% of rat liver protein was replaced every 4–5 days from the dietary source. Such replacement cannot represent serum proteins (such as albumin), since the steady-state level of such proteins in liver is of the order of only 1–2% of total liver protein.

7.5.2 Turnover is largely intracellular

The life span of hepatic cells is of the order of 160–400 days[43-45] and hence the extensive turnover occurring in 4–5 days precludes cell replacement as the explanation for the turnover observed in liver.

7.5.3 There is a marked heterogeneity of rates of replacement of different proteins (enzymes)

Table 7.1 provides a representative listing of rates of degradation of various specific proteins and cell organelles of rat liver, as well as the methodology employed for measurement of such rates. More extensive listings of various proteins are given by Schimke and Doyle[3] and Rechcigl[2]. The wide range of half-lives for these specific proteins is remarkable, ranging from 11 min for ornithine decarboxylase to 16 days for LDH_5. In addition there is no necessary relationship between half-lives and metabolic functions of the enzymes. For example, glucokinase and LDH_5, both involved in carbohydrate metabolism, have markedly different half-lives (30 h *v.* 16 days), as do tyrosine amino-transferase and arginase (1.5 h *v.* 4–5 days), both involved in amino acid catabolism. Perhaps more striking is the lack of correlation between the cell fraction or organelle and the rate of turnover of specific proteins. For example, δ-aminolevulinate synthetase, a mitochondrial enzyme, has a

Table 7.1 Half-lives of specific enzymes and sub-cellular fractions of rat liver

Enzymes	*Half-life*	*Ref.*	*Method of measure*
Ornithine decarboxylase (soluble)	11 min	154	Loss of activity after puromycin
δ-Aminolevulinate synthetase (mitochondria)	70 min	155	Loss of activity after puromycin
Alanine-aminotransferase (soluble)	0.7–1.0 days	156	Time course of enzyme change
Catalase (peroxisomal)	1.4 days	157	Recovery of activity after irreversible inhibition of activity
Tyrosine aminotransferase (soluble)	1.5 h	23	Isotope decay*
Tryptophan oxygenase (soluble)	2 h	16	Isotope decay*
Glucokinase (soluble)	1.25 days	158	Change of enzyme activity
Arginase (soluble)	4–5 days	24	Isotope uptake and decay*
Glutamic-alanine transaminase	2–3 days	11	Time course of enzyme change
Lactate dehydrogenase isozyme-5	16 days	53	Isotope uptake*
Cytochrome c reductase (endoplasmic reticulum)	60–80 h	159	Isotope decay
Cytochrome b_5 (endoplasmic reticulum)	100–120 h	159	Isotope decay
Hydroxymethylglutaryl CoA reductase (endoplasmic reticulum)	2–3 h	160, 161	Activity decay after cycloheximide and isotope decay
Acetyl CoA carboxylase (soluble)	2 days	49	Isotope decay*
Cell fractions			
Nuclear	5.1 days	106	Isotope decay
Supernatant	5.1 days	106	Isotope decay
Endoplasmic reticulum	2.1 days	106	Isotope decay
Plasma membrane	2.1 days	102	Isotope decay
Ribosomes	5.0 days	162	Isotope decay
Mitochondria	4–5 days	156	Isotope decay

* Denotes use of immunoprecipitation techniques.

half-life of 1 h, whereas the overall (or mean) rate of turnover of mitochondrial protein is 4–5 days. Of current interest is the relatively rapid turnover of cellular membranes of rat liver (half-lives of 2–3 days for endoplasmic reticulum and plasma membranes). Yet for specific enzymes of the endoplasmic reticulum, there is again a remarkable heterogeneity of turnover rates.

7.6 REGULATION OF ENZYME LEVELS IN INTACT ANIMALS

7.6.1 Tryptophan oxygenase

As shown originally by Knox and Mehler[46] and studied extensively by Knox and Piras[47], as well as by Feigelson *et al.*[48], the activity of tryptophan oxygenase can be increased by the administration of either hydrocortisone and other glucocorticoids, or by tryptophan and certain tryptophan analogues. Schimke *et al.*[16] obtained an antibody to the enzyme and showed that the difference in enzyme activity resulted from differences in immunogically reactive protein. The results of Figure 7.1 showing the time course of accumulation of enzyme activity suggested that the hormone increased the rate of enzyme synthesis, and that tryptophan resulted in a decreased rate of enzyme degradation. Table 7.2 shows experiments involving short-term incorporation of isotope into total protein and immunoprecipitable tryptophan oxygenase designed to assess the effect of these agents on the rate of enzyme synthesis. Hydrocortisone or tryptophan, or both, were administered to rats for varying time periods, followed by [^{14}C]leucine. Extracts of livers were prepared 40 min after isotope administrations, and enzyme was

Table 7.2 Incorporation of (40 min) [^{14}C]leucine into rat liver tryptophan oxygenase*

		[^{14}C]*Leucine incorporation*	
Treatment	*Enzyme activity* (units (g *of liver*)$^{-1}$)	*Tryptophan oxygenase* (total c/min)	*Supernatant protein* (c/min mg^{-1})
None	4.2	1368	1190
Hydrocortisone			
4 h	13.6	5640	1320
12 h	31.4	6502	1491
Tryptophan			
4 h	8.2	1620	1564
12 h	14.1	1670	1165
Hydrocortisone + tryptophan			
4 h	28.3	7680	1491
12 h	72.0	7280	1018

* Rats were given repeated doses of hydrocortisone or L-tryptophan, or both, at 4 h intervals for the times indicated. Each rat was given, 40 min before death, a single intraperitoneal injection of 20 μCi [^{14}C]leucine in 1.0 ml of 0.85% NaCl. Results of [^{14}C]leucine incorporation into tryptophan oxygenase are reported as total net counts per minute in the precipitate from the total DEAE-cellulose extract of two rats. See reference 16 for details.

isolated by immunoprecipitation. As shown in Table 7.2 hydrocortisone increases isotope incorporation into tryptophan oxygenase by approximately four- to five-fold, whereas isotope incorporation into total protein increases by only 30%. Tryptophan does not increase the extent of isotope incorporation, although its administration results in enzyme accumulation (Figure 7.1). These findings are consistent with the proposal that hydrocortisone increases the rate of enzyme synthesis.

That tryptophan does affect the rate of enzyme degradation is shown by the results shown in Figure 7.2, in which isotope is administered to animals

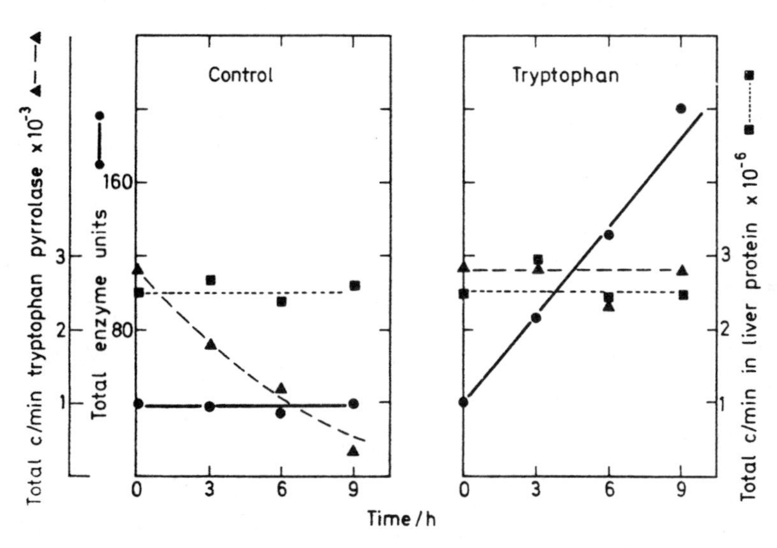

Figure 7.2 Effect of L-tryptophan administration on the loss of tryptophan oxygenase (pyrrolase) prelabelled with [^{14}C]leucine. Rats were given single injections of 20 μCi [^{14}C]leucine. Sixty minutes later, two animals were killed. The remainder were given 10 ml of 0.85% NaCl or 10 ml of 0.85% NaCl containing 150 mg L-tryptophan. These injections were repeated in the remaining animals after 4 and 8 h. At the times specified, the livers of two animals in each group were removed and frozen immediately. At the end of the experiment extracts of the livers were prepared, and the radioactivity that was incorporated into tryptophan oxygenase and protein was determined. The values given are for totals of combined extracts of two animals. ●——●, enzyme activity; ▲— —▲, total radioactivity in protein precipitated by the tryptophan oxygenase antiserum; ■- - - -■, radioactivity in total cellular protein. See Ref. 16 for details

1 h prior to the indicated zero time[16]. At the end of this time isotope in the free amino acid pool is negligible, and hence new proteins will be synthesised largely from unlabelled amino acids[16]. Figure 7.2 shows that in control animals, where enzyme levels remain constant, there is a progressive loss of radioactivity in the immunoprecipitate, with an estimated half-life of 3–4 h. This shows turnover of the enzyme in the basal state. In contrast, when tryptophan is administered to the animals, the existing enzyme does not decay, indicating that tryptophan stabilises the enzyme *in vivo*.

7.6.2 Acetyl CoA carboxylase

In addition to hormonal and substrate effects on synthesis and degradation of specific proteins as exemplified by the studies with tryptophan oxygenase, the nutritional status of the animal has profound effects on both synthesis and degradation. This pattern of independent regulation of rates of synthesis and degradation is illustrated by studies by Majerus and Kilburn with acetyl CoA carboxylase[49]. Fatty-acid synthesis is inhibited by starvation and is accelerated when rats are placed on fat-free diets. Numa *et al.*[50] have found that the first enzyme in the fatty-acid pathway of fatty-acid synthesis, acetyl CoA carboxylase, is rate limiting, and its activity as measured by assays *in vitro* fluctuates widely during fasting and re-feeding. Majerus and Kilburn[49] have extended these studies by combined use of radioisotopic and immunological techniques to assess the role of synthesis and degradation in the regulation of this enzyme. As summarised in Table 7.3 and Figure 7.3, when rats are fed

Table 7.3 Synthesis and degradation in nutritional control of acetyl CoA carboxylase*

Nutritional status	*Enzyme activity* (milliunits (mg protein)$^{-1}$)	*Relative* K_s†	*Half-life* h‡
Fat-free diet	14.0	3.0	48
Purina rat chow	6.0	(1.3)§	48
Change from fat-free diet to fasting 48 h	1.9	0.3	18

* Data from Majerus and Kilburn[49].

† Relative rate of synthesis = total c/min precipitated by specific antiacetyl CoA carboxylase antibody c/min mg^{-1} total protein. [^{3}H]leucine was injected intraperitoneally and animals were killed after 40 min.

‡ Half-lives determined as described in Figure 7.3.

§ Relative K_s calculated from known enzyme level and K_d as per equation 7.3.

a fat-free diet for 48 h, the liver activity of acetyl CoA carboxylase is 14.0 milliunits mg^{-1} protein, whereas this value is 0.6 milliunits mg^{-1} protein for rats maintained on a diet containing 45% vegetable oil (not shown in Table 7.3). Maintenance on Purina rat chow results in 6.0 milliunits mg^{-1} protein, and animals fasted after maintenance on a fat-free diet for 60 h have 0.9 milliunits mg^{-1} protein of acetyl CoA carboxylase activity. The rates of enzyme synthesis were markedly affected by the diet. Thus animals maintained on the fat-free diet synthesised acetyl CoA carboxylase at a rate ten times greater than animals starved for 48 h relative to total protein (Table 7.3). However, the precipitous decline in enzyme activity that occurs during fasting is due not only to a decrease in the rate of enzyme synthesis, but also to an acceleration in the rate of enzyme degradation. Thus, as shown in Figure 7.3, prelabelled enzyme decays with a half-life of 48 h in the fat-free dietary state, whereas during starvation, the enzyme is inactivated far more rapidly

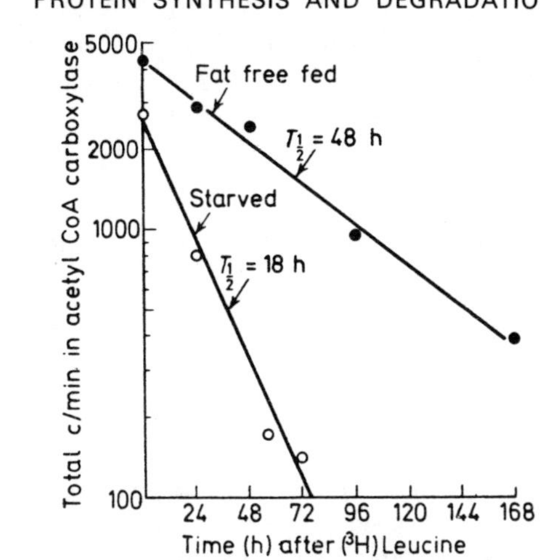

Figure 7.3 Acetyl CoA carboxylase turnover. Rats weighing 150–170 g were used in these experiments. Rats fed a fat-free diet for 48 h were given intraperitoneal injections of 250 μCi [^{3}H]leucine, and were subsequently killed in groups of three, 40 min after injection (zero time), and at the subsequent times indicated. In a second experiment, rats fasted 48 h were given intraperitoneal injections of 500 μCi [^{3}H]leucine; the rats were killed in groups of four at the times indicated. Acetyl CoA carboxylase was purified on DEAE-cellulose columns, and the radioactivity incorporated into the enzyme was determined by precipitation with a specific antibody. The results are reported as the total radioactivity incorporated into acetyl CoA carboxylase in each group. See Majerus and Kilburn[49] for details

(half-life of 18 h). The difference in enzyme level between fat-free diet and Purina rat chow under steady-state conditions (14.0 *v.* 6.0 milliunits mg^{-1} protein) would appear to result from an effect only on enzyme synthesis, since Majerus and Kilburn found, in experiments comparable to Figure 7.3, that the half-life of the enzyme is the same (48–50 h) with both the Purina chow and fat-free diets.

7.6.3 LDH isoenzymes

In many instances there are remarkable differences in the specific activities of enzymes catalysing the same reaction in different tissues. In certain cases, particularly in comparing liver with tissues such as muscle or kidney, this is based on the fact that they represent completely different proteins with different regulatory properties, for example the hexokinase isozymes, including the liver-specific glucokinase[51], and the pyruvate kinase isozymes[52]. One particularly important study is that of Fritz *et al.*[53] who have determined the rates

Table 7.4 Steady-state levels of LDH_5 and rates of synthesis and degradation*

Tissue	*Enzyme level* (pM g^{-1} tissue)	K_s (pmol g^{-1} day^{-1})	K_d (day^{-1})	*Half-life*†
Heart	5.5	2.2	0.400	1.6 (1.0)
Muscle	295	65.2	0.018	31 (22.0)
Liver	1600	65.0	0.040	16 (2 2)

* From Fritz *et al.*[53].
† Parentheses indicate values for total soluble protein.

of synthesis and degradation of lactate dehydrogenase isozyme-5 in rat liver, heart muscle and skeletal muscle as summarised in Table 7.4. These workers have found that the tissue differences in enzyme levels were not related solely to rates of synthesis, but that the rate of LDH-5 degradation was also markedly different in the different tissues. Thus the half-lives of this isozyme were 16, 1.6 and 31 days in liver, heart muscle and skeletal muscle respectively, compared to mean half-lives of 2.2, 1.0 and 22 days for total soluble protein in these same tissues. Thus the same isozyme in different tissues may be degraded at markedly different rates. Hence one cannot assume that differences in enzyme levels from tissue to tissue result from differences in rates of synthesis.

7.6.4 Genetic regulation of enzyme levels

The lack of ability to manipulate genetics has limited the rate of progress and depth of knowledge concerning enzyme regulation in animal cells. It is encouraging to note the increasing attempts to obtain and investigate mutations in animal systems that regulate enzyme levels. A number of recent studies have indicated the feasibility of using inbred mouse strains to study enzyme regulation. Several of these studies are outlined briefly here.

Rechcigl and Heston[54] and Ganschow and Schimke[27] have described a mutation which regulates the rate of degradation of liver catalase in two closely related inbred mouse strains, C57B1/6 and C57B1/Ha (or C57B1/He). In addition these latter workers found that the catalase activity of these two strains is catalytically altered compared to a number of other mouse strains. Their results are summarised in Figure 7.4. Comparing three inbred mouse strains, DBA/2, C57B1/6, and C57B1/Ha, the C57B1/6 strain has 60% of the catalase activity of DBA/2. Studies based on enzyme purification, as well as the use of immunological techniques, showed that DBA/2 and C57B1/6 strains contained the same number of immunologically-reactive molecules, whereas the C57B1/Ha strain had twice as many enzyme molecules. Enzyme purified to homogeneity from both of the C57B1 strains has only 60% of the activity of that purified to homogeneity from the DBA/2 strain. Thus the difference between the three strains results from two independent mutations, one affecting the catalytic activity, the other affecting the rate constant of degradation. The mutation affecting degradation appears to be relatively specific for catalase among liver proteins, since the rate of degradation of total liver protein was not affected by this mutation, nor was urate oxidase, another peroxisomal protein[27].

The nature of the mutation affecting catalase degradation remains to be established. Enzyme purified from the two C57Bl strains appears to be similar by a number of criteria, including electrophoresis, sedimentation, heat, and trypsin stability[27]. The mutation affecting catalytic activity is inherited as a co-dominant trait[55], consistent with the probability that it affects the structure of the catalase protein. In contrast, the mutation decreasing degradation is inherited as a recessive trait, in other words, rapid turnover is dominant[55]. Ganschow and Schimke[27] provided some evidence that the

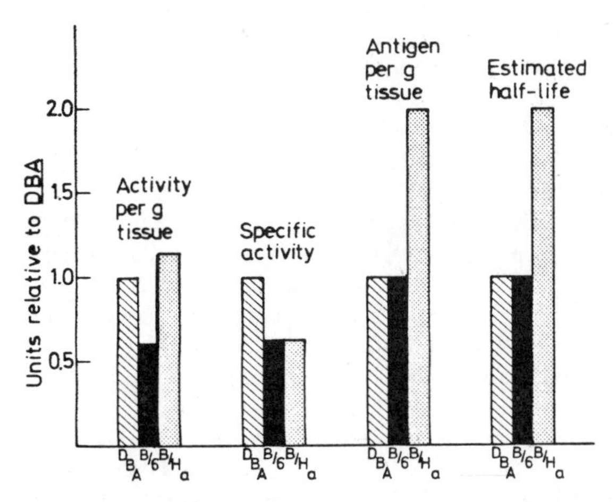

Figure 7.4 Summary of liver catalase phenotypes. See Ref. 27 for details

mutations affecting catalytic activity and the rate of degradation were not linked, based on the finding that backcrosses of F_1 hybrids of DBA/2 and C57Bl/Ha strains resulted in progeny with higher catalase levels than are found on any of the parental strains. Such animals would contain the catalytically more active catalase (characteristic of DBA/2 mice), with the slower rate of degradation (characteristic of the C57Bl/Ha strain).

Not only do mutations affect the rate of enzyme degradation, they also affect the rate of enzyme synthesis. For instance Doyle and Schimke[28] have studied the regulation of γ-aminoaevulinate dehydratase as controlled by two alleles at the laevulinate (Lv) locus, first described by Russell and Coleman[56]. Mouse strains homozygous for the Lv^a allele (AKR) have high liver enzyme activity compared to strains homozygous for the Lv^b allele (C57Bl/6). Heterozygotes have intermediate levels. Coleman[57] showed the enzyme purified from liver of both homozygous genotypes was indistinguishable by certain physical, chemical and enzymatic criteria. Doyle and Schimke[28] have shown by immunochemical methods that the Lv locus controls the amount, and not the activity of δ-aminolaevulinate dehydratase. In addition they have shown that the next enzyme in the pathway of haem synthesis, uroporphyrinogen-I synthetase, is not affected by the Lv locus. Using isotopic techniques they have shown that this locus regulates the rate of enzyme synthesis without affecting the rate constant of enzyme degradation. In the heterozygote Lv^{ab}, the rate of enzyme synthesis is also intermediate. More recently, Doyle[61] purified enzyme from the two homozygous states and determined that the molecular weight, amino acid composition and tryptic peptides, as partially separated by chromatography on Dowex-50 columns, are similar. This suggests that the mutation affecting the rate of synthesis is not a structural mutation.

There are a number of examples of differences in enzyme levels in human genetic defects with altered rates of synthesis or degradation. Mutations of catalase[62] and glucose 6-phosphate dehydrogenase[63] also appear to affect the rate of enzyme degradation, by making the enzyme more labile *in vivo* in red cells. Another interesting example is the elevated levels of AMP-pyrophosphorylase which occur in the red cells of patients lacking the entirely separate enzyme, IMP-pyrophosphorylase[64]. This phenomenon results from stabilisation of the AMP-pyrophosphorylase during red cell ageing[65].

Perhaps the most significant aspect of these studies is the fact that mutations affecting enzyme levels are readily found, both in the mouse and human population. A systematic screen of more than 100 inbred mouse strains would undoubtedly reveal a large number of mutations affecting enzyme levels. The possibility of use of mutagenic agents to facilitate the isolation of mutants is also feasible, in view of the finding of a number of structural mutations of catalase obtained by Feinstein *et al.*[66] from among some 10 000 progeny of an x-irradiated male mouse.

7.7 REGULATION OF ENZYME LEVELS IN CULTURED CELLS

In the preceding section representative examples were described of the manner in which various experimental variables, including hormones, substrates, nutrition, tissue specificity and mutations can affect enzyme levels. Typical of these examples is the finding that both rates of synthesis and degradation can be altered. Enzyme levels in cultured cells likewise undergo marked variations in activity, as a function of growth phase[67], phase of the cell cycle[68], or additions of hormones, and alterations in specific substrate levels[31,102a]. Several examples will be discussed briefly here.

7.7.1 Tyrosine aminotransferase (TAT)

The regulation of tyrosine aminotransferase levels in cultured hepatoma cells has been studied extensively, largely by Tomkins and his colleagues[30] and by Kenney[37]. Such cell lines were originally reported by Pitot *et al.*[58], and were derived from minimal deviation hepatomas. The HTC and Ruber H-35 cell lines have been studied most extensively, and both respond to the addition of glucocorticoids with an approximately tenfold increase in synthesis of TAT[30,37]. It is of interest that these cell lines, although derived from liver, contain no assayable tryptophan oxygenase, and there is no stimulation of its synthesis by glucocorticoids[58]. As in the intact rat, insulin stimulates TAT synthesis in the Ruber H-35 cells, whereas glucagon has no effect[69-71]. This latter difference between the intact animal and liver explants can be ascribed to a lack of adenyl cyclase in the cultured cells[69]. Following addition of the glucocorticoid to the medium, there is an increase in enzyme activity to a new steady-state value ten times that of the basal state. This results from a tenfold increase in isotope incorporation into specific tyrosine aminotransferase immunoprecipitates. Thus one can conclude from equation (7.3) that the hormone affects only the rate of synthesis. When the cells are washed free of hormone, enzyme activity falls to basal levels with a half-life of *ca.* 2 h., a value similar to that in intact animals[23].

As is the case with every steroid hormone-mediated stimulation of protein synthesis, the administrations of inhibitors of RNA synthesis at the time of hormone administration abolishes the rise of enzyme activity. In cell culture systems there is no increase in bulk RNA synthesis[72], as opposed to the finding in the intact animal, where steroid hormone administration results in the accumulation of ribosomal and transfer RNA species[73, 74]. This finding suggests that ribosomal RNA synthesis is not necessary for the hormone-mediated response. That a potentially unique species of RNA accumulates following glucocorticoid administration in cultured cells has been demonstrated in a series of interesting experiments by Peterkofsky and Tomkins[72]. They have found that if the accumulation of TAT is prevented by the inclusion in the medium of an inhibitor of protein synthesis, but in the absence of an inhibitor of RNA synthesis, and after approximately 1 h the medium is changed such that protein synthesis can proceed, now in the absence of RNA synthesis (addition of actinomycin D), then the accumulation of TAT can proceed, and does so without any lag. Recalling that if actinomycin D is present from time zero no increase in TAT is seen, the conclusion to be made is that the hormone results in the accumulation of an induction-specific RNA that can be translated subsequently. Granner *et al.*[75] have found that there is a critical time period of *ca.* 20 min following hormone addition to the medium before there is any demonstrable accumulation of capacity for increased TAT synthesis.

An additional phenomenon that has been studied extensively in the hepatoma cells, and which remains a subject of controversy, concerns the effect of actinomycin D addition to the medium at a time when the cells are actively synthesising TAT. Thus, under such conditions, TAT accumulation is greater than in the presence of hormone alone. The explanation for this phenomenon, for which the term 'superinduction' has been coined, remains controversial. Reel and Kenney[70] have provided evidence in the Ruber H-35 cell line, using combined immunological and immunoprecipitation techniques to measure both the rate of synthesis and the rate of degradation of TAT, that actinomycin D decreases both the rate of synthesis and the rate of degradation of the enzyme. The decrease in rate of degradation is greater than the decrease in rate of synthesis, and hence the result is a further accumulation of enzyme. An effect of actinomycin D on inhibition of enzyme degradation is in keeping with a number of enzymes in other systems[26, 35, 36]. On the other hand,

Martin *et al.*[76] and Thompson *et al.*[77], studying a comparable cell line (HTC) have concluded that actinomycin D has no effect on the inactivation of TAT. Using immunological techniques, Thompson *et al.*[77] have reported an increase in the rate of TAT synthesis. The differences between the conclusions of this group may be the result of: (a) different labelling times to measure rates of synthesis; this is an important consideration since the half-life of tyrosine aminotransferase is short [2–3 h[49, 77]], and hence a long 'pulse' relative to the half-life of the enzyme will reflect degradation as well as synthesis. Reel and Kenney[70] used 15 min pulses, whereas Thompson *et al.*[77] used a pulse of 6 h; (b) differences in culture conditions: Hershko and Tomkins[79] have reported that protein degradation in these cell lines is sensitive to the amino acid concentration of the medium. Reel and Kenney[70] used a medium with a low amino acid concentration for these studies, a medium which Hershko and Tomkins[79] have found to retard TAT inactivation; (c) inadequate specificity of immunoprecipitation of TAT; neither of these groups showed that the antibodies they use specifically precipitated tyrosine aminotransferase from radioactive homogenates or extracts (see comments on methodology above).

Tomkins *et al.*[30] have elaborated a model to explain 'super-induction' which involves a repressor that acts at the translational level. They propose that the mRNA for TAT is synthesised continually, and that it can interact with ribosomes to direct the synthesis of enzyme or it can interact reversibly with a repressor molecule, resulting in an inactive complex which is subject to degradation. The repressor is postulated to have a rapid turnover. Therefore an inhibition of its synthesis by actinomycin D will result in a decreased concentration of the repressor, and lead to more mRNA available for translation. Indirect support for this model comes from the observation that there are marked differences in the effectiveness of actinomycin D and glucocorticoids during phases of the cell cycle[30].

An alternative explanation for 'superinduction' in another hormonally responsive system, i.e. chick oviduct, will be discussed in Section 7.8.6.

7.7.2 Folate reductase and methotrexate

A perplexing clinical problem has been the development in patients of tumour cells that are resistant to the folate analogues methotrexate and aminopterin[81]. The mechanism of action involves the irreversible binding to, and inhibition of, folate reductase[81]. The development of resistance, as studied in cultured cells, can result from various mechanisms, including alterations in folate reductase such that the methotrexate binding constant is decreased[82, 164] as well as increased in activity of folate reductase[83-86, 165]. Only the last phenomenon is of concern in the context of this chapter.

Two general types of increased activity are observed. In one type, the addition of methotrexate to cells results in a progressive increase in enzyme activity of *ca.* three- to five-fold[87]. When the drug is removed from the medium, activity returns rapidly to basal levels. It should be noted that enzyme activity is assayed under conditions in which the drug–enzyme complex is dissociated, thereby allowing for an estimate of total enzyme activity. Hillcoat *et al.*[87]

have analysed this phenomenon and find that the addition of methotrexate stabilises the enzyme against intracellular degradation, and propose that the accumulation of enzyme results from decreased degradation. It is of interest to find that methotrexate stabilises partially purified folate reductase against proteolytic[88] and heat inactivation[89a].

A second type of increased folate reductase activity develops when cells are cultured continuously in the presence of the drug for long periods of time[29]. When such cells are now grown in the absence of methotrexate, activity does not fall. Thus there appears to be a permanent change in the capacity of such cells to synthesise enzyme. This problem has been studied most recently by Nakamura and Littlefield[29] using hamster kidney cell lines; one grown continuously in methotrexate had a 125-fold increase in folate reductase activity. They have prepared an antibody to folate reductase, purified to homogeneity using an affinity chromatography step which utilises the high binding constant for methotrexate to the enzyme. They have demonstrated that the differences in assayable enzyme activity are associated with a comparable difference in immunological protein. In preliminary studies they have also shown that there is a *ca.* 20-fold increased rate of enzyme synthesis between the two cell lines. The results are not sufficiently definitive to know whether the entire 125-fold differences can be accounted for by the altered rate of synthesis, or whether altered degradation also contributes to the differences in enzyme levels. The mechanism for the altered rate of enzyme synthesis in this most interesting system is unknown. Littlefield[89], using somatic-hybridisation techniques, demonstrated that there is no dominant cytoplasmic regulator of folate reductase synthesis. Thus one must consider among the possibilities selective gene amplification, increased mRNA synthesis, decreased mRNA degradation, and so on. This system is particularly interesting, since both (presumably) genetic and epigenetic factors affect the synthesis and degradation of this particular enzyme.

7.7.3 Glutamine synthetase regulation by hormones and substrate

Glutamine synthesis activity is regulated by both hormones and substrate. The original studies on hormone regulation were performed with developing chick retinal tissues. Chick retina glutamine synthetase accumulates at day 14–15 of chick development[90]. Moscona and Piddington[91], as well as Reif-Lehrer and Amos[92] demonstrated that retinal explants in culture could be induced to synthesise glutamine synthetase in response to a factor present in chick serum, which was found to be hydrocortisone. Moscona and Kirk[93] also found that addition of glutamine to the medium in which the retinal explants were incubated partially suppressed the hormone-mediated increase in synthetase activity.

In cultured cells, glutamine synthetase activity is also regulated by glutamine and hydrocortisone. The original observation was that of De Mars[94] who found that culturing HeLa cells in the absence of glutamine gave five- to ten-fold higher levels of glutamine synthetase activity than in its presence. This result was interpreted in terms of an affect of glutamine in 'repressing' the synthesis of enzyme. Subsequently, Paul and Fottrell[95] concluded that glutamine was accelerating the rate of degradation of the enzyme, rather than affecting its synthesis. The effect of glutamine on glutamine synthetase has been studied more recently in Chinese Hamster cells[96] and in HTC cells[97]. The latter cell type is also sensitive to glucocorticoid, as is the retinal explant system, and in this system glutamine suppresses the accumulation of hormone-mediated enzyme accumulation. Furthermore hydrocortisone addition and removal of glutamine from the medium led to a greater accumulation of enzyme than either regulator alone. These studies have not been carried to the point where the mechanisms can be interpreted in terms of altered rates of synthesis and degradation. However, the problem appears to be very similar to that of the effect of hormone and tryptophan on tryptophan oxygenase, except that the substrate, glutamine, may destabilise the enzyme, rather than the situation with tryptophan oxygenase, where substrate stabilises the active enzyme.

7.8 MOLECULAR MECHANISMS OF ENZYME REGULATION

Preceding sections of this review have dealt with certain techniques to determine rates of synthesis and degradation of proteins, and with representative examples of alterations of both rates by various experimental variables. In the following section, potential mechanisms will be discussed, and in particular in relation to methodology available for their study.

7.8.1 Regulation of enzyme (protein) synthesis

The question of what regulates the synthesis of a specific protein in animal tissues need not, and in all likelihood does not, have a single universal answer. Depending on the developmental, hormonal, physiological or growth state, one of any number of steps in protein synthesis may be rate limiting, including (a) the number of copies of the specific gene[98]; (b) the rate of transcription of that gene as regulated by DNA-associated proteins[4]; (c) the packaging and transport of specific mRNA into the cytoplasm[99]; (d) utilisation of mRNA, as regulated by steps in initiation, elongation, or release; (e) modified stability of the mRNA; and (f) assembly of the released peptide chains into a biologically functioning unit, whether associated with other proteins (such as multimeric enzymes), phospholipids (membranes), or nucleic acids (ribosomes and chromosomes).

Discussion of the potentially rate limiting steps in protein synthesis, and concepts concerning their regulation, have been presented in detail by various authors[3, 100, 101], and it seems unnecessary to detail them here, in particular since little information is actually available on whether such proposed mechanisms are operative in any given system.

Indirect experiments using various inhibitors, e.g. actinomycin D, puromycin, cycloheximide, etc. cannot provide definitive answers, and at best can be used only to conclude that RNA and/or protein synthesis is required for the observed effect. They do not, *per se*, demonstrate that it is specific mRNA or specific protein synthesis that is required for the observed effect. Clearly what is necessary for 'solving' the equation of regulation of specific protein synthesis is the ability to quantify and isolate each of the components involved in protein synthesis, including specific mRNA, polysomes, and the specific gene(s). Several years ago it seemed unlikely that such techniques would be available; however, in the past 2 years, such possibilities are not only possible, but in some instances are a reality.

7.8.2 Quantification of the amount of mRNA

An increasing number of animal cell mRNAs have been successfully translated in heterologous protein synthesising systems, including reticulocyte lysates[168], Krebs ascites tumour cells[78,125], as well as the frog oocyte[105]. The mRNAs thus far successfully translated include, globin[78,168,169], ovalbumin[104], alpha crystallines[103], conalbumin*, rat serum albumin† and light chain of immunoglobulin[107], among others. The criteria for synthesis of the specific product

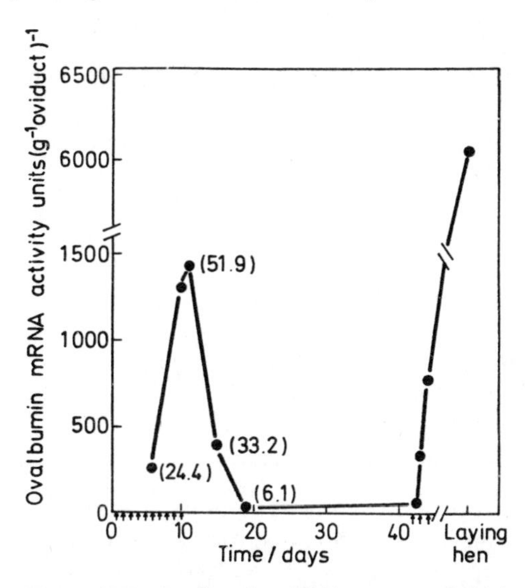

Figure 7.5 Ovalbumin mRNA content of chick oviduct tissue during primary oestrogen stimulation (1–10 d), withdrawal (11–41 d), and secondary stimulation (42–44 d). Oestrogen (1 mg) was administered to chicks on days designated with arrows. Oviducts were removed and total mRNA activity determined. At several points portions of the same tissue used for the preparation of nucleic acid were incubated in culture for 1 h with tritiated amino acids to determine the relative rate of ovalbumin synthesis. This is expressed as a percentage of total protein synthesis (numbers in parentheses). See Ref. 108 for details

* Gonzalez, C. and Schimke, R. T. (to be published).

† Taylor, J. M. and Schimke, R. T. in press.

include specificity of immunoprecipitation, the size of the product synthesised, chromatographic behaviour of the protein synthesised, and similarity of peptides between the authentic protein and that synthesised in the heterologous system. Assays of mRNA content can be made quantitative, and therefore can be used to quantify the amount of specific mRNA[108].

Figure 7.5 shows a comparison of the amount of assayable ovalbumin mRNA as a function of hormonal stimulation of chick oviduct[108]. In this system, the administration of oestrogen to chicks over a period of 10 days results in the accumulation of the capacity to synthesise ovalbumin such that at the end of 10 days, *ca.* 50% of the protein synthesised is ovalbumin. Following withdrawal of hormone, ovalbumin synthesis, expressed as a percentage of total protein synthesis, declines. Ovalbumin synthesis can be restimulated in the chick by a secondary series of oestrogen stimulations[109] (also see Figure 7.10). Figure 7.5 shows that the amount of assayable ovalbumin mRNA is directly proportional to the percentage of protein synthesised as ovalbumin. Thus in this system there is a direct correlation between ovalbumin mRNA content, and the rate of ovalbumin synthesis. Similar conclusions have been made by Means *et al.* in this same system[110]. Thus in this case no evidence can be found for a masked form of mRNA. It should be emphasised, however, that if the mRNA were in a form that was not translatable in the heterologous system, it would not be detected. In a subsequent section, the use of a complementary DNA product to determine the existence of untranslatable mRNA sequences will be discussed.

7.8.3 Quantification of the number of polysomes synthesising a specific protein

The identification of polysomes synthesising a specific protein can be accomplished by one of several techniques. If the protein synthesised is of a unique size, the polysomes can be identified by their unique size class. For instance, myosin polysomes can be identified in embryonic chick muscle tissues by their large size[111]. However, in general, a protein is not likely to be of a unique size, and hence other methods must be employed. Immunological techniques based on the reaction of antibody with growing nascent chains is one such technique that is applicable, both to polysome quantification, as well as isolation of specific mRNA (see next section). Figure 7.6 shows the specificity of binding of an antiovalbumin antibody to hen polysomes[112]. Panel (a) shows the binding of 30 μg of [^{125}I]antiovalbumin, and indicates that antibody is bound to polysomes with a peak at a 12-membered region. The binding is saturable, as shown in Panel (b), where the polysomes were first incubated with a 15-fold excess of unlabelled antiovalbumin prior to addition of the [^{125}I]antiovalbumin. Panel (c) shows that the blocking of binding of the [^{125}I]antiovalbumin is specific, since the addition of antibovine serum albumin does not block the binding of [^{125}I]antiovalbumin. Panel (d) shows that [^{125}I]antiserum to bovine albumin does not bind to the polysomes from hen.

The binding of [^{125}I]antiovalbumin at saturation can be used to quantify the number of polysomes that synthesise ovalbumin, as shown in Figure 7.7.

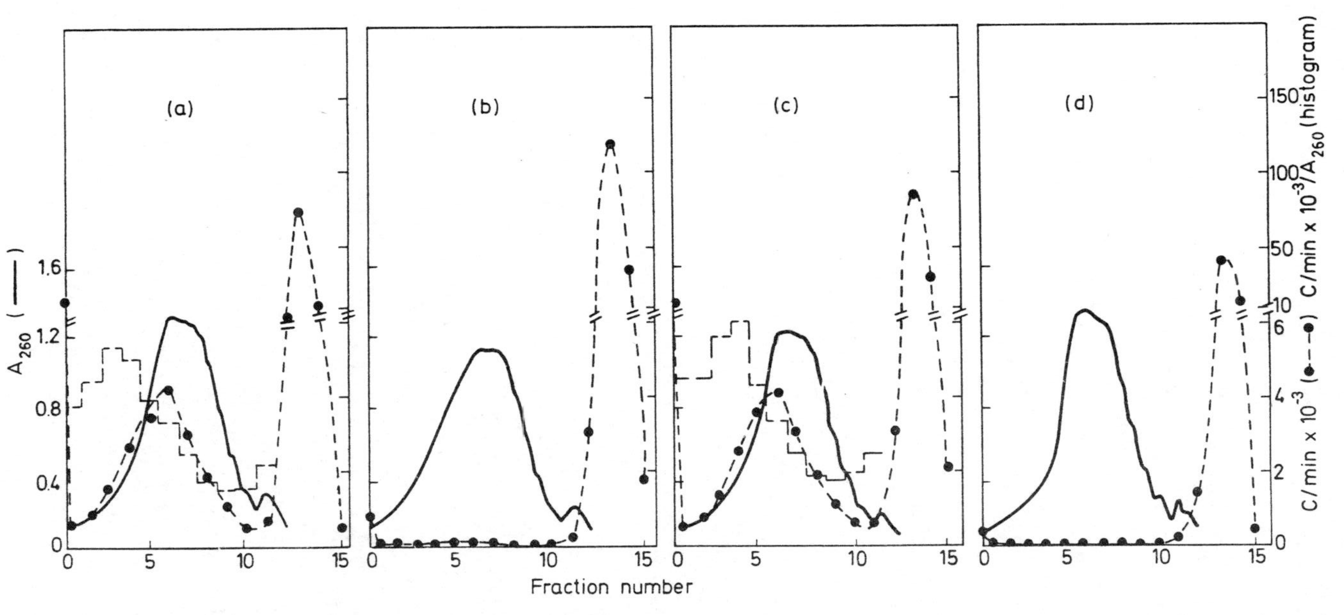

Figure 7.6 Binding of [^{125}I]anti-ovalbumin and [^{125}I]anti-BSA to hen oviduct polysomes. Polysomes (10A_{260} units in 1.0 ml) were incubated at 4°C with (a) 30 ug of [^{125}I] anti-ovalbumin for 30 min; (b) 500 µg of unlabelled anti-ovalbumin for 30 min followed by 30 µg of [^{125}I]anti-ovalbumin for 30 min more; (c) 500 µg of unlabelled anti-BSA for 30 min followed by 30 µg of [^{125}I]anti-ovalbumin for 30 min; and (d) 30 µg of [^{125}I]antiBSA for 30 min. After incubation the polysomes were layered over a continuous sucrose gradient and centrifuged. Fractions (1.0 ml) were collected to measure specific activity and radioactivity. See Ref. 112 for details

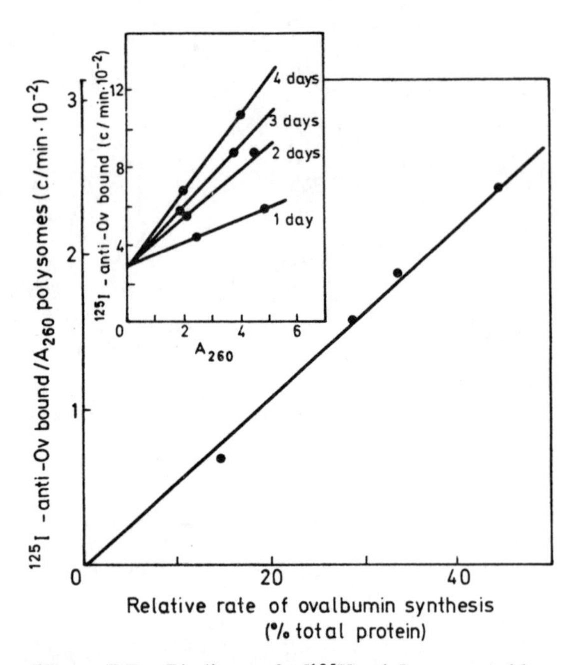

Figure 7.7 Binding of [^{125}I]anti-Ov to oviduct ribosomes from chicks given 1–4 days of secondary stimulation with oestrogens. Chicks were treated with oestrogen (2 mg per day, secondary stimulation). Oviduct magnum was isolated and total ribosomes were prepared. Different amounts of ribosomes from each preparation were incubated with [^{125}I]anti-Ov for 1 h at 4 °C (inset), and amount bound determined. [^{125}I]anti-Ov bound per A_{260} of polysomes was determined from the slope of the line in inset and corrected for the percentage of total ribosomes which sediment as polysomes (determined by analysing an aliquot of homogenate) on sucrose gradients; in all preparations 70–80% of the ribosomes sedimented as polysomes. The relative rate of ovalbumin was determined by incubating pieces of oviduct magnum in Hanks medium with [^{3}H]amino acids and then determining the percentage of the total protein synthesised which was precipitable with anti-Ov. See Ref. 113 for details

Polysomes were prepared from chicks to which oestrogen had been administered for varying amounts of time, resulting in oviducts in which ovalbumin constituted varying proportions of the total protein synthesised. Figure 7.7 shows that there is a good correlation between the proportion of ovalbumin synthesis, and the degree of binding of the ovalbumin antibody per unit of total polysomes.

7.8.4 Immunoisolation of specific mRNA

Central to the study of regulation of protein synthesis is the isolation of specific mRNAs. In the case of the globin chain mRNA, the presence of a cell type that synthesises large amounts (95%) of a single group of proteins of a uniquely small size allows for mRNA isolation on the basis of size only. In the case of the silk fibroin mRNA, its unique base composition allows for separation from other RNAs[59,114] by density-centrifugation techniques. However, generally mRNAs will not be of unique size or base composition. An immunological approach appears most promising[112,115-117]. The specific reaction of antibody with growing nascent chains allows for isolation of specific polysomes, and hence their mRNA. Figure 7.8 shows the isolation of ovalbumin mRNA based on the reaction, of the specific antibody with polysomes, followed by reaction of the antibody (bivalent) with a matrix of glutaraldehyde fixed ovalbumin[118]. The ovalbumin matrix is large, can be readily centrifuged, and contains sufficient antigenic determinants on its surface to react with the bivalent antibody bound to the polysomes. The polysome-matrix can be washed to remove the non-reacted polysomes, followed by release of the specific polysomes in a washing medium containing EDTA. Figure 7.8 (inset) shows the pattern of elution of D. 260 by repeated washes of the polysome-matrix in 0.5 M sucrose, containing 0.15 M NaCl and 1% each of Triton X-100 and desoxycholate followed by elution with a wash of 0.01 Tris-Cl pH 7.5 containing 0.05 M EDTA. Also shown is the capacity of the isolated RNA to direct ovalbumin synthesis in a reticulocyte lysate system. The specifically adsorbed polysomes are enriched sevenfold for ovalbumin mRNA content. Since the tissue from which the polysomes were isolated synthesised only 16% ovalbumin, a sevenfold enrichment of mRNA is all that can be expected. In this experiment 100% of mRNA activity was recovered.

This immunoadsorbed RNA fraction containing ovalbumin mRNA also contains ribosomal RNA. The ovalbumin mRNA can be separated from rRNA by use of the millipore binding procedure of Brawerman *et al.*[119], or the poly dT cellulose technique of Nakazato and Edmonds[120], and thus presumably contains sequences of poly rA, as has been proposed for a large number of mRNAs from animal tissues[121,122].

We believe that this sequence of techniques can be used to isolate any specific mRNA, provided that an antibody can be obtained that is reactive with nascent chains. The major limitation of this technique at the present time is the amount of non-specific trapping of polysomes in the process of precipitation or adsorption of polysomes. This amounts to *ca.* 1–2% with the various matrices which we have employed[118]. The trapping is not a

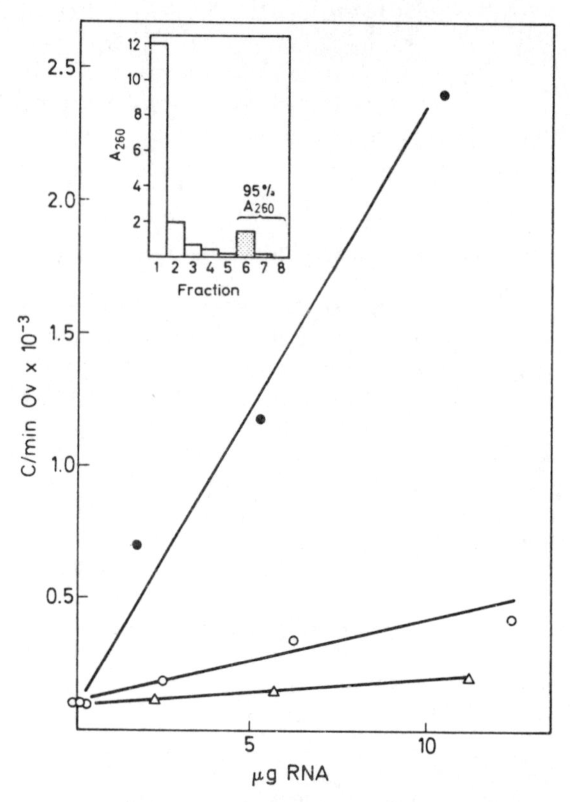

Figure 7.8 *In vitro* synthesis of ovalbumin by RNA extracted from immunoadsorbed chick oviduct polysomes. Chicks received a secondary stimulation with oestrogen for 18 h. Oviduct magnum was isolated and the relative rate of ovalbumin synthesis was measured by incubating explants of the oviduct. Ovalbumin was 17% of the total protein synthesised. Polysomes prepared from the same oviduct (20 A_{260} units) were treated for immunoadsorption. The A_{260} of the different fractions during the treatment was measured as in Figure 7.1 and is presented in the inset. Shaded fractions represent the eluted polysomes. RNA was extracted from total (○- - - -○), nonadsorbed (△- - - - △), and adsorbed polysomes (●- - - -●), and was assayed at different concentrations in the reticulocyte lysate system for the synthesis of ovalbumin. See Ref. 118 for details

function of the immunological reaction, and apparently involves the interaction of polysomes with the granular matrix, whether it be cross-linked protein, agarose, sepharose or acrylamide beads. In the case of ovalbumin, which constitutes 60% of the polysomes, this minor contamination is not important. However, if a protein constitutes only 1% of the protein synthesised, then, the contamination obviously becomes a significant problem.

7.8.5 Synthesis of a DNA product complementary to the ovalbumin mRNA, and an analysis of the number of copies of the ovalbumin gene

The presence of poly(rA) in the ovalbumin mRNA, presumably at the 3′ end[122], allows for the synthesis of a DNA product complementary to the ovalbumin mRNA. Details of the synthesis of this probe, and evidence that it hybridises only with ovalbumin mRNA, are presented by Sullivan *et al.*[123]. Figure 7.9 shows an experiment which indicates that there is only one copy of the ovalbumin gene per haploid genome of the chick, and that there is no amplification in the oviduct. In this experiment chick fibroblast DNA was labelled with [^{3}H]thymidine, and the single copy DNA was obtained. To this DNA, serving as a marker for single copy material, was added a small amount of the single stranded [^{14}C]labelled DNA product complementary to the ovalbumin mRNA, and 10 mg of cell DNA, either from chick liver, or chick oviduct. The question to be asked is whether the [^{14}C]DNA product, which is essentially hybridising with the cell DNA, does so along a C_0t curve of single copy DNA, or at a lower C_0t value, indicative of hybridisation to multiple copy (reiterated) DNA. Figure 7.9 shows that the probe hybridises along the C_0t curve of the marker single copy DNA. For more details consult Sullivan *et al.*[123].

It is of interest that although gene amplification has been demonstrated for ribosomal genes in frog amphibians[98], differential gene amplification does not appear to be a general mechanism whereby specific cell types are capable of synthesising large amounts of a specific protein[59, 80, 124].

7.8.6 Analysis of actinomycin D 'superinduction'

The techniques available for quantifying the amount of specific mRNA, and identification and quantification of specific polysomes has been used in the oviduct to analyse the perplexing problem of the so-called 'superinduction', whereby the administration of actinomycin D to a tissue that is actively synthesising a protein actually leads to a greater accumulation of that protein. Figure 7.10 shows the pattern of accumulation in the capacity to synthesise ovalbumin as a function of hormone administration, its withdrawal and re-administration[126]. Noteworthy is the fact that if actinomycin D is administered to chicks and, 4 h later, oviduct fragments are studied for ovalbumin synthesis (compare open and closed circles), that ovalbumin synthesis now constitutes a greater proportion of protein synthesised. This is formally similar to the superinduction as studied in HTC cells, and found in many experimental systems[127]. This problem has been studied extensively by Palmiter and Schimke[126] with the following findings:

(a) There is no decrease, or only a moderate decrease, in the absolute rate of protein synthesis following actinomycin D administration to chicks.

(b) Following incubation of oviduct explants in the presence of concentrations of actinomycin D that totally inhibit RNA synthesis, the capacity for synthesis of ovalbumin increases (as it does when the drug is administered to the intact animal), and that capacity does not decay with time compared

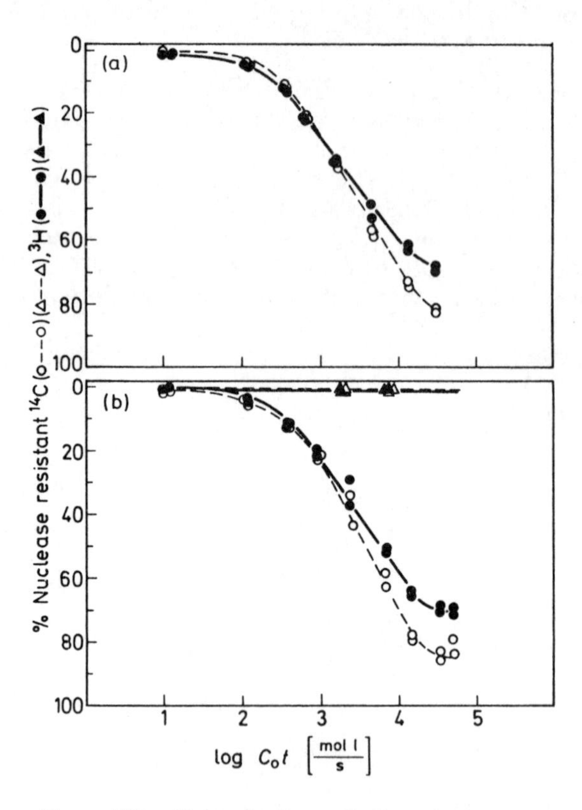

Figure 7.9 Determination of the absolute copy number of ovalbumin sequences in the chicken genome. [^{3}H]-single stranded DNA product at 1 ng ml^{-1}, [^{14}C]-unique sequence chicken DNA at 300 ng ml^{-1} and unlabelled cellular DNA from chicken liver, chicken oviduct or calf thymus at 10 mg ml^{-1} were mixed together, then melted and re-annealed. At different times aliquots were taken and assayed for resistance to S_1 nuclease. The data are plotted relative to the C_0t of the unlabelled cellular DNA. Panel (a): [^{14}C]-unique sequence DNA, ○, and [^{3}H]ovalbumin specific DNA product, ●, reassociated with chicken liver DNA. Panel (b): [^{14}C]-unique sequence DNA, ○, and [^{3}H]ovalbumin specific DNA product, ●, reassociated with chicken oviduct DNA; [^{14}C]-unique sequence DNA, △, and [^{3}H] ovalbumin specific DNA, ▲, reassociated with calf thymus DNA. See Ref. 129 for details

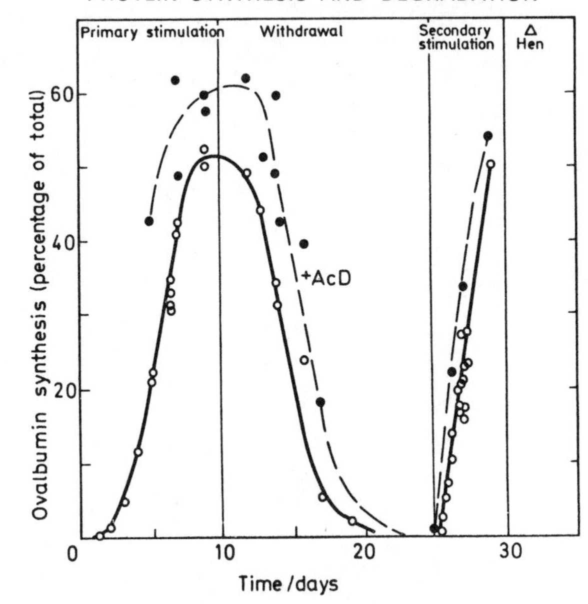

Figure 7.10 Effect of oestrogen and actinomycin D on the relative rate of synthesis of ovalbumin in chick oviduct during primary stimulation, withdrawal and secondary stimulation. Immature chicks 4 days old were injected intramuscularly with 1 mg oestradiol benzoate daily (primary stimulation), and after 10 days without oestrogen administration (withdrawal) administration was resumed (secondary stimulation). Chicks in groups of 2–4 were injected with actinomycin D (5 mg kg^{-1}) for 4–5 h prior to isolating oviducts. Fragments of oviduct were then incubated in Hanks' salt solution for 1 h with [^{3}H]amino acids (10 μCi ml^{-1}). Following homogenisation and centrifugation at 100 000 × g for 1 h, ovalbumin was precipitated from the supernatant using a specific antibody. Results are presented as a percentage of total acid-precipitable radioactivity in supernatant that is precipitated immunologically with anti-ovalbumin antibody. Details are given in Ref. 126. ○——○, oestrogen; ●- - -●, oestrogen plus actinomycin D 4 h before killing

to the untreated control explants. On the other hand, the capacity to synthesise other protein (not specified) declines with a half-life of 4–5 h. Whether this indicates that some mRNAs are unstable relative to others (the old interpretation of such experiments), or whether actinomycin D simply renders some mRNAs untranslatable by an unknown mechanism, as suggested by recent studies of Singer and Penmann[21] and Murphy and Attardi[20], is unknown.

(c) The combination of increased proportional synthesis of ovalbumin and no or little decrease of total protein synthesis indicates that the absolute rate of ovalbumin synthesis has increased. Parenthetically, it should be stated that both conalbumin and lysozyme are also 'superinduced' in this system. This

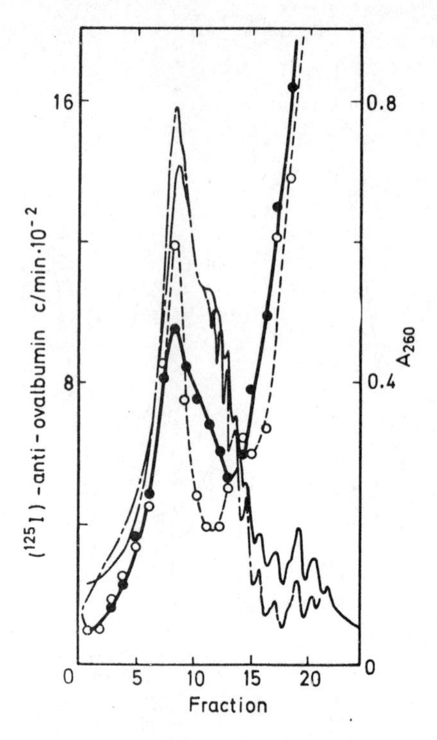

Figure 7.11 Effect of actinomycin D on the size of ovalbumin-synthesising polysomes during oestrogen stimulation. Magnum polysomes were isolated from chicks treated as in Figure 7.10. They were incubated with [^{125}I]antiovalbumin and displayed on sucrose gradients. Sedimentation is from right to left, monosomes are in Fraction 19. [^{125}I]antibody binding and A_{260} profile of polysomes from actinomcyin D treated chicks; (○- - - -○, - - - -); that of controls, (●——●, ——). Details are given in Ref. 126

increase in absolute rate of ovalbumin synthesis could result from an increase in translatable mRNA, an increase in the number of polysomes synthesising ovalbumin, or an increased rate of translation of the existing ovalbumin mRNA. Studies of Palmiter and Schimke[126], and of Rhoads *et al.*[108] indicate that there is no increase in translatable mRNA following actinomycin D, nor is there redistribution of ovalbumin mRNA between polysomes and other cell fractions. Hence we conclude that there is no alteration in mRNA content or distribution. Quantification of the number and size distribution of ovalbumin synthesising polysomes based on [^{125}I]antiovalbumin binding is shown in Figure 7.11. Integration of the amount of radioactivity bound to polysomes from control and actinomycin D treated chicks shows that essentially the same number of nascent peptide chains are present in both tissues, and hence the number of polysomes synthesising ovalbumin is similar in both instances. Figure 7.11 shows one more finding of significance, and that is that following actinomycin D treatment, the size distribution of ovalbumin-synthesising polysomes has shifted to a far sharper modality about 12-membered polysomes, indicating that on the average there are more ribosomes per mRNA.

That the polysomes synthesising ovalbumin (as well as conalbumin) are actually functioning more rapidly is shown in Figure 7.12, which provides estimates of transit times as determined by the technique of Fan and Penman[128]. In such an experiment oviduct fragments are incubated in a labelled

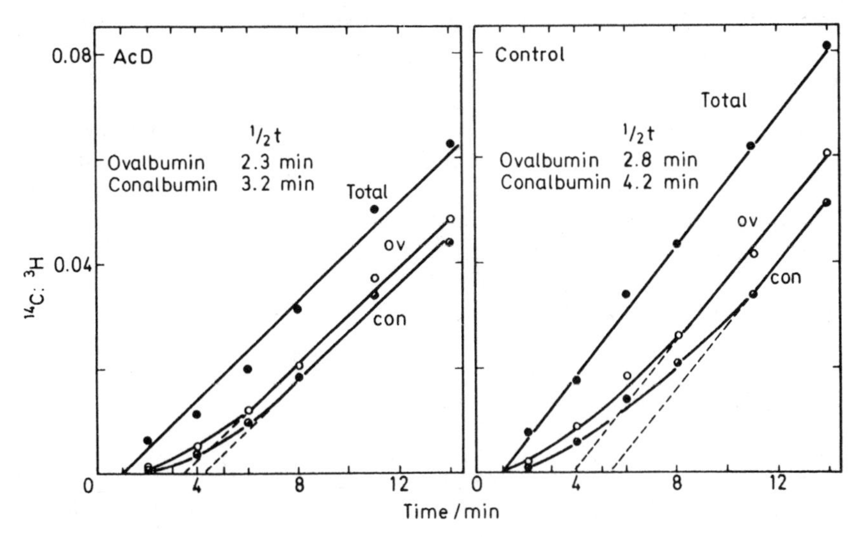

Figure 7.12 Effect of actinomycin D on the average mRNA transit time. Magnum explants from chicks which had received 5 days of primary stimulation were incubated at 33°C for 5 h; one set included actinomycin D at 10 μg ml^{-1}. [^{3}H]amino acids were added after 3 h and [^{14}C]amino acids were added at 5 h; then, at the intervals indicated, samples were removed for the determination of radioactivity in total and released proteins. The horizontal distance between the two curves is an estimate of the mean half-transit time ($\frac{1}{2}T$). See Ref. 126 for details

amino acid mixture, and at varying times tissue is homogenised and analysed for total incorporation which includes both nascent chains and released proteins, as well as for the radioactivity present in a 100 000 × g supernatant following treatment of the tissue with detergents, which is precipitable with specific antibodies against conalbumin and ovalbumin, i.e. release protein only. The difference between these two curves, projected to the base line, indicates the half-transit times of the released proteins. As shown in Figure 7.12 in the control fragments, the half-transit time for ovalbumin is shorter than that for conalbumin, which is in keeping with the molecular weights of the two proteins of 43 000 and 73 000 daltons respectively. Following treatment of the chicks with actinomycin D, the half-transit times are characteristically reduced some 30%. This 30% decrease in transit time accounts for the increased rate of ovalbumin synthesis produced by actinomycin D.

We therefore propose that in this system the mechanism for so-called superinduction requires only that some factor is rate-limiting for overall protein synthesis, and that following actinomycin D administration, some mRNAs are rendered untranslatable. The mRNA that remain can now be translated more efficiently. Clearly under these circumstances the rate of chain initiation must increase, and this, coupled with the finding of Figure 7.11 that following actinomycin D, the size of ovalbumin synthesising polysomes increases, suggested that, indeed, initiation is the rate-limiting step in protein synthesis that is partially relieved by the removal of some mRNAs. Thus it is unnecessary to invoke specific regulatory models for the superinduction phenomenon in this system. Such experiments leave unresolved the question of why some mRNAs are stable relative to others, or some are rendered untranslatable following actinomycin D treatment.

The foregoing series of experiments, and experimental techniques are capable of a finer dissection of the regulation of protein synthesis than has hitherto been available, and their application to other systems and other proteins should provide far more definitive answers to the question of what controls, the absolute rate of synthesis of a given protein depending on the state of the cells. Under various conditions, different factors appear to be rate-limiting.

7.8.6 Regulation of protein degradation

Although the continual degradation of cell proteins is extensive, it has been difficult to determine specific molecular mechanisms, both because the intermediate reaction products of specific protein degradation in physiological systems have essentially not been identified, and also because the properties of protein degradation as found in intact cells systems have not been adequately mimicked in broken cell systems[129]. Any understanding of the molecular mechanisms or control of protein degradation must take into account certain fundamental properties of such turnover: (a) the degradation appears to be random, inasmuch as the loss of labelled protein during a chase period, or fall of enzyme activity following elevation to a high level, follows first-order kinetics; (b) there is a marked heterogeneity of turnover rates of individual proteins (see Table 7.1); (c) the rate constant of degradation is in many cases characteristic of a given protein, but in other cases can be altered markedly.

We may consider two general types of mechanisms for the regulation of protein degradation.

7.8.6.1 Properties of the protein molecule as a substrate for degradation

Protein molecules can exist in various conformational states of varying degrees of detection. Thus a protein molecule might be subject to degradation only when it assumes certain conformations[143]. Thus a heterogeneity of degradation would occur among different proteins (enzymes), depending on the number and nature of particularly labile peptide bonds exposed in certain conformations. Interactions of proteins with various ligands, including other proteins, co-factors, prosthetic groups, lipids, etc. can alter such conformations and thereby alter the proteins as substrates for inactivation[130]. The model that emerges from this concept is one in which the protein molecules are individually available to a degradative process that is present at all times. Shifting concentrations of substrates, co-factors, etc. as occur under various hormonal and physiological conditions would lead to a variety of effects on specific enzymes, either to stabilise them (tryptophan and tryptophan oxygenase), or to labilise them (as proposed for glutamine and glutamine synthetase). Effects of ligands to alter physical and enzymatic inactivation of numerous enzymes are well known[131]. Consistent with this hypothesis is the demonstration by Bond of a general correlation between known rates of degradation of proteins *in vivo* and their rate of inactivation by trypsin and chymotrypsin *in vitro*[132]. In addition, Goldberg[133] has recently shown that when amino acid analogues are incorporated into the proteins of *E. coli*, protein degradation is accelerated, either as studied *in vivo*, or *in vitro*, using added proteases. Dice, Bradley and Schimke have made similar observations with amino acid analogues in animal cells in culture*. Dice, Dehlinger and Schimke[134] have also demonstrated that there is a correlation between *in vivo* rates of turnover of rat liver cytoplasmic proteins and their susceptibility to digestion by various proteases, including trypsin, chymotrypsin and pronase.

Such a concept could also explain the development of heterogeneity of rate constants of degradation. Taking a cue from mutations in *E. coli* which decreases the stability of the lac repressor[163], and arginine tRNA synthetase[135] and B-galactosidase[136], one can readily envisage an evolutionary process in which certain mutations in the amino acid sequence of proteins that either increase or decrease stability of a protein would be advantageous to an organism, depending on whether rapid or slow turnover would be advantageous.

A series of papers from Schimke and his collaborators[134, 137-140] have provided evidence that there is a correlation between the rate of degradation of proteins, whether associated with membranes, ribosomes, or cytoplasmic proteins, and their molecular size. This correlation is found for proteins of various tissues, and for cytoplasmic, membrane and ribosomal proteins. In addition, Dehlinger *et al.*[141] have shown that the different subunits of the multienzyme complex of rat liver, fatty acid synthetase, are turning over at different rates, and that the larger subunits are turning over at a greater rate than the smaller subunits. This has led Dehlinger and Schimke[137] to propose that macromolecular complexes are in a continual state of association dissociation, and that degradation occurs primarily in the dissociated state, a proposal in keeping with the suggestion of Fritz *et al.*[142]. Thus one of the rate limiting parameters for degradation that should be considered is the facility of dissociation of a protein. The requirement for dissociation of a protein into subunits prior to degradation could explain the discrepancy between the general correlation of the size of protein and its rate of turnover, and the fact

* Dice, J. F., Bradley, M. and Schimke, R. T., to be published.

that three specific enzymes, with roughly the same molecular weight, i.e. LDH_5[166], arginase, and tyrosine aminotransferase[162], have half-lives *in vivo* of 16 days[53], 4–5 days[167], and 1.5 h[23] respectively, if the rate of dissociation of these three proteins into subunits were markedly different.

Considerations of enzymatic mechanism are hampered by lack of suitable mutants in the degradative process itself. The lack of such mutants is curious and may indicate that continual degradation is so important to cell viability that such mutants would be lethal. Another problem in studying degradation of proteins involves the identification of the products of specific protein degradation once a protein has lost enzymatic activity or immunological reactivity. One of the few examples in which intermediates have been demonstrated is the study by Goldschmidt, who studied a mutant in B-galactosidase in growing *E. coli*[136]. He identified mutations in this enzyme, which were point mutations in the structural gene, for which no B-galactosidase activity could be demonstrated in induced cells. Using extremely brief pulses of radioactivity he could demonstrate the synthesis of enzyme, with subsequent appearance of large-sized fragments of the enzyme within a few seconds after removal of the pulse. This study is instructive in that it demonstrates that intermediates in the degradation of this protein appear to be generated by endopeptidase attack, and at relatively specific sites since specific-sized intermediates in degradation were observed. Furthermore, such degradation products are extremely short-lived. This would be in keeping with the concept that initial peptide bond-cleavage of the protein leads to unfolding of the protein, and subsequent rapid peptidase cleavage of the remaining fragments[143]. These experiments also point out the difficulties in demonstrating intermediate reaction products. Goldschmidt's experiments support the concept that the rate of degradation of a protein is dependent on the structure of the protein as a substrate for a non-specific degradative process.

7.8.6.2 Alterations in activity of a degradative process

In the preceding paragraph the degradative system was considered to be non-specific with respect to the protein molecules degraded, and to be potentially active at all times. It is also conceivable, if not plausible, that the rate of degradation of total or specific proteins may be controlled by activation, inhibition, translocation within a cell, or *de novo* synthesis of degrading enzymes. In addition to the problem in studying enzyme degradation discussed in the preceding paragraph, several curious observations are of note that should be explained in the formulation of suitable mechanism(s) for degradation. In both animal and bacterial systems, inhibition of energy production and protein synthesis inhibits protein degradation (see Ref. 6 for a detailed review). Various explanations have been offered for such observations, including co-factor requirements[144], necessity for maintaining structural integrity of organelles, most specifically lysosomes[145], and a requirement for continued synthesis of degradative enzymes that are turning over rapidly[135]. The evidence for these explanations is highly indirect, and is based on the fact that various drugs that inhibit energy generation, or protein or RNA synthesis prevent the inactivation of enzymes, or release of prelabelled protein to a soluble form.

Since the consequences of lack of ATP, and inhibition of protein synthesis on the intracellular concentrations of many small molecules are vast, it is equally plausible that such drug effects result from accumulation of amino acids, tRNA species, etc. which may regulate by ligand interaction the activity of degradative enzymes or of proteins as substrates for degradation.

One obvious candidate for a degradative system is the lysosome, which occurs in virtually all cells[145]. Lysosomes are intracellular organelles that contain acid hydrolases and are currently conceived as being involved in the autophagy of discrete areas of cytoplasm. It is most difficult to conceive that lysosomes are involved in that protein degradation whose properties involve randomness and heterogeneity of degradation rate constants among different proteins, whether so-called 'soluble' proteins or those associated with membranes or ribosomes. Thus some mechanisms such as acetylation, formylation, or, as recently suggested, deamidation[146] would be required for the recognition of whether a protein molecule were to be degraded, perhaps involving transport in a lysosome. It seems reasonable to this author to propose that the system of lysosomes is important where cell involution or gross changes in rates of protein degradation occur, such as starvation and cell death, whereas the degradation that occurs in normal steady-state conditions, involves a system(s) not clearly understood at present[147]. This could involve lysosomes, but acting as a sieve, rather than in an 'all or none' fashion.

Another possibility is that there are specific degrading enzymes for specific proteins. There are examples of proteins or enzymes which appear to inhibit or inactivate specific enzymes[60, 148-150]. In addition, studies by Tata[151] and by Gross and Lapiere[152] on degeneration of amphibian tail during the thyroxine-induced metamorphosis suggest that *de novo* synthesis of specific proteolytic enzymes is required for these instances of tissue involution.

One particularly instructive study is that of Katanuma and his colleagues who have purified an endopeptidase that appears to be specific for pyridoxal containing enzymes, e.g. ornithine transaminase[153]. They have found that this protease activity specifically attacks the enzyme in the apo form, but not the holoenzyme. The degrading enzyme splits the enzyme into smaller sized fragments, which are then rapidly hydrolysed by endogenous peptidases. Thus this system involves a specific degrading enzyme and modification of the substrate ligand interaction (prosthetic group). In addition the activity of the enzyme is not constant, but increases under conditions of a pyridoxal deficient diet.

At one extreme, then, we might propose that the degradation of each protein requires a specific protein. This, however, is impossible, since the continual replacement of essentially all proteins would require that there exist a protein to degrade a protein . . . *ad infinitum*. It is most likely that, just as there are a number of different enzymes that hydrolyse RNA in an organism such as *E. coli*, there are also a number of different functional tasks at different sites and times, the sum total of which results in continual protein degradation.

References

1. Knox, W. E., Auerbach, V. H. and Lin, E. C. C. (1956). *Physiol. Rev.*, **36**, 164
2. Rechcigl, M. Jr. (1971). *Enzyme Synthesis and Degradation in Mammalian Systems*, p. 236 (M. Rechcigl, Jr., editor) (Baltimore: University Park Press)

3. Schimke, R. T. and Doyle, D. (1970). *Annu. Rev. Biochem.*, **39,** 929
4. Epstein, W. and Beckwith, J. R. (1968). *Annu. Rev. Biochem.* **37,** 411
5. Jacob, F. and Monod, J. (1961). *J. Molec. Biol.*, **3,** 318
6. Schimke, R. T. (1970). *Mammalian Protein Metabolism*, Vol. 4, p. 177 (H. N. Munro, editor) (New York: Academic Press)
7. Schoenheimer, R. (1942). *The Dynamic State of Body Constituents* (Cambridge, Mass.: Harvard University Press)
8. Coon, H. G. (1966). *Proc. Nat. Acad. Sci.* (*U.S.A.*), **53,** 66
9. Yasumura, Y., Tashjian, A. H. and Sato, G. H. (1966). *Science*, **154,** 1186
10. Schubert, D. (1968). *Proc. Nat. Acad. Sci.* (*U.S.A.*), **60,** 683
11. Segal, H. L. and Kim, Y. S. (1963). *Proc. Nat. Acad. Sci.* (*U.S.A.*), **50,** 912
12. Berlin, C. M. and Schimke, R. T. (1965). *Molec. Pharmacol.*, **1,** 149
13. Freedland, R. A. and Szepesi, B. (1971). *Enzyme Synthesis and Degradation in Mammalian Systems*, p. 103 (M. Rechigl, Jr., editor) (Baltimore: University Park Press)
14. Swick, R. W. (1957). *J. Biol. Chem.*, **231,** 751
15. Rudack, D., Chrisholm, E. M. and Holten, D. (1971). *J. Biol. Chem.*, **246,** 1249
16. Schimke, R. T., Sweeney, E. W. and Berlin, C. M. (1965). *J. Biol. Chem.*, **240,** 322
17. Soeiro, R. and Amos, H. (1966). *Biochim. Biophys. Acta*, **129,** 406
18. Revel, M., Hiatt, H. H., and Revel, J. P. (1964). *Science*, **146,** 1311
19. Laszlo, J., Miller, D. S., McCarty, K. S. and Hochstein, P. (1966). *Science*, **151,** 1007
20. Murphy, L. and Attardi, G. (1973). *Proc. Nat. Acad. Sci.* (*U.S.A.*), **70,** 115
21. Singer, R. and Penman, S. (1972). *Nature* (*London*), **240,** 101
22. Kenney, F. T. (1962). *J. Biol. Chem.*, **237,** 1610
23. Kenney, F. T. (1967). *Science*, **156,** 525
24. Schimke, R. T. (1964). *J. Biol. Chem.*, **239,** 3808
25. Jost, J. P., Khairallah, E. A. and Pitot, H. C. (1968). *J. Biol. Chem.*, **243,** 3057
26. Sellinger, O. Z., Lee, K. L. and Fesler, K. W. (1966). *Biochim. Biophys. Acta*, **124,** 289
27. Ganschow, R. E. and Schimke, R. T. (1969). *J. Biol. Chem.*, **244,** 4649
28. Doyle, D. and Schimke, R. T. (1969). *J. Biol. Chem.* **244,** 5449
29. Nakamura, H. and Littlefield, J. W. (1972). *J. Biol. Chem.*, **247,** 179
30. Tomkins, G. M., Gelehrter, T. D., Granner, D., Martin, D., Jr., Samuels, H. H. and Thompson, E. B. (1969). *Science*, **166,** 1474
31. Schimke, R. T. (1967). *Nat. Cancer Inst. Monog.*, **27,** 301
32. Silpananta, P. and Goodridge, A. (1971). *J. Biol. Chem.*, **246,** 5754
33. Cohen, P. P. (1966). *Harvey Lectures*, Series 60, 119
34. Balinsky, J. B., Shambaugh, G. E. and Cohen, P. P. (1970). *J. Biol. Chem.*, **245,** 128
35. Grossman, A. and Mavrides, C. (1967). *J. Biol. Chem.*, **242,** 1398
36. Shambaugh, G. E., III, Balinsky, J. B. and Cohen, P. P. (1969). *J. Biol. Chem.*, **244** 5295
37. Kenney, F. T. (1970). *Mammalian Protein Metabolism*, Vol. 4 (H. N. Munro, editor) (New York: Academic Press)
38. Loftfield, R. B. and Harris, A. (1956). *J. Biol. Chem.*, **219,** 151
39. Gan, J. C. and Jeffay, H. (1967). *Biochim. Biophys. Acta*, **148,** 448
40. Righetti, P., Little, E. P. and Wolf, G. (1971). *J. Biol. Chem.*, **246,** 5724
41. Poole, B. (1971). *J. Biol. Chem.*, **246,** 188
42. Schimke, R. T. (1973). *Methods in Enzymology*, in the press
43. Buchanan, D. L. (1961). *Arch. Biochem. Biophys.*, **94,** 500
44. MacDonald, R. A. (1961). *Arch. Int. Med.*, **107,** 335
45. Swick, R. W., Koch, A. L. and Handa, D. T. (1956). *Arch. Biochem. Biophys.*, **63,** 226
46. Knox, W. E. and Mehler, A. H. (1951). *Science*, **113,** 237
47. Knox, W. E. and Piras, M. M. (1967). *J. Biol. Chem.*, **242,** 2959
48. Feigelson, P., Feigelson, M. and Greengard, O. (1962). *Recent Progr. Hormone Res.*, **18,** 491
49. Majerus, P. W. and Kilburn, E. (1969). *J. Biol. Chem.*, **244,** 6254
50. Numa, S., Matsuhashi, M. and Lynen, F. (1961). *Biochem. Z.*, **334,** 203
51. Schimke, R. T. and Grossbard, L. (1968). *Annu. Rev. N.Y. Acad. Sci.*, **151,** 322
52. Tanaka, T., Harano, Y., Sue, F. and Morimura, H. (1967). *J. Biochem.* (*Tokyo*), **62,** 71
53. Fritz, P. J., Vesell, E. S., White, E. L. and Pruitt, K. M. (1969). *Proc. Nat. Acad. Sci.* (*U.S.A.*), **62,** 558
54. Rechcigl, M., Jr. and Heston, W. E. (1963). *J. Nat. Cancer Inst.*, **30,** 855

55. Heston, W. E., Hoffman, H. A. and Rechcigl, M., Jr. (1965). *Genetic Res.*, **6,** 387
56. Russell, R. L. and Coleman, D. L. (1963). *Genetics*, **48,** 1033
57. Coleman, D. L. (1966). *J. Biol. Chem.* (1966). **241,** 5511
58. Pitot, H. C., Peraine, C., Morse, P. A., Jr. and Potter, V. R. (1966). *Natl. Cancer Inst. Monograph*, **13,** 229
59. Suzucki, Y., Gage, L. O. and Brown, D. D. (1972). *J. Molec. Biol.*, **70,** 637
60. Gancedo, C. and Holzer, H. (1968). *Eur. J. Biochem.*, **4,** 190
61. Doyle, D. (1971). *J. Biol. Chem.*, **246,** 4965
62. Matsubara, S., Suter, H. and Aebi, H. (1967). *Humangenetik*, **4,** 29
63. Yoshida, A., Stamatoyannopoulos, G. and Motulsky, A. (1967). *Science*, **155,** 97
64. Seegmiller, J. E., Rosenbloom, F. M. and Kelley, W. N. (1967). *Science*, **155,** 1682
65. Rubin, C. S., Balis, M. E., Piomelli, S., Berman, P. H. and Dancis, J. (1969). *J. Lab. Clin. Med.*, **74,** 732
66. Feinstein, R. N., Howard, J. B., Braun, J. T. and Seaholm, J. E. (1966). *Genetics*, **53,** 923
67. Krooth, R. S., Howell, R. S. and Hamilton, H. B. (1962). *J. Exp. Med.*, **115,** 313
68. Forrest, G. L. and Klevecz, R. R. (1972). *J. Biol. Chem.*, **247,** 3147
69. Lee, K-L., Reel, J. R. and Kenney, F. T. (1970). *J. Biol. Chem.*, **245,** 5806
70. Reel, J. R. and Kenney, F. T. (1968). *Proc. Nat. Acad. Sci.* (*U.S.A.*), **61,** 200
71. Reel, J. R., Lee, K-L. and Kenney, F. T. (1970). *J. Biol. Chem.*, **245,** 5800
72. Peterkofsky, B. and Tomkins, G. M. (1968). *Proc. Nat. Acad. Sci.* (*U.S.A.*), **60,** 222
73. Greenman, D. L., Wicks, W. D. and Kenney, F. T. (1965). *J. Biol. Chem.*, **240,** 4420
74. Wicks, W. D., Greenman, D. L. and Kenney, F. T. (1965). *J. Biol. Chem.*, **240,** 4414, 4420
75. Granner, D. K., Thompson, E. B. and Tomkins, G. M. (1970). *J. Biol. Chem.*, **245,** 1472
76. Martin, D., Jr., Tomkins, G. M. and Granner, D. (1969). *Proc. Nat. Acad. Sci.* (*U.S.A.*) **62,** 248
77. Thompson, E. B., Granner, D. K. and Tomkins, G. M. (1970). *J. Molec. Biol.*, **54,** 159
78. Ross, J., Aviv, H., Scolnick, E. and Leder, P. (1972). *Proc. Nat. Acad. Sci.* (*U.S.A.*), **69,** 264
79. Hershko, A. and Tomkins, G. M. (1971). *J. Biol. Chem.*, **246,** 710
80. Bishop, J. O., Pemberton, R. and Baglioni, C. (1972). *Nature New Biol.*, **235,** 231
81. Friedkin, M. (1963). *Annu. Rev. Biochem.*, **32,** 185
82. Sirotnak, F. M., Kurita, S. and Hutchison, D. J. (1968). *Cancer Res.*, **28,** 75
83. Kashket, E. R., Crawford, E. J., Friedkin, M., Humphreys, S. R. and Goldin, A. (1964). *Biochemistry*, **3,** 1928
84. Perkins, J. P., Hillcoat, B. L. and Bertino, J. R. (1967). *J. Biol. Chem.*, **242,** 4771
85. Raunio, R. P. and Hakala, M. T. (1967). *Molec. Pharmacol.*, **3,** 279
86. Misra, D. K., Humphreys, S. R., Friedkin, M., Goldin, A. and Crawford, E. J. (1961). *Nature* (*London*), **189,** 39
87. Hillcoat, B. L., Sweet, V. and Bertino, J. R. (1967). *Proc. Nat. Acad. Sci.* (*U.S.A.*), **58,** 1632
88. Hakala, M. T. and Zakrzrewski, S. F. (1966). *Molec. Pharmacology*, **2,** 432
89. Littlefield, J. W. (1969). *Proc. Nat. Acad. Sci.* (*U.S.A.*), **62,** 88
89a. Hakala, M. T., Zakrzrewski, S. F. and Nichol, C. A. (1961). *J. Biol. Chem.*, **236,** 952
90. Kirk, D. L. and Moscona, A. A. (1963). *Develop. Biol.*, **8,** 341
91. Moscona, A. A. and Piddington, R. (1967). *Science*, **158,** 496
92. Reif-Lehrer, L. and Amos, H. (1965). *Biochem. J.*, **106,** 425
93. Moscona, A. A. and Kirk, D. L. (1965). *Science*, **148,** 519
94. De Mars, R. (1958). *Biochem. Biophys. Acta*, **27,** 435
95. Paul, J. and Fottrell, P. F. (1963). *Biochem. Biophys. Acta*, **67,** 334
96. Tiemeier, D. C. and Milman, G. (1972). *J. Biol. Chem.*, **247,** 5722
97. Kulka, R. G., Tomkins, G. M. and Crook, R. B. (1972). *J. Biol. Chem.*, **54,** 175
98. Dawid, I. B. and Brown, D. M. (1968). *Science*, **160,** 272
99. Georgeiv, G. P. (1967). *Progr. Nucleic Acid Res. Molec. Biol.*, **6,** 259
100. Gross, P. R. (1968). *Annu. Rev. Biochem.*, **37,** 631
101. Wainwright, S. D. (1972). *Control Mechanisms and Protein Synthesis* (New York: Columbia University Press)
102. Hirsch, C. A. and Hiatt, H. H. (1966). *J. Biol. Chem.*, **241,** 5936

102a. Schimke, R. T. (1969). *Axenic Mammalian Cell Reactions*, p. 181 (G. L. Tritsch, editor) (New York: Manel Dekker)
103. Clayton, R. M., Truman, D. E. S. and Campbell, J. C. (1972). *Cell Differentiation*, **1,** 25
104. Rhoads, R. E., McKnight, G. S. and Schimke, R. T. (1971). *J. Biol. Chem.*, **246,** 7407
105. Gurdon, J. B., Lane, C. D., Woodland, H. R. and Marbaix, G. (1971). *Nature* (*London*), **233,** 177
106. Arias, I. M., Doyle, D. and Schimke, R. T. (1969). *J. Biol. Chem.*, **244,** 3303
107. Stavnezer, J. and Huang, R. C. C. (1971). *Nature* (*London*), **230,** 172
108. Rhoads, R. E., McKnight, G. S. and Schimke, R. T. (1973). *J. Biol. Chem.*, **247,** in the press
109. Palmiter, R. D., Christensen, A. K. and Schimke, R. T. (1970). *J. Biol. Chem.*, **245,** 833
110. Means, A. R., Comstock, J. P., Rosenfeld, G. C. and O'Malley, B. W. (1972). *Proc. Nat. Acad. Sci.* (*U.S.A.*), **69,** 1146
111. Heywood, S. M., Dowben, R. M. and Rich, A. (1967). *Proc. Nat. Acad. Sci.* (*U.S.A.*), **57,** 1002
112. Palacios, R., Palmiter, R. D. and Schimke, R. T. (1972). *J. Biol. Chem.*, **247,** 2316
113. Palmiter, R. D., Palacios, R. and Schimke, R. T. (1972). *J. Biol. Chem.*, **247,** 3296
114. Suzuki, Y. and Brown, D. D. (1972). *J. Molec. Biol.*, **63,** 409
115. Allen, E. R. and Terrence, C. F. (1968). *Proc. Nat. Acad. Sci.* (*U.S.A.*), **60,** 1209
116. Holme, G., Boyd, S. L. and Sehon, A. H. (1971). *Biochem. Biophys. Res. Commun.*, **45,** 240
117. Takagi, M. and Ogata, K. (1971). *Biochem. Biophys. Res. Commun.*, **42,** 125
118. Palacios, R., Sullivan, D., Summers, N. M., Kiely, M. L. and Schimke, R. T. (1973). *J. Biol. Chem.*, **248,** 540
119. Brawerman, G., Mendecki, J. and Lee, S. Y. (1972). *Biochemistry*, **11,** 637
120. Nakazato, H. and Edmonds, M. (1972). *J. Biol. Chem.*, **247,** 10, 3365
121. Adesnik, M., Salditt, M., Thomas, W. and Darnell, J. E. (1972). *J. Molec. Biol.*, **71,** 21
122. Darnell, J. E., Wall, R. and Tushinski, (1971). *Proc. Nat. Acad. Sci.* (*U.S.A.*), **68,** 1321
123. Sullivan, D., Palacios, R., Stavnezer, J., Taylor, J. M., Faras, A. J., Kiely, M. L., Summers, N. M., Bishop, J. M. and Schimke, R. T. (1973). *J. Biol. Chem.*, Nov. 10th issue
124. Harrison, P. R., Hell, A., Birnie, G. D. and Paul, J. (1972). *Nature New Biol.*, **239,** 219
125. Packman, S., Aviv, H., Ross, J. and Leder, P. (1972). *Biochem. Biophys. Res. Commun.*, **49,** 813
126. Palmiter, R. D. and Schimke, R. T. (1973). *J. Biol. Chem.*, **248,** 1502
127. Tomkins, G. M., Levinson, B. B., Baxter, J. D. and Dethlefsen, L. (1972). *Nature New Biol.*, **239,** 88
128. Fan, H. and Penman, S. (1970). *J. Molec. Biol.*, **50,** 655
129. Schimke, R. T., Sweeney, E. W. and Berlin, C. M. (1965). *J. Biol. Chem.*, **240,** 4609
130. Grisolia, S. (1964). *Physiol. Rev.*, **44,** 657
131. Green, N. M. and Neurath, H. (1954). *The Proteins*, Ch. 2B, p. 1059 (H. Neurath and K. Bailey, editors) (New York: Academic Press)
132. Bond, J. S. (1971). *Biochem. Biophys. Res. Commun.*, **43,** 333
133. Goldberg, A. L. (1972). *Proc. Nat. Acad. Sci.* (*U.S.A.*), **69,** 422
134. Dice, F., Dehlinger, P. J. and Schimke, R. T. (1973). *J. Biol. Chem.*, **248,** in the press
135. Williams, L. S. and Neidhardt, F. C. (1969). *J. Molec. Biol.*, **43,** 529
136. Goldschmidt, R. (1970). *Nature* (*London*), **228,** 1151
137. Dehlinger, P. J. and Schimke, R. T. (1970). *Biochem. Biophys. Res. Commun.*, **49,** 1473
138. Dehlinger, P. J. and Schimke, R. T. (1971). *J. Biol. Chem.*, **246,** 2574
139. Dehlinger, P. J. and Schimke, R. T. (1972). *J. Biol. Chem.*, **297,** 1257
140. Dice, F. and Schimke, R. T. (1972). *J. Biol. Chem.*, **247,** 98
141. Dehlinger, P. J., Tweto, J. and Larrabee, A. R. (1972). *Biochem. Biophys. Res. Commun.*, **48,** 1371
142. Fritz, P. J., White, E. L., Vessell, E. S. and Pruitt, K. M. (1971). *Nature* (*London*), **230,** 119
143. Linderstrom-Lang, K. (1950). *Cold Spring Harbor Symposium Quant. Biol.*, **14,** 117
144. Penn, N. W. (1961). *Biochim. Biophys. Acta*, **53,** 490
145. De Duve, C. and Wattiaux, R. (1966). *Annu. Rev. Physiol.*, **28,** 432

146. Robinson, A. B., McKerrow, J. H. and Cary, P. (1970). *Proc. Nat. Acad. Sci.* (*U.S.A.*), **66,** 753
147. Hartley, B. S. (1960). *Annu. Rev. Biochem.*, **29,** 45
148. Bonsignore, A., DeFlora, A., Mangiarotti, M. A., Lorenzoni, I. and Alema, S. (1968). *Biochem. J.*, **106,** 147
149. Dvorak, H. F., Anraku, Y. and Heppel, L. A. (1966). *Biochem. Biophys. Res. Commun.*, **24,** 628
150. Messenguy, F. and Wiame, J. (1969). *FEBS Lett.*, **3,** 47
151. Tata, J. R. (1966). *Devel. Biol.*, **13,** 77
152. Gross, J. and Lapiere, C. M. (1962). *Proc. Nat. Acad. Sci.* (*U.S.A.*), **48,** 1014
153. Kominani, E., Koyayashi, K., Kominani, S. and Katanuma, N. (1972). *J. Biol. Chem.*, **247,** 6848
154. Russell, D. and Snyder, S. H. (1968). *Proc. Nat. Acad. Sci.* (*U.S.A.*), **60,** 1420
155. Marver, H. S., Collins, A., Tschudy, D. P. and Rechcigl, M., Jr. (1966). *J. Biol. Chem.*, **241,** 4323
156. Swick, R. W., Rexroth, A. K. and Stange, J. L. (1968). *J. Biol. Chem.*, **243,** 3581
157. Price, V. E., Sterling, W. R., Tarantola, V. A., Hartley, R. W., Jr. and Rechcigl, M., (1962). *J. Biol. Chem.*, **237,** 3468
158. Niemeyer, H. (1966). *Nat. Cancer Inst. Monog.*, **27,** 29
159. Kuryama, Y., Omura, T., Siekevitz, P. and Palade, G. E. (1969). *J. Biol. Chem.*, **244,** 2017
160. Higgins, M., Kawachi, T. and Rudney, H. (1971). *Biochem. Biophys. Res. Commun.*, **45,** 138
161. Shapiro, D. J. and Rodwell, V. W. (1971). *J. Biol. Chem.*, **246,** 3210
162. Auricchio, F., Valierote, F., Tomkins, G. and Riley, W. (1970). *Biochim. Biophys. Acta*, **221,** 307
163. Platt, T., Miller, J. H. and Weber, K. (1970). *Nature* (*London*), **228,** 1154
164. Blumenthal, G. and Greenberg, O. M. (1970). *Oncology*, **24,** 223
165. Bertino, J. R., Silber, R., Freeman, M., Alenty, A., Albrecht, M., Gabrio, B. W. and Huennekens, F. M. (1963). *J. Clin. Invest.*, **42,** 1899
166. Castellino, F. J. and Barker, R. (1968). *Biochemistry*, **7,** 2207
167. Hirsh-Kolb, H. and Greenberg, D. M. (1969). *J. Biol. Chem.*, **243,** 6123
168. Labrie, F. and Korner, A. (1968). *J. Biol. Chem.*, **243,** 1120
169. Lockard, R. E. and Lingrel, J. B. (1969). *Biochem. Biophys. Res. Commun.*, **37,** 204
170. Molloy, G. R., Sitz, T. O. and Schmidt, R. R. (1973). *J. Biol. Chem.*, **248,** 1970
171. Segal, H. L., Rosso, R. G., Hopper, S. and Weber, M. M. (1962). *J. Biol. Chem.*, **237,** PC3303

8
mRNA Stability and the Control of Specific Protein Synthesis in Highly Differentiated Cells

F. C. KAFATOS
Harvard University and University of Athens, Greece

and

R. GELINAS
Harvard University

8.1 INTRODUCTION

8.1.1 Comments on highly differentiated cells

We may define as highly differentiated those cells which, at some point in their career, become specialised for the large-scale production of one or a few specialised protein products. Many of these cells are secretory: the pancreatic acinar cells, the giant salivary gland cells of flies, the tubular gland cells of chick oviduct and the cocoonase-producing cells of silkmoths are well known examples. Other highly differentiated cells do not secrete their specialised products, but retain them intracellularly: vertebrate red blood cells, muscle cells and lens fibres belong in this category.

Highly differentiated cells represent only an extreme in differentiation for specific protein synthesis. Many developmental phenomena (e.g. production of embryonic inducers) involve synthesis of rather minor amounts of specific proteins, or of other products. Thus, highly differentiated cells cannot be considered representative of all differentiated systems, either in descriptive or in mechanistic terms. Still, there are so many highly differentiated cell types and their methodological advantages are so numerous, that intensive study is both justified and productive.

One of the most important attributes of highly differentiated cells is the ease with which their specialised macromolecular product can be assayed. Because of the small number of proteins produced, and the high rate of their production, in many cases the proteins have been completely purified and extensively studied. Thus, reliable chemical or immunochemical methods for their estimation have been developed. This in turn has permitted quantitative description of the kinetics of specific protein synthesis during differentiation, with greater reliability than is possible for relatively minor tissue components, such as enzymes of intermediary metabolism. The single-minded specialisation of these cells has similar advantages at the RNA level, as indicated by the recent progress in specific mRNA isolation. It is not unreasonable to expect that such systems will permit the first analysis of differentiation at the level of specific mRNA and gene metabolism.

When the development of highly differentiated cells is described in terms of *translational specialisation*, i.e. in terms of changing absolute or relative rates of cell-specific protein synthesis over time, a rather common pattern is often observed[1,2]. After an initial 'predifferentiated phase', during which the specific product is not produced (or at least is produced at a rate of no greater than that observed in other cells of the body), the period of differentiation becomes apparent by a heightened rate of specific protein synthesis. This period, in turn, can be separated into three phases. In Phase I (protodifferentiated phase), the rate of specific synthesis is relatively low. In Phase II (fully differentiated phase), specific synthesis increases dramatically and continues to accelerate for a significant period of time. Finally, in Phase III the rate of synthesis either stabilises or is further modulated; in time-limited cells, such as reticulocytes, synthesis usually decays rapidly, whereas in cells which remain functional synthesis is modulated upwards or downwards by hormones or other factors.

The three transition points between the various phases are obvious targets for study. An additional significant feature is the continuous acceleration of cell-specific protein synthesis during Phase II of differentiation. The high degree of translational specialisation which defines these cells is attained gradually, rather than as a step function. In that respect, differentiation is not adequately described by models of alternative, mutually exclusive states, but must take into account the actual kinetics of developmental changes. How, then, is the progressive change in translational specialisation controlled?

8.1.2 A simple postulate

An analysis of the control of cell-specific protein synthesis can be attempted by means of a very simple postulate: the kinetics of protein synthesis during Phase II of differentiation reflect the accumulation of specific mRNA, resulting from rapid transcription from a single gene and an unusual mRNA stability.

In this paper the implications and predictions of this postulate will be examined, both in theoretical terms and in the light of experimental evidence. The postulate will be put forth not as an established fact but as a tentative generalisation, useful if it has heuristic value. It will serve as a probe for consideration of multiple levels of developmental regulation. While the major focus will be on the importance of mRNA stability, we will also attempt an evaluation of the general significance of gene multiplicity, transcriptional control (especially the rate of RNA chain initiation), post-transcriptional controls other than mRNA stability (e.g. mRNA storage, variations in translational productivity, etc.).

In addition, this analysis will help focus attention on an important biological parameter: the time constant of changes in protein synthetic rate. In the systems mentioned above, the duration of Phase II is of the order of days. However, there are systems in which major changes in specific protein synthesis occur over a period of hours or even minutes. Such short time-constant changes are well known in early embryogenesis, but also occur in some highly differentiated cells. To what extent does the time constant of the system influence the choice of methods for regulation of specific protein synthesis?

8.2 ON PROPORTIONALITY BETWEEN SPECIFIC mRNA CONTENT AND PROTEIN SYNTHESIS

8.2.1 Some clarifications

If one attempts to understand changing rates of protein synthesis in terms of production, use and degradation of the corresponding mRNA, a key question is whether a proportionality exists between the specific mRNA content of the cells and the rate of corresponding protein synthesis. Clearly, proportionality is the foundation of any strict transcriptional control model. This proportionality should hold in an absolute sense, i.e. over time (or even in different cell types) if mRNA is the factor limiting protein synthesis and no 'masked mRNA' pool exists in the cells. However, specific protein synthesis can be evaluated not only in absolute terms, but also relative to the total protein synthesis in the cells at that time. Indeed, for methodological reasons, relative rates can often be determined much more accurately than absolute protein synthetic rates[2]. It is important to recognise that the requirements are much less stringent for proportionality between the *relative* rate of specific protein synthesis (as a percentage of the total) and the *relative* mRNA content (content of a specific mRNA as a percentage of total mRNA in the cell) than for absolute proportionality. In relative terms, proportionality will hold even if protein synthesis becomes limited by translational or 'masking' factors, provided these factors do not discriminate among mRNAs. Thus, '*absolute*' proportionality demands that mRNA be rate limiting; '*relative*' proportionality merely demands non-specificity of factors controlling mRNA availability and translational productivity (yield of polypeptide chains per minute per active mRNA molecule).

In terms of our simple postulate (Section 8.1.2), it seems probable that, in Phase II of differentiation, at least relative proportionality is the rule in long time-constant systems. In short time-constant systems and at other phases of differentiation, proportionality may or may not apply.

8.2.2 Evidence supporting proportionality between mRNA content and specific protein synthesis

Until recently, it was impossible to obtain rigorous evidence regarding proportionality between mRNA content and protein synthesis in eukaryotes. However, we have now entered a period of specific mRNA isolation. We can predict confidently that rigorous evaluation of the proportionality will be made in a number of systems in the near future, as we learn how to assay for

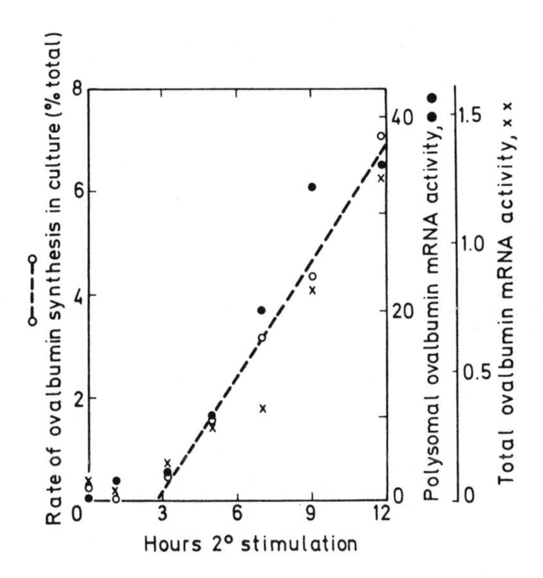

Figure 8.1 Palmiter's data on the proportionality between rate of ovalbumin synthesis and ovalbumin message content, after secondary oestrogen stimulation. The rate of synthesis was determined as in Ref. 14. Message content was determined with a reticulocyte cell-free protein synthesising system[138]. Determinations for polysome-bound and for total cellular ovalbumin message were performed separately; the activity units used were arbitrary and different. (From R. Palmiter, unpublished results.)

specific mRNAs quantitatively, both in chemical terms (e.g. hybridisation) and in functional terms (translation in cell-free systems or in frog oocytes).

An exemplary study of this type has been made in the hormone-stimulated chick oviduct[3, 13-16]. In this tissue, following secondary oestrogen stimulation, the rate of ovalbumin synthesis increases dramatically and with nearly linear kinetics, after an initial lag period of *ca.* 3 h. Quantitative assays reveal a strict proportionality between the rate of ovalbumin synthesis and the cellular content of ovalbumin mRNA, both total and polysome-bound (Figure 8.1).

A second line of support for proportionality comes from the studies of Gurdon and his colleagues on the synthesis of haemoglobin by frog oocytes in response to injected haemoglobin mRNA. They observed that the rate of haemoglobin synthesis increases linearly with amount of message injected, over a substantial range, and that the injected message does not compete with endogenous mRNA, i.e. does not depress the background level of protein synthesis[5, 6]. Non-linear response and competition occur only with substantial haemoglobin message inputs (10 μg ml^{-1} oocyte cytoplasm)[6, 7]. The results clearly indicate that, at least with this mRNA and this cell, message content is rate-limiting.

An additional line of support can be obtained from the agreement between kinetics of specific protein synthesis and predictions of transcriptional control models. For example, as will be discussed in Section 8.3.3, a reasonable agreement with the predictions is observed in the case of histone synthesis in the S phase of HeLa cells.

8.2.3 Evidence against absolute proportionality

Quite obviously, factors other than mRNA content can be made rate limiting by an appropriate experimental design. A trivial case would be withdrawal of an essential amino acid. In this discussion we shall not treat the large body of literature on effects of culture conditions, but only 'natural' regulatory processes.

Deviations from proportionality can be caused either by a pool of untranslated mRNA or by changes in the productivity of mRNA translation. *Translational productivity* P_r is the number of polypeptide molecules produced per unit time per molecule of polysome-bound mRNA[2] and can be described by equation (8.1):

$$P_r = r/t \qquad (8.1)$$

where r is the number of ribosomes associated with a message, and t the time that each requires to traverse it. An alternative expression is equation (8.1a):

$$P_r = v/s \qquad (8.1a)$$

where v is the rate of peptide chain elongation (codons translated per unit time) and s the average spacing of ribosomes on the message (codons per ribosome). The two types of deviation from proportionality may or may not be linked. Generally, a decrease in initiation rate (which will be manifested as a decrease in polysome size if elongation and termination remain constant) will create a significant pool of ribosome-free message only if the probability of initiation is substantially lower for free mRNA than for mRNA which already has ribosomes attached.

In a number of systems, circumstantial evidence suggests that ribosome-free mRNA exists in the cytoplasm during Phase III, when protein synthesis is modulated downwards. In reticulocytes, it is reported that as much as 20% of the total translatable α-globin mRNA can be isolated from the polysome-free (post-80S) supernatant[8, 9, 112]. Similarly, in the silk gland of *Bombyx mori* the full amount of chemically defined fibroin mRNA (labelled by long exposure to ^{32}P) can still be detected even after a substantial decline in the rate of fibroin synthesis[10]. Although these results are suggestive, a cautionary remark is in order. Rigorous evaluation of mRNA content is most likely to be obtained by a combination of both functional and chemical methods. Chemical criteria such as size or hybridisation cannot distinguish between irreversibly inactivated mRNA (e.g. molecules in which quantitatively minor initiation sequences have been degraded by nuclease attack) and potentially functional mRNA untranslated because of post-transcriptional regulation. Conversely, translation in a non-homologous system may not discriminate between forms of an mRNA which are distinguished in the actual cell. The ultimate desirability of combined functional and chemical criteria should also be borne in mind elsewhere in this paper.

Translational productivity may be depressed in rabbit reticulocytes as a result of progressive IF-3 depletion[11]. This initiation factor promotes the recycling of ribosomal subunits into polysomes (and blocks the formation of inactive single ribosomes[12]). In lysates, its depletion appears to affect non-specifically the translation of several mRNAs[11].

The synthesis of ovalbumin and other secretory products decreases after withdrawal of oestrogen from the previously-stimulated chick oviduct[13-15], and increases again upon re-stimulation. During the withdrawal period (Phase III), minor changes occur in the translational parameters. The rate of peptide chain elongation appears to be depressed to 70–80% of the stimulated value[15]. In the course of 4 days of withdrawal, the spacing of ribosomes on the ovalbumin polysomes increases by a factor of 1.5, as indicated by a change in the average size of ^{125}I-antiovalbumin-binding polysomes, from 12 to 8 ribosome aggregates[15]. This effect seems to be due to a limitation upon peptide chain initiation, since inhibitors of chain elongation increase the polysome size during oestrogen withdrawal, but not during stimulation. The effects on elongation and spacing indicate that overall translational productivity decreases to 50% (equation (8.1a)). A somewhat more pronounced decrease is suggested by the statement that after complete withdrawal the average polysome size declines by a factor of 2–3 (with only a 10% decrease in the average size of the proteins synthesised); however, no quantitative data are presented in this case. It appears that the change in translational productivity is general rather than specific for particular proteins. There is no evidence for the accumulation of a pool of untranslated mRNA during withdrawal[15, 16].

8.2.4 Evidence regarding specific translational factors

As discussed above, even relative proportionality breaks down when *specific* 'masking', initiation, or other translational factors operate *in vivo*. The best documented case of this type is the masking of mRNA during oogenesis and its apparently specific, programmed unmasking during early embryogenesis (see Chapter 3). For example, by use of a double-labelling procedure, Terman has documented shifts in the types of proteins synthesised at various stages of early sea urchin development, and has shown that some of these changes occur even in the absence of RNA synthesis, in embryos developing in saturating levels of actinomycin D (e.g. shifts between profiles typical of zygotes and 8 h embryos or blastulae)[17]. In the clam *Spisula* the profile of proteins synthesised changes significantly between fertilisation and first cleavage[18]. It is not unlikely that these rapid changes are controlled by specific unmasking. Although the evidence is incomplete and in some ways ambiguous, it may well be that specific translational regulation operates in early embryos as well as in other systems where development occurs after temporary arrest (e.g. germinating seeds).

In a number of systems, the onset of Phase II of differentiation has been interpreted as activation of pre-existing, translationally repressed mRNA. Thus, actinomycin experiments suggested that the mRNA for various enzymes of carbohydrate metabolism were synthesised at variable periods prior to their use during slime-mould development[19] (see also Ashworth, Chapter 1). However, current work on this problem in another laboratory suggests that actinomycin alone does not completely suppress mRNA synthesis in the slime-mould[20]. New mRNA synthesis can be completely blocked only when actinomycin is used in conjunction with another inhibitor of transcription such as donamycin. Under these conditions there is only a brief (*ca.* 2 h) lag between treatment and complete cessation of specific enzyme synthesis. This result suggests that the earlier studies in which transcription was incompletely inhibited do not provide conclusive evidence for translational control at the level of mRNA activation[20].

Similar circumstantial evidence for translational activation at the onset of Phase II has been presented in the case of glutamine synthetase induction in the vertebrate retina[21]. In the cocoonase-producing cells of the silkmoth galea, treatment with saturating levels of actinomycin D for a day prior to the beginning of Phase II still permits a substantial increase in the relative rate of cocoonase synthesis, and even in the absolute rate of [^{3}H]leucine incorporation into cocoonase; however, careful analysis of the results indicates that this phenomenon is due to the differential stability of cocoonase mRNA and to an unexplained increase in the specific radioactivity of intracellular leucine in the presence of actinomycin. No specific translational 'unmasking' need be postulated[2].

A case of variation in translational productivity within a single cell has been well documented in reticulocytes. Hunt *et al.* observe that polysomes synthesising β-globin are larger than those making α-chains[22, 23], despite the similar molecular weight of the two proteins. Yet reticulocytes synthesise nearly equal amounts of α- and β-globin. Hunt *et al.* interpreted their results in terms of equal productivities, by marshalling evidence that the time necessary for a ribosome to traverse the β-message is correspondingly longer than for α-message (parallel changes in r and t in equation (8.1))[22, 23]. However, more recent evidence suggests that the translation times (or chain elongation rates) are indeed the same. Apparently, the overall equality of α- and β-synthesis is attained by a lower translational productivity on α-mRNA (initiation only 0.65 times as frequent as on β-mRNA, identical elongation) which is exactly balanced by a 1.5 × excess content of α-message[24, 25]. The functional significance of these differences is unclear.

It should be remarked that major variations in translational parameters are not encountered even between different systems. A wide variety of eukaryotic systems from insects to mammals have rather uniform average ribosome spacing and peptide chain elongation rates (Tables 8.1 and 8.2); in combination these observations also imply uniformity of initiation rates. The spacing seems constant regardless of temperature, possibly because initiation and elongation have similar temperature coefficients; it is the same as in prokaryotes. In mammalian cells at 37 °C, properly calculated chain elongation rates are consistently about half the bacterial rate. Insect rates are comparable to mammalian when the necessary temperature correction is applied (see Refs. 22, 23 or 15 for a proper coefficient). In the highly differentiated follicular

Table 8.1 Rates of polypeptide chain elongation in eukaryotes

System (proteins)	*Temperature °C*		*v**	*Ref.*
Rabbit reticulocytes (Hb)	37	$\bar{v} = 8.5$	6–10	113
Rabbit reticulocytes (Hb)	37		7–10	114
Chinese hamster cells (general)	37		7	115
Growing HeLa cells (general)	37		11	116
Oestrogen-stimulated chick oviduct (ovalbumin)	41		5	15
Polyphemus follicular cells (chorion proteins)	23.5		1.3	26
(Non-specific proteins)			1.9	26

* $\bar{v}$ = codons translated per ribosome per second.

Table 8.2 Average spacing of ribosomes on active mRNA

System (proteins)	*Temperature °C*		*s**	*Ref.*
Rat liver (general)	37	$\bar{s} = 32$	30±3	117
Rabbit reticulocytes	37			
(general Hb)			29	118–120
(α-chain)			34–40	23, 24
(β-chain)			27–32	23, 24
Embryonic chick skeletal muscle (myosin)	37		29	121
Embryonic chick (collagen)	37		36	122
Chick oviduct (ovalbumin)	41		33	15, 138
Polyphemus follicular cells (chorion proteins)	25		37	26
Polyphemus galea (cocoonase zymogen)	25		30	26
E. coli (β-galactosidase)	37		32	124

* s = codons per ribosome.

cells of silkmoths, no difference greater than experimental error could be documented between the elongation rates of differentiation-specific and 'general' proteins[26].

The paradoxical phenomenon of 'superinduction' has been used as an argument for specific translational regulation. This term refers to the increased rate of specific protein synthesis after actinomycin treatment[27]; it should be distinguished from apparent superinduction, which is merely an increase in the specific radioactivity of the intracellular precursor[28]. In the chick oviduct, superinduction can be interpreted in non-specific terms; as mRNA decays in the absence of transcription, there is a lessening of competition for non-specific rate-limiting translational factors, and as a consequence a greater translational rate can be supported by the mRNAs for the secretory products, since these RNAs are differentially stable and are enriched in the actinomycin-treated cells[16]. However, this interpretation may not apply to the super-induction of tyrosine aminotransferase in hepatoma cells or liver[27,29]. In this case, superinduction is substantial and the half-life of TAT message in the absence of the inducer, dexamethasone, or in the absence of RNA synthesis inhibitors such as actinomycin and mercaptopyridethyl benzimi-dazole (MPB) is extremely short. Tomkins postulates that a rapidly turning-over specific translational repressor, the 'R protein', controls the level of TAT synthesis by continually binding to TAT mRNA and thereby speeding its degradation. In this view, dexamethasone can induce and maintain TAT synthesis by binding to and rendering inactive (or otherwise reducing the level of) the R protein[110]. But, when the inducer is removed, the level of TAT synthesis drops precipitously because the inducer is no longer present to prevent the sequestration by R protein of TAT mRNA and its eventual enhanced degradation. Tomkins' model endows the R protein with a very short -half-life which implies that its synthesis should be quite sensitive to inhibition of transcription. In fact, TAT mRNA 'rescue' experiments have been performed in which the addition of actinomycin or MPB to de-induced cultures can prevent the usual loss of TAT synthetic capacity. Tomkins has accumulated an impressive amount of circumstantial evidence in favour of his model. Still, direct proof is lacking, since the mRNA has not yet been isolated and quantified directly. In the present context, it may be remarked that the evidence for the model comes primarily from de-induction experiments, and may be more relevant to Phase III than to Phase II regulation.

Specific initiation factors have been postulated on the basis of evidence for differential mRNA translation in cell-free systems[30]. The *in vitro* phenomenon itself has been challenged, in view of the successful translation of a variety of exogenous mRNAs by reticulocyte lysates[31-36], the Krebs ascites cell-free systems[37-42], and even living frog oocytes[5,6,43]. In the ascites system, successful translation has been accomplished even with mRNAs from invertebrates: sea urchin histone mRNA[44] and silkmoth eggshell mRNA[45]. The postulated stage-specific initiation factors from insects[46] have not yet been sufficiently characterised. Regardless of how the controversy is resolved about specific factors assayable in cell-free systems, it will still be necessary to demonstrate that such factors play a regulatory role *in vivo*.

In summary, the evidence for and against *specific* translational regulation (other than through mRNA stability) is still sketchy. Reasonable evidence for translational regulation exists in early embryos, and somewhat weaker in Phase III systems, but in the latter it is not certain that regulation is specific, rather than non-specific as in the oviduct. Direct mRNA assay in the oviduct during secondary stimulation—the first rigorous test with a Phase II system—showed strict proportionality between mRNA content and rate of *in vivo* specific protein synthesis (Figure 8.1). Thus, while recognising that specific translational regulation may exist (especially in newly activated, Phase III, or short-time constant systems), it seems reasonable to explore regulatory models emphasising the importance of changing mRNA levels, controlled through the interplay of transcription and specific mRNA stability.

8.3 RATES OF mRNA SYNTHESIS AND DEGRADATION AND THEIR EFFECTS ON THE KINETICS OF SPECIFIC PROTEIN SYNTHESIS

8.3.1 A basic equation

A detailed treatment of effects of mRNA synthesis and degradation on specific protein synthesis has been presented elsewhere[2,47]. Here we will review the salient points.

This treatment is based on certain assumptions about the nature of mRNA synthesis and degradation. First, mRNA synthesis is assumed to be a zero-order reaction with respect to message content. That is, mRNA synthesis is not controlled by direct feed-back from mRNA content. Secondly, mRNA degradation is assumed to be a stochastic process, i.e. random with respect to the age of any particular mRNA molecule. A stochastic process would lead to first-order or exponential kinetics. Although non-random, 'ageing' models of RNA degradation have been presented[48], the bulk of the quantitative evidence in the literature supports the suggestion of first-order mRNA decay, both in bacteria[49] and in eukaryotes[50,51]. In our experience with insect tissues, the kinetics of both total[28] and specific protein synthesis[52] after actinomycin treatment fit the exponential model well.

These two assumptions are completely analogous to those supported by extensive evidence in the case of protein synthesis and breakdown (see Schimke, Chapter 7). As in the case of protein turnover, the assumptions lead directly to the following basic equation (8.2) which describes the rate of change in mRNA content as a function of synthesis and breakdown:

$$\frac{\mathrm{d}M}{\mathrm{d}t} = S - DM \qquad (8.2)$$

where M is the amount of mRNA, S the mRNA synthetic rate, and D the first-order mRNA decay constant. D is related to the half-life T by equation (8.3):

$$D = (\ln 2)/\mathrm{T} \qquad (8.3)$$

Whenever mRNA content is proportional to the rate of specific protein synthesis, equation (8.2) is the basic equation for the kinetics of specific protein synthesis as a function of mRNA synthesis and degradation. In other words, this is the equation for regulation exerted through transcription and mRNA stability. Under the term 'transcription' we assume any processing that may be necessary to convert the direct gene transcript to potentially functional mRNA.

8.3.2 A case of simple transcriptional control: constant rate of mRNA synthesis

According to equation (8.2), what are the consequences of the simplest transcriptional control, i.e. the setting of the 'transcription thermostat' to a constant value of mRNA production, S? Quite clearly, if $S = DM$ we are dealing with a steady state, and the rate of the specific protein synthesis itself should be constant. By contrast, if S is set at a value greater than DM, the specific message will begin to accumulate according to equation (8.4):

$$M_t = \frac{S}{D} - \left(\frac{S}{D} - M_0\right) e^{-Dt} \qquad (8.4)$$

where M_0 is the initial amount of specific mRNA and M_t is the amount at time t. Accumulation will continue until the plateau value, $M = S/D$, is approached asymptotically.

Obviously, as equation (8.4) indicates, the absolute rate of message accumulation as well as the equilibrium value, will depend on both synthesis and degradation. However, as in the analogous case of protein turnover, the *shape* of the message accumulation curve will be determined by the mRNA stability alone. Thus, if two mRNAs differ by a factor of two in their stability, in a developmental change in which the mRNA content will ultimately increase x-fold for both, as a result of an x-fold increase in synthetic rate, the time necessary for the half maximal change to be attained will be twice longer for the stable as compared with the less stable of the two.

This consideration emphasises that the stability of differentiation-specific mRNA must be substantial if it is to play a significant role in the determination of Phase II kinetics. In this connection, it is immaterial whether differentiation-specific mRNA is equally or more stable than other mRNAs in the same cell. The greater the absolute stability of mRNA, the longer the period during which accumulation can occur as a result of a step change in transcription. In long time-constant systems, changes in translational specialisation may be ascribed to a simple transcriptional change ('resetting the transcription thermostat' to a new level at the beginning of Phase II) only if the mRNA half-life is comparable to the time constant.

Equation (8.4) describes a curve with a monotonically changing slope (Figure 8.2a). Actual Phase II kinetics are usually sigmoidal (Figure 8.2b,c)[2]. The sigmoid nature can be largely ascribed to DNA synthesis which continues during the early portion of Phase II but subsequently ends: mRNA synthesis per gene could remain constant while the overall rate of synthesis accelerates as long as multiplication of the genome continues. Moreover, it would not be unreasonable to allow for a short transient in an otherwise simple transcriptional model, by assuming either a non-instantaneous readjustment of the transcription thermostat, or, less probable, a translational readjustment during

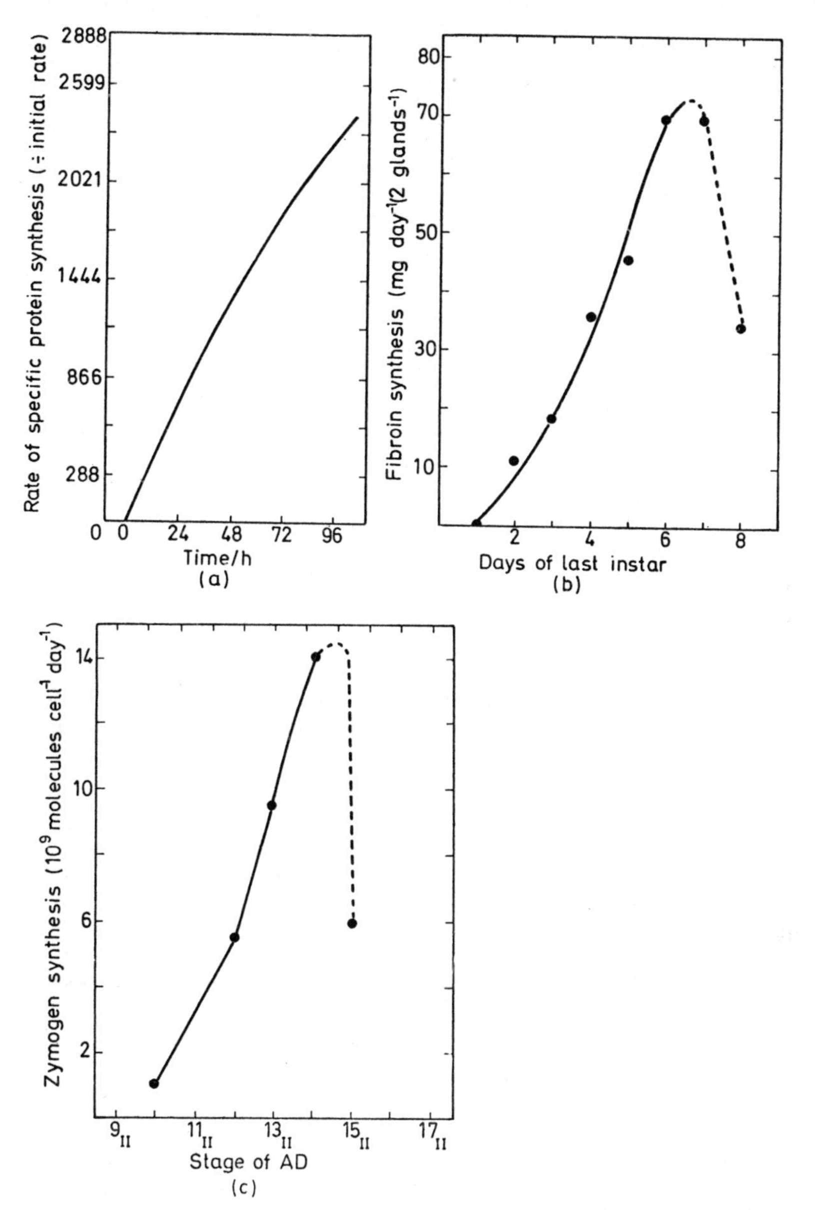

Figure 8.2 Rates of specific protein synthesis during Phase II of differentiation. (a) A theoretical curve for kinetics resulting from constant transcription (at a rate 10.4% of total mRNA synthesis) of a differentiation-specific message which has a half-life of 100 h. Non-specific mRNA synthesis and decay (half-life 2.5 h) were assumed to be constant, as was the number of specific gene copies. (From Ref. 47 by courtesy of Karolinska Institutet.) (b) Kinetics of fibroin synthesis in the silk gland of *Bombyx mori* during the last larval stage. Phase II is shown in solid line, Phase III in dotted line. (From Ref. 103.) (c) Kinetics of cocoonase zymogen synthesis in the galea of the silkmoth *Antheraea polyphemus*, during the second half of adult development (days 9–15). The data were obtained from the derivative of the zymogen accumulation curve. (From Ref. 2, courtesy of Academic Press, Inc.)

the transition from Phase I to Phase II. The essence of the model is that during most of Phase II a continuously *changing* rate of specific protein synthesis is driven by *constant* rapid transcription of stable mRNA.

A simple transcriptional model can be postulated even for somewhat more complicated developmental systems. If each gene in a coordinately controlled gene battery produces mRNA of a different half-life, the kinetics of gene expression would appear independent for each gene even though the transcription thermostats of all were readjusted simultaneously. In this connection, it is important to realise that highly differentiated tissues often produce a number of products, often each with distinct kinetics. Examples are embryonic rat pancreas[1], and the vertebrate lens in which different crystallins are produced in precise succession[54]; this possibility raises the question of how uniform mRNA half-lives are in any one cell type. The possibility of coordinately controlled but non-coordinately expressed genes is particularly attractive for hormone responsive tissues, since hormones commonly set in motion a complex sequence of events. As a result of diversity in mRNA half-lives, a single transcriptional regulatory step could be obscured in a bewildering array of seemingly independent changes. This explanation could be evaluated, since it predicts translational kinetics differing in slope rather than in lag phase.

8.3.3 A more complex transcriptional control: changing rates of mRNA production driving changing rates of protein synthesis

If transcriptional control is widespread it may be expected to act in many cases as continuously variable, rather than step regulation. What, then, are the consequences of varying transcription on the rate of protein synthesis, when mRNA half-life remains constant? To explore this question, we will turn to the synthesis of histones in synchronised HeLa cells during the DNA synthetic phase of the cell cycle (S).

In a classic paper, Borun *et al.*[55] identified an 8S species of rapidly-labelled RNA as the putative message for the A class of small histones. Variable amounts of newly-synthesised 8S RNA were found associated with polysomes at the time of histone synthesis, and at no other time. This RNA selectively disappeared upon treatment of the cells with the DNA synthesis inhibitor, cytosine arabinoside, in parallel with the cessation of histone synthesis. The RNA also disappeared from polysomes upon actinomycin treatment, with the same half-life as the decay of histone A synthesis (*ca.* 1 h). Finally, the RNA was only associated with polysomes which were of the expected small size and which were synthesising histones, as shown by the characteristic size distribution and amino acid ratio (high lysine:tryptophan) of the nascent polypeptides. The identification of 8S RNA as histone message, in sea urchin as well as HeLa, has been confirmed in several recent studies[4, 44, 56, 57, 161].

Figure 8.3 shows the time course of histone A synthesis, as evaluated by pulse-labelling with [^{14}C]lysine (20 min) at various stages of the cell cycle, extracting nuclear histones, and analysing on SDS-polyacrylamide gels.

Figure 8.3 also shows the amount of 8 S RNA newly synthesised and associated with polysomes at different stages (30 min [^{3}H]uridine pulse). Assuming no variations in permeability to isotope, and postulating transcriptional control (i.e. that the mRNA appearing in polysomes is proportional to the mRNA transcribed, and that protein synthesis is proportional to message content), the two curves in Figure 8.3 represent changes in the rate of mRNA

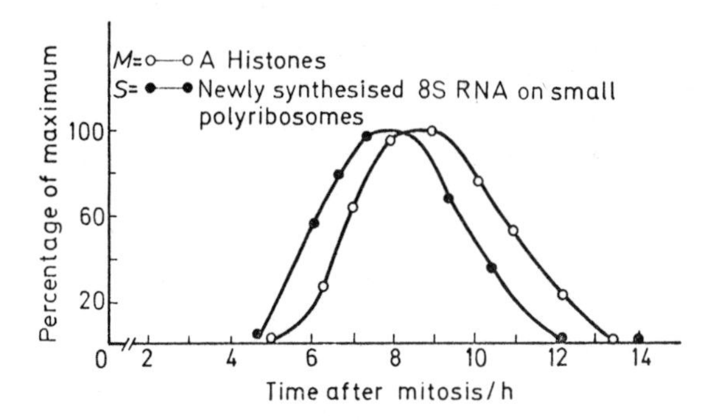

Figure 8.3 Histone message and histone protein synthesis in HeLa cells. (Data from Ref. 55.) The synthesis of message (●——●) was estimated from the appearance of labelled 8S RNA on small polysomes, following a 30 min pulse of [^{3}H]uridine at the indicated times of the cell cycle. The synthesis of A type histones (○——○) was estimated from the appearance of labelled A histones in the nucleus, following a 20 min pulse of [^{14}C]lysine at the indicated times

synthesis, *S*, and in the amount of mRNA accumulated, *M* (8S RNA labelling and histone labelling, respectively). Can curve *M* be derived mathematically from curve *S*? Are the kinetics of histone synthesis driven by changing rates of transcription?

Equation (8.2) can be rearranged to yield:

$$\frac{S}{M} = \left(\frac{dM}{dt}\right)\frac{1}{M} + D \tag{8.5}$$

Thus, our model would predict that a plot of $\frac{S}{M}$ *v.* $\frac{dM}{Mdt}$ should yield a straight line, with *D* as the *y*-intercept. Figure 8.4 shows that the data are indeed consistent with the expected linearity (although no quantitative evaluation of the goodness of fit is possible, since the data are insufficient and second hand).

During each cell cycle, each HeLa cell must synthesise an amount of histones equal to its DNA content, *ca.* 15.4 pg[53]; according to the data of Borun *et al.*[55], *ca.* 67.6% of that, or 10.4 pg, must be A histones, of average mol. wt. 13 000. Thus, 4.8×10^{8} molecules of histone A must be produced.

This number must equal the integral of curve M, multiplied by the translational productivity, P. Thus, assuming the value of P_r as 16 polypeptides $mRNA^{-1}$ min^{-1} (the usual mammalian value at 37 °C; see Tables 8.1 and 8.2 and Ref. 2), an absolute scale can be assigned to curve M. M_{max} is estimated as 1.2×10^5 mRNA molecules $cell^{-1}$, a value comparable with the maximum specific mRNA content in several highly differentiated cell types[2].

With the value of M defined, equation (8.5) can be used to calculate either D or S, when one of the two is known. No absolute data on S exist. However, it seems reasonable to assign to D the value 0.65 h^{-1}. This is calculated from the *ca.* 1 h half-life of both histone A synthesis and polysomal 8S RNA label upon actinomycin treatment. We may assume that this half-life is not an actinomycin artefact, since even in the normal cells M decays with a similar

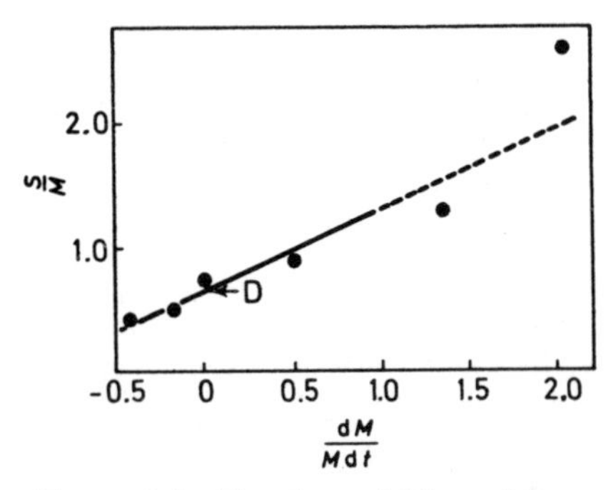

Figure 8.4 The data of Figure 8.3, replotted according to equation (8.5). See the text for details

half-life when S becomes negligible (Figure 8.3)[50]. This permits us to calculate the maximum rate of histone mRNA synthesis per cell as 1.4×10^3 mRNA molecules min^{-1}. We shall return to this number in a later section.

In sum, transcriptional control may be inferred when adequate translational (M) and transcriptional (S) data can be fitted by equation (8.5). Provided we can measure the value of translational productivity, knowledge of the mRNA synthetic rate defines the mRNA degradation constant, and vice versa. In the special case of constant mRNA synthesis, both S and D can be determined simultaneously by plotting $\frac{1}{M}$ v. $\frac{dM}{Mdt}$ in a manner analogous to the determination of K_m and V_{max} by the Lineweaver–Burk plot. This emphasises the importance of obtaining quantitative kinetics of specific protein synthesis (or, better still, message accumulation) during Phase II of differentiation—especially since these kinetics are easier to obtain than direct measurements of S and, as we shall see next, are more reliable than values of D determined by present methods.

8.4 EXPERIMENTAL EVIDENCE ON mRNA HALF-LIVES

8.4.1 Introduction

A large body of evidence, mostly circumstantial, indicates that mRNA species of widely varying stability exist in eukaryotic cells. The stability may change radically during cell differentiation, and mRNA of vastly different stabilities may co-exist in the same cell. It may be that the variation of mRNA half-life is of fundamental importance to the generally long-lived eukaryotic cell, being a key element in the quantitative and temporal control over the rate of specific protein synthesis. In marked contrast, nearly all the mRNA of prokaryotic cells has a uniformly short half-life[58-60]. It is especially interesting that bacterial cells may stabilise certain mRNAs when committed to specialised synthetic functions such as sporulation[61], synthesis of penicillinase or flagellin[62,63], or production of some viruses[64].

In this section we shall discuss some representative experimental evidence on mRNA half-lives in eukaryotic cells: first the evidence on overall mRNA half-lives, as determined both by the kinetics of labelling and by actinomycin D experiments, and then the evidence for highly stable, differentiation-specific mRNAs and for diversity of mRNA half-lives.

8.4.2 Average mRNA half-lives

8.4.2.1 Determination by kinetics of labelling

(a) *Sea urchins*—Without the use of any drugs or metabolic inhibitors, Brandhorst and Humphreys obtained evidence for two classes of RNA, with half-lives of 5–10 min and 60–90 min, in sea urchin embryos[65]. By disregarding the very slight amount of stable RNA which accumulates during this stage, Brandhorst and Humphreys argued that RNA synthetic rates and degradation rates were equal. When the accumulation curves for labelled adenosine were subjected to standard first-order decay analysis, the data were consistent with the existence of these two classes of short and moderately long half-life, in the approximate proportion 2:1. Brandhorst and Humphreys later fractionated sea urchin embryos and found that the shorter half-life RNA is strictly nuclear while the RNA having a 60–90 min average half-life could be purified from cytoplasmic polysomes[50]. An important fraction of the labelled polysomal RNA is undoubtedly histone mRNA, since sea urchin blastulae and plutei, which increase their cell number exponentially, are actively engaged in histone synthesis[44]. Other experiments on sea urchins[57], and HeLa and mouse L cells[4,55,66], are consistent with an average half-life of 60–90 min for histone mRNA—substantially shorter than the average mRNA half-life in these cells (see also Section 8.3.3).

(b) *Cultured mammalian cells*—The discovery that a substantial fraction of eukaryotic mRNA contains poly A had led to a technology of mass mRNA isolation by binding to immobilised poly U, poly dT or nitrocellulose. With poly A content as the sole criterion, the metabolism of mRNA can be studied irrespective of mRNA translation.

Average mRNA stabilities have now been measured in HeLa and mouse L cells by this method. It appears that poly-A containing RNA is much more stable than previously suspected, having an average half-life close to the generation time of the cell.

In a manner similar to Brandhorst and Humphreys, Greenberg followed in exponentially growing mouse L cells the approach to steady-state labelling of an operationally-defined mRNA class (polysomal RNA which binds to Millipore filters under ionic and pH conditions specific for molecules containing poly A[51]). The data fit a single-component decay curve, with a half-life of 10 h, or slightly less than the 15 h doubling time of the L cells. It was verified that the specific activity of RNA precursor pools remained essentially constant throughout the experiment.

Singer and Penman have measured the kinetics of labelling of whole cytoplasmic RNA retained on poly U filters[67,68]. Their data cannot be reconciled with a single-component decay, and have been interpreted in terms of two components, with apparent half-lives of 7 and 24 h[68]. In a double isotope, pulse- and long-term labelling experiment, 'old' and 'new' mRNA can be shown to differ in average sedimentation coefficient.

Recently, the development of a low-temperature chase technique has permitted direct study of the decay of polysomal, poly A-containing mRNA in HeLa cells[69]. A 3 h exposure to 4 °C in the presence of 0.01 M unlabelled uridine was shown to reduce the specific activity of the intracellular UTP pool by 90 %, irrespective of the duration of prior labelling. Upon rewarming to 37 °C, normal protein and DNA synthesis were resumed within minutes, although RNA synthesis showed a slight depression which was gradually overcome within the next 3 h. The [^{3}H]mRNA, labelled prior to the 'cold chase' for periods between 30 min and 24 h, was observed under these conditions to decay with a half-life of at least 2 days regardless of the labelling time.

8.4.2.2 Determination by actinomycin experiments

Early determinations of the average stability of eukaryotic mRNA were based on the use of high levels of actinomycin, hopefully to end effectively and specifically all transcription. The decay of polysomes, protein synthetic ability, or labelled polydisperse RNA were then used as an index of mRNA decay. It now seems that this method underestimates both the absolute stability and the diversity of mRNA half-lives. A better approximation of the average stability may be obtained by drug-independent methods, as detailed in the previous section. An indication of diversity may be obtained from the relative (actinomycin-based) half-lives of different mRNA species in the same tissue and at the same time (Section 8.4.3.3).

(a) *Cultured mammalian cells*—The earliest estimate of bulk mRNA half-life in HeLa cells was 3–4 h, based on two polysome profiles, one partially and the other completely decayed after the cells had been exposed to a high level (5 μg ml^{-1}) of actinomycin for 3 and 7 h, respectively[70]. This assessment of mRNA stability in HeLa cells, despite its relative dearth of data, remained the standard literature reference for many years. Cheevers and Sheinin obtained essentially similar results with mouse 3T3 cells growing in monolayers[71]. They pre-incubated cells in 0.01 μg ml^{-1} actinomycin for 30 min (under which conditions rRNA synthesis was shown to be depressed to 'undetectable' levels) and measured the kinetics of entry of [^{3}H]uridine labelled nascent RNA into polysomes. They interpreted the observed biphasic accumulation curve as meaning that much (80%) of the labelled mRNA synthesised in 2 h after the addition of [^{3}H]uridine was unstable. The remainder appeared not to turn over within 6 h, the longest labelling time of the experiment.

Hodge *et al.* concluded that mRNA present in the cytoplasm of HeLa cells persisted through mitosis[72]. Treatment of cells before metaphase with 2 μg ml^{-1} actinomycin prevented neither reformation of polysomes nor a progressive increase in the rate of overall protein synthesis nor synthesis of certain electrophoretically distinct proteins for nearly 1 h after the termination of metaphase. Craig *et al.* attempted to measure the relative lifetimes of mRNAs coding for ribosomal proteins, histones, and other proteins in L cells, by measuring the synthesis of these proteins *in vitro* on polysomes derived from cells exposed to actinomycin for various times[73]. Polysomes from which labelled ribosomal proteins could be purified, as well as the remainder of the non-histone synthesising polysomes, decayed with a half-life of 2.5–3 h. Polysomal histone synthesis was much more sensitive to actinomycin and decayed within 1 h. However, the scarcity of data and their substantial scatter make the quantitative conclusions of this study of doubtful reliability. A short half-life for histone message was also estimated by Borun *et al.* but denied by Gallwitz *et al.*[74]. In the latter study, the levels of actinomycin or MPB used were probably insufficient to inhibit much of mRNA production (95% inhibition of total RNA synthesis).

(b) *Mammalian liver*—Average mRNA stabilities were studied intensively in mammalian liver soon after actinomycin became available. Early studies of rat liver polysomes suggested that perhaps as much as 50–80% of the polysomal mRNA breaks down within 4–8 h of actinomycin injection[75]. Other early work suggested that bulk liver mRNA was quite stable[76, 77]. Using a 3 h pulse of [^{14}C]leucine *in vivo*, Revel and Hiatt observed as much protein synthetic capacity in livers from rats exposed to actinomycin for 40 h as in control livers. All of these early experiments used levels of actinomycin high enough to kill the experimental animals within hours but different laboratories could not agree on whether actinomycin-induced artefacts were leading to abnormally long[78, 79] or short half-lives[77]. Trakatellis *et al.* found that labelled 4.5S to 20S RNA prepared from liver polysomes, as well as the capacity of liver ribosomes to incorporate radioactive leucine, decayed in parallel, with a half-life of 8–12 h after administration of actinomycin[79]. The same workers observed that in the absence of drugs the specific activity of their operationally defined 'mRNA fraction' reached a plateau after 4 h of labelling. They argued that the true half-life of rat liver mRNA was about 2 h and that exposure of liver to actinomycin resulted in artefactually increased half-lives. There was no proof that the 4.5S to 20S fraction is representative, or that it contains exclusively mRNA. Wilson *et al.* also inferred average mRNA stability from actinomycin-induced decay of rat liver polysomes[81]. Their decay curve showed two components, suggesting that about two-thirds of the mRNA has a half-life of 3–5 h and the remainder a half-life of 80 h. These authors did include controls to show that actinomycin was not inducing or unmasking any new ribonuclease activity in the liver and that no presumptive mRNA of heterogeneous sedimentation coefficient (excluding 4.5S RNA) could be purified from the post-80S regions of polysome gradients after exposure to actinomycin.

(c) *Studies on an insect tissue: the silkmoth galea*—In a series of experiments designed to measure the stability of the mRNA for the differentiation-specific product of the silkmoth galea, the zymogen of the proteolytic enzyme cocoonase, Kafatos and co-workers have also obtained information about the non-specific mRNAs present in galea cells[2]. The evidence for mRNA stability is indirect, since it was inferred from the resistance of protein synthesis to actinomycin, but is consistent for all three techniques employed.

One type of experiment used quantitative autoradiography to assess the rate of protein synthesis[82]. After various culture periods, matched control and actinomycin treated (60 μg ml^{-1}) galae were labelled for 1 h with [^{3}H]amino acids and chased in non-radioactive media for 4 h, a period sufficient for transport of newly-synthesised prococoonase into the large zymogen storage vacuole. The rates of synthesis of zymogen and general cellular proteins were inferred from the grains over the vacuole and the remaining cytoplasm, respectively. The synthesis of general cellular proteins was strongly inhibited by actinomycin and decayed with a 2 h half-life. Actinomycin at a level high enough to inhibit more than 99 % of all RNA synthesis had almost no effect on the synthesis of zymogen. Zymogen synthesis persisted at 70 % of the control rate after even 2 days of exposure to actinomycin.

A different type of experiment allowed for the determination of the average mRNA stability as well as the relative diversity of non-zymogen mRNA[2, 47, 83]. Again, pairs of galeae were cultured for various periods from 4.4 to 7.7 h with and without actinomycin. Control tissues were then labelled with [^{14}C]leucine and actinomycin-treated tissues with [^{3}H]leucine. Then the tissues were combined, homogenised, and their proteins analysed by SDS-gel electrophoresis. The ratio of ^{3}H to ^{14}C in any particular region of the gel was used as an indication of the corresponding mRNA stability: the higher the ratio the greater the mRNA stability (more directly, the greater the resistance of protein synthesis to actinomycin). The $^{3}H/^{14}C$ ratio over the zymogen region of the gel was always so high that it could be adopted as an internal standard corresponding to effectively infinite mRNA stability. This circumvented the difficulties inherent in determining absolute rates of protein synthesis[2]. By dividing the $^{3}H/^{14}C$ ratio of any non-zymogen region of the gel by the isotope ratio at the zymogen peak, a value (R_t) could be obtained which was a measure of how low that specific non-zymogen protein synthesis had become as a result of exposure to actinomycin for time t. By assuming that mRNA decay was exponential, the half-life T could be calculated from equation (8.6):

$$T = -\frac{t \ln 2}{\ln R_t} \tag{8.6}$$

which may be derived from the definition of half-life. In 11 separate experiments the overall mRNA half-life for all mRNAs coding for proteins larger than 33 000 daltons was 2.5 ± 0.24 h (95 % confidence limits).

In a third type of experiment both control and actinomycin-treated tissues were labelled with [^{3}H]leucine and were processed separately so the specific activity of intracellular leucine pools and presumably the absolute rates of protein synthesis could be calculated[28]. Synthesis of zymogen and non-zymogen protein was determined from samples electrophoresed in SDS-acrylamide gels. Incorporation data were divided by the corresponding precursor specific activity to calculate the actual rates of protein synthesis. Relative to the controls, the actinomycin-treated tissues showed a progressive, approximately exponential, decrease in the rate of [^{3}H]leucine incorporation into non-zymogen proteins. The results were consistent with a zymogen mRNA half-life of 99 h (justifying the methodology adopted in the second type of experiments) and an average half-life of 3–6 h for all non-zymogen messengers. Thus, all three types of experiments gave the same general result: in the galea of the silkmoth, production of the differentiation-specific product takes place on stable mRNA with a half-life of the order of 100 h, whereas the average non-zymogen messenger is unstable, with a half-life of only 2–3 h.

(d) *Studies on insect wing*—Actinomycin-induced decay of protein synthesis was also used to measure average mRNA stabilities and the diversity of mRNA half-lives in the developing silkmoth wing[52]. Wing tissue from three different stages of development (days 1–2, 5–6 and 8–9) was cultured in the presence or absence of actinomycin for periods up to 50 h and then labelled with ^{3}H or ^{14}C amino acids for 1 h. The results showed a good fit to an exponential decay model, and average half-lives of 10 h, 13 h and 13 h were measured for tissue from the various developmental stages. At the longer exposures to actinomycin the decay of incorporation levelled off somewhat. This would be consistent with the existence of more stable mRNA, but the authors caution against over-interpretation because of the relatively few samples taken and the low levels of incorporation attained after the long exposures to the drug.

Correlation coefficients were calculated for both exponential and linear decay models. The exponential decay model consistently showed a somewhat better fit, when the data for synthesis of either total or electrophoretically purified proteins were analysed.

(e) *Chick oviduct*—Additional evidence for the presence of at least two classes of mRNA with different stabilities in a single tissue comes from the work of Palmiter and colleagues[3,14,16,85]. Synthesis of differentiation-specific secretory proteins in chick oviducts was initiated by secondary administration of oestradiol-17β. After induction of synthesis, chick oviducts were cultured *in vitro* in the presence and absence of actinomycin, and the synthetic rates determined for the major secretory protein, ovalbumin, and for the total non-secretory protein. In actinomycin, generalised protein synthesis decayed with a half-life of 5 h while ovalbumin synthesis decayed with a half-life of 14 h.

8.4.2.3 Evaluation of average half-life studies

Much of the work to date has taken one of two approaches. In the actinomycin approach, it has been assumed that the rate of protein synthesis is proportional (in an absolute or relative sense; see Section 8.2) to mRNA content, and the decay of protein synthesis in the presence of actinomycin has been followed in an attempt to infer the persistence of functional mRNA after cessation of transcription, which is hopefully brought about by exposure to the drug. In the kinetic approach, exposure to any drug has been avoided; a chemical or biological criterion has been adopted for mRNA (size, poly A content, cytoplasmic or polysomal localisation, synthesis at a time of negligible ribosomal RNA production) and the kinetics of label accumulation in the presumed mRNA have been treated as the reverse of a decay curve to yield estimates of mRNA stability.

The actinomycin approach has the obvious and serious drawback of depending on the unproven assumption of proportionality between protein synthesis and mRNA content. In particular, absolute half-lives determined by this method are very suspect, since they assume absolute proportionality between mRNA content and protein synthesis, and since they are very sensitive to side effects of actinomycin, such as the change in the specific radioactivity of the intracellular amino acid precursor[28]. By contrast, relative half-lives determined by this method are substantially more meaningful. These are half-lives of different mRNA classes within a single tissue and only depend on the assumption of relative proportionality between mRNA content and protein synthesis, i.e. on the assumption that any rate limiting factors other than mRNA are non-specific (Section 8.2). Double-label methodologies, in which control and actinomycin-treated samples are analysed together, are an additional improvement since they minimise experimental artefacts.

The kinetic approach is very important in providing cross checks independent of the assumption of proportionality; its increasing use should certainly be encouraged. At the same time, it must be recognised that this approach also has its limitations. First, it suffers from the danger of unrecognised re-incorporation of precursor, which has long played the analogous studies of protein degradation. The extremes to which Murphy *et al.* have resorted to accomplish a meaningful chase situation attest to this problem; in their study, it is likely that the long period required to establish chase conditions permitted the most rapidly turning over mRNAs to decay, biasing the results towards abnormally long half-lives[69]. Secondly, it should be borne in mind that the conclusions are no more definitive than the criterion used to define the mRNA. For example, use of polysomal localisation as a criterion for globin message would introduce an error to the extent that globin message can be found in the post-ribosomal supernatant[8, 9, 112]. As another example, although the current opinion is that essentially all mRNAs in eukaryotic cells have poly A fragments[86], it should be borne in mind that this opinion may have to be revised; in at least one case, histone message, strong evidence exists that poly A is absent[4, 161]. Since the indications are that histone message is relatively short-lived, as well as having other metabolic peculiarities[4, 66, 86, 111, 161], it is possible that over-reliance on the poly A criterion may lead to an over-estimate of mRNA half-life.

On a more technical level, it is difficult to make many full comparisons and criticisms of many of the half-life determinations based on decay of polysomes or polysomal RNA. In addition to variations in polysome yield which can be expected with different procedures, different investigators probably measured different features of the polysome profile[70, 73, 81]. For example, some workers specifically avoided measuring active disomes and active single ribosomes in their analysis of polysome gradients. These species may increase drastically in response to depressed rates of polypeptide chain initiation, which may be caused by actinomycin[67]. It is also likely that some mRNA was lost in older procedures utilising phenol for deproteinisation of polysomal RNA. It is now known that poly A-containing mRNA may specifically partition to the organic phase and/or interface during phenol extraction, unless the aqueous phase has high pH and low monovalent cation concentration[87, 88] or unless the extracting medium contains chloroform in addition to phenol from the start[4, 89].

It is clear that a rigorous study of mRNA metabolism would include a combination of physical (presence in polysomes or cytoplasmic fractions), chemical (poly A content, lack of methylation) and biological criteria (EDTA or puromycin releaseability, ability to program a cell-free protein synthesising system). Purification and quantification of specific mRNAs along the lines of the pioneering work of Palmiter and Schimke[108, 133-138] and O'Malley and co-workers[105-107] is clearly a major goal for the future.

8.4.3 Evidence regarding diversity of mRNA half-lives

8.4.3.1 Average half-lives and diversity of half-lives

It is important to recognise that experiments which evaluate average half-lives are by their very nature inadequate to provide any measure of the diversity of half-lives except in the most qualitative sense. When two or more classes of mRNA exist, of widely different average half-lives (as in the galea), it is possible to detect them with simple decay experiments of the type discussed previously, provided conditions (e.g. duration of the experiment) are chosen appropriately. However, when stabilities are distributed across a continuous spectrum the experimental error may be sufficient to give the appearance of single component or two component decay. In this case, Occam's razor may be misapplied. Thus, it becomes important to review the evidence which specifically documents the unusually high stability of some mRNA species and the diversity of mRNA half-lives corresponding to individual 'non-specific' proteins in any one eukaryotic tissue.

8.4.3.2 The high stability of differentiation-specific mRNA

Long before the formulation of the 'central dogma' of molecular biology, incontrovertible evidence on the ability of anucleate cells to survive and produce protein was at hand. After nuclear extrusion, the immediate red blood cell precursors (reticulocytes) are capable of haemoglobin synthesis. Anucleate amoebae live and feed for hours. Perhaps most impressive was the pioneering work of Hammerling on *Acetabularia*[90]. These single-celled marine algae can live for months without nuclei: large anucleate parts of the rhizoid contain in their cytoplasm the information necessary to program dramatic morphological changes.

(a) *Review of older studies*—Early in the last decade the study of cell differentiation and specialisation had progressed to the biochemical level, and the general protein and nucleic acid metabolism of differentiated systems came under intense scrutiny. Many highly differentiated tissues or cells were seen to have common features: large amounts of often a small number of proteins are synthesised over temporally defined periods; the specialised cells eventually convert most of their dry mass into differentiation-specific product and may then die; synthesis of the differentiation-specific product is relatively insensitive to actinomycin.

Synthesis of crystalline proteins in lens tissue from chick embryos[91,92] and in calf lens fibre cells[93,94] was shown to be resistant to actinomycin. Similar results were obtained for production of feather proteins (keratins) by chick skin cells[91] and for haemoglobin synthesis in mammalian reticulocytes[95,96] or in chicken embryos[97]. Yaffee and Feldman showed that differentiated muscle cells were relatively resistant to actinomycin and would remain functional for at least 48 h after *in vitro* culture with the drug[98]. Wilson *et al.* were able to detect synthesis of albumin proteins in the actinomycin-treated rat liver[99].

(b) *Pancreas*—Rutter and his colleagues have studied differentiation-specific protein synthesis in the embryonic rat pancreas[1,100]. Pancreatic rudiments were studied during short-term organ culture *in vitro* on days 13–20 of gestation. During this period, as the pancreatic cells are passing through the protodifferentiated and fully differentiated phases (see Section 8.1.1) a group of specific digestive enzymes, the secretory products of the pancreas, increase in concentration by 3–5 orders of magnitude. This change in enzyme concentration apparently reflects the fact that the rates of synthesis for these specific secretory proteins have increased *ca.* $1–5 \times 10^4$ times[2].

In a typical actinomycin experiment, pancreas cultures were treated *in vitro* with 10^{-8} M actinomycin for 48 h periods at progressive times during a 7 day culture period, after which the extent of [^{3}H]leucine incorporation into specific secretory proteins was measured. In general, the later in Phase II the cultures were exposed to actinomycin, the smaller the reduction of specific protein synthesis. Essentially all protein synthesis could be blocked by actinomycin prior to the onset of Phase II. Once Phase II had started, selective inhibition of protein synthesis could be observed: the synthesis of certain proteins became insensitive to actinomycin earlier than others. Late in Phase II, actinomycin had no effect on the synthesis of any of the pancreatic products studied. These results were consistent with the hypothesis that mRNA for the exocrine proteins of the pancreas is stable once it is synthesised, and accumulates during Phase II. The initiation of Phase II occurs at different times for different proteins, implying independent regulation. The differential acquisition of insensitivity to the drug was consistent with the independent kinetics of synthesis normally demonstrated for several different products.

(c) *Insect galea*—Production of large amounts of secretory proteins can also occur on apparently very stable mRNAs in insects. Several experiments were cited above (Section 8.4.2.2) which document the high differential stability of the mRNA coding for cocoonase zymogen in the modified epidermal cells of the galea (apparent half-life of *ca.* 100 h). In contrast, mRNAs in the same cells coding for non-secretory proteins have an average half-life of the order of 2–3 h.

(d) *Chick oviduct*—The stabilities of mRNAs for differentiation-specific secretory and for non-secretory proteins have also been measured for the chick oviduct[16]. The differentiation of the tubular gland cells of the chick oviduct can be primed by a suitable regimen of primary hormone stimulation and withdrawal, so that secondary administration of oestrogen will precipitate the rapid formation of polysomes[85] on newly-synthesised mRNA. The mRNA synthesised in response to secondary oestrogen treatment codes largely for the secretory proteins of the oviduct: ovalbumin, conalbumin, ovomucoid and lysozyme[14]. The relative rate of synthesis of these four secretory proteins increased after administration of actinomycin, which would be expected if these proteins were synthesised on mRNAs of greater than average stability[14]. Ovalbumin mRNA specific activity was actually shown to increase in polysomes from oviduct tissue previously exposed to actinomycin. By measuring the decay of the rates of ovalbumin and conalbumin synthesis, mRNA half-lives were determined to be 14 and 8 h respectively[16]. Similar half-lives were inferred in the absence of actinomycin: the synthesis of secretory (but not of other) proteins decays in stimulated and explanted tissue, presumably because of specific termination of transcription. The synthesis of non-secretory proteins in the presence of actinomycin decayed with an apparent average half-life of *ca.* 5 h (see Section 8.4.2.2e).

(e) *Erythropoiesis in vertebrates*—The study of the control of haemoglobin synthesis during erythropoiesis in the higher vertebrates has provided evidence that haemoglobin synthesis is programmed by stable mRNAs. There are fundamental similarities in the nature of erythropoiesis in birds, frogs, mice and man: in each case erythropoiesis involves an ontogenetic shift from primitive to definitive cell lines. The cell lines often originate at different sites in the animal[101]. The embryonic cell lines are particularly suitable for study, since they often develop synchronously, as cohorts. Just as in Phase II specific protein synthesis in the oviduct, pancreas and galea, the relative rate of haemoglobin synthesis (computed as a percentage of total protein synthesis per cell) increases during erythroid development. Marks and his colleagues have shown that after haemoglobin synthesis has begun in either yolk sac or foetal liver erythroid cells, it can continue for extended periods in the presence of actinomycin[102]. Non-haemoglobin protein synthesis decays rapidly in the presence of actinomycin. As a result, and as in other highly specialised cell types, actinomycin increases the proportion of total protein synthesis devoted to the differentiation-specific product.

(f) *Fibroin synthesis by silkmoths*—The mRNA for fibroin, by indirect evidence at least, also appears to be highly stable. Fibroin, the major protein of silk, is produced by the posterior silk gland of the caterpillars of *Bombyx mori*, primarily at the end of the larval stage[103]. The full amount of chemically-recognisable fibroin mRNA can be isolated as long as 48 h after the peak of fibroin synthesis; it is not known whether this mRNA is still biologically functional[104]. During preparation of ^{32}P-labelled mRNA for sequencing studies, it became apparent that the best yield of mRNA of the highest specific activity was possible after 4–5 days of *in vivo* labelling of silkmoth larvae with ^{32}P. These empirical findings are consistent with the fibroin mRNA having a high stability and a low initial rate of transcription (see Section 8.5).

(g) *Evaluation*—Differentiation-specific protein synthesis in many embryonic and mature tissues of vertebrates and insects shows strikingly constant features. Specific protein synthesis starts gradually as cells pass from Phase I to Phase II of development (Section 8.1). Then, the rate of specific protein synthesis may increase by as much as several orders of magnitude. During late Phase I and throughout Phase II, the synthesis of non-specialised proteins is sensitive to inhibition of RNA synthesis with actinomycin, whereas differentiation-specific protein synthesis is relatively refractory. The most parsimonious interpretation of the results is that differentiation-specific mRNA is differentially stable, although the actinomycin results could also be interpreted in terms of *specific* translational factors (but *not* non-specific; see Section 8.2.1). In such tissues (oviduct[105-108,133-138], reticulocytes[36,109], silk gland[84] moth follicle[158]) it has recently become possible to isolate specific mRNAs late in Phase II or Phase III. Initial, as yet not rigorous, indications from direct studies of mRNA metabolism tend to substantiate that differentiation-specific mRNA is highly stable[3]. This conclusion is in agreement with the older, purely biological indication that differentiation-specific protein synthesis survives the loss of the nucleus.

8.4.3.3 Diversity in 'general' mRNA

Some evidence has already been reviewed for diversity of half-lives of 'general' mRNAs of a single tissue. The two-component kinetic decay curves in HeLa and the unusually short half-life of histone message are two examples. Here we shall discuss attempts to quantify the variability of half-lives around the average value.

(a) *Studies on moth wing*—In the moth wing system described above (Section 8.4.2.2d) Kafatos *et al.* again applied a double-label method to partially fractionated proteins[52]. [^{14}C]labelled control proteins and [^{3}H]proteins synthesised after various exposures to actinomycin were coelectrophoresed on SDS gels. Diversity of half-lives was inferred from the diversity of the isotope ratio in different gel slices. Internal standardisation of the experiment in this way minimised artefacts resulting from variations in gel slicing, exact reproducibility of electrophoretic conditions, or differential losses of proteins during processing. Since the specific activity of the precursor pool was not determined, the absolute value of the average mRNA half-life (against which individual half-lives were standardised) was based on incorporation data alone. The apparent mRNA half-lives of six recognisable classes of proteins in the gels (9.7, 10, 15, 18, 41 and 59 h) varied by a factor of six in relative magnitude. The rate of incorporation into individual components, as well as the total incorporation, decayed exponentially with time, suggesting that mRNA decay is indeed stochastic.

(b) *Studies on moth galea*—In the silkmoth galea, the diversity of mRNA half-lives for non-zymogen proteins was measured by a similar technique. It was recognised that when a total tissue extract is electrophorised in a single gel each gel slice contains a large number of different proteins, quite probably dissimilar. Thus the half-life diversity inferred from a single gel represents a minimal estimate, and can be made more accurate by additional purification of the proteins, which should reduce overlap and dampening of the variation.

In this case, whole galeal protein was divided into ten different ammonium sulphate fractions prior to electrophoretic analysis[2, 47, 83]. This pre-fractionation step increased the number of ratio peaks which could be identified on the SDS gels by an order of magnitude. A substantial diversity in half-life was found for those messengers present in large enough amounts to yield recognisable protein peaks under the experimental conditions (*ca.* 0.02% of the total protein synthesis). In a sample of 50 partially purified proteins, 49 mRNA half-lives were in the range of 1.09–4.48 h, and one was 9.8 h; the mean was 2.37 h and the standard deviation 1.26 h. This sort of analysis showed that the stability of zymogen mRNA was well above the range of the bulk of the non-zymogen messengers.

(c) *Evaluation and conclusions*—It now appears undeniable that substantial diversity of mRNA half-lives exist—quite analogous to the diversity of protein half-lives[124]. Although the experiments that have specifically attempted quantitative evaluation of this diversity suffer from the drawbacks of inhibitor use, even drug-independent experiments give some indication of mRNA half-life diversity (Section 8.4.2.1). In Brandhorst and Humphreys's kinetic study of sea urchin mRNA stability[50] the 95% confidence limits indicate that one-quarter of the radioactive RNA could have a half-life more than 0.28 or 1.9 times different than the mean, one-half could lie beyond 0.66 or 1.33 times the mean, and three-quarters could be beyond 0.85 or 1.15 times the mean. The drug-dependent estimate of diversity[2, 52, 83] indicates a sixfold range of half-lives for non-specific mRNA; this is undoubtedly an underestimate, since dampening of the variation should be expected in experiments of this type, which use only partially-purified components. When the exceedingly stable differentiation-specific message is considered, the range of half-life diversity is extended to two orders of magnitude[2, 82, 83]. Even in the unlikely case that these results reflect primarily specificity of translational factors (e.g. specific initiation factors), the observation of apparent half-life diversity is important since it documents extensive post-transcriptional control.

Although bacterial message is typically short-lived, even in *E. coli* half-life diversity can be observed. In T4-infected *E. coli*, individual mRNAs have half-lives varying from less than 80 s to more than 20 min; phage message is the more stable fraction[125].

The actual mechanism of differential mRNA stability remains a mystery. An appealing hypothesis is that non-specific degradation machinery exists in the cell acting on diverse messages which vary in susceptibility, ultimately because of their primary structure: the primary structure leads to association with specific protective proteins, or specifies secondary or tertiary structure which in turn regulates resistance to nuclease attack. In the case of protein breakdown, some evidence exists for a major effect of primary sequence on stability[126, 127]. In addition, it is possible that the degradation mechanism involves more than one type of nuclease.

Sequence studies of the small RNA phages of *E. coli* have shown that their primary sequence permits a great deal of secondary structure, such as hair-pin loops and cloverleaf forms[128, 129]. Assuming that the secondary structure of the picorna viruses is related to stability, it could help account for the very high stability of these viral RNAs during infection of *E. coli*[130] (half-lives of the order of 90 min).

Simultaneous with its discovery, speculations were advanced that poly A is in some way involved with mRNA stability. The poly A sequence may be the attachment point for at least some of the protein that is associated with mRNA as it leaves the nucleus and travels to the cytoplasmic polysomes[131,132]. It is intriguing that the size of poly A seems to decrease with message age. More generally, the untranslated regions known to exist in mRNA may have as one of their functions the control of mRNA stability.

8.5 IMPLICATIONS OF DIVERSITY IN mRNA HALF-LIVES

In Section 8.3 we concluded that Phase II kinetics can potentially be generated with *constant* transcription rates, so long as the mRNA half-life is comparable to the duration of Phase II. In Section 8.4.3 we saw that circumstantial but strong evidence exists that the mRNAs coding for differentiation-specific proteins are generally long-lived, and that they tend to be differentially stable, i.e. to have a half-life at or above the high end of a range encountered in the cells. If mRNAs are diverse in stability and differentiation-specific mRNAs are unusually stable, implications about regulation can be drawn from the transcriptional control hypothesis beyond those presented in Section 8.3.

First, the transcriptional load on differentiation-specific genes is lightened. This is an obvious conclusion, since the level of any mRNA represents a balance between synthesis and degradation.

Second, Phase II protein synthesis kinetics can be generated with a relatively *low* (as well as constant) rate of mRNA synthesis[47]. For example, let us assume that the half-life of a specific message is of the order of 100 h, and is approximately 40 times longer than the average half-life of other mRNAs in the cell, as seems to be the case in the silkmoth galea. It can be shown (see equations 8.9, 8.10 and 8.10a below) that the rate of specific protein synthesis can increase by more than two orders of magnitude within 4 days (while other protein synthesis remains constant), and that the relative protein synthesis rate can become 10%, if specific mRNA synthesis remains constant at only 0.55% of the rate of total mRNA production[47]. With the same conditions, if specific mRNA synthesis is 10% of the total, the rate of specific protein synthesis can increase *ca.* 2900-fold and become 70% of the total in a period of time equal to the mRNA half-life of 4 days (Figure 8.2a)[47]. If one takes into account that mRNA production is a small fraction of total transcription, because of the rapid turnover of Hn RNA, it is clear that only a very small fraction of total transcription need be supported by the differentiation-specific gene to generate Phase II kinetics. This leads to the prediction that differentiation-specific mRNA is unlikely to be recognised after a brief pulse-labelling period. Such failures are a common experience of workers in the field. In the case of the silk gland, in which fibroin synthesis accounts for nearly all protein synthesis, fibroin mRNA cannot be distinguished as a labelled species except with long labelling periods, longer than 6 h[84]. In general, a long-pulse short-chase regimen is optimal for recognising essentially stable specific mRNAs.

The transcription of specific mRNA, the mRNA half-life, and the duration of Phase II are variables which can be manipulated individually to yield

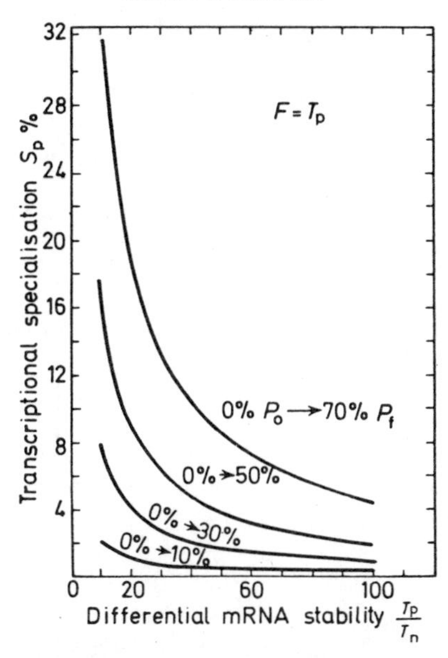

Figure 8.5 The relationships between differential mRNA stability (ratio of specific and non-specific half-lives, T_p/T_n transcriptional specialisation (% of total message production which is devoted to the specific message, S_p%), and the degree of translational specialisation which can be attained in a period of time F, equal to the specific mRNA half-life T_p (rate of specific protein synthesis, relative to the total, initially set as zero, and finally equal to the indicated value P_f). The calculations were based on equation (8.9), which assumes constant transcription and constant non-specific protein synthesis. For further details see the text

Phase II kinetics. More specifically, the interrelationships of these parameters can be analysed as follows.

Let us define more precisely as *translational specialisation* the fraction P of total protein synthesis which is devoted to a specific product. If protein synthesis is proportional to mRNA content,

$$P = \frac{M_p}{M_p + M_n} \tag{8.7}$$

where M_p is the accumulated amount of mRNA coding for the specific product and M_n is the amount of mRNAs for all other proteins produced by the cell the proportion of the non-specific protein synthesis N will be

$$N = 1 - P = \frac{M_n}{M_p + M_n} \tag{8.8}$$

If for simplicity we consider that the absolute rate of non-specific protein synthesis remains constant during Phase II, M_n is constant and presumably S_n and D_n as well. If we are dealing with Phase II kinetics which are controlled by constant transcription and differential stability of mRNA, we can deduce from equations (8.2)–(8.4) that the necessary rate of specific mRNA synthesis S will be

$$S = \frac{M_n D(P_f N_o - P_o N_f \, e^{-DF})}{N_f N_o - N_o N_f \, e^{-DF}} \tag{8.9}$$

where D is the decay constant of the specific message, F the duration of Phase II, and the subscripts 0 and f indicate initial and final translational values*.

Figures 8.5 and 8.6 apply equation (8.9) and show the interrelationships between the duration of Phase II (F), the change in percentage translational specialisation during Phase II ($\%P_0 \rightarrow \%P_f$), the absolute (T_p) and differential (T_p/T_n) stability of the specific message, and the percentage transcriptional specialisation (the percentage of total mRNA production which must be devoted to the specific message, $\%S_p$). Figure 8.5 shows that when the duration of Phase II is comparable to the half-life of the specific message, attainment of 70% translational specialisation (which is near the upper limit of the changes observed in real systems) requires either a very high degree of transcriptional specialisation or a substantial differential stability of the specific message. As the degree of the translational specialisation which must be attained is reduced, the levels of transcriptional specialisation and differential stability that are necessary decrease, but their inverse relation persists.

Figure 8.6 shows that the cost in terms of transcriptional specialisation needed for attainment of a particular translational level change is reduced as the duration of Phase II increases. In general, the transcriptional demands approach an asymptotic minimum when Phase II has a minimum duration within a factor of 2 of the specific mRNA half-life. The actual level of transcriptional specialisation is dictated by the differential mRNA stability, and the magnitude of the translational change (Figures 8.6a and 8.6b).

The above treatment is in agreement with intuitively obvious generalisations. The greater the differential mRNA stability, the lesser the transcriptional load on the specific gene. For any given differential mRNA stability, the greater the duration of Phase II (the longer the time-constant of the system), the lesser the transcriptional load on the specific gene and the greater the translational specialisation which can be attained. Beyond making these generalisations concrete, equation (8.9) and Figures 8.5 and 8.6 are helpful in indicating that the relationships between these parameters are non-linear. If the number of specific gene copies is limited (Section 8.6), extreme transcriptional specialisation may have to be avoided. The non-linearity of the curves then imposes guidelines on the other parameters. The

*The time-course of change in translational specialisation will be described by

$$P_t = \frac{X}{X+1} \tag{8.10}$$

where

$$X = \left(\frac{P_f N_0 - P_o N_f \, e^{-DF}}{N_f N_o - N_o N_f \, e^{-DF}}\right) \left(1 - \frac{P_f \, e^{-DT} - P_o \, e^{-DT}}{P_f - P_o \, e^{-DF}}\right) \tag{8.10a}$$

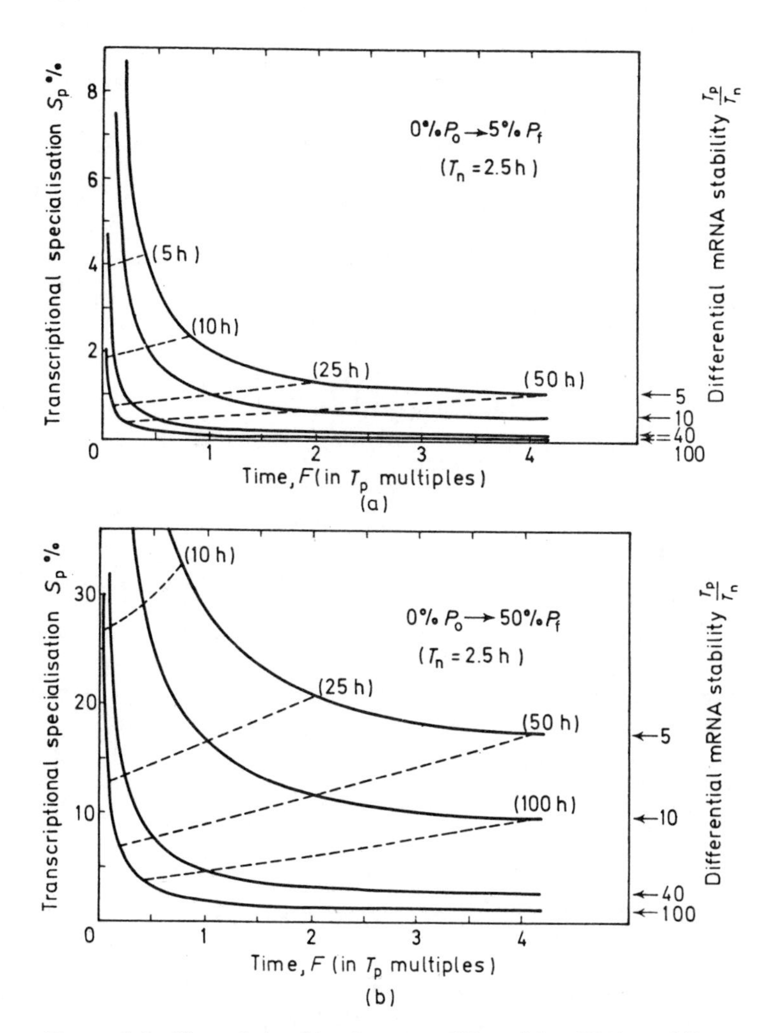

Figure 8.6 The relationships between differential mRNA stability, transcriptional specialisation, and the time-constant of the system (F, duration of Phase II, during which the indicated translational specialisation, $\%P_o \rightarrow \%P_f$, is attained). The calculations were based on equation (8.9), which assumes constant transcription and constant non-specific protein synthesis. As a concrete example, actual values of the time-constant are given in parentheses and positioned in the solid curves by means of the dotted lines (these time-constants correspond to a uniform value of 2.5 h for the average non-specific mRNA half-life T_n, and to a value for the specific mRNA half-life T_p given by the product of 2.5 h times T_p/T_n, the differential mRNA stability). The positions of the 10 and 25 h dotted lines indicate that the time constant must be of the order of a day or more if transcriptional specialisation is to approach the asymptotic minimum. For further details, see Figure 8.5 and the text. (a) Specific protein synthesis changing from zero to 5% of the total protein synthesis in time F. (b) Specific protein synthesis changing from zero to 50% of the total in time F

minimum time-constant of the system should be within about a factor of 2 of the specific mRNA half-life (i.e. the minimum duration of Phase II should be approximately 1–4 days if the specific mRNA has a half-life 5–100 times greater than an average non-specific mRNA half-life of 2.5 h; Figure 8.6). The specific mRNA should be 10–100 times more stable than the average non-specific mRNA if the cells are to specialise within that time so as to devote 10–70% of their total protein synthesis to the specific product (Figure 8.5).

If the actual behaviour of the system markedly deviates from these guidelines, we may suspect that the kinetics of Phase II are not controlled by constant contemporaneous transcription of differentially stable mRNA. In particular, rapid-response systems are good candidates for control through activation of existing mRNA—or through gene redundancy.

8.6 IS GENE MULTIPLICITY NECESSARY?

8.6.1 Long time-constant systems

The upper limit to transcriptional specialisation is obviously related to the number of specific gene copies present in the cell. We may, then, inquire whether the transcriptional demands of highly differentiated cells can be accommodated with a single copy of the specific gene per genome, or whether we must postulate gene multiplicity—either redundancy in the basic genome[139,140] or amplification in this particular cell line[141].

First we must estimate the maximum productivity of a single gene copy. As in the case of translation, the productivity P will be given by equation (8.1a). In this case, P is the number of message molecules produced per gene copy per min, v the rate of RNA chain elongation (nucleotides transcribed per polymerase per min) and s the average distance of RNA polymerases on the DNA (nucleotides per polymerase).

The rate v has been measured repeatedly in bacteria, and shown to have a rather constant value, the same for both mRNA and rRNA. This value changes with temperature, as expected for an enzymatic reaction, being approximately 2.7×10^3 nucleotides per polymerase per min at 37 °C and *ca.* 1.0×10^3 at 25 °C. Yet it is essentially unaffected by even major changes in cellular growth rates[142-145]. Even in eukaryotes it seems to be comparable. It has been roughly estimated as $<5 \times 10^3$ for ribosomal precursor RNA synthesis in HeLa[146] and as $\geq 4 \times 10^3$ for polio virus synthesis[147]. Its uniformity suggests that RNA chain elongation is not an important level of regulatory control. Thus, the high degree of transcriptional regulation known in bacteria and surmised for eukaryotes must operate through s. This conclusion is in agreement with the highly variable polymerase spacing on bacterial and eukaryotic chromosomes visualised by electron microscopy[148-150]. Clearly, the spacing is affected by RNA chain elongation, initiation and termination rates. If elongation is not very variable, we must conclude that initiation of transcription (and, less likely, termination) is of major regulatory significance. Initiation is apparently the level of both positive (e.g. polymerase-mediated) and negative (repressor-mediated) control in bacteria[58, 159, 160].

Although s is variable, the maximal reasonable transcription rate can be calculated on the basis of the minimal value of s encountered. It is interesting that three quite disparate systems extremely active in transcription, *E. coli.* ribosomal DNA, *Triturus* amplified rDNA, and *Triturus* lampbrush chromosomes, all show very similar tight packing of polymerases when examined under the electron microscope[148, 149]. The values of s are 74, 79 and 97 nucleotide pairs per polymerase, respectively. The average (84) is just over twice the physically-possible minimal value (a eukaryotic polymerase molecule has a diameter of 125 Å, equivalent to 38 base pairs). Thus, if in equation (8.1a) we set $s = 84$ and $v = 3.4 \times 10^3$ (a value half-way between the established bacterial rate and the lowest estimated eurkaryotic rate), we can estimate that the maximal reasonable transcriptional productivity is 40 mRNA molecules per gene copy per min at 37 °C. By extrapolation, the value at 25 °C can be estimated as 15.

Knowing the values of the translational parameters (Tables 8.1 and 8.2), we can calculate the translational productivity. Hence, when the kinetics of specific protein synthesis are accurately known, we can deduce the kinetics of accumulation of *active* mRNA simply by dividing the protein synthesis rates by the translational productivity. If we are dealing with long time-constant Phase II systems, we may assume that no significant pool of stored, inactive mRNA exists (Section 8.2), and hence we have an estimate of the kinetics of *total* mRNA accumulation. Can these kinetics be accommodated with a single copy of the specific gene per genome?

The rate of mRNA production in highly-specialised long time-constant Phase II systems can be estimated in several ways from the kinetics of accumulation (Table 8.3). The minimal estimate is the average rate, calculated by dividing the maximal attained value of M by the duration of Phase II, M_{max}/f, or more generally $\Delta M/\Delta t$. A better estimate is obtained from the steepest portion of the message accumulation curve (maximum $\mathrm{d}M/\mathrm{d}t$). Although the stability of mRNA is not accurately known, we may attempt a refinement by correcting $\mathrm{d}M/\mathrm{d}t$ for message decay, according to equation (8.2), with the reasonable assumption that the mRNA half-life is of the order of 2 days (Section 8.4.3.2). The three types of estimate for specific mRNA synthesis (per haploid DNA equivalent) agree within a factor of 2–3 (Table 8.3). It is remarkable that these estimates, for several diverse long time-constant Phase II systems, are just lower than the calculated maximal transcription rate per gene copy. The only exception is the silk gland, for which the estimates are suspect because of uncertain values for genome size, molecular weight of the protein product and translational productivity. In this system, a direct estimate of $\Delta M/\Delta t$ can be obtained from Suzuki and Brown's study of chemically-defined silk message[84, 104]; this estimate is in good agreement with the maximal transcriptional productivity. In general, the data available to date indicate that long time-contant Phase II systems could sustain their prodigious rates of specific protein synthesis with a single copy of the specific gene per genome, provided this gene is almost maximally active in transcription and provided the mRNA it synthesises is stable. Unstable mRNA (with a half-life of a few hours) is incompatible with one specific gene copy per genome[2].

Within the past year, independent and direct evidence has accumulated against gene multiplicity in highly specialised cell types[151-154]. Thus, it appears

Table 8.3 Calculated rates of specific mRNA synthesis

System and temperature	$\frac{\Delta M}{\Delta t}$ (mRNA HGE^{-1} min^{-1})	*maximum* $\frac{dM}{dt}$ (mRNA HGE^{-1} min^{-1})	$S = \frac{dM}{dt} + \frac{M \text{ Ln } 2}{T}$ (mRNA HGE^{-1} min^{-1})	*Expected maximal reasonable transcription rate per gene copy* (mRNA gene^{-1} min^{-1})
Ovalbumin (chick oviduct; 41 °C)	21	26	37	40
Chymotrypsinogen (embryonic rat pancreas; 37 °C)	4.4	7.3	10	40
Haemoglobin y (embryonic mouse primitive erythroblasts; 37 °C)	13	—	—	40
Cocoonase zymogen (polyphemus galea; 25 °C)	3.5	6.4	10	15
Silk fibroin (*B. mori* silk glands; 25 °C)	20 (5)*	35	63	15

HGE = haploid genome equivalent, i.e. amount of DNA equal to the haploid level. For details, see the text and Ref. 2

* Calculated from the direct data on fibroin mRNA[84]

Values for M were determined from rates of protein synthesis using translational productivity factors of 3.8, 16, 16, 2.8, and 2.8 polypeptides per mRNA per minute respectively, for synthesis of ovalbumin, chymotrypsinogen, haemoglobin y, cocoonase zymogen, and silk fibroin as described in the legends to Tables II and III in reference 2. $\Delta M/\Delta t$ is calculated from the total amount of M accumulated and the duration of Phase II. dM/dt is the maximum rate of increase in M during Phase II. S includes a correction for message decay. T, the effective trancription time, was taken as 1.6, 3, 1.1, 5, and 5.5 days for ovalbumin, chymotrypsinogen, haemoglobin y, cocoonase zymogen, and silk fibroid mRNAs, respectively. See Ref. 2 for more details that transcriptional maximisation and mRNA stability are the mechanisms used by highly-differentiated long time-constant Phase II systems, rather than gene amplification or redundancy. This generalisation leads to the prediction that differentiation-specific genes in such systems could be visualised as chromatin regions of unusually close polymerase packing—distinguished from ribosomal cistrons because of the latter's characteristic length and repeating nature[148-150].

8.6.2 Short time-constant systems

The situation may be entirely different for short time-constant systems. We have already seen that very rapid changes in specific protein synthesis may reflect activation of existing mRNA (Section 8.2.4). In other systems, rapid changes may require gene multiplicity.

A case in point is histone synthesis. We have seen that HeLa cells may synthesise as many as 1.4×10^3 A histone mRNA molecules/min (Section 8.3.3). If there are five types of histone and HeLa is diploid with respect to the histone gene loci, even if all histones are produced equally the mRNA synthetic rate is three times higher than the calculated maximum productivity per gene copy. Thus, the kinetics shown in Figure 8.3 are somewhat above the limit possible with a single gene copy per locus. The kinetics of histone synthesis during early embryogenesis are substantially faster, indicating that considerable gene multiplicity must exist.

During cleavage, production of nuclei is exponential and hence the number of total active histone message molecules per embryo, M_e, can be assumed to double during each division cycle

$$\frac{\mathrm{d}M_e}{\mathrm{d}t} \approx \frac{M_e}{\tau} \tag{8.11}$$

where τ is the (brief) duration of the cleavage cycle. Although some of the histone message molecules used in early cleavage seem maternal in origin, much histone message synthesis occurs on the embryonic genome[44] and undoubtedly becomes by far the dominant source of the message in late cleavage, when histone synthesis (on a per embryo basis) is several orders of magnitude more rapid than immediately after fertilisation. From equations (8.2) and (8.11) it follows that

$$S = (\frac{1}{\tau} + D)M \tag{8.12}$$

But

$$M = \frac{h}{P_r \tau} \tag{8.13}$$

where h is the number of histone protein molecules per genome (approximately equal to the haploid DNA weight multiplied by 6×10^{23} and divided by 1.3×10^4 the average histone mol. wt.). P_r is the translational productivity, which we can take to be *ca.* 2.8 polypeptides mRNA^{-1} min^{-1} at 25°C. On the basis of the half-life histone message, as estimated by the use of actinomycin, we can set $D = 1.2 \times 10^{-2}$ min^{-1}; in any case, since $D \ll 1/\tau$ for cleavage, the precise value of D is not important. Thus

$$S = \frac{h(1 + D\tau)}{P_r \tau^2} \tag{8.14}$$

For *Drosophila*, $\tau = 10$ min and $h \approx 8.3 \times 10^6$ molecules. Therefore, $S = 3.3 \times 10^4$ histone message molecules produced per min; if transcriptional productivity is 15 (Table 8.3), there must be more than 2×10^3 total histone genes. Similarly, for the sea urchin *Paracentrotus* $\tau = 24$ min and $h = 3.2 \times 10^7$. Therefore $S = 2.5 \times 10^4$, and again there must be *ca.* 2×10^3 total histone genes, as a minimum. These estimates are in good agreement with the experimental evidence on histone gene reiteration[155]. In *Drosophila*, *in situ* hybridisation experiments suggest that histone genes have a substantial reiteration frequency[156]. In *Paracentrotus*, it has been roughly estimated by DNA-excess hybridisation that each of three different histone message classes hybridise to DNA reiterated *ca.* 400-fold.

We conclude that histone gene redundancy *in the basic genome* is dictated by the need for very rapid histone message production by the progenitors of all somatic and germ cells, during early embryogenesis. Even with their apparently large number of copies, histone genes must approach the upper limit of transcriptional productivity during cleavage. In other systems, very rapid protein synthesis kinetics may be limited to a single line of differentiated somatic cells. In such systems, control may reside in the activation of existing mRNA. Alternatively, if differentiation is controlled by contemporaneous transcription, gene multiplicity may again be mandatory, and in this case may be attained by specific gene amplification in that particular somatic cell line, rather than by gene redundancy in the basic genome. A good example of such highly-differentiated short time-constant systems are the follicular cells of silkmoths, which synthesise massive amounts of eggshell proteins[26] in succession, according to a precise programme of overlapping synthetic phases lasting a few hours. We have recently identified on polyacrylamide slab gels the messages coding for eggshell proteins[158] and have established for them an assay based on translation in a Krebs ascites cell-free system[45], preparatory to a direct study of message metabolism and gene reiteration.

8.7 CONCLUSIONS AND SUMMARY

In this paper, consideration of mRNA stability was undertaken in the broader framework of the flow of genetic information, from gene to protein. The multitude of levels at which this information flow can be controlled is well known. From our perspective it is important to understand which of these levels are actually used in major regulatory processes, and especially which ones are most important in eukaryotic differentiation.

It is clear that we are not yet at the point where we can answer these questions with certainty. Much of our information on important variables (such as transcriptional and translational parameters, mRNA stability, mRNA content, gene number) is not rigorous, being indirect, average rather than specific, and at best semi-quantitative. With the evidence on hand, it is only possible to reach certain tentative generalisations—and to begin to pinpoint some biological parameters, such as the time-constant of developmental change, which may dictate which of the possible regulatory levels are used in fact.

In Section 8.2 we review the evidence on whether the rate of specific protein synthesis can be assumed to bear a proportionality to the content of the corresponding message. It is clear that proportionality fails in certain systems, especially those in which storage of masked mRNA permits subsequent abrupt activation (fertilisation, seed germination), and perhaps in highly differentiated systems during 'shut off'. At the same time, it is notable that the empirically-measurable translational parameters (rate of polypeptide chain elongation and ribosomal spacing on the mRNA) vary only within narrow ranges; thus, the translational productivity of those messages which are active (as distinct from message 'unmasking' or activation) seems to be of limited regulatory importance. In the specific case of Phase II or differentiation in the normal long time-constant systems, it appears that the kinetics of the changing rate of specific protein synthesis reflect, to a first approximation at least, the kinetics of specific mRNA accumulation.

In Section 8.3, a theoretical treatment emphasises that the mRNA decay constant is equally as important as the mRNA synthetic rate in determining the kinetics of message accumulation. We point out that even a constant rate of mRNA synthesis can generate the accumulation kinetics characteristic of long time-constant Phase II systems, provided the specific mRNA stability is high. On the other hand, a particular mRNA decay constant, in combination with changing rates of mRNA production, may dictate the kinetics of message accumulation and disappearance; this may be the case for histone synthesis during the S phase of the cell cycle.

The experimental evidence on mRNA stability is reviewed in Section 8.4. The evidence has long been indirect, and as a result the field is now in a major flux, as newer more direct methods yield unexpected results. While absolute values of mRNA half-lives are quite suspect, relative half-lives may be more meaningful. As of now, it appears that a considerable diversity in mRNA half-lives exists within each eukaryotic cell, and that the message for the characteristic protein products of highly differentiated cells is very stable, most probably unusually so in comparison to other messages in the same cell.

In Section 8.5 we explore the implications of the high and differential stability of the differentiation-specific message. Continuing the theoretical treatment of Section 8.3, we find that Phase II kinetics can be generated with a *low*, as well as constant, rate of mRNA production, because of the specific message stability. More generally, transcription of specific mRNA, mRNA half-life and duration of Phase II are variables which can be manipulated individually to yield Phase II protein synthesis kinetics. The interrelationships of these variables are non-linear, and thus specific transcription can be limited to a relatively low rate only if the other two parameters follow certain guidelines.

In Section 8.6, the 'relatively low rate of specific transcription' is evaluated more precisely. The data are consistent with the hypothesis that Phase II protein synthesis kinetics are controlled by constant contemporaneous transcription of differentially stable mRNA, on a *single copy* of the specific gene per genome—a copy which is almost fully 'turned-on', so as to attain near maximal transcriptional productivity with a near maximal initiation rate. Direct evidence now exists against multiplicity of the genes coding for products of long time-constant highly differentiated cells. The analysis can thus be reversed, and we can conclude, given the limitation of one gene copy per genome, that both differential mRNA stability and transcriptional maximisation are important in this type of differentiation.

By contrast, in short time-constant systems, characterised by exceedingly rapid changes in protein synthesis, differential mRNA stability and transcriptional maximisation on a single gene copy may be inadequate. One alternative control is activation of existing message, as discussed above. Another is gene multiplicity. It is shown that the high reiteration frequency of histone genes is dictated by the extremely short time-constant of embryonic development (doubling of the rate of histone synthesis every few minutes, with each cleavage cycle). It is proposed that somatic gene amplification may occur only in short time-constant highly differentiated cells.

Our conclusions are non-rigorous, because of the indirect nature of much of the evidence. This fact emphasises the importance of shifting from the protein to the nucleic acid level in molecular studies of differentiation. The recent passion for specific mRNA isolation is a most welcome trend in this field. In order to test our ideas about major regulatory levels in differentiation, one immediate goal is clear: to use the new methodologies of specific mRNA quantitation, in a direct attempt to correlate the kinetics of specific protein synthesis with the metabolism of the corresponding mRNA.

Acknowledgements

We thank Linda S. Lawton for her expert secretarial assistance. We are grateful to Dr. R. D. Palmiter for communicating and allowing us to use his unpublished data (Figure 8.1), as well as for helpful discussions and Drs. P. Gage, J. Greenberg, W. Jellinek, R. Singer, B. Brandhorst, R. Williamson, and H. Lodish for communicating their results. We acknowledge the support of NIH training grant 1 TO1 HD00415 (R. Gelinas) and NIH grant 5 R01 HD04701, NSF grant GB-35608X, and a grant from the Rockefeller Foundation (F. C. Kafatos).

References

1. Rutter, W. J., Kemp, J., Bradshaw, W., Clark, W., Ronzio, R. and Sanders, T. (1968). *J. Cell. Physiol.*, **72,** 1
2. Kafatos, F. C. (1972). *Current Topics in Developmental Biology*, Vol. 7, 125 (A. A. Moscona and A. Monroy, editors) (New York: Academic Press)
3. Palmiter, R. D. (1973). *J. Biol. Chem.* (in press)
4. Adesnik, M. and Darnell, J. (1972). *J. Molec. Biol.*, **67,** 397
5. Lane, C. D., Marbaix, G. and Gurdon, J. (1971). *J. Molec. Biol.*, **61,** 73
6. Gurdon, J. B., Lane, C., Woodland, H. and Marbaix, G. (1971). *Nature (London)*, **233,** 177
7. Moar, V. A., Gurdon, J., Lane, C. and Marbaix, G. (1971). *J. Molec. Biol.*, **61,** 93

8. Gianni, A. M., Giglioni, B., Ottolenghi, S., Comi, P. and Guidotti, G. (1972). *Nature New Biol.*, **240,** 183
9. Jacobs-Lorena, M. and Baglioni, C. (1972). *Proc. Nat. Acad. Sci.* (*U.S.A.*), **69,** 1425
10. Gage, P. (1972). Personal communication
11. Kaempfer, R. (1973). Personal communication
12. Kaempfer, R. and Kaufman, J. (1972). *Proc. Nat. Acad. Sci.* (*U.S.A.*), **69,** 3317
13. Chan. L., Means, A. R. and O'Malley, B. W. (1973). *Proc. Nat. Acad. Sci.*, (*U.S.A.*), **70,** 1870
14. Palmiter, R. D. (1972). *J. Biol. Chem.*, **247,** 6450
15. Palmiter, R. D. (1972). *J. Biol. Chem.*, **247,** 6770
16. Palmiter, R. D. and Schimke, R. (1973). *J. Biol. Chem.*, **248,** 1502
17. Terman, S. A. (1970). *Proc. Nat. Acad. Sci.* (*U.S.A.*), **65,** 985
18. Nadel, M., Carroll, A. and Kafatos, F. (1973). In preparation
19. Roth, R., Ashworth, J. and Sussman, M. (1966). *Proc. Nat. Acad. Sci.* (*U.S.A.*), **59,** 1235
20. Firtel, R., Baxter, L. and Lodish, H. (1973). *J. Molec. Biol.* (in the press)
21. Moscona, A. A., Moscona, M. and Saenz, N. (1968). *Proc. Nat. Acad. Sci.* (*U.S.A.*), **61,** 160
22. Hunt, T., Hunter, T. and Munro, A. (1968). *J. Molec. Biol.*, **36,** 31
23. Hunt, R. T., Hunter, A. and Munro, A. (1968). *Nature* (*London*), **220,** 481
24. Lodish, H. F. (1971). *J. Biol. Chem.*, **246,** 7131
25. Lodish, H. F. (1972). *J. Biol. Chem.*, **247,** 3622
26. Paul, M., Goldsmith, M., Hunsley, J. and Kafatos, F. (1972). *J. Cell Biol.*, **55,** 653
27. Tomkins, G. M., Levinson, B., Baxter, J. and Dethlefsen, L. (1972). *Nature New Biol.*, **239,** 9
28. Regier, J. C. and Kafatos, F. (1971). *J. Biol. Chem.*, **246,** 6480
29. Tomkins, G. M., Gelehrter, T., Granner, D., Martin, D., Samuels, H. and Thompson, E. (1969). *Science*, **166,** 1474
30. Heywood, S. M. (1970). *Proc. Nat. Acad. Sci.* (*U.S.A.*), **67,** 1782
31. Rhoads, R. E., McKnight, G. and Schimke, R. (1971). *J. Biol. Chem.*, **246,** 7407
32. Prichard, P. M., Picciano, D., Laycock, D. and Anderson, W. (1971). *Proc. Nat. Acad. Sci.* (*U.S.A.*), **68,** 2752
33. Berns, A. J., Strous, G. and Bloemendal, H. (1972). *Nature New Biol.*, **236,** 7
34. Heywood, S. M. (1969). *Cold Spring Harbor Symp. Quant. Biol., Synthesis of Myosin on Heterologous Ribosomes*, **34,** 799 (Cold Spring Harbor Lab. 1969: Cold Spring Harbor, L.I., N.Y.)
35. Stavnezer, J. and Huang, R. (1971). *Nature New Biol.*, **230,** 172
36. Pemberton, R. E., Housman, D., Lodish, H. and Baglioni, C. (1972). *Nature New Biol.*, **235,** 99
37. Aviv, H., Boime, I. and Leder, P. (1971). *Proc. Nat. Acad. Sci.* (*U.S.A.*), **68,** 2303
38. Aviv, H., Boime, I., Loyd, B. and Leder, P. (1972). *Science*, **178,** 1293
39. Housman, D., Pemberton, R. and Taber, R. (1971). *Proc. Nat. Acad. Sci.* (*U.S.A.*), **68,** 2716
40. Mathews, M. B., Osborn, M. and Lingrel, J. (1971). *Nature New Biol.*, **233,** 177
41. Mathews, M. B., Osborn, M., Berns, A. and Bloemendal, H. (1972). *Nature New Biol.*, **236,** 5
42. Swan, D., Aviv, H. and Leder, P. (1972). *Proc. Nat. Acad. Sci.* (*U.S.A.*), **69,** 1967
43. Stevens, R. H. and Williamson, A. (1972). *Nature* (*London*), **239,** 143
44. Gross, K., Ruderman, J., Jacobs-Lorena, M., Baglioni, C. and Gross, P. (1973). *Nature New Biol.*, **241,** 272
45. Hunsley, J., Gelinas, R. and Kafatos, F. (1973). In preparation
46. Ilan, J. and Ilan, J. (1971). *Develop. Biol.*, **25,** 280
47. Kafatos, F. C. (1972). *Acta Endocrinol. Suppl.* **168,**
48. Sussman, M. (1970). *Nature* (*London*), **225,** 1245
49. Jacquet, M. and Kepes, A. (1971). *J. Molec. Biol.*, **60,** 453
50. Brandhorst, B. P. and Humphreys, T. (1972). *J. Cell Biol.*, **53,** 474
51. Greenberg, J. (1972). *Nature* (*London*), **240,** 102
52. Yund, M. A., Kafatos, F. and Regier, J. (1973). *Develop. Biology*, **33,** 362
53. *Handbook of Biochemistry*, 2nd edn., H-107 (1970) (H. A. Sober, editor) (Cleveland, Ohio: Chemical Rubber Co.)

54. Clayton, R. M. (1970). *Current Topics in Developmental Biology*, Vol. 5, 115 (A. A. Moscona and A. Monroy, editors) (New York: Academic Press)
55. Borun, T. W., Scharff, M. and Robbins, E. (1967). *Proc. Nat. Acad. Sci.* (*U.S.A.*), **58,** 1977
56. Kedes, L. H., Gross, P., Cognetti, G. and Hunter, A. (1969). *J. Molec. Biol.*, **45,** 337
57. Moav, B. and Nemer, M. (1970). *Biochemistry*, **10,** 881
58. Geiduschek, E. P. and Haselkorn, R. (1969). *Annu. Rev. Biochem.*, **38,** 647
59. Levinthal, C., Keynan, A. and Higa, A. (1962). *Proc. Nat. Acad. Sci.* (*U.S.A.*), **48,** 1631
60. Salser, W., Janin, J. and Levinthal, C. (1968). *J. Molec. Biol.*, **31,** 237
61. del Valle, R. and Aronson, A. I. (1962). *Biochem. Biophys. Res. Commun.*, **9,** 421
62. Pollock, M. R. (1963). *Biochim. Biophys. Acta*, **76,** 80
63. Martinez, R. J. (1966). *J. Molec. Biol.*, **17,** 10
64. Jaenish, R., Jacob, E. and Hofschneider, P. H. (1970). *Nature* (*London*), **227,** 59
65. Brandhorst, B. P. and Humphreys, T. (1971). *Biochemistry*, **10,** 877
66. Schochetman, G. and Perry, R. P. (1972). *J. Molec. Biol.*, **63,** 591
67. Singer, R. and Penman, S. (1972). *Nature* (*London*), **240,** 100
68. Singer, R. and Penman, S. (1973). Personal communication
69. Murphy, W. and Attardi, G. (1973). *Proc. Nat. Acad. Sci.* (*U.S.A.*), **70,** 115
70. Penman, S., Scherrer, K., Becker, Y. and Darnell, J. (1963). *Proc. Nat. Acad. Sci.* (*U.S.A.*), **49,** 654
71. Cheevers, W. and Sheinin, R. (1970). *Biochim. Biophys. Acta*, **204,** 449
72. Hodge, L., Robbins, E. and Scharff, M. (1969). *J. Cell Biol.*, **40,** 497
73. Craig, N., Kelly, D. and Perry, R. (1971). *Biochim. Biophys. Acta*, **246,** 493
74. Gallwitz, D. and Mueller, G. (1969). *J. Biol. Chem.*, **244,** 5947
75. Staehelin, T., Wettstein, F. and Noll, H. (1963). *Science*, **140,** 180
76. Hiatt, H. (1962). *J. Molec. Biol.*, **5,** 217
77. Revel, M. and Hiatt, H. (1963). *Proc. Nat. Acad. Sci.* (*U.S.A.*), **51,** 810
78. Endo, Y., Tominaga, H. and Natori, Y. (1971). *Biochim. Biophys. Acta*, **240,** 215
79. Trakatellis, A., Axelrod, A. and Montjar, M. (1964). *Nature* (*London*), **203,** 1134
80. Trakatellis, A., Axelrod, A. and Montjar, M. (1964). *J. Biol. Chem.*, **239,** 4237
81. Wilson, S. and Hoagland, M. (1967). *Biochem. J.*, **103,** 556
82. Kafatos, F. and Reich, J. (1968). *Proc. Nat. Acad. Sci.* (*U.S.A.*), **60,** 1458
83. Kafatos, F. and Moore, P. (1973). In preparation
84. Suzuki, Y. and Brown, D. (1972). *J. Molec. Biol.*, **63,** 409
85. Palmiter, R. D., Oka, T. and Schimke, R. (1971). *J. Biol. Chem.*, **246,** 724
86. Adesnik, M., Salditt, M., Thomas, W. and Darnell, J. (1972). *J. Molec. Biol.*, **71,** 21
87. Brawerman, G., Mendecki, J. and Lee, S. (1971). *Biochemistry*, **11,** 637
88. Lee, S., Mendecki, J. and Brawerman, G. (1971). *Proc. Nat. Acad. Sci.* (*U.S.A.*), **68,** 1331
89. Perry, R., Latorre, J., Kelley, D. and Greenberg, J. (1972). *Biochim. Biophys., Acta*, **262,** 220
90. Hammerling, J. (1966). *Develop. Biology*, p. 23 (R. Flickinger, editor) (Dubuque, Iowa: Brown and Co.)
91. Scott, R. and Bell, E. (1964). *Science*, **145,** 711
92. Reeder, R. and Bell, E. (1965). *Science*, **150,** 71
93. Papaconstantinou, J. (1967). *Science*, **156,** 338
94. Stewart, J. and Papaconstantinou, J. (1967). *J. Molec. Biol.*, **29,** 357
95. Marks, P., Burka, E. and Schlessinger, D. (1962). *Proc. Nat. Acad. Sci.* (*U.S.A.*), **48,** 2163
96. Reich, E., Franklin, R., Shatkin, A. and Tatum, E. (1962). *Proc. Nat. Acad. Sci.* (*U.S.A.*), **48,** 1238
97. Wilt, F. (1965). *J. Molec. Biol.*, **12,** 331
98. Yaffee, D. and Feldman, J. (1964). *Develop. Biol.*, **9,** 347
99. Wilson, S., Hill, H. and Hoagland, M. (1967). *Biochem. J.*, **103,** 567
100. Kemp, J. D., Walther, B. T. and Rutter, W. (1972). *J. Biol. Chem.*, **247,** 3941
101. Marks, P. and Rifkind, R. (1972). *Science*, **175,** 955
102. Fantoni, A., de la Chapelle, A., Rifkind, R. and Marks, P. (1968). *J. Molec. Biol.*, **33,** 79
103. Tashiro, Y., Morimoto, T., Matsuura, S. and Nagata, S. (1968). *J. Cell Biol.*, **38,** 574

104. Gage, P. (1972). Personal communication
105. O'Malley, B., Rosenfeld, G., Comstock, J. and Means, A. (1972). *Nature New Biol.*, **240,** 45
106. Means, A., Comstock, J., Rosenfeld, G. and O'Malley, B. (1972). *Proc. Nat. Acad. Sci.* (*U.S.A.*), **69,** 1146
107. Rosenfeld, G., Comstock, J., Means, A. and O'Malley, B. (1972). *Biochem. Biophys. Res. Commun.*, **47,** 387
108. Palacios, R., Sullivan, D., Summers, N., Keily, M. and Schimke, R. (1973). *J. Biol. Chem.*, **248,** 540
109. Williamson, R., Morrison, M., Lanyon, G., Eason, R. and Paul, J. (1971). *Biochemistry*, **10,** 3014
110. Baxter, J., Rousseau, G., Benson, M., Garcea, R., Ito, J. and Tomkins, G. (1972). *Proc. Nat. Acad. Sci.* (*U.S.A.*), **69,** 1892
111. Penman, S., Vesco, C. and Penman, M. (1968). *J. Molec. Biol.*, **34,** 49
112. Spohr, G., Imaizumi, T., Stewart, A. and Scherrer, K. (1972). *FEBS Lett.*, **28,** 165
113. Knopf, P. and Lamfrom, H. (1965). *Biochim. Biophys. Acta*, **95,** 398
114. Hunt, T., Hunter, T. and Munro, A. (1969). *J. Molec. Biol.*, **43,** 123
115. Fan, H. and Penman, S. (1970). *J. Molec. Biol.*, **50,** 655
116. Vaughan, M., Pawlowski, P. and Forchhammer, J. (1971). *Proc. Nat. Acad. Sci.* (*U.S.A.*), **68,** 2057
117. Staehelin, T., Wettstein, F., Oura, H. and Noll, H. (1964). *Nature* (*London*), **201,** 264
118. Warner, J., Rich, A. and Hall, C. (1962). *Science*, **138,** 1399
119. Warner, J., Knopf, P. and Rich, A. (1963). *Proc. Nat. Acad. Sci.* (*U.S.A.*), **49,** 122
120. Marks, P., Rifkind, R. and Danon, D. (1963). *Proc. Nat. Acad. Sci.* (*U.S.A.*), **50,** 336
121. Herrmann, H., Heywood, S. and Marchok, A. (1970). *Current Topics in Developmental Biology*, Vol. **5,** 181 (A. A. Moscona and A. Monroy, editors) (New York: Academic Press)
122. Lazarides, E. and Lukens, L. (1971). *Nature New Biol.*, **237,** 37
123. Kiho, Y. and Rich, A. (1964). *Proc. Nat. Acad. Sci.* (*U.S.A.*), **51,** 111
124. Schimke, R. T. (1969). *Curr. Top. Cell. Regul.*, **1,** 77
125. Marrs, B. L. and Yanofsky, C. (1971). *Nature* (*London*), **234,** 168
126. Goldberg, A. L. (1971). *Nature New Biol.*, **234,** 51
127. Goldberg, A. L. (1972). *Proc. Nat. Acad. Sci.* (*U.S.A.*), **69,** 422
128. DeWachter, R., Merregaert, J., Vandenberghe, A., Contreras, R. and Fiers, W. (1971). *Eur. J. Biochem.*, **22,** 400
129. Lodish, H. F. (1970). *J. Molec. Biol.*, **50,** 689
130. Igarashi, S. J., Elliott, J. and Bisonnette, R. (1970). *Can. J. Biochem.*, **48,** 47
131. Kwan, S. and Brawerman, G. (1972). *Proc. Nat. Acad. Sci.* (*U.S.A.*), **69,** 3247
132. Blobel, G. (1973). *Proc. Nat. Acad. Sci.* (*U.S.A.*), **70,** 924
133. Haines, M. E., Carey, N. H. and Palmiter, R. D. (1973). *J. Biol. Chem.* (in press)
134. Palmiter, R. D. and Smith, L. T. (1973). *Mol. Biol. Reports* (in press)
135. Palacios, R., Palmiter, R. and Schimke, R. (1972). *J. Biol. Chem.*, **247,** 2316
136. Palmiter, R. D., Palacios, R. and Schimke, R. (1972). *J. Biol. Chem.*, **247,** 3296
137. Rhoads, R. E., McKnight, G. and Schimke, R. (1973). *J. Biol. Chem.*, **248,** 2031
138. Palmiter, R. D. (1973). *J. Biol. Chem.*, **248,** 2095
139. Thomas, C. A. (1971). *Annu. Rev. Genetics*, **5,** 237
140. Lee, S. and Thomas, C. (1973). *J. Molec. Biol.*, (In the press)
141. Brown, D. and Dawid, I. (1968). *Science*, **160,** 272
142. Rose, J. K., Mosteller, R. and Yanofsky, C. (1970). *J. Molec. Biol.*, **51,** 541
143. Manor, H., Goodman, D. and Stent, G. (1969). *J. Molec. Biol.*, **39,** 1
144. Bremer, H. and Yuan, D. (1968). *J. Molec. Biol.*, **38,** 163
145. Jacquet, M. and Kepes, A. (1971). *J. Molec. Biol.*, **60,** 453
146. Greenberg, H. and Penman, S. (1966). *J. Molec. Biol.*, **21,** 527
147. Darnell, J. E., Girard, M., Baltimore, D., Summers, D. and Maizel, J. (1967). *The Molecular Biology of Viruses* (S. J. Colter and W. Paranchych, editors), p. 375. (New York: Academic Press)
148. Miller, O. L., Beatty, B., Hamkalo, B. and Thomas, C. (1970). *Cold Spring Harbor Symp. Quant. Biol.*, **35,** 505
149. Miller, O. L., Hamkalo, B. and Thomas, C. (1970). *Science*, **169,** 393
150. Miller, O. L. and Bakken, A. (1972). *Acta Endocrinol. Supp.*, **168,** 155

151. Bishop. J. O., Pemberton, R. and Baglioni, C. (1972). *Nature New Biol.*, **235,** 231
152. Harrison, R. P., Hill, A., Birnie, G. and Paul, J. (1972). *Nature* (*London*), **239,** 219
153. Suzuki. Y., Gage, L. and Brown, D. (1972). *J. Molec. Biol.*, **70,** 637
154. Schimke, R. (1972). Personal communication
155. Kedes, L. and Birnstiel, M. (1971). *Nature New Biol.*, **230,** 165
156. Pardue, M. L. (1973). Private communication
157. Paul, M. and Kafatos, F. (1973). In preparation
158. Gelinas, R. E. and Kafatos, F. (1973). *Proc. Nat. Acad. Sci., U.S.A.* (in press)
159. Calendar, R. (1970). *Annu. Rev. Microbiology*, **24,** 241
160. Losick, R. (1972). *Annu. Rev. Biochemistry*, **41,** 409
161. Greenberg, J. R. and Perry, R. (1972). *J. Molec. Biol.*, **72,** 91

Editor's Comments

The specific nature of the mechanisms controlling degradation in differentiated eukaryotic cells has yet to yield to experimentation. A great deal more has been learned about possible controls of transcription which may operate in eukaryotes. Since in prokaryotes, DNA-dependent RNA polymerases are intimately involved in the regulation of transcription through the mediation of a number of special factors, it was natural to expect that similar enzymes might play an important part in the specific regulation of transcription in eukaryotic cells. The discovery by Roeder and Rutter (1969) a few years ago of multiple polymerases in the sea urchin, which has been extended by other investigators to many other species, has engendered a great deal of interest in these enzymes and the current situation is discussed in detail in the next essay.

On the other hand, it has become rather clear in recent years that in eukaryotes mechanisms may operate which do not exist in prokaryotes. Many years ago, Stedman and Stedman (1950) suggested that the histones might be regulators of gene activity. In the early 1960s, a number of investigators showed that they did in fact inhibit the transcription from DNA by RNA transcriptase. This observation initiated the very extensive research into nucleoproteins and chromatin which has followed. Largely as a result of observations by Paul and Gilmour, attention has shifted from the histones to the non-histone proteins. In their original studies, Paul and Gilmour (1968) used bacterial RNA polymerase to transcribe RNA from chromatin and then tested this RNA by hybridisation to DNA. They provided evidence that only a limited set of sequences in DNA was accessible to the polymerase, that the accessible DNA differed from one organ to another and that the accessibility was mediated by non-histone proteins. These original studies employed methods which enabled only the more repetitive sequences of the DNA in the genome to be studied. More recently, however, Gilmour and Paul (1973) have reinvestigated the problem using a DNA copy from globin messenger RNA to estimate the extent to which the globin gene is transcribed in chromatin from erythropoietic and non-erythropoietic tissue. They have shown that in chromatin from mouse foetal liver (which is an erythropoietic organ) the globin gene is accessible to bacterial RNA polymerase whereas in chromatin from mouse brain it is not. They have, moreover, obtained evidence that a specificity of transcription is again mediated through non-histone proteins. Almost identical observations have been made independently by Axel, Cedar and Felsenfeld (1973), in the duck reticulocyte.

References

Axel, R., Cedar, H. and Felsenfeld, G. (1973). *Cold Spring Harbor Symp. Quant. Biol.*, **38** (in press)
Gilmour, R. S. and Paul, J. (1973). *Proc. Nat. Acad. Sci.* (in press)
Paul, J. and Gilmour, R. S. (1968). *J. Mol. Biol.*, **34**, 305
Roeder, R. G. and Rutter, W. J. (1969). *Nature* (*London*), **224**, 234
Stedman, E. and Stedman, E. (1950). *Nature* (*London*), **166**, 780

9
RNA Polymerase and Transcriptional Regulation in Physiological Transitions*

W. J. RUTTER, M. I. GOLDBERG and J. C. PERRIARD
University of California

---

* This work was supported by NIH USPHS grant HDO4617, National Science Foundation #3–5256, the Swiss National Foundation (to J.-C.P) and the Giannini Foundation for Medical Research (to M.G.)

9.1 INTRODUCTION

Selective gene transcription from a complex genome implies a modality in the specificity of the transcribing system[1]: the transcriptive system (including polymerase and whatever regulatory elements influence transcription) must recognise specific DNA sequences for initiation. In contrast, the process of elongation involves an indiscriminate copying of the DNA template. Termination may also involve recognition of a specific DNA sequence. The initiating and terminating modes require specificity-determining elements, either in the chromosomal template, the RNA polymerase, or in other components of the transcribing system. Three distinct, but not mutually exclusive, models to explain such specificity have been proposed (Figure 9.1):

1. The structure of the template may restrict transcription by sequestering regions of the DNA from an indiscriminate RNA polymerase. Several mechanisms of restricting template transcription have been considered.

(a) Histones may be general transcription inhibitors[2,3]. Selective placement or selective 'derepression' would then provide an 'open' template for transcription.

(b) Specific negative effectors (repressors) may operate at the gene or operon level[4]. The lac[5] and lambda[6] repressors are paradigms for repressor function.

(c) Specific positive effectors may activate transcription. As a precedent for positive regulation, the arabinose C gene product is known to effect transcription of the arabinose operon[7].

(d) Conformational modulations of the template may affect transcription. Superhelix formation, for example, may result in the formation of transcriptively active single-stranded regions and also perhaps transcriptively inactive regions by physical exclusion[8,9].

2. Polymerase-associated factors may confer initiation and termination specificity for particular genes. This intriguing possibility was suggested by the discovery that a protein (σ factor) associated with *E. coli* polymerase is required for efficient transcription of T4 DNA.[10] A general model based on

Possible mechanisms of control
of transcriptive specificity

Template restriction

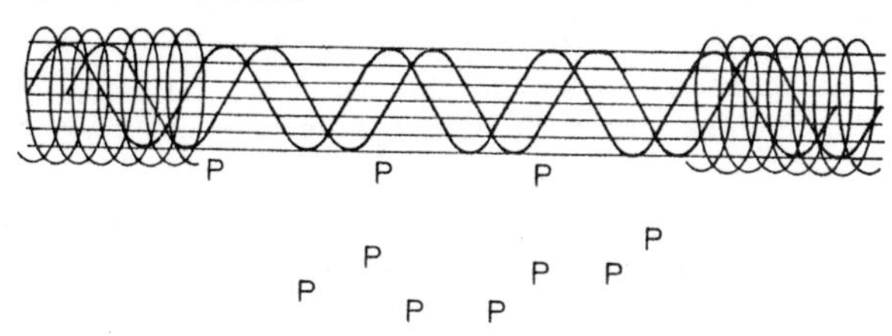

Multiple selector subunits

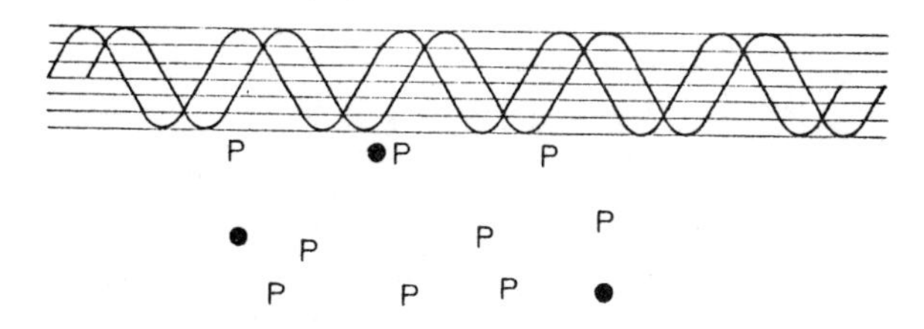

Multiple RNA polymerases

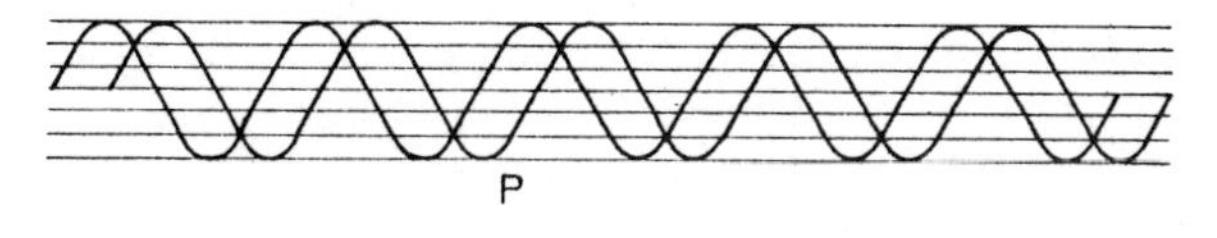

Figure 9.1 Possible mechanisms of control of transcriptive specificity. Requirements for transcriptive specificity

Recognition of specific DNA sequence	: Initiating mode
Indiscriminate reading of DNA	: Elongating mode
Dissociation of template/RNA polymerase complex	: Terminating mode

Hypotheses:

1. Template restricts transcription	= Indiscriminate polymerase (elongating mode only)
2. Multiple selector subunits	= Indiscriminate polymerase (elongating mode only)
3. Multiple RNA polymerases with inherent specificity	= Polymerases have initiating, elongating, terminating modes

this hypothesis suggests that 'σ' factors interacting with the polymerase are the determinants of specificity.

3. Multiple RNA polymerases, each with a different transcriptive function, may play a role in selective gene expression. The extension of this hypothesis to the expression of a large number of individual genes seems unlikely. However, it is an attractive means for the control of sets of genes. The model requires that the enzymes themselves have an initiation mode (a termination mode may or may not be necessary). Multiple RNA polymerases have been detected in the nuclei of a variety of organisms spanning the eucaryotic domain[11]. There is good evidence that some of these enzymes have different functions[12].

In this review we will summarise the current status of knowledge concerning the polymerases and the regulation of their activity. The major question is whether changes in the specificity of the RNA polymerase are causally linked to alterations in the RNA synthetic program during physiological transitions.

9.2 PROKARYOTIC TRANSCRIPTION

There are several transcriptive functions in prokaryotic systems. These include the synthesis of RNAs associated with the translational apparatus (rRNA, 5S and transfer RNA), the synthesis of mRNA for specific proteins, and the synthesis of RNA associated with the initiation of DNA synthesis. Independent regulation of these various transcriptive functions requires a method for selective expression of particular genes. All of these functions are apparently carried out by a single DNA-dependent RNA polymerase. The qualitative and quantitative control of transcription by this enzyme is effected through a combination of positive and negative regulatory mechanisms. We discuss here the structure of the procaryotic RNA polymerase and the mechanism of its interaction with the template. We will also describe how the polymerase structure is modified and activity regulated during particular physiological transitions.

9.2.1 Prokaryotic RNA polymerase

RNA polymerase (holoenzyme) is a complex molecule that can be reversibly dissociated into a 'core' enzyme and an additional subunit, termed σ. The core polymerase is comprised of three polypeptides, a, β, and β' in the combination (a_2, β, β'). The holoenzyme contains these polypeptides plus σ (a_2, β, β', σ). It has an aggregate mass of 500 000 daltons[13]. The role of the different subunits has not yet been completely elucidated, but it is known that β' (165 000 daltons) binds to DNA, while rifampicin, an inhibitor of initiation of transcription, binds to the β subunit (155 000 daltons)[14]. Purified a, β and β' subunits (molar ratio 2:1:1) can be recombined to yield the core enzyme activity[15].

9.2.2 Phage RNA polymerase

After infection of *E. coli* by one of the T bacteriophages, the host transcription machinery is pre-empted for the production of viral RNA. This change in transcriptional specificity is correlated with, and is probably mediated by, the alteration of subunits of the host RNA polymerase (T-even phages)[16] or the appearance of a new viral polymerase (T-odd phages)[17]. Early after infection, transcription in T7-infected *E. coli* cells becomes rifampicin-resistant[18]. This resistance is accounted for by the synthesis of a phage-specific RNA polymerase which carried out transcription of late T7 genes[17]. The T7 polymerase has a single polypeptide chain of molecular weight 110 000 daltons. Its activity is not affected by antisera against *E. coli* RNA polymerase. Thus, the T7 polymerase bears little structural relationship to the host enzyme.

A similar low molecular weight polymerase has been isolated from T3-infected cells[19]. Each phage enzyme exhibits a preference for its homologous template. The existence of phage polymerases of low molecular weight indicates that a complex subunit structure is not required for the transcription of specific genes. This suggests that some of the subunits of the host polymerase have non-catalytic functions, such as the determination of specificity and regulation of activity.

9.2.3 Mechanism of polymerase action

9.2.3.1 For initiation

The holoenzyme is required for the transcription of a specific template. It interacts with a relatively few, selected regions (promoters) on the DNA to form a stable initiation complex, which is not dependent on substrate[20]. In contrast, the core enzyme binds to DNA in a readily reversible manner and transcribes readily from non-specific templates such as calf thymus DNA or denatured DNA. The binding of the core enzyme seems to be an interaction with the deoxyribose phosphate backbone of the DNA, since it is prevented either by polyanions, such as heparin, or by high ionic strength[21].

The specific reaction of the holoenzyme with the promoter appears to be a stepwise process. After binding to the DNA, the enzyme may drift along the template until it reaches the start signal for transcription[22, 23]. The σ subunit may act as an allosteric effector in the process. After initiation, σ is released and can be re-used in a new initiation step[24, 25].

9.2.3.2 Elongation

Chain elongation is carried out more efficiently in the absence of σ because the core enzyme has a higher affinity than the holoenzyme for non-promoter DNA[20]. The rate of elongation may not be constant and may reflect the nucleic acid sequence of secondary structure of the template[26].

9.2.3.3 Termination

Termination of RNA chains seems to occur at specific DNA sequences and requires the presence of the termination factor, ρ[27,28]. At high ionic strength, the polymerase is able to terminate[29,30]. Since the RNA molecules produced in high salt are longer than those made in the presence of ρ, it is assumed that the polymerase stops transcription at different sites in the absence of ρ[27].

9.2.4 Prokaryotic polymerases and regulation of transcription

In many cases the pattern of transcription depends upon the extracellular environment and is readily reversible. Under certain conditions, however, there is a commitment to a series of essentially irreversible events. Such a program, once initiated continues despite environmental alterations. In bacterial systems, there are two well-characterised examples of programmed transcription: bacteriophage development and bacterial sporulation. These processes are similar in some aspects to differentiation. Therefore, studies of control of transcription in these processes may produce models of regulation that are relevant to eukaryotic systems. Studies of transcriptive regulation in phage λ show how the expression of specific genes is regulated by other gene products. Such control leads to a means of temporal regulation of transcription. In both T4 bacteriophage development and *B. subtilis* sporulation significant changes that are coordinated with the developmental program occur in the polymerase molecule and associated proteins. In *B. subtilis* there is evidence suggesting that these changes may be causally related to the onset of sporulation.

9.2.4.1 Phage λ

Upon infection, the linear genome of λ circularises and can either lysogenise (covalently insert into the *E. coli* DNA) or grow lytically[31,32]. During lysogeny, a single gene (cl) is expressed[33], and its product, the λ repressor, inhibits all further transcription of λ DNA by binding to two operator regions, O_L and O_R[34]. When the repressor is inactivated or absent, as occurs during lytic growth, the host RNA polymerase transcribes two cistrons to the left of P_L and right of P_R[35]. The products of these genes (known as N and tof respectively) then mediate the further transcription of the λ DNA[36].

In an *in vitro* system, *E. coli* polymerase holoenzyme, in the presence of ρ factor, produces several transcripts from λ DNA. Among them are two RNAs (12S and 7S)[27], that correspond in size to the message expected from the P_L and P_R cistrons. It appears from genetic studies that the product of the N gene is an 'anti-terminator' protein that permits continued transcription past a ρ-dependent termination sequence[37]. One of the products of the P_R cistron is a repressor that binds to the operator for the cl gene, thus preventing the transcription of the λ repressor gene. Addition of λ repressor before RNA polymerase inhibits the transcription of the 12S and 7S RNA[34]. Binding of

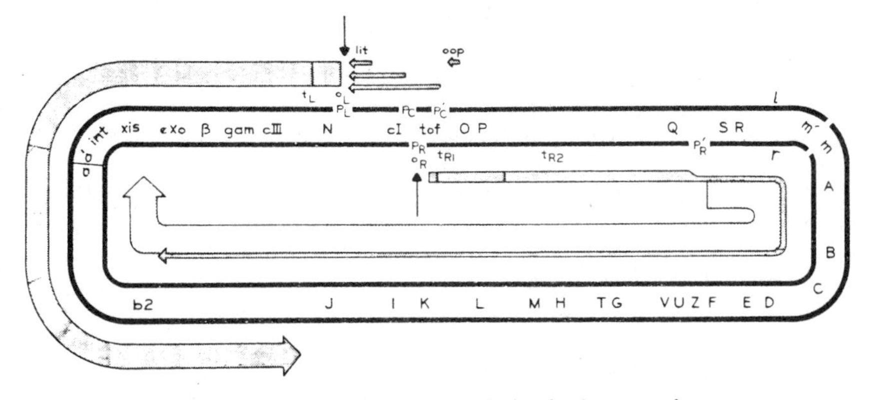

Figure 9.2 Schematic representation of transcription in the phage λ

Transcription from the circularised (vegetative) genome is depicted. The major transcripts occur leftward on the L-strand from the relevant promoter P_L (see arrow) and rightward on the r strand from the promoter P_R (see arrow). The relative magnitude of transcription is indicated by the width of the arrow. Note that the rightward transcription is enhanced with the activation of the promoter P'_R. The genes N, C_1 (leftward), tof (rightward) are involved in the regulation of initiation of transcription (see text). The minor leftward transcripts oop, lit, C are not discussed here (Diagram: courtesy of Dr W. Szybalski, University of Wisconsin.)

repressor inhibits formation of an initiation complex[34], perhaps by blocking binding of the enzyme to the promoter region[38].

9.2.4.2 Bacteriophage T4

The bacteriophage T4 infects *E. coli.* Subsequently host RNA synthesis is inhibited, phage-specific macromolecular synthesis is initiated and the host cells are lysed[39, 40]. A temporal sequence for the synthesis of phage-specific mRNAs can be detected[41]. A first class of mRNA appears immediately after infection (immediate early), does not require protein synthesis, and hybridises exclusively to one strand of T4 DNA. The synthesis of the second class of RNA (delayed early) begins approximately two minutes after infection. Delayed early RNA hybridises to the same DNA strand as immediate early but contains other specific sequences. Phage-directed protein synthesis may be required for delayed early RNA synthesis. The transcription of the late T4 RNA genes begins 12 min after infection. These genes apparently code for viral coat protein. Their transcription requires both protein and DNA synthesis. Late RNA hybridises to the opposite DNA strand.

There is no known T4 RNA polymerase, and the subunits of host core polymerase are conserved during infection[16, 42]. Therefore, if the sequence of transcription is mediated by changes in RNA polymerase, alteration of the host polymerase must be involved. Bautz and Dunn[43] have reported that σ factor disappears soon after infection. The loss of σ factor is presumably responsible for restriction of the transcription of the host genome. Shortly after the loss of host σ factor, the subunits of *E. coli* core polymerase are modified by the covalent attachment of AMP or ADP to the α subunit and by addition of new amino acids to the enzyme protein[16]. Schachner and Zillig[44] have shown that all the subunits of the core enzyme are modified. Stevens[45] has demonstrated that the polymerase also acquires four small T4-specific proteins of 10 000–25 000 daltons, These polypeptides are not synthesised until 5 min after infection, suggesting they are coded for by delayed early genes.

None of these alterations has yet been shown to mediate synthesis of specific classes of T4 RNA. Thus the attractive hypothesis that transcription specificity in these systems is induced by modifications in the polymerase or associated factors, is not yet proven.

9.2.4.3 **B. subtilis** ***sporulation***

The transition from vegetative cell to spore in *B. subtilis* involves the production of stage-specific gene products[46]. Although the complete sequence of events leading to spore formation has not been elucidated, many changes in levels of protein coincident with sporulation have been detected. Among these is the induction of synthesis late in logarithmic growth of an alkaline protease. Studies have shown that this protease is involved in sporulation[47], since temperature-sensitive protease mutants remain in their vegetative form at non-permissive temperatures[48].

Several observations suggest there are specific changes in the structure of *B. subtilis* polymerase during sporulation. The bacteriophage φ-e reproduces well in vegetative, but not sporulating, *B. subtilis*[49]. In *in vitro* experiments vegetative polymerase can effectively synthesise rRNA from *B. subtilis* DNA and transcribe φ-e DNA[50]. The spore polymerase is relatively inactive on both *B. subtilis* DNA and φ-e DNA[51,52,229], The subunit structure of *B. subtilis* vegetative polymerase is similar to that of *E. coli*[53]. The purified sporulating polymerase lacks σ factor and contains a modified β subunit of lower molecular weight (110 000 daltons) than the vegetative enzyme (155 000 daltons)[52,54]. At present it is not clear whether these changes are artifacts caused by degradation during the isolation procedure[55]. In addition, a new polymerase subunit of 60 000–70 000 daltons appears[56,57]. It has been presumed that one or more of these changes is related to the alterations in transcriptive specificity that occur during sporulation.

The results of genetic studies also support a specific role of polymerase in the sporulation process. Hussey *et al.*[52] have isolated mutants which contain an altered polymerase and which fail to sporulate. The isolated mutant enzyme is resistant to rifampicin and is presumed to have an altered β subunit. This enzyme shows no obvious changes in subunit structure and retains the ability to synthesise rRNA. Leighton[58] has described another mutant that contains a rifampicin-resistant polymerase. This enzyme loses its ability to synthesise rRNA, and proceeds through the early stages of sporulation as in the wild type. However, the later stages of sporulation are blocked. These observations suggest that some secondary change in the polymerase is required for the completion of sporulation.

The evidence from both biochemical and genetic studies is consistent with the hypothesis that changes in the subunit structure of RNA polymerase mediate the changes in transcription associated with *B. subtilis* sporulation. Proof of this hypothesis must await the demonstration of sporulation-specific transcription by an altered polymerase.

9.3 EUKARYOTIC TRANSCRIPTIVE SYSTEMS

Sometime after the first reports of RNA polymerase activity in higher organisms[59], Widnell and Tata[60] discovered that the type of RNA synthesised in nuclei was dramatically altered by the ionic conditions. In the presence of magnesium at low ionic strength, the RNA synthesised had a high G + C content reminiscent of ribosomal RNA, while in the presence of manganese at higher ionic strength, the base composition of the RNA resembled that of the DNA. Subsequently others[61,62] showed that in magnesium-low salt, RNA synthesis occurred preferentially in the nucleoli, whereas in manganese-higher salt, the labelled RNA was dispersed throughout the nucleus. These results were consistent with the existence of RNA polymerases having different salt and divalent metal ion requirements, but were also compatible with the selective action of these ions on the structure of chromatin, with a consequent effect on transcription by a single polymerase.

Roeder and Rutter[11] succeeded in solubilising and stabilising the polymerase activity and were able to resolve three RNA polymerase activities by DEAE-Sephadex chromatography. These activities exhibited different catalytic characteristics, including magnesium and manganese requirements and ionic strength optima. The variant properties of these enzymes explained the earlier observations. Chambon *et al.*[63] subsequently confirmed the presence of two polymerases in calf thymus, which they termed A and B. Multiple polymerases have since been found in nuclei of all eukaryotic organisms examined (see CSHSQB vol. 35, 1970 and Ref. 64). Polymerase I and II are invariably present. Polymerase III is less readily detected, although it represents a substantial fraction of the activity in the nuclei from sea urchin and *Xenopus* embryos and yeast[11,65]. In nuclear extracts of rat liver, however, polymerase III activity represents no more than a few per cent of the total activity and is very labile[12].

It is significant that the properties originally described for the rat liver and sea urchin enzymes, with few exceptions, hold true for the activities found in the nuclei of other organisms. This implies that these characteristics are related to a function which is evolutionarily perpetuated and hence is probably biologically meaningful.

Polymerase II is specifically inhibited by α-amanitin, a toxin from the mushroom *Amanita phalloides;* neither polymerase I nor polymerase III nor the procaryotic polymerases are inhibited by this compound[66-68]. Rifampicin, the inhibitor of prokaryotic polymerases, has no effect on eucaryotic enzymes[63] However, a number of rifamycin derivatives inhibit RNA polymerase[69]. This effect is apparently not specific since these derivatives also inhibit DNA polymerases and reverse transcriptase[70].

The nuclear enzymes can be adequately classified according to the general characteristics listed in Table 9.1. An alternative nomenclature has been proposed by Chambon *et al.*[63], based solely on α-amanitin sensitivity. The

Table 9.1 Nuclear RNA polymerases: properties and transcriptive specificity Data was compiled after Roeder and Rutter[11], Blatti *et al.*[12], and Jacob[64]. For further explanation see text

Nuclear RNA polymerases

Properties	I	II	III
	(a) (b)	(a) (b)	
DEAE Elution ~$M(NH_4)_2SO_4$	0.05–0.15	~0.25	~0.35
Mn^{2+}/Mg^{2+} activity ratio	1	5–50	2–3
Ionic strength optimum	0.05	0.10–0.15	0–0.2
Rifampicin inhibition %	0	0	0
α-amanitin inhibition %	0	100	0

Transcriptive specificity:	I	III	II
Nuclear localisation	Nucleolus	(Nucleoplasm)	Nucleoplasm
Transcriptive range	<10%		>90%
Transcriptive products	Ribosomal precursor		HnRNA
	5S RNA		mRNA
	tRNA		

enzymes which are insensitive to α-amanitin are the A type, while the enzymes sensitive to α-amanitin, are B-type polymerases. This limits enzyme classification to two categories and does not take into account fundamental differences in structure between nuclear and organellular enzymes. For this reason we prefer our original terminology. Eventually, it will be more appropriate to develop a classification which associates each type with a particular function (the genes transcibed) rather than with any arbitrary property.

Heterogeneity in polymerase was observed in the earliest studies[11,71]. Further investigation has shown that there are two (a and b) and perhaps more different structural forms of polymerase II[72-76]. Two chromatographically different forms of polymerase I have been demonstrated in the tissues of several higher organisms[77-80] (Goldberg and Perriard, unpublished). It is not known at this time whether these subspecies represent allelic variants with different transcriptional specificities, different regulatory forms of the same enzyme, or degraded or modified forms not necessarily related to any specific function.

9.3.1 Structural homologies between eukaryotic and prokaryotic polymerases

Because of the basic differences in structure of eukaryotic and prokaryotic chromosomes, it is by no means certain that the transcribing enzymes in higher organisms are related either in structure or in the mechanism of regulation to the procaryotic polymerases. Studies on the subunit composition are important, therefore, to both characterise and distinguish the eukaryotic polymerases and establish the extent of structural homology with the prokaryotic enzymes. Polymerase II has been purified to homogeneity from several organisms[74,81-85]. Two forms of polymerase II (IIa and IIb) have been identified in rat liver and calf thymus[72-74, 81]. Each enzyme has four components that are resolved by electrophoresis in sodium dodecyl sulphate. The sum of the molecular weights of the four components approximates the molecular weight of the native molecule. Thus it is assumed that they are subunits present stoichiometrically in the polymerase. Polymerases IIa and IIb are distinguished by a different high molecular weight subunit. The proportions of polymerases IIa and IIb isolated from rat liver appear to be influenced by the presence of a proteolytic inhibitor, suggesting that one may be derived from the other by proteolytic conversion[74]. Polymerase I has also been isolated in essentially pure form and is composed of two high molecular weight and several lower molecular weight subunits[86-88]. The exact number of subunits is still in doubt because most polymerase I preparations have a low specific activity and perhaps are not homogeneous.

It is not possible to rigorously demonstrate that the components detected by electrophoresis in SDS gels are actual subunits of the molecule, since reconstitution experiments similar to those performed by Heil and Zillig[15] on the prokaryotic enzymes have not been carried out. It seems virtually certain that the two high molecular weight components of both polymerases I and II are indeed subunits, since they are invariably present and occur in stoichiometric amounts. The eukaryotic nuclear polymerases have a subunit structure resembling that of the prokaryotic enzyme (Table 9.2). This structural similarity suggests a functional homology of the eukaryotic and prokaryotic polymerases and implies that the mechanism of transcription is fundamentally similar. This analogy may be extended to include the possibility that transcription from 'true' initiation sites by eukaryotic polymerases is also dependent upon the action of σ-like factors.

Table 9.2 Apparent structural homologies between prokaryotic and eukaryotic RNA polymerases Molecular weight of subunits of *E. coli* RNA polymerase[13], rat liver nuclear polymerase II[74] and nucleolar polymerase I[88] were estimated by SDS-acrylamide gel electrophoresis[230]. Molecular weights of the native enzyme molecules were estimated from their sedimentation coefficients in glycerol density gradients[12,13]

Subunit	E. coli *core*	II		I *Nucleolar*
		(A)	(B)	
β'	$(155\,000)_1$	$(190\,000)_1$		$(200\,000)_1$
β''			$(170\,000)_1$	
β	$(145\,000)_1$	$(150\,000)_1$	$(150\,000)_1$	$(125\,000)_1$
a	$(40\,000)_2$	$(35\,000)_1$	$(35\,000)_1$	(60 000)
a'		$(25\,000)_1$	$(25\,000)_1$	(44 000)
		(16 000)	(16 000)	~(20 000)
Molecular weight	380 000	400 000	400 000	350–400 000
Structure	$a_2\beta\beta'$	$aa'\beta\beta'$	$aa'\beta\beta''$	

9.3.2 Polymerase I and ribosomal RNA transcription

A number of observations now support the contention that a major function of polymerase I is the transcription of rRNA:

1. Under conditions where the activity of polymerase I is favoured (Mg-low ionic strength plus α-amanitin to block polymerase II activity), the RNA produced by isolated nuclei resembles rRNA. Its base composition is similar to rRNA precursor, and unlabelled rRNA competes with the hybridisation of this RNA with nuclear DNA[12]. In such nuclei, transcription of the ribosomal cistrons has been shown to occur only from the H strand of the DNA, as in the intact cell, and to be insensitive to α-amanitin[89].

2. Under the same conditions, the synthesis of RNA is localised in the nucleolus[61, 62], known to be the site of rRNA synthesis[90].

3. Polymerase I is found largely in nucleoli and can be isolated in high yield from this organelle[88, 91]. The fact that I is still present in the cells of a *Xenopus* mutant which has no nucleolus does not mitigate against a nucleolar localisation of this enzyme in normal cells[65]. This result does indicate, however, that the synthesis of I is not coupled to the synthesis of rRNA or to the presence of rRNA cistrons.

Although these results strongly imply that a major function of polymerase I is the synthesis of rRNA, they do not rule out other transcriptive roles for this enzyme. Neither do they eliminate the possibility that another α-amanitin-resistant polymerase (polymerase III) may also transcribe rDNA.

A study of the mechanism of action of polymerase I would be greatly aided by a specific inhibitor of this enzyme. Cycloheximide has been reported to be a specific inhibitor *in vitro*[92, 93] but, as other workers have failed to confirm this finding[94], it cannot be a generally useful inhibitor. Recently, Shields and Tata[95] have shown that polymerase I is more heat-labile than polymerase II, and perhaps polymerase III. The differences in sensitivity might help to discriminate among the various enzymes.

There has been no demonstration of the specific synthesis of rRNA by polymerase I *in vitro*. Roeder *et al.*[65] have found that both polymerases I and II from *Xenopus*, as well as *E. coli* polymerase, transcribe rDNA symmetrically. In contrast, Hecht and Birnstiel[96] have demonstrated that *Micrococcus luteus* polymerase is capable of synthesising rRNA from purified *Xenopus* ribosomal genes. The synthesis of this specific product seems to be a function of the size of the rDNA. The studies of Roeder *et al.*[65] were carried out on rDNA smaller than the ribosomal gene. We have found that polymerase I from rat liver can transcribe nucleolar DNA (enriched in ribosomal genes) four times more efficiently than polymerase II[88]. These results are compatible with a site-specific reaction of polymerase I with the ribosomal genes. Tocchini-Valentini and Crippa[97] have also reported preliminary studies suggesting that the nucleolar polymerase can bind ribosomal DNA more effectively than total DNA. In accord with these observations is the finding that polymerase I and Polymerase II apparently bind to and transcribe from different sites on rat liver and calf thymus DNA[76, 98].

9.3.3 Polymerase II and the transcription of heterogeneous and messenger RNAs

A number of observations support the contention that most heterogeneous RNA and mRNA is synthesised by polymerase II.

1. In the presence of Mn-high ionic strength (0.4 M $AmSO_4$), the RNA

produced in isolated nuclei is found broadly dispersed throughout the nucleoplasm[61] and has a G + C content similar to that of total nuclear DNA. The addition of α-amanitin inhibits the appearance of this RNA[99].

2. RNA synthesised in the absence of α-amanitin hybridises to nuclear DNA and is competitive with unlabelled nuclear RNA, but not with rRNA[12].

3. Analysis of the RNA synthesised *in vitro* by nuclei isolated from adenovirus-infected cells indicates that virus-specific mRNA is also α-amanitin-sensitive, and is most effectively transcribed under the same conditions responsible for host cell, heterogeneous RNA synthesis[100, 101].

4. Polymerase II can be isolated from a nucleoplasmic fraction and is inhibited by low concentrations of α-amanitin[66, 91].

There is as yet no evidence implicating polymerase II in the transcription of specific genes. However, polymerase I and II do transcribe synthetic and natural templates at different rates and the RNAs produced have different base ratios[12, 63, 102]. Only polymerase II transcribes mammalian virus SV40 DNA to produce RNA corresponding in size to the virus-specific RNA[103]. These results imply that each enzyme has some capacity to recognise different DNA sequences. On the other hand, they do not indicate which aspects of specificity are determined by the polymerase itself, and which are prescribed by the template or other regulatory factors.

9.3.4 Polymerase III and other transcriptive activities

The transcriptive function of polymerase III has yet to be elucidated.

1. Polymerase III might be a precursor of polymerase I and/or polymerase II. High levels of polymerase III are found in the mature eggs of sea urchin and *Xenopus* oocytes. In the sea urchin, total polymerase activity remains constant during early embryological development. During this period polymerase III activity declines and polymerase II and I activities increase[98, 104, 105] (Morris and Rutter, unpublished).

2. Polymerase III may synthesise both the 5S RNA found in ribosomes and transfer RNA. Price and Penman[106] have shown that in isolated nuclei the synthesis of these RNA classes requires Mg^{2+}, is relatively independent of ionic strength, and is not inhibited by α-amanitin. These conditions are similar to the requirements of polymerase III[11, 65]. An α-amanitin insensitive polymerase activity (possibly polymerase III) has been found in the nucleoplasm where the synthesis of 5S and tRNA is known to occur[107].

3. A role of polymerase III in the transcription of rRNA has not been ruled out.

9.3.5 Another transcription function: The initiation of DNA synthesis

It has recently been demonstrated in bacteriophages[108], bacteria[109] and in higher organisms[110] that DNA synthesis requires an RNA primer that is presumably synthesised by an RNA polymerase at a specific initiation site. The polymerase responsible for the initiation of DNA synthesis in eukaryotes has not been defined. It is possible that the same polymerases transcribing the genes may be associated with their replication. However, the initiation of DNA synthesis has substantially different characteristics than transcription (see Ref. 111). It is symmetric rather than asymmetric and is bidirectional rather than unidirectional. Furthermore, DNA and RNA synthesis occur at different times in the cell cycle. The existence of a separately regulated enzyme with this function appears likely to us.

9.3.6 Organellular transcription systems

It was demonstrated in the 1960s that several organelles contained DNA and carried out independent RNA synthesis[112, 113]. It has since been shown that the translational system in mitochondria (and probably that in chloroplasts) resembles that found in prokaryotic organisms and is structurally different from that found in the cytoplasm[114]. This finding has provided support for the hypothesis that such organelles arose from a prokaryote existing in a symbiotic or saprophytic relationship with a nucleated cell[114]. Whether the RNA polymerase found in these organelles is related to the prokaryotic enzymes is relevant to this hypothesis. Preliminary tests of the sensitivity of mitochondrial transcription to rifampicin have been inconclusive[115-119].

Isolation and determination of the mitochondrial polymerase subunit structure have been carried out from a number of organisms. Küntzel and Schäfer[116] first reported that the *Neurospora* mitochondrial polymerase is composed of a single low molecular weight subunit (64 000 daltons). A similar finding has been reported for the *Xenopus* mitochondrial enzyme[119]. Should these observations be confirmed, then the characteristics of the transcription system in mitochondria will be substantially different from both those in the nucleus and prokaryotes. In fact, these enzymes resemble the bacteriophage T7 andT3 polymerases, rather than the multi-subunit bacterial and nuclear polymerases[17, 19]. The remarkable difference between the nuclear and mitochondrial enzymes strongly supports the contention of independent genetic systems. However, the interesting report that the mitochondrial polymerase is apparently coded for by nuclear genes and its mRNA is translated on cytoplasmic ribosomes[120] emphasises the integration of the nuclear and mitochondrial systems within the cell.

The characterisation of the mitochondrial and chloroplast polymerases is difficult because of possible nuclear contamination. We have attempted a different approach to the study of transcription in organelles. A transcription complex consisting of organellular DNA, an endogenous RNA polymerase and other DNA-binding proteins has been isolated from chloroplasts of *Euglena gracilis*[121]. RNA synthesis by this complex is indifferent to added exogenous DNA (including chloroplast DNA). The RNA product is complementary to chloroplast DNA, but it is not a symmetric copy of it. These observations open up the way to a more comprehensive analysis of chloroplast transcription.

9.3.7 Cytoplasmic RNA polymerases and putative cytoplasmic transcription systems

It has recently been suggested that RNA polymerase is present in the soluble fraction of the cytoplasm[122]. The existence of such a cytoplasmic polymerase would be significant because of the proposals by Bell[123] of the involvement of cytoplasmic informational DNA in the synthesis of cell-specific proteins. Bell and Brown[122] claim to have detected two cytoplasmic RNA polymerases. The possibility that these enzyme activities may be contributed by nuclear or even mitochondrial enzymes has not been eliminated. Seifart *et al.*[82]. have partially purified a polymerase activity from the postmitochondrial supernatant of rat liver. It appears to chromatograph differently from both nuclear polymerases I and II. Its co-factor optima are similar to those of nuclear polymerase II, but it has a reduced sensitivity to α-amanitin and cannot be stimulated by a protein factor specific for polymerase II. Whether this polymerase represents a separate enzyme with a discrete function, is a derivative, or a complex of polymerase II with other components, is uncertain.

9.4 REGULATION OF TRANSCRIPTION

The regulation of transcription may be affected by control of any of the components involved in the RNA synthesis: the template, the polymerase, associated factors and other environmental conditions.

9.4.1 Template

The major problem in studying eukaryotic transcription is associated both with the genetic complexity and the physical and chemical integrity of the template. Faithful transcription requires a high molecular weight, intact DNA[96]. DNA prepared by standard procedures contains both single and double-stranded breaks and single-stranded regions[124,125]. Such regions represent artificial initiation sites[126] for the *E. coli* polymerase and probably function for the eukaryotic enzymes in the same way. Higher order structure, trace quantities of proteins binding to DNA, and ions affecting the DNA structure all may exert an additional influence on the regions transcribed and the rate of transcription[127].

Attempts to circumvent these problems through the use of 'purified' genes or simplified gene systems can provide only partial answers to the fundamental questions. Neither DNA highly enriched in certain genes[128,129], or complementary DNA prepared from mRNA with reverse transcriptase[130], is an optimal template for studies of specificity. These DNA preparations contain possible artificial initiation sites, additional deoxynucleotide sequences, and, even more seriously, lack the putative promoter and operator regions which must interact specifically with the polymerase.

Although studies of the transcription of small animal viruses, such as SV40 or adenovirus, can be intrinsically informative, they may provide little information about the presumably complex regulation of transcription in nuclei. However, these relatively simple viral DNAs contain 'true' initiation sites and will be very useful in establishing the basic transcriptive characteristics of the eukaryotic enzymes. Comparative studies of the viral and phage systems will elucidate the relationship between prokaryotic and eukaryotic polymerases. *E. coli* holoenzyme, for example, can transcribe SV40 DNA asymmetrically[131]. In contrast, eukaryotic polymerases, as isolated, are not competent to transcribe T4 DNA[12, 63, 83]. Addition of a protein factor stimulates transcription on phage λ and adenovirus DNA[84, 98]; this or a similar factor may exert a σ-like action on T4 transcription as well.

A related approach involves the study of transcription of the organellular genomes (mitochondria, chloroplasts) which have many of the advantages of the viral systems and interact with the nuclear genome[120, 132]. The availability of a large number of mutants should facilitate analysis of the functional components of these transcriptive systems.

The situation becomes even more complex when the transcription of chromatin is considered. A major problem in such studies concerns the possible mobility of chromosomal proteins during isolation or manipulation of the chromatin in various media. Exchange of chromosomal proteins under various conditions has been demonstrated[133, 134]. The addition of histones to chromatin usually restricts both the transcriptive activity and capacity in a non-specific fashion; however some non-histone proteins appear to reverse this non-specificity[135, 136]. Because of the diversity of methods used in the isolation of chromatin, it is unlikely that preparations from the same source have identical properties[136]. Improved methods minimising chromosomal protein exchange may circumvent many of these difficulties[137, 138].

The endogenous polymerase activity in chromatin preparations has different properties from that of the solubilised exogenous enzymes transcribing DNA. Typically the addition of salt to chromatin preparations produces a biphasic increase in the rate of transcription[139-141]. These observations are difficult to interpret, since they may be related to changes in the chromatin structure, the removal of chromatin proteins, or an effect on the polymerases themselves. Studies involving the addition of exogenous RNA polymerases are also subject to considerable uncertainty[142, 143]. In most chromatin preparations, endogenous polymerases participate predominantly in chain elongation, and there is relatively little chain initiation[139, 140, 144]. Initiation by exogenous enzymes may occur at artificial rather than 'true' initiation sites. Furthermore, the enzyme preparations frequently employed in these experiments may contain nuclease activities and hence the template may be modified significantly. Finally, added exogenous enzymes may influence the activity of endogenous polymerase.

Despite these technical problems, the aim of elucidation of the specific aspects of transcriptive control may be realised through studies in these complex systems. The specificity of transcription detectable in isolated nuclei is at least partially evident in chromatin[135, 145-147]. It might be noted in this regard that only polymerase II is apparently able to transcribe chromatin[139]. Continued development of techniques for isolating or synthesising specific regions of the genome will dramatically extend the feasibility of reconstruction of the chromosome from its components. Of crucial importance to the analysis of these systems is the development of procedures for studying the transcription of specific genes. In particular, the method for isolation of pure mRNA and synthesis of complementary DNA provides a direct means for assaying specific transcripts[130].

Initial studies on specific synthesis of globin mRNA from duck reticulocyte chromatin using *E. coli* polymerase as a transcribing agent have been reported by Axel *et al.*[148]. These results have been confirmed and extended by Paul *et al.* (personal communication). A significant level of RNA complementary to globin cDNA was found in the transcripts of reticulocyte chromatin, while none was detected in the transcripts of liver or brain chromatin or in that from purified reticulocyte DNA. Furthermore, reconstitution of chromatin producing globin mRNA could be achieved provided that reticulocyte non-histone proteins are employed. The simplistic interpretation from these experiments is that specificity is determined by the template and not by the polymerase. However, it would be hasty to conclude that there are not elements of specificity associated with the polymerases themselves. Homologous proteins from higher organisms and bacteria frequently show general similarities in their mechanism of action, although they also exhibit significant differences in specificity. Thus, it is not particularly surprising that a prokaryotic polymerase can transcribe mRNA sequences from a eukaryotic genome, when the inherent structural similarities of the enzymes and the similar basic structure of DNA in both types of organisms are taken into account. These factors may also explain why the efficiency of transcription of the globin gene by the *E. coli* polymerase appears low.

9.4.2 Protein factors which affect transcription

Several protein factors isolated from bacteria have been shown to influence transcription, including those like σ that bind to the polymerase[10] and others, such as the cAMP binding-protein required for expression of the lac genes, that bind to the template[149].

The search in eukaryotic organisms for similar factors has just begun. There are several experimental difficulties to be dealt with, including both the problem of developing assay procedures for specific transcription and the possibility of non-specific effects. DNAses, for example, may create new initiation sites[124, 126, 150]. RNAses may inhibit the reaction by destroying the RNA product[151]. Other proteins may also influence transcription non-specifically by protecting the enzyme or by interacting with the DNA. Bovine serum albumin at relatively high concentrations stimulates eukaryotic polymerases significantly[94, 98].

Studies of the effect of prokaryotic transcription factors on eukaryotic systems have not yielded conclusive information. σ factor does not stimulate transcription by eukaryotic enzymes on T4 DNA or on a number of other DNA templates[12, 63, 152]. However, high concentrations of σ inhibit transcription, suggesting the factor can interact with the enzyme. These results do not rule out the possibility that σ may influence the formation of initiation complexes at true initiation sites on eukaryotic templates, since in most of the DNA preparations used transcription probably occurs from artificial initiation sites. For this reason the preliminary report that σ stimulates rRNA transcription when injected into amphibian oocytes[97] deserves further study. The termination factor, ρ, also has had no measurable effect on the activity of polymerases I and II when assayed on different templates[63, 152].

A number of factors which affect transcription have been detected in eukaryotic systems. Stein and Hausen[151] and subsequently Seifart[153] isolated a protein fraction from whole cell homogenates that stimulated only polymerase II activity on native DNA. At low ionic strength, the degree of stimulation of polymerase II by these and similar factors may be as high as 20-fold[154]. Some of these factors have been shown to act at the level of initiation[84, 155], while others affect the chain elongation[156]. The proteins in these fractions may either interact directly with the template[157] or may bind weakly to the polymerase[84, 151, 158, 159]. Partial or complete loss of such proteins during purification may provide one explanation for the low recoveries usually reported with these enzymes.

Factors selectively stimulating polymerase I [71, 88, 160] or stimulating both polymerases I and II have also been reported[155]. We have recently shown that a protein fraction resolved from highly purified polymerase I is required for activity on native calf thymus DNA, but not for activity on denatured DNA or polydeoxycytidylic acid. Recombination of the fraction with the modified polymerase results in recovery of the original activity on native DNA[88]. The action of this factor superficially resembles that of the σ factor. Preliminary studies suggest that similar proteins may be required by polymerase II for the reaction with double-stranded DNA (Rutter and Weinberg, unpublished). Other putative factors include hormone binding proteins[161, 162], cAMP binding proteins[163] and the acidic proteins isolated from non-histone chromatin proteins[164, 165]. Whether any of these molecules control the expression of a specific transcript can only be discerned after their isolation has been achieved.

The search for specific factors should continue and perhaps be intensified now that methods exist for the assay of specific gene transcripts[130, 148].

9.4.3 Enzyme levels

The quantity and quality of RNA synthesis can, in principle, depend either on the concentration of polymerase or on the level of activity of the enzyme. In growing *E. coli* cells, there is a pool of excess RNA polymerase[166] and the polymerase subunits seem to be synthesised at the same rate as other cellular proteins[167]. The regulation of growth in this system, therefore, does not depend upon the concentration of RNA polymerase itself, but rather upon modulation of its activity. This is particularly evident in regulation of rRNA synthesis in *E. coli*. Although the rRNA genes account for only about 0.4% of the genome[168], rRNA accounts for 40% of the total RNA synthesised in growing cells and is therefore preferentially synthesised[169]. The rRNA genes are transcribed at a rate 100-fold higher than the remainder of the genome. It has been suggested that transcription of the ribosomal genes by the polymerase is controlled by the protein co-factors and a small molecular weight molecule (ppGpp), whose concentration depends on the physiological state of the cells[170, 171]. A similar mechanism may regulate rRNA synthesis in HeLa cells[172].

On the other hand, inhibition of host-cell transcription after bacteriophage infection presumably occurs by destruction or alteration of the host polymerase[16, 17]. Similar effects on host RNA synthesis have been found in viral infections in mammalian cells[173, 174]. This may be analogous to some instances of terminal differentiation where the loss of nuclear function is involved[175].

The level of polymerase activity varies considerably. Polymerase II is typically high in rapidly growing cells (32 000 molecules per cell) and lower in non-growing cells (8000 molecules per cell) in *Dictyostelium discoideum*[85]. The concentrations detected may vary from 800 molecules per cell in some thymus preparations[74] to about 50 000 molecules per cell in rapidly growing tumour cells in tissue culture[176]. There is also considerable variation depending upon the nutritional or hormonal status of the tissues in question. For example, the yield of polymerase II in calf thymus has varied over a 100-fold range[74, 76]. This difference may be due to a drastic lowering of the polymerase activity during regression of calf thymus.

Efforts to determine the level of polymerase I in eucaryotic tissues have been less successful. Polymerase I is apparently less stable than polymerase II and turns over more rapidly. Yu and Feigelson[177] have estimated the half-life of polymerase I (or that of a factor essential for its activity) to be approximately 1.5 h. These experiments also emphasise the sensitivity of the rRNA system to fluctuations in the metabolic state.

9.4.4 Physiological transitions

In every differentiative or reversible metabolic transition thus far studied, significant changes in the RNA polymerase activities have been detected, although in no instance is it yet known whether these changes are causally related to the onset of such transitions. A dramatic change in polymerase level accompanies the initiation of embryonic development. In reversible transitions initiated by hormones and other humoral factors, nutrients, mitogenic drugs, etc., modest changes in polymerase I and sometimes in polymerase II activity have been demonstrated.

9.4.4.1 Oocyte maturation, fertilisation and embryogenesis

During maturation, amphibian and echinodermal oocytes accumulate reserves of macromolecules and nutrients. Early embryological development in these organisms is then independent of external nutrients. The amplification of the ribosomal genes and the accumulation of rRNA is one aspect of this process[128, 178]. The level of RNA polymerase in mature oocytes is sufficient to provide polymerase during the early stages of embryogenesis (up to several thousand cells)[105]. Thus the egg has the characteristics of a 'preloaded' system.

The types of polymerase present in mature oocytes vary in different organisms. In amphibian oocytes, there are high levels of both polymerase I and III. In addition, modified polymerase II is present. In *Xenopus* larvae, polymerase III disappears and only RNA polymerase I and IIb are found. The level of total RNA polymerase activity per cell decreases dramatically from the unfertilised egg to the swimming tadpole stage, when it is similar to that found in tissue culture cells or adult liver in the same organism[105]. This finding is in agreement with the decrease of RNA synthesis on a per cell basis during early development[179] and suggests that the concentration of the enzymes is not limiting for transcription.

Sea urchin oocytes contain polymerase III almost exclusively (greater than 90%); the remaining activity is polymerase II. Polymerase I is not detectable. After fertilisation, there is a decrease in level of polymerase III coincident with an increase in the levels of polymerase II and I. The total polymerase level remains constant until about gastrulation[98, 104].

The high level of polymerase III in oocytes could be responsible for the synthesis of rRNA. However, after fertilisation, no significant amounts of rRNA[180, 181] are made although polymerase I and III are present. In later stages of embryogenesis (gastrula), polymerase I is relatively enriched[71] concomitant with the appearance of nucleoli in the cells[182]. The activation of polymerase I or its sequestration in the nucleolar structure might be necessary for the initiation of rRNA synthesis.

9.4.4.2 *Other developmental systems*

Activation of spores from the fungus, *Rhizopus stolonifer*, results in a pattern of changes in polymerase activity similar to that found after fertilisation and during embryogenesis of sea urchin and *Xenopus*[183]. *Rhizopus* spores contain only polymerases I and III. Three hours after the induction of germination, polymerase II activity is detectable. Polymerase I may be modified during this transition since the enzyme exhibits different chromatographic behaviour and co-factor optima. The appearance of polymerase II is almost coincident with the maximum in RNA synthesis 4 h after the onset of germination[183].

Little is known of the changes in polymerase levels during other specific differentiative transitions. A general decline in the level of polymerase activity has been noted during the transition from an active to an inactive state. The maturation of avian red blood cells is accompanied by inactivation of nuclear RNA synthesis. There is a temporal sequence involving cessation of rRNA synthesis before the decrease in heterogeneous nuclear RNA synthesis[175]. In accord with these observations, the blood of anaemic chickens enriched in immature blood cells contains higher levels of polymerase II than blood from untreated animals[184]. Longuere and Rutter (unpublished) have shown that polymerase I and II can be detected only in the early stages of red blood cell differentiation. Subsequently polymerase I and then polymerase II are lost from the cells.

In the slime mould, *Dictyostelium discoïdeum*, the differentiation of the vegetative amoeba to fruiting body is accompanied by a reduction in both polymerase I and polymerase II activity[85]. At this time, new RNA species are produced as revealed by hybridisation to unique sequences of DNA[185]. Pong and Loomis[85] have been unable to find changes in the structure of polymerase II during this transition.

During encystment of the protozoan, *Acanthamoeba castellani* (induced by removal of the nutrients), it has been reported that polymerase I and II activities drop to very low levels[186]. This process is superficially similar to sporulation in *B. subtilis* and an analysis of possible changes in polymerase structure and specificity should be performed.

9.4.4.3 Steroid hormones

Steroid hormones are thought to penetrate the cell and bind with a specific cytoplasmic receptor, which then migrates to the nucleus[187-189]. The transformed complex of steroid and receptor in turn interacts with the chromosome to induce the response. In all tissues examined, steroid hormone induction is accompanied by changes in polymerase activity[12, 190-192]. These changes may represent a secondary temporal response since they are observed subsequent to a number of initial events elicited by the hormones. This is particularly evident in the effects of oestrogen on the rat uterus and the chick oviduct. Within 5 min after injection of oestrogen, the hormone is taken up by the cells and apparently bound to the cytoplasmic receptor[188]. Concurrently, there is an early effect of steroid hormones on cell permeability and substrate pool sizes which appears to anticipate and probably facilitate the subsequent increases in biosynthetic activity[193-195].

The rate of RNA synthesis in the oviduct begins to increase slightly after 1 h and is obvious between 2 and 4 h[196-198]. The major product appears to be 45S ribosomal precursor RNA[195]. Ovalbumin mRNA can be detected within 3 h of oestrogen treatment[199]. These synthetic events appear to occur subsequent to the production of an oestrogen-induced protein which is thought to appear within 30 min[200, 201]. The synthesis of the mRNA for this protein may occur within 15 min after hormone administration and cannot be suppressed by addition of protein inhibitors concomitantly with the hormone[202]. Thus the synthesis of this induced protein may be the primary biosynthetic response to oestrogen.

Measurements of polymerase activity show that polymerase I increases within 2 h of oestrogen treatment[12, 140, 201, 203]. Subsequently the activities of both polymerases I and II are enhanced. Increased accumulation of RNA is evident 6 h after hormone administration, and precedes increased protein synthesis (12 h) and DNA synthesis (18–24 h)[204]. These events are part of the secondary biosynthetic response in which growth and replication occur. Whether the steroid-cytoplasmic receptor complex is directly involved in these later events is not known. The results of several experiments suggest that the increased levels of polymerase activity[144, 205] found at short times after hormone administration may be due to a modulation in the activity of the enzymes, while at later times the further increase may reflect an increment in the number of polymerase molecules[12, 140, 203]. It has been proposed that the cytosol steroid-binding proteins stimulate polymerase activity in the early responses[188, 189, 206].

9.4.4.4 *Polypeptide hormones*

Most, if not all, protein hormones are thought to interact with specific receptors on the cell surface[207]. These in turn influence the synthesis of a second messenger (cAMP or cGMP) which induces the physiological response. The general response to polypeptide hormones is increased RNA and protein synthesis within a few hours after administration[208-211]. As in the case with steroid hormones, the earliest observable effect is the stimulation of 45S rRNA precursor synthesis[212, 213]. This increase may be correlated with an increase in polymerase I activity. Presumably these changes are cAMP-mediated. Varrone *et al.*[163] have reported the isolation of a cytoplasmic protein, which, when preincubated with cAMP and DNA, stimulated polymerase activity 200%. Thus it seems possible that some of the putative polymerase effects may be caused by other protein factors.

9.4.4.5 *Growth and regeneration*

Contact-inhibited tissue culture cells synthesise less rRNA than growing cells Growth is induced in confluent fibroblast monolayers by addition of fresh medium[215]. A similar effect is observed when Ehrlich ascites cells are transferred to a medium enriched in amino acids[216]. Polymerase activity increases in two phases, suggesting that the enzyme first may be activated, and subsequently the level of the polymerase may increase.

Growth is also stimulated in regenerating tissues. After partial hepatectomy, polymerase I activity increases several-fold while the increase in polymerase II activity is smaller and occurs later[12, 203, 217]. These results are in good agreement with the finding that 45S rRNA precursor synthesis is enhanced in regenerating liver tissue[218]. A minimal deviation rat hepatoma cell line contains almost nine times as much extractable polymerase I as rat liver tissue[219].

These findings support the contention that rapidly growing tissues contain a high content of polymerase I. It is not known whether these higher levels are a cause or an effect of growth.

9.4.4.6 *Lymphocyte transformation by PHA*

Lymphocytes stimulated with the mitogenic drug PHA produce an increase of polymerase I within 15 min, while polymerase II increases only after a lag period of 1 h[220]. These changes correlate with an increase in ribosomal and total RNA synthesis[221, 222]. It has been suggested that the early response involves 'chromatin' activation[220] while the later effects (24 h) may represent an increase in polymerase concentration[223].

9.5 MECHANISMS OF TRANSCRIPTIVE SPECIFICITY

In prokaryotes, three transcriptive functions [(1) the synthesis of the RNAs associated with the translational system, rRNA, tRNA and 5S RNA, (2) synthesis of mRNAs, and (3) the synthesis of initiator RNA involved in DNA replication] are apparently carried out by the same enzyme, while in eukaryotes there is a division of labour among multiple RNA polymerases. It seems likely that at least one polymerase is associated with each of the major functions.

It is unlikely there will be changes in the characteristics of the enzymes associated with functions 1 and 3 during developmental transitions, since these processes are common to all cells in the organism. The significant question then is whether such transitions involving the expression of new sets of genes are accompanied by a novel polymerase II. We believe this possibility is not great. Many mRNAs are common to all cell types. During a physiological transition some new mRNAs are found but there is no evidence for the production of an entirely new set of mRNAs[224]. Perhaps the most attractive possibilities for regulation by polymerase II modification are in terminal differentiation and in programmed cell death, where synthesis of RNA is essentially eliminated[1].

For the resolution of this problem it is crucial to define whether transcriptive specificity is identical in polymerase II isolated from different tissues. Although no structural differences are obvious in the enzymes isolated from thymus and liver[74, 76], discriminating tests especially with respect to specificity have not been made. The degree of multiplicity of polymerase II in a particular tissue could correspond to the number of genes being transcribed, but the number of forms found thus far have not been associated with distinct functions[76].

It seems more likely that the transcription of specific genes is associated with the production of specific molecules regulating the transcribing system. We believe such specificity factors will not interact directly with polymerase II. This conclusion derives from a consideration of the kinetics of formation of the putative ternary complex, consisting of a factor, RNA polymerase and the DNA template. Two basic bimolecular pathways can be envisaged (Figure 9.3). The two pathways are of course thermodynamically equivalent, but the second pathway does have certain kinetic disadvantages. The transcriptive range of a particular polymerase molecule is constrained by the formation of a factor-polymerase complex. Therefore the kinetic problem of polymerases reaching their specific transcription sites becomes complex, since both bimolecular parts of the reaction are repeated with each round of transcription (this assumes that the factor is only required for specific

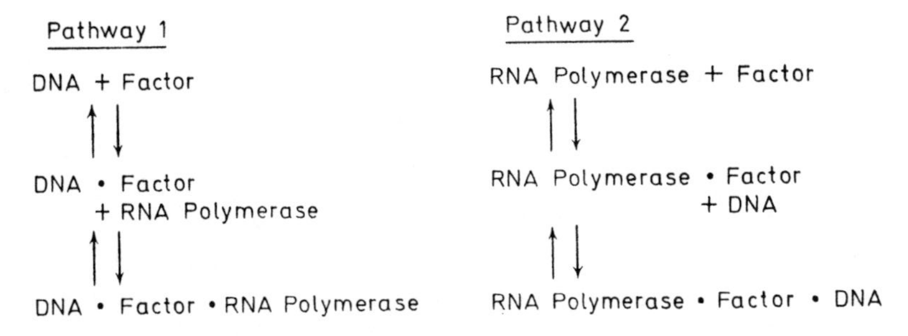

Figure 9.3 Possible reaction sequences for factors with DNA or RNA polymerases in the initiation of specific transcription

initiation and is released as soon as chain elongation begins). In contrast, in pathway 1, interaction of the specificity factor with the DNA can form a stable complex that is not affected by the transcription processes. Only the second bimolecular reaction is required with each transcriptive cycle. We believe therefore that the main elements of transcriptive determination are likely to reside with the template rather than with the polymerase itself.

Specific translation factors are possible candidates for this group of regulatory proteins. In prokaryotes, the protein synthesis elongation factors (Tu and Ts) are thought to be involved in the regulation of rRNA synthesis[171]. There is increasing evidence for the existence of specific translation factors in eukaryotes that might act by binding at a defined sequence of the mRNA[225,226]. Such factors might recognise the complementary DNA and free the correct strand of the DNA for transcription by polymerase.

Soluble factors binding to the enzymes have the disadvantage of requiring a high enzyme concentration in order to saturate specific sites on the template. The RNA polymerase of the transcriptive systems carrying out only the synthesis of a small number of RNAs (like transcription systems I and III) could be directed to its gene sets by such factors. One method for increasing the effective concentration of the polymerase might be through the physical or chemical sequestration of transcriptive systems, as in the case of localisation of ribosomal genes and polymerase I in the nucleolus. Sequestration might have two consequences: it could restrict general diffusion of polymerase and other nucleolar proteins and it could also form a matrix for the optimal interaction of polymerase I and the ribosomal genes. Disturbance of nucleolar morphology leads to the increased appearance of polymerase I in the nucleoplasm and decreased rRNA synthesis[227,228]. These findings suggest that the nucleolar organisation is required for the assembly of ribosomes and for the efficient production of ribosomal and related RNAs.

The precedent, however, is for template-related regulation. A limited degree of specificity seems to be exhibited through the multiplicity of polymerases operating in different transcriptive systems. If these conclusions are true, then what is the future of studies involving RNA polymerase? It is largely in the reconstruction of transcriptional specificity where the homologous enzyme will be most useful in interacting with the specific template and where the nature of controls is most likely to be discerned.

References

1. Rutter, W. J., Pictet, R. L. and Morris, P. W. (1973). Toward molecular mechanism of developmental processes. *Ann. Rev. Biochem.*, **42,** 601
2. Huang, R. C. C. and Bonner, J. (1962). Histone, a supressor of chromosomal RNA synthesis *Proc. Nat. Acad. Sci.*, **48,** 1216
3. Allfrey, V. G., Littau, V. C. and Mirsky, A. E. (1963). On the role of histones in regulating ribonucleic acid synthesis in the cell nucleus. *Proc. Nat. Acad. Sci.*, **49,** 414
4. Jacob, F. and Monod, J. (1961). Genetic regulatory mechanism in the synthesis of proteins. *J. Mol. Biol.*, **3,** 318
5. Gilbert, W. and Muller-Hill, B. (1966). Isolation of the lac repressor. *Proc. Nat. Acad. Sci.*, **56,** 1891
6. Ptashne, M. (1967). Specific binding of the λ phage repressor to λ DNA. *Nature (London)*, **214,** 232

7. Sheppard, D. and Englesberg, E. (1966). Positive control in the L-arabinose gene-enzyme complex of *Escherichia coli* B/r as exhibited with stable merodiploids. *Cold Spring Harbor Symp. Quant. Biol.*, **31,** 345
8. Pardon, J. F. and Wilkins, M. H. F. Jr (1972). A supercoil model for nucleohistone. *J. Mol. Biol.*, **68,** 115
9. Crick, F. (1971). General model for the chromosomes of higher organisms. *Nature* (*London*), **234,** 25
10. Burgess, R. R., Travers, A. A., Dunn, J. J. and Bautz, E. K. F. (1969). Factor stimulating transcription by RNA polymerase. *Nature* (*London*), **221,** 43
11. Roeder, R. G. and Rutter, W. J. (1969). Multiple forms of DNA-dependent RNA polymerase in eucaryotic organisms. *Nature* (*London*), **224,** 234
12. Blatti, S. P., Ingles, C. J., Lindell, T. J., Morris, P. W., Weaver, R. F., Weinberg, F. and Rutter, W. J. (1970). Structure and regulatory properties of eucaryotic RNA polymerase. *Cold Spring Harbor Symp. Quant. Biol.*, **35,** 649
13. Burgess, R. R. (1969). Separation and characterisation of the subunits of ribonucleic acid polymerase. *J. Biol. Chem.*, **244,** 6168
14. Rabussay, D. and Zillig, W. (1969). A rifampicin resistant RNA polymerase from *E. coli* altered in the β-subunit. *FEBS Lett.*, **5,** 104
15. Heil, A. and Zillig, W. (1970). Reconstitution of bacterial DNA-dependent RNA polymerase from isolated subunits as a tool for the elucidation of the role of the subunits in transcription. *FEBS Lett.*, **11,** 165
16. Goff, C. G. and Weber, K. (1970). A T4 induced RNA polymerase α subunit modification. *Cold Spring Harbour Symp. Quant. Biol.*, **35,** 101
17. Chamberlin, M., McGrath, J. and Waskell, J. (1970). New RNA polymerase from *Escherichia coli* infected with bacteriophage T7. *Nature* (*London*), **228,** 227
18. Summers, W. C. and Siegel, R. B. (1969). Control of template specificity of *E. coli* RNA polymerase by a phage coded protein. *Nature* (*London*), **223,** 1111
19. Dunn, J. J., Bautz, F. A. and Bautz, E. K. F. (1971). Different template specificities of phage T3 and T7 RNA polymerase. *Nature New Biol.*, **230,** 94
20. Hinkle, D. C. and Chamberlain, M. J. (1972). Studies of the binding of *Escherichia coli* RNA polymerase to DNA. 1. The role of sigma subunit in site selection. *J. Mol. Biol.*, **70,** 157
21. Zillig, W., Zechel, K., Rabussay, D., Schachner, M., Sethi, V. S., Palm, P., Heil, A. and Seiffert, W. (1970). On the role of different subunits of DNA dependent RNA polymerase from *E. coli* in the transcription process. *Cold Spring Harbor Symp. Quant. Biol.*, **35,** 47
22. Blattner, F. R., Dahlberg, J. E., Boettiger, J. K., Fiandt, M. and Szybalski, W. (1972). Distance from a promoter mutation to an RNA synthesis startpoint on bacteriophage λ DNA. *Nature New Biol.*, **237,** 232
23. Schafer, R., Zillig, W. and Zechel, K. (1973). A model for the initiation of transcription by DNA-dependent RNA polymerase from *Escherichia coli*. *Eur. J. Biochem.*, **33,** 207
24. Travers, A. A. and Burgess R. R. (1969). Cyclic re-use of the RNA polymerase sigma factor. *Nature* (*London*), **222,** 537
25. Krakow, J. S., Daley, K. and Karstadt, A. (1969). *Azotobacter vinlandii* RNA polymerase. VII. Transitions during unprimed r(I-C) synthesis. *Proc. Nat. Acad. Sci.*, **62,** 432
26. Manor, H., Goodman, D. and Stent, G. S. (1969). RNA chain growth rates in *E. coli* *J. Mol. Biol.*, **39,** 1
27. Roberts, J. W. (1969). Termination factor for RNA synthesis. *Nature* (*London*), **224,** 1168
28. Gaff, C. G. and Minkley, E. G. (1970). The RNA polymerase sigma factor: a specificity determinant. *Lepetit Colloq. on RNA Polymerase and Transcription*, 124 (L. Silvestri, editor) (North Holland Publ. Co.)
29. Millette, R. L., Trotter, C. D., Herrlich, P. and Schweiger, M. (1970). *In vitro* synthesis termination and release of active messenger RNA. *Cold Spring Harbor Symp. Quant. Biol.*, **35,** 135
30. Maitra, U., Lockwood, A. H., Dubnoff, J. S. and Guha, A. (1970). Termination, release and reinitiation of RNA chains from DNA templates by *Escherichia coli* RNA polymerase. *Cold Spring Harbor Symp. Quant. Biol.*, **35,** 143

31. Young, E. T. and Sinscheimer, R. L. (1964). Novel intra-cellular forms of lambda DNA. *J. Mol. Biol.*, **10,** 562
32. Echols, H. (1971). Lysogeny: viral repression and site specific recombination. *Ann. Rev. Biochem.*, **40,** 827
33. Ptashne, M. and Hopkins, N. (1968). The operators controlled by the λ phage repressors. *Proc. Nat. Acad. Sci.*, **60,** 1282
34. Steinberg, R. A. and Ptashne, M. (1971). *In vitro* repression of RNA synthesis by purified λ phage repressor. *Nature New Biol.*, **230,** 76
35. Kourilsky, P., Bourguignon, M. F., Bouquet, M. and Gros, F. (1970). Early transcription controls after induction of prophage λ. *Cold Spring Harbor Symp. Quant. Biol.*, **35,** 305
36. Szybalski, W., Boure, K., Fiandt, M., Hayes, S., Hradecna, Z., Kumar, S., Lozeron, H. A., Nijkamp, H. J. J. and Stevens, W. F. (1970). Transcriptional units and their control in *Escherichia coli* phage λ: operons and scriptons. *Cold Spring Harbor Symp. Quant. Biol.*, **35,** 341
37. Roberts, J. W. (1970). The ρ factor: termination and anti-termination in lambda. *Cold Spring Harbor Symp. Quant. Biol.*, **35,** 121
38. Chadwick, P., Pirrotta, V., Steinberg, R., Hopkins, N. and Ptashne, M. (1970). The λ and 434 phage repressor. *Cold Spring Harbor Symp. Quant. Biol.*, **35,** 283
39. Monod, J. and Wollman, E. (1947). L'inhibition de la croissance et de l'adaptation enzymatique chez les bactéries infectees par le bactériophage. *Ann. Inst. Pasteur*, **73,** 937
40. Calendar. R. (1970). The regulation of phage development. *Ann. Rev. Microbiol.*, **24,** 241
41. Salser, W., Bolle, A. and Epstein, R. (1970). Transcription during bacteriophage T4 development: A demonstration that distinct subclasses of the 'early' RNA appear at different times and some are 'turned off' at late times. *J. Mol. Biol.*, **49,** 271
42. Stevens, A. (1970). An isotopic study of DNA-dependent RNA polymerase of *E. coli* following T4 phage infection. *Biochem. Biophys. Res. Commun.*, **41,** 367
43. Bautz, E. K. F. and Dunn, J. (1969). DNA-dependent RNA polymerase from phage T4 infected *E. coli*: An enzyme missing a factor required for transcription of T4 DNA. *Biochem. Biophys. Res. Commun.*, **34,** 230
44. Schachner, M. and Zillig, W. (1970). Fingerprint maps of tryptic peptides from subunits of *Escherichia coli* and T4-modified DNA-dependent RNA polymerases *Eur. J. Biochem.*, **22,** 513
45. Stevens, A. (1972). New small polypeptides associated with DNA dependent RNA polymerase of *Escherichia coli* after infection with bacteriophage T4. *Proc. Nat. Acad. Sci.*, **69,** 603
46. Yamakawa, T. and Doi, R. H. (1971). Preferential transcription of *Bacillus subtillis* light deoxyribonucleic acid strands during sporulation. *J. Bacteriol.*, **106,** 305
47. Schaeffer, P. (1969). Sporulation and the production of antibiotics, exoenzymes and exotoxins, *Bact. Rev.*, **33,** 48
48. Leighton, T. J., Doi, R. H., Warren, R. A. J. and Kelln, R. A. (1973). The relationship of serine protease activity to RNA polymerase modification and sporulation in *Bacillus subtilis. J. Mol. Biol.*, **76,** 103
49. Sonenshein, A. L. and Roscoe, D. H. (1969). The course of phage phi-e infection in sporulating cells of *Bacillus subtilis* strain 3610. *Virology*, **39,** 265
50. Hussey, C., Pero, J., Shorenstein, R. G. and Losick, R. (1972). *In vitro* synthesis of ribosomal RNA by *Bacillus subtilis* RNA polymerase. *Proc. Nat. Acad. Sci.*, **69,** 407
51. Losick, R. and Sonenshein, A. L. (1969). Change in the template specificity of RNA polymerase during sporulation. *Nature* (*London*), **224,** 35
52. Hussey, C., Losick, R. and Sonenshein, A. L. (1971). Ribosomal RNA synthesis is turned of during sporulation of *Bacillus subtilis. J. Mol. Biol.*, **57,** 59
53. Losick, R., Sonenshein, A. L., Shorenstein, R. G. and Hussey, C. (1970). Role of RNA polymerase in sporulation. *Cold Spring Harbor Symp. Quant. Biol.*, **35,** 443
54. Losick, R., Shorenstein, R. G. and Sonenshein, A. L. (1970). Structural alteration of RNA polymerase during sporulation. *Nature* (*London*), **227,** 910
55. Linn, T. G., Greenleaf, A. R., Shorenstein, R. G. and Losick, R. (1973). The loss of the sigma activity of RNA polymerase of *Bacillus subtilis* during sporulation. *Proc. Nat. Acad. Sci.*, **70,** 1865

56. Greenleaf, A. L., Linn, T. G. and Losick, R. (1973). Isolation of a new RNA polymerase binding protein from sporulating *Bacillus subtilis*. *Proc. Nat. Acad. Sci.*, **70,** 490
57. Holland, M. J. and Whiteley, H. R. (1973). A new polypeptide associated with RNA polymerase from *Bacillus subtilis* during late stages of vegetative growth (submitted to *Biochem. Biophys. Res. Commun.*)
58. Leighton, T. J. (1973). An RNA polymerase mutation causing temperature-sensitive sporulation in *Bacilus subtilis*. *Proc. Nat. Acad. Sci.*, **70,** 1179
59. Weiss, S. B. and Gladstone, L. (1959). A mammalian system for the incorporation of cytidine triphosphate into ribonucleic acid. *J. Amer. Chem. Soc.*, **81,** 4118
60. Widnell, C. C. and Tata, J. R. (1966). Studies on the stimulation by ammonium sulfate of the DNA dependent RNA polymerase of isolated rat liver nuclei. *Biochim. Biophys. Acta*, **123,** 478
61. Maul, G. G. and Hamilton, T. H. (1967). The intranuclear localisation of two DNA-dependent RNA polymerase activities. *Proc. Nat. Acad. Sci.*, **57,** 1371
62. Pogo, A. O., Littau, V. C., Allfrey, V. G. and Mirsky, A. E. (1967). Modification of ribonucleic acid synthesis in nuclei isolated from normal and regenerating liver: some effect of salt and specific divalent cations. *Proc. Nat. Acad. Sci.*, **57,** 743
63. Chambon, P., Gissinger, F., Mandel, J. L., Kedinger, C., Gniazdowski, M. and Meilhac, M. (1970). Purification and properties of calf thymus DNA-dependent RNA polymerases A and B. *Cold Spring Harbor Symp. Quant. Biol.*, **35,** 693
64. Jacob, S. T. (1973). Mammalian RNA polymerases. *Prog. Nucl. Acid. Res. and Mol. Biol.*, **13,** 93
65. Roeder, R. G., Reeder, R. H. and Brown, D. D. (1970). Multiple forms of RNA polymerase in *Xenopus laevis:* their relationship to RNA synthesis *in vivo* and their fidelity of transcription *in vitro*. *Cold Spring Harbor Symp. Quant. Biol.*, **35,** 727
66. Lindell, T. J., Weinberg, F., Morris, P. W., Roeder, R. G. and Rutter, W. J. (1970). Specific inhibition of nuclear RNA polymerase II by α-amanitin. *Science*, **170,** 447
67. Meilhac, M., Kedinger, C., Chambon, P., Faulstich, H., Govindan, M. V. and Wieland T. (1970). Amanitin binding to calf thymus RNA polymerase B. *FEBS Lett.*, **9,** 258
68. Novello, F., Fiume, L. and Stirpe, F. (1970). Inhibition by alpha-amanitin of ribonucleic acid polymerase solubilised from rat liver nuclei. *Biochem. J.*, **116,** 177
69. Meilhac, M., Tysper, Z. and Chambon, P. (1972). Animal DNA-dependent RNA polymerases 4. Studies on inhibition by rifamycin derivatives. *Eur. J. Biochem.*, **28,** 291
70. Riva, S. and Silvestri, L. G. (1972). Rifamycins: a general view. *Ann. Rev. Microbiol.*, **26,** 199
71. Roeder, R. G. and Rutter, W. J. (1970). Multiple ribonucleic acid polymerases and ribonucleic acid synthesis during sea urchin development. *Biochemistry*, **9,** 2543
72. Chesterton, C. J. and Butterworth, P. H. W. (1971). Purification of the rat liver from B DNA-dependent RNA polymerases. *FEBS Lett.*, **15,** 181
73. Mandel, J. L. and Chambon, P. (1971). Purification of RNA polymerase B activity from rat liver. *FEBS Lett.*, **15,** 175
74. Weaver, R. F., Blatti, S. P. and Rutter, W. J. (1971). Molecular structures of DNA dependent RNA polymerases (II) from calf thymus and rat liver. *Proc. Nat. Acad. Sci.*, **68,** 2994
75. Kedinger, C., Nuret, P. and Chambon, P. (1971). Structural evidence for two α-amanitin sensitive RNA polymerases in calf thymus. *FEBS Lett.*, **15,** 169
76. Chambon, P., Gissinger, F., Kedinger, C., Mandel, J. L., Meilhac, M. and Nuret, P. (1972). Structural and functional properties of three mammalian nuclear DNA-dependent RNA polymerases. *Gene Transcription in Reproductive Tissue*, 222 (E. Diczfalusy, editor), Karolinska Symp. on Research Methods in Reproductive Endocrinology
77. Chesterton, C. J. and Butterworth, P. H. W. (1971). Studies on the origin of the form Ib mammalian DNA-dependent RNA polymerase. *FEBS Lett.*, **13,** 275
78. Chesterton, C. J. and Butterworth, P. H. W. (1971). A new form of mammalian DNA-dependent RNA polymerase and its relationship to the known forms of the enzyme. *FEBS Lett.*, **12,** 301
79. Chesterton, C. J. and Butterworth, P. H. W. (1971). Selective extraction of form I DNA-dependent RNA polymerase from rat liver nuclei and its separation into two species. *Eur. J. Biochem.*, **19,** 232

80. Adman, R., Schultz, L. D. and Hall, B. D. (1972). Transcription in yeast: separation and properties of multiple of RNA polymerases. *Proc. Nat. Acad. Sci.*, **69,** 1702
81. Kedinger, C. and Chambon, P. (1972). Animal DNA-dependent RNA polymerases. 3. Purification of calf thymus BI and BII enzymes. *Eur. J. Biochem.*, **28,** 283
82. Seifart, K. H., Benecke, B. J. and Juhasz, P. P. (1972). Multiple RNA species from rat liver tissue: possible existence of a cytoplasmic enzyme. *Arch. Biochem. Biophys.*, **151,** 519
83. Dezelee, S. and Sentenac, A. (1973). Role of DNA-RNA hybrids in yeast. Purification and properties of yeast RNA polymerase B. *Eur. J. Biochem.*, **34,** 41
84. Sugden, B. and Keller, W. (1973). Mammalian deoxyribonucleic acid-dependent ribonucleic acid polymerases. *J. Biol. Chem.*, **248,** 3777
85. Pong, S. S. and Loomis, W. F. (1973). Multiple nuclear RNA polymerases during development of *Dictyostelium discoideum. J. Biol. Chem.*, **248,** 3933
86. Gissinger, F. and Chambon, P. (1972). Animal DNA-dependent RNA polymerases. 2. Purification of calf-thymus Al enzyme. *Eur. J. Biochem.*, **28,** 277
87. Ponta, H., Ponta, U. and Wintersberger, E. (1972). Purification and properties of DNA-dependent RNA polymerases from yeast. *Eur. J. Biochem.*, **29,** 110
88. Goldberg, M. I., Perriard, J. C., Hager, G., Hallick, R. B. and Rutter, W. J. (1973). Transcriptional systems in eucaryotic cells. *International Symposium on the Control of Transcriptions*, 12–15 (Calcutta, India: Bose Institute, Feb.)
89. Reeder, R. H. and Roeder, R. G. (1972). Ribosomal RNA synthesis in isolated nuclei. *J. Mol. Biol.*, **67,** 433
90. Attardi, G. and Amaldi, F. (1970). Structure and synthesis of ribosomal RNA. *Ann. Rev. Biochem.*, **39,** 183
91. Roeder, R. G. and Rutter, W. J. (1970). Specific nucleolar and nucleoplasmic RNA polymerases. *Proc. Nat. Acad. Sci.*, **65,** 675
92. Horgen, P. A. and Griffin, D. H. (1971). Specific inhibitors of the three RNA polymerases from the aquatic fungus *Blastocladiella emersonii*, *Proc. Nat. Acad. Sci.*, **68,** 338
93. Timberlake, W. E., Hagen, G. and Griffin, D. H. (1972). Rat liver DNA-dependent RNA polymerase I is inhibited by cycloheximide. *Biochem. Biophys. Res. Commun.*, **48,** 823
94. Higashinakagawa, T. and Muramatsu, M. (1972). In vitro effect of cycloheximide on the nucleolar and extranucleolar nuclear RNA polymerases of rat liver. *Biochem. Biophys. Res. Commun.*, **47,** 1
95. Shields, D. and Tata, J. R. (1973). Differential thermal sensitivities of eucaryotic DNA-dependent RNA polymerases. *FEBS Lett.*, **31,** 209
96. Hecht, R. M. and Birnstiel, M. L. (1972). Integrity of the DNA template, pre-requisite for the faithful transcription of *Xenopus* rDNA *in vitro. Eur. J. Biochem.*, **29,** 489
97. Tocchini-Valentini, G. P. and Crippa, M. (1970). RNA polymerases from *Xenopus laevis. Cold Spring Harbor Symp. Quant. Biol.*, **35,** 737
98. Rutter, W. J., Morris, P. W., Goldberg, M., Paule, M. and Morris, R. W. (1973). RNA polymerases and transcriptive specificity in eukaryotic organisms. *The Biochemistry of Gene Expression in Higher Organisms*, 89 (J. K. Pollak and J. W. Lee, editors) (Sidney: Australia and New Zealand Book Co.)
99. Egyhazi, E., D'Monte, B. and Edstrom, J. E. (1972). Effects of α-amanitin on *in vitro* labelling of RNA from defined nuclear components in salivary gland cells from *Chironomus tentans. J. Cell. Biol.*, **53,** 523
100. Price, R. and Penman, S. (1972). Transcription of the adenovirus genome by an alpha-amanitin sensitive ribonucleic acid polymerase in HeLa cells. *J. Virology*, **9,** 621
101. Wallace, R. D. and Kates, J. (1972). State of adenovirus 2 deoxyribonucleic acid in the nucleus and its mode of transcription: studies with isolated viral deoxyribonucleic acid-protein complexes and isolated nuclei. *J. Virol.*, **9,** 627
102. Smuckler, E. A. and Tata, J. R. (1972). Nearest neighbor base frequency of the RNA formed by rat liver DNA-dependent RNA polymerase A and B with homologous DNA. *Biochem. Biophys. Res. Commun.*, **49,** 16
103. Mandel, J. L., Kedinger, C., Gissinger, F. and Chambon, P. (1973). Size of the RNAs synthesised by purified calf thymus DNA-dependent RNA polymerases on SV40 DNA. *FEBS Lett.*, **29,** 109
104. Rutter, W. J., Ingles, C. J., Weaver, R. F., Blatti, S. P. and Morris, P. W. (1972).

RNA polymerases and transcriptive specificity in eucaryotes. *Molecular Genetics and Developmental Biology*, 143 (M. Sussman, editor) (Prentice-Hall, Inc. Englewood Cliffs, N. J.)

105. Roeder, R. G. (1972). RNA polymerases during amphibian development. *Molecular Genetics and Developmental Biology*, 163 (M. Sussman, editor) (Prentice-Hall, Inc. Englewood Cliffs, N. J.)
106. Price. R. and Penman, S. (1972). A distinct RNA polymerase activity, synthesising 5.5S, 5S and 4S RNA in nuclei from adenovirus 2-infected HeLa cells. *J. Mol. Biol.*, **70,** 435
107. Zylber, E. and Penman, S. (1971). Products of RNA polymerases in HeLa cell nuclei. *Proc. Nat. Acad. Sci.*, **68,** 2861
108. Brutlag, D., Schekman, R. and Kornberg, A. (1971). A possible role for RNA polymerase in the initiation of M13 DNA synthesis. *Proc. Nat. Acad. Sci.*, **68,** 2826
109. Lark, K. G. (1972). Evidence for the direct involvement of RNA in the initiation of DNA replication in *E. coli.* 15T. *J. Mol. Biol.*, **64,** 47
110. Keller, W. (1972). RNA-primed DNA synthesis *in vitro. Proc. Nat. Acad. Sci.*, **69,** 1560
111. Klein, A. and Bonhoeffer, F. (1972). DNA replication. *Ann. Rev. Biochem.*, **41,** 301
112. Neubert, D. and Helge, H. (1965). Studies on nucleotide incorporation into mitochondrial RNA. *Biochem. Biophys. Res. Commun.*, **18,** 600
113. Spencer, D. and Whitfield, P. R. (1966). Ribonucleic acid synthesising activity of spinach chloroplasts and nuclei. *Arch. Biochem. Biophys.*, **121,** 336
114. Borst, P. (1972). Mitochondrial nucleic acids. *Ann. Rev. Biochem.*, **41,** 333
115. Scragg, A. H. (1971). Mitochondrial DNA-directed RNA polymerase from *Saccharomyces cerevisae* mitochondria. *Biochem. Biophys. Res. Commun.*, **45,** 701
116. Küntzel, H. and Schäfer, K. P. (1971). Mitochondrial RNA polymerase from *Neurospora crassa. Nature New Biol.*, **231,** 265
117. Tsai, M. J., Michaelis, G. and Criddle, R. S. (1971). DNA-dependent RNA polymerase from yeast mitochondria. *Proc. Nat. Acad. Sci.*, **68,** 473
118. Wintersberger, E. (1972). Isolation of a distinct rifampicin-resistant RNA polymerase of yeast, *Neurospora* and liver. *Biochem. Biophys. Res. Commun.*, **48,** 1287
119. Wu, G. J. and Dawid, I. B. (1972). Purification and properties of mitochondrial deoxyribonucleic acid dependent ribonucleic acid polymerase from ovaries of *Xenopus laevis. Biochemistry*, **11,** 3589
120. Barath, Z. and Kuntzel, H. (1972). Induction of mitochondrial RNA polymerase in *Neurospora crassa. Nature New Biol.*, **240,** 195
121. Hallick, R. B., Richards, O. C. and Rutter, W. J. (1973). Isolation of an RNA polymerase-chloroplast DNA transcription complex from chloroplasts of *Euglena gracilis.* (submitted to *Proc. Nat. Acad. Sci.*)
122. Bell, E. and Brown, J. (1972). RNA polymerase activities in the cytoplasm of differentiating chick muscle cells. *Biochim. Biophys. Acta*, **269,** 237
123. Bell, E. (1971). Informational DNA synthesis distinguished from that of nuclear DNA by inhibitors of DNA synthesis. *Science*, **174,** 603
124. Dausse, J.-P., Sentenac, A. and Fromageot, P. (1972). Interaction of RNA polymerase from *Escherichia coli* with DNA. Influence of DNA scissions on RNA polymerase binding and chain initiation. *Eur. J. Biochem.*, **31,** 394
125. Gross-Bellard, M., Audet, P. and Chambon, P. (1973). Isolation of high-molecular weight DNA from mammalian cells. *Eur. J. Biochem.*, **36,** 32
126. Vogt, V. (1969). Breaks in DNA stimulate transcription by core polymerase. *Nature (London)*, **223,** 854
127. von Hippel, P. H. and McGhee, J. D. (1972). DNA-protein interactions. *Ann. Rev. Biochem.*, **41,** 231
128. Birnstiel, M. L., Chipchase, M. and Speirs, J. (1971). The ribosomal RNA cistrons. *Prog. Nucl. Acid. Res. and Mol. Biol.*, **11,** 351
129. Shih, T. Y. and Martin, M. A. (1973). A general method of gene isolation. *Proc. Nat. Acad. Sci.*, **70,** 1697
130. Ross, J., Aviv, H., Scolnick, E. and Leder, P. (1972). *In vitro* synthesis of DNA complementary to purified rabbit globin mRNA. *Proc. Nat. Acad. Sci.*, **69,** 264
131. Westphal, H. (1970). SV40 strand selection by E. coli RNA polymerase. *J. Mol. Biol.*, **50,** 407

132. Schafer, K. P., Bugge, G., Grandi, M. and Kuntzel, H. (1971). Transcription of mitochondrial DNA *in vitro* from *Neurospora crassa*. *Eur. J. Biochem.*, **21**, 473.
133. Clark, R. J. and Felsenfeld, G. (1971). Structure of chromatin. *Nature New Biol.*, **229**, 101
134. Itzhaki, R. (1970). Structure of deoxyribonucleoprotein as revealed by its binding to polylysine. *Biochem. Biophys. Res. Commun.*, **41**, 25
135. Paul, J. and Gilmour, R. S. (1968). Organ-specific restriction of transcription in mammalian chromatin. *J. Mol. Biol.*, **34**, 305
136. Paul, J., Carroll, D., Gilmour, R. S., More, J. A. R., Threlfall, G., Wilkie, M. and Wilson, S. (1972). Functional studies on chromatin. *Gene Transcription in Reproductive Tissue*, 277 (E. Diczfalusy, editor), Karolinska Symp. on Research Methods in Reproductive Endocrinology
137. Cedar, H. and Felsenfeld, G. (1973). Transcription of chromatin *in vitro*. *J. Mol. Biol.*, **77**, 237
138. Bhorjee, J. S. and Pedersen, T. (1973). Chromatin: Its isolation from cultured mammalian cells with particular reference to contamination by nuclear ribonucleoprotein particles. *Biochemistry*, **12**, 2766
139. Butterworth, P. H. W., Cox, R. F. and Chesterton, C. J. (1971). Transcription of mammalian chromatin by mammalian DNA-dependent RNA polymerases. *Eur. J. Biochem.*, **23**, 229
140. Cox, R. F., Haines, M. E. and Carey, N. H. (1973). Modification of the template capacity of chick, oviduct chromatin for form B RNA polymerase by estradiol. *Eur. J. Biochem.*, **32**, 513
141. Wintersberger, U., Smith, P. and Letnansky, K. (1973). Yeast chromatin: preparation from isolated nuclei, histone composition and transcription capacity. *Eur. J. Biochem.*, **33**, 123
142. Silverman, B. and Mirsky, A. E. (1973). Accessibility of DNA in chromatin to DNA polymerase and RNA polymerase. *Proc. Nat. Acad. Sci.*, **70**, 1326
143. Tsai, M. J. and Saunders, G. F. (1973). Transcription of chromatin by human polymerase. *Biochem. Biophys. Res. Commun.*, **51**, 756
144. Barry, J. and Gorski, J. (1971). Uterine ribonucleic acid polymerase. Effects of estrogen on nucleotide incorporation in 3′ chain termini. *Biochemistry*, **10**, 2384
145. Bekhor, I., Kung, G. M. and Bonner, J. (1969). Sequence-specific interaction of DNA and chromosomal protein. *J. Mol. Biol.*, **39**, 351
146. Huang, R. C. C. and Huang, P. C. (1969). Effect of protein-bound RNA associated with chick embryo chromatin on template specificity of the chromatin. *J. Mol. Biol.*, **39**, 365
147. Maryanka, D. and Gould, H. (1973). Transcription of rat liver chromatin with homologous enzyme. *Proc. Nat. Acad. Sci.*, **70**, 1161
148. Axel, R., Cedar, H. and Felsenfeld, G. (1973). Synthesis of globin ribonucleic acid from duck reticulocyte chromatin *in vitro*. *Proc. Nat. Acad. Sci.*, **70**, 2029
149. de Crombrugghe, B., Chen, B., Anderson, W., Nissley, P., Gottesman, M. and Pastan, I. (1971). lac DNA, RNA polymerase and cyclic AMP receptor protein, cyclic AMP, lac repressor and inducer are the essential elements for controlled lac transcription. *Nature New Biol.*, **231**, 139
150. Gniazdowski, M., Mandel, J. L., Gissinger, F., Kedinger, C. and Chambon, P. (1970). Calf thymus RNA polymerases exhibit template specificity. *Biochem. Biophys. Res. Commun.*, **38**, 1033
151. Stein, H. and Hausen, P. (1970). Factors influencing the activity of mammalian RNA polymerase. *Cold Spring Harbor Symp. Quant. Biol.*, **35**, 709
152. Sugden, B. and Sambrook, J. (1970). RNA polymerase from HeLa cells. *Cold Spring Harbor Symp. Quant. Biol.*, **35**, 663
153. Seifart, K. H. (1970). A factor stimulating the transcription on double stranded DNA by purified RNA polymerase from rat liver nuclei. *Cold Spring Harbor Symp. Quant. Biol.*, **35**, 719
154. Lentfer, D. and Lezius, A. G. (1972). Mouse myeloma polymerase B: Template specificities and the role of a transcription-stimulating factor. *Eur. J. Biochem.*, **30**, 278
155. diMauro, E., Hollenberg, C. P. and Hall, B. D. (1972). Transcription in yeast: a factor that stimulates yeast RNA polymerases. *Proc. Nat. Acad. Sci.*, **69**, 2818
156. Seifart, K. H., Juhasz, P. P. and Benecke, B. J. (1973). A protein factor from rat liver

tissue enhancing the transcription of native templates by homologous RNA polymerase B. *Eur. J. Biochem.*, **33,** 181

157. Mondahl, H., Ganguly, A., Das, A., Mandal, R. K. and Biswas, B. B. (1972). Ribonucleic acid polymerase from eucaryotic cells. *Eur. J. Biochem.*, **28,** 143
158. Goldberg, M. L. and Moon, H. D. (1970). Partial purification of RNA polymerase from bovine thymus. *Arch. Biochem. Biophys.*, **141,** 258
159. Lee, S. C. and Dahmus, M. E. (1973). Stimulation of eucaryotic DNA-dependent RNA polymerase by protein factors. *Proc. Nat. Acad. Sci.*, **70,** 1383
160. Higashinakagawa, T., Onishi, T. and Murmatsu, M. (1972). A factor stimulating the transcription by nucleolar RNA polymerase in the nucleolus of rat liver. *Biochem. Biophys. Res. Commun.*, **48,** 937
161. Arnaud, M., Beziat, Y., Borgna, J. L., Guilleux, J. C. and Mousseron-Canet, M. (1971). Le recepteur de l'oestradiol, l'amp cyclique et la RNA polymerase nucleolaire dans l'uterus de genisse. Stimulation de la biosynthese de RNA *in vitro. Biochem. Biophys. Acta*, **254,** 241
162. DeSombre, E. R., Mohla, S. and Jensen, E. V. (1972). Estrogen-independent activation of the receptor protein of calf uterine cytosol. *Biochem. Biophys. Res. Commun.*, **48,** 1601
163. Varrone, S., Ambesi-Impiombato, F. S. and Macchia, V. (1972). Stimulation by cyclic 3′-5′-adenosine monophosphate of RNA synthesis in a cell free system. *FEBS Lett.*, **21,** 99
164. Teng, C. S., Teng, C. T. and Allfrey, V. G. (1971). Studies of nuclear acidic proteins, evidence for their phosphorylation, tissue specificity, selective binding to deoxyribonucleic acid, and stimulatory effects on transcription. *J. Biol. Chem.*, **246,** 3597
165. Shea, M., and Kleinsmith, L. J. (1973). Template-specific stimulation of RNA synthesis by phosphorylated non-histone chromatin proteins. *Biochem. Biophys. Res. Commun.*, **50,** 473
166. Dalbow, D. G. (1973). Synthesis of RNA polymerase in *Escherichia coli* B/r growing at different rates. *J. Mol. Biol.*, **75,** 181
167. Matzura, H., Hansen, B. S. and Zeuthen, J. (1973). Biosynthesis of the β and β' subunits of RNA polymerase in *Escherichia coli. J. Mol. Biol.*, **74,** 9
168. Yanofsky, S. A. and Spiegelman, S. (1962). The identification of the ribosomal RNA cistron by sequence complementarity. II. Saturation of the competitive interaction at the RNA cistron. *Proc. Nat. Acad. Sci.*, **48,** 1465
169. Lazzarini, R. A. and Winslow, R. M. (1970). The regulation of RNA synthesis during growth rate transitions and amino acid deprivation in *E. coli. Cold Spring Harbor Symp. Quant. Biol.*, **35,** 383
170. Trayers, A., Kamen, R. I. and Schlief, R. F. (1970). Factor necessary for ribosomal RNA synthesis. *Nature* (*London*), **228,** 748
171. Travers, A., Baillie, D. L. and Pedersen, S. (1973). Effect of DNA conformation on ribosomal RNA synthesis *in vitro. Nature New Biol.*, **243,** 161
172. Smulson, M. E. and Thomas, J. (1969). Ribonucleic acid biosynthesis of human cells during amino acid deprivation. *J. Biol. Chem.*, **244,** 5309
173. Mahy, H. W. J., Hastie, N. D. and Armstrong, S. J. (1972). Inhibition of influenza virus replication by alpha-amanitin: mode of action. *Proc. Nat. Acad. Sci.*, **69,** 1421
174. Rott, R. and Scholtissek, C. (1970). Specific inhibition of influenza replication by alpha-amanitin. *Nature* (*London*), **228,** 56
175. Attardi, G., Parnas, H. and Attardi, B. (1970). Pattern of RNA synthesis in duck erythrocytes in relationship to the stage of cell differentiation. *Exp. Cell Res.*, **62,** 11
176. Keller, W. and Goor, R. (1970). Mammalian RNA polymerase: structural and functional properties. *Cold Spring Harbor Symp. Quant. Biol.*, **35,** 671
177. Yu, F. L. and Feigelson, P. (1972). The rapid turnover of RNA polymerase of rat liver nucleolus and of its messenger RNA. *Proc. Nat. Acad. Sci.*, **69,** 2833
178. Sconzo, G., Bono, A., Albanese, I. and Giudice, G. (1972). Studies on sea urchin oocytes, II. Synthesis of RNA during oogenesis. *Exp. Cell Res.*, **72,** 95
179. Claycomb, W. C. and Villee, C. A. (1971). Synthesis of RNA by nuclei isolated from embryos of *Xenopus laevis* at different developmental stages. *Exp. Cell Res.*, **69,** 430
180. Nemer, M. and Infante, A. A. (1967). Ribosomal ribonucleic acid of the sea urchin egg and its fate during embryogenesis. *J. Mol. Biol.*, **27,** 73
181. Emerson, C. P. and Humphreys, T. (1970). Regulation of DNA-like RNA and the

apparent activation of ribosomal RNA synthesis in sea urchin embryos: quantitative measurements of newly synthesised RNA. *Develop. Biol.*, **23,** 86

182. Karasaki, S. (1968). The ultrastructure and RNA metabolism of nucleoli in early sea urchin embryos. *Exp. Cell Res.*, **52,** 13
183. Gong, C. S. and van Etten, J. L. (1972). Changes in soluble ribonucleic acid polymerases associated with the germination of *Rhizopus stolonifer* spores. *Biochim. Biophys. Acta*, **272,** 44
184. van der Westhuyzen, D. R., Boyd, M. D. C., Fitschen, W. and von Holt, C. (1973). DNA-dependent RNA polymerase in maturing avian erythrocytes. *FEBS Lett.*, **30,** 195
185. Firtel, R. A. (1972). Changes in the expression if single-copy DNA during development of the cellular slime mold *Dictyostelium discoideum*. *J. Mol. Biol.*, **66,** 363
186. Rudick, V. L. and Weisman, R. A. (1973). DNA-dependent RNA polymerase from trophozoites and cysts of *Acanthamoeba castellani*. *Biochim. Biophys. Acta*, **299,** 91
187. Tata, J. R. (1968). Hormonal regulation of growth and protein synthesis. *Nature* (*London*), **219,** 331
188. Jensen, E. V. and DeSombre, E. R. (1972). Mechanism of action of the female sex hormones. *Ann. Rev. Biochem.*, **41,** 203
189. Mueller, G. C., Vonderhaar, B., Kim, U. H. and Le Mahieu, M. (1972). Estrogen action: an inroad to cell biology. *Rec. Prog. Hor. Res.*, **28,** 1
190. Mainwaring, W. I. F., Mangan, F. R. and Peterken, B. M. (1971). Studies on the solubilised ribonucleic acid polymerase from rat ventral prostate gland. *Biochem. J.*, **123,** 619
191. Glasser, S. R., Chytil, F. and Spelsberg, T. C. (1972). Early effects of oestradiol-17β on the chromatin and activity of the deoxyribonucleic acid-dependent ribonucleic acid polymerases (I and II) of the rat uterus. *Biochem. J.*, **130,** 947
192. Yu, F. L., and Feigelson, P. (1971). Cortisone stimulation of nucleolar RNA polymerase activity. *Proc. Nat. Acad. Sci.*, **68,** 2177
193. Billing, R. J., Barbiroli, B. and Smellie, R. M. S. (1969). The mode of action of oestradiol. I. The transport of RNA precursors into the uterus. *Biochim. Biophys. Acta*, **190,** 52
194. Munns, T. W. and Katzman, P. A. (1971). Effect of estradiol on uterine ribonucleic acid metabolism: *in vitro* uptake and incorporation of ribonucleic acid precursors. *Biochemistry*, **10,** 4941
195. Knowler, J. T. and Smellie, R. M. S. (1971). The synthesis of ribonucleic acid in immature rat uterus responding to oestradiol-17β. *Biochem. J.*, **125,** 605
196. McGuire, W. L. and O'Malley, B. W. (1968). Ribonucleic acid polymerase activity of the chick oviduct during steroid-induced synthesis of a specific protein. *Biochim. Biophys. Acta*, **157,** 187
197. Hamilton, T. H. (1968). Control by estrogen of genetic transcription and translation. *Science*, **161,** 649
198. Luck, D. N. and Hamilton, T. H. (1972). Early estrogen action: stimulation of the metabolism of high molecular weight and ribosomal RNA's. *Proc. Nat. Acad. Sci.*, **69,** 157
199. O'Malley, B. W., Rosenfield, G. C., Comstock, J. P. and Means, A. R. (1972). Steroid hormone induction of a specific translatable messenger RNA. *Nature New Biol.*, **240,** 45
200. Barnea, A. and Gorski, J. (1970). Estrogen induced protein: Time course of synthesis. *Biochemistry*, **9,** 1899
201. Baulieu, E. E., Wira, C. R., Milgrom, E. and Raynaud-Jammet, C. (1972). Ribonucleic acid synthesis and oestradiol action in the uterus. *Gene Transcription in Reproductive Tissue* 396 (E. Diczfalusy, editor) Karolinska Symp. on Research Methods in Reproductive Endocrinology
202. DeAngelo, A. B. and Gorski, J. (1970). Role of RNA synthesis in the estrogen induction of a specific uterine protein. *Proc. Nat. Acad. Sci.*, **66,** 693
203. Lindell, T. J., Morris, P. W., Weinberg, F. and Rutter, W. J. (1973). Differential stimulation of rat hepatic and uterine RNA polymerase activities following hormone treatment or partial hepatectomy. (submitted to *Endocrinology*)
204. Billing, R. J., Barbiroli, B. and Smellie, R. M. S. (1969). The mode of action of oestradiol. II. The synthesis of RNA. *Biochim. Biophys. Acta*, **190,** 60

205. Benecke, B. J., Ferencz, A. and Seifart, K. H. (1973). Resistance of hepatic RNA polymerases to compounds effecting RNA and protein synthesis *in vivo*. *FEBS Lett.*, **31,** 53
206. Mohla, S., DeSombre, E. R. and Jensen, E. V. (1972). Tissue specific stimulation of RNA synthesis by transformed estradiol-receptor complex. Biochem. *Biophys. Res. Commun.*, **46,** 661
207. Behrens, O. K. and Grinnan, E. L. (1969). Polypeptide hormones. *Ann. Rev. Biochem.*, **38,** 83
208. Adiga, P. R., Murthy, P. U. N. and McKenzie, J. M. (1971). Stimulation by thyrotropin, long acting thyroid stimulator, and dibutyryl 3′-5′-adenosine monophosphate of protein and ribonucleic acid synthesis and ribonucleic acid polymerase activities in porcine thyroid *in vitro*. *Biochemistry*, **10,** 702
209. Akamatsu, N., Kamiya, T., Maeda, H. R., Endo, N., Fukui, N. and Miura, Y. (1971). Effect of growth hormone on the RNA polymerase and deoxythymidine kinase activities of regenerating liver in hypophysectomised rats. *J. Biochem.* (*Japan*), **69,** 1091
210. Smuckler, E. A. and Tata, J. R. (1971). Changes in hepatic nuclear DNA-dependent RNA polymerase caused by growth hormone and triiodothyronine. *Nature* (*London*), **234,** 37
211. Jungmann, R. A. and Schweppe, J. S. (1972). Mechanism of action of gonadotropin. II. Control of ovarian nuclear ribonucleic acid polymerase activity and chromatin template activity. *J. Biol. Chem.*, **247,** 5543
212. Oravec, M. and Korner, A. (1971). Stimulation of synthesis of DNA-like and ribosomal RNA by growth hormone. *J. Mol. Biol.*, **58,** 489
213. Salaman, D. F., Betteridge, S. and Korner, A. (1972). Early effects of growth hormone on nuclear and nucleoplasmic RNA synthesis and RNA polymerase activity in normal rat liver. *Biochim. Biophys. Acta*, **272,** 382
214. Emerson, C. P. (1971). Regulation of the synthesis and the stability of ribosomal RNA during contact inhibition of growth. *Nature New Biol.*, **232,** 101
215. Rovera, G., Farber, J. and Baserga, R. (1971). Gene activation in WI-38 fibroblasts stimulated to proliferate: requirement for protein synthesis. *Proc. Nat. Acad. Sci.*, **68,** 1725
216. Franze-Fernandez, B. and Pogo, A. O. (1971). Regulation of the nucleolar DNA-dependent RNA polymerase by amino acids in Ehrlich ascites tumor cells. *Proc. Nat. Acad. Sci.*, **68,** 3040
217. Novello, F. and Stirpe, F. (1970). Simultaneous assay of RNA polymerase I and II in nuclei isolated from resting and growing rat liver with the use of alpha-amanitin. *FEBS Lett.*, **8,** 57
218. Rizzo, A. J. and Webb, T. E. (1972). Regulation of ribosome formation in regenerating rat liver. *Eur. J. Biochem.*, **27,** 136
219. Chesterton, C. J., Humphrey, S. M. and Butterworth, P. H. W. (1972). Comparison of the multiple deoxyribonucleic acid-dependent ribonucleic acid polymerase forms of whole rat liver and a minimal-deviation rat hepatoma cell line. *Biochem. J.*, **126,** 675
220. Pogo, B. G. T. (1972). Early developments in lymphocyte transformation by phytohemagglutinin. I. DNA-dependent RNA polymerase activities in isolated lymphocyte nuclei. *J. Cell Biol.*, **53,** 635
221. Cooper, H. L. (1969). Ribosomal ribonucleic acid production and growth regulation in human lymphocytes. *J. Biol. Chem.*, **244,** 1946
222. Cooper, H. L. (1969). Ribosomal ribonucleic acid wastage in resting and growing lymphocytes. *J. Biol. Chem.*, **244,** 5590
223. Cooke, A. and Brown, M. (1973). Stimulation of the activities of solubilized pig lymphocyte RNA polymerases by phytohemagglutinin. *Biochem. Biophys. Res. Commun.*, **51,** 1042
224. Church, R. B. and McCarthy, B. J. (1967). Ribonucleic acid synthesis in regenerating and embryonic liver. *J. Mol. Biol.*, **23,** 459–475
225. Heywood, S. M. (1970). Specificity of mRNA binding factor in eukaryotes. *Proc. Nat. Acad. Sci.*, **67,** 1782
226. Nudel, U., Lebleu, B. and Revel, M. (1973). Discrimination between messenger ribonucleic acids by a mammalian translation initiation factor. *Proc. Nat. Acad. Sci.*, **70,** 2139

227. Fiume, L. and Stripe, F. (1966). Decreased RNA content in mouse liver nuclei after intoxication with alpha-amanitin. *Biochim. Biophys. Acta*, **123**, 643
228. Sekeris, C. E. and Schmid, W. (1972). Action of alpha-amanitin *in vivo* and *in vitro*. *FEBS Lett.*, **27**, 41
229. Sonenshein, A. L. and Losick, R. (1970). RNA polymerase mutants blocked in sporulation. *Nature* (*London*), **227**, 906
230. Shapiro, A. L. Vinuela, E. and Maizel, J. V. (1967). Molecular weight estimation of polypeptide chains by electrophoresis in SDS-polyacrylamide gels. *Biochem. Biophys. Res. Commun.*, **37**, 965

10
The Role of Chromosomal Proteins as Gene Regulators

A. J. MACGILLIVRAY and D. RICKWOOD
The Beatson Institute for Cancer Research, Glasgow

10.1 INTRODUCTION

Much of our current knowledge of the mechanisms involved in differentiation and development has come from investigations of the control of protein synthesis in both prokaryotes and eukaryotes. Thus it is generally recognised that the same basic rules apply to mammalian as well as bacterial cells, namely that the genetic information for protein synthesis is carried in the DNA and that this is transcribed into messenger RNA. Such RNA is in turn translated into polypeptide chains on ribosomes by a process which also involves activating enzymes, transfer RNA and other factors. Such a mechanism for protein synthesis obviously provides numerous levels at which control could be applied and indeed there is evidence to suggest that such controls can operate at all levels during the formation of the differentiated state. In this chapter we will consider regulation at a single level, namely that of transcription, since there is much evidence to indicate that during differentiation, controls of considerable importance are applied at this stage.

The investigations into gene regulation in prokaryotes have pointed to two forms of control, the 'positive' function of factors which stimulate the transcription of DNA by RNA polymerase (e.g. CAP factor) and the 'negative' effects of repressors of transcription (e.g. *lac* repressor). However, when dealing with gene regulation in eukaryotes a number of important features are encountered. Firstly, the eukaryotic genome is very much larger than that of, say, bacteria being often 100–1000 times as great. In addition, only a few genes are present in prokaryotic DNA as multiple copies, while in eukaryotes a considerable proportion of the genome consists of 'repetitious' DNA; the remainder is present as 'unique' sequences in amounts far greater than that found in the entire genome of many bacteria[1].

Another major feature of eukaryotic cells concerns the fact that the DNA is permanently associated with large amounts of protein, whereas in prokaryotes protein is associated only transiently with the DNA. The eukaryotic chromosomal material (termed nucleoprotein or chromatin) is housed inside the nuclear membrane, separate from the protein-synthesising machinery of the ribosomes, and consists of the loosely packed 'euchromatin' and tightly condensed 'heterochromatin'. Euchromatin is generally accepted as containing those regions of DNA which are actively transcribed[2-4], but it appears to contain a mixture of repetitious and unique DNA[5]. Heterochromatin, on the other hand, has been recognised as consisting of a number of components, namely 'facultative heterochromatin' containing DNA which at other times is found in euchromatin, and 'constitutive heterochromatin' which is probably never transcribed[6] and which also includes a fraction of highly repetitious DNA, termed 'satellite DNA' which is found in the heterochromatic centromeric region[7,8]. Thus it appears that in many eukaryotic chromosomes much of the DNA is incapable of being transcribed throughout the lifetime of the cell.

During replication there are distinct differences between the properties of prokaryotic and eukaryotic chromosomes. The former are simply separated after duplication into daughter cells. Thereafter, most genes respond to changes in the environment. In eukaryotes, replication is, of course, an essential part of the process of differentiation leading to the development of body organs and tissues. After replication, the eukaryotic chromosomes become tightly condensed prior to division involving the complex mechanisms of meiosis or mitosis.

Other features of the differentiated state are also relevant to the present discussion on gene control. These include the evidence that all differentiated cells in an organism contain the same genetic information[9] and the investigations which have shown the remarkable stability of the differentiated state, even under *in vitro* conditions[10-13].

It would, therefore, be reasonable to expect that, in addition to the transcriptional controls already described in prokaryotes, there may be other, possibly more complicated, forms of gene control in eukaryotes. The possible involvement of the chromosomal proteins in such mechanisms was first proposed by the Stedmans[14] and during the past 20 years many attempts have been made to verify this hypothesis. In the following sections we will review the structural and functional roles of the chromosomal proteins and describe some current views on eukaryotic gene regulation.

10.2 CHROMATIN

10.2.1 Chromosomes and the isolation of chromatin

Morphological studies of interphase chromosomes, e.g. the giant chromosomes of insects and the lampbrush chromosomes of oocytes, indicate that the chromosomal unit is a fibre of 100–200 Å in diameter. Ris[15] first of all suggested that this consisted of two molecules of DNA forming a double fibre but recent evidence has supported the view of Dupraw[16] that the structure represented a nucleoprotein fibre of 30 Å in diameter which formed a supercoiled strand of some 200 Å in diameter. This latter proposal was supported by x-ray diffraction studies which indicated a primary nucleoprotein fibre of diameter 35 Å which was coiled into a helical structure[17, 18]. More recent studies of calf thymus nucleohistone, using the electron microscope and physical criteria, show an elongated structure of diameter 80–120 Å which it has been suggested might represent a DNA double helix irregularly supercoiled with a pitch of 45 Å[19]. Other electron microscopical work on calf thymus chromatin shows knobby threads of diameter 200–300 Å[20]. Structures with dimensions similar to those described above have also been observed in sea urchin sperm[21], human lymphocytes[22] and in nucleated erythrocytes[23, 24]. Although we know that the chromosomal material consists mostly of inactive condensed heterochromatin together with some active diffuse euchromatin, there is little information available concerning the organisation of these fractions along the chromosome. Only Miller's group (see Miller and Bakken[25]) appear to have made any contribution to this field. Their evidence from electron microscope studies of amphibian and mammalian genes actively involved in RNA synthesis, show that ribosomal genes are repetitious and are separated by inactive 'spacer regions'. Gradients of closely spaced fibrils run from each RNA polymerase locus giving a 'Christmas tree' effect. During the synthesis of non-ribosomal RNA in amphibia, the polymerases are closely spaced along loops extending from the active chromosomes, with long RNA containing fibrils leading from each locus. In contrast, in HeLa cells the polymerases are widely separated, at irregular intervals and the RNA-fibrils are much shorter.

An essential prerequisite for the characterisation of chromosomal proteins is their preparation from chromosomes which have been isolated so as to retain their native components as far as possible. Although metaphase chromosomes can be isolated with relative ease[26], during interphase the chromosomes are more extended and hydrated. Hence they are usually isolated as 'chromatin', the nucleoprotein complex remaining after removal of cytoplasmic and soluble nuclear components[27, 28]. Analyses of a number of chromatins are given in Table 10.1, expressed as ratios of the DNA content.

Apart from DNA itself the other major constituents of chromatin consists of two types of proteins, the histones which are extremely basic and the non-histone proteins. Although such a classification was originally a somewhat arbitrary one based on the relative solubilities of the two groups of proteins in dilute acid, more recent procedures for the fractionation of chromatin

Table 10.1 Chemical composition of chromatins

Source	*DNA*	*RNA*	*Histone*	*Non-histone protein*	*Ref.*
Pig cerebellum	1	0.13	1.6	0.5	185
Pig pituitary	1	0.11	1.56	0.45	185
Rat liver	1	0.04	1.15	0.95	176
Rat kidney	1	0.06	0.95	0.70	176
Calf thymus	1	0.05	0.89	0.21	458
Chick liver	1	0.03	1.17	0.88	176
Chick erythrocyte	1	0.02	1.08	0.54	176
Pea bud	1	0.05	1.10	0.41	176
Sea urchin sperm	1	0.04	1.02	0.13	109
Algae (*A. cylindrica*)	1	0.04	0.02	0.07	86
Algae (*E. cohnii*)	1	0.09	0.08	0.48	87
Fungus (*M. gypseum*)	1	0.05	0.03	1.02	88
Yeast (*S. cerevisiae*)	1	0.1	1.3	0.38	91
Slime mould (*P. polycephalum*)	1	—	1.1	9.1	94

utilising ion-exchange chromatography have confirmed the presence of these two distinct groups of proteins in a number of chromatins[28-31]. In general there is an approximately 1:1 ratio between the DNA and histone contents of most chromatins. However, some species, e.g. certain algae and fungi, have been found to contain little histone-like protein. On the other hand the amount of non-histone protein in chromatin varies from tissue to tissue (Table 10.1).

10.2.2 Histones

Much of the background to the isolation and chemistry of histones has been adequately reviewed by others[32-35]. Briefly, the histones of the somatic cells of most higher organisms have been found to consist of five main groups, namely histones F1, F2b, F2a1, F2a2 and F3 according to the nomenclature of Johns[34]. It is obvious from their analyses that these five components are all highly basic proteins of a limited molecular weight range (11 000–22 000). Their content of basic amino acid residues lies between 20 and 40 mol % with the lysine and arginine contents varying from one fraction to the next, such that the former is a major constituent of the F1 histone fraction. Cysteine is found only in the F3 histone, whilst tryptophan is totally lacking in any of the histone fractions. In three fractions (F1, F2a1, F2a2) the *N*-terminal serine is blocked by an acetyl group. With the exception of the very lysine-rich F1 histone, individual histone fractions appear to consist of one molecular species[34, 36]. During the past few years virtually all of the histone fractions have been isolated in pure form and their primary structure determined. Complete and partial amino acid sequences of the five histone fractions are given in Table 10.2 and the properties of each of these proteins are summarised later.

Table 10.2 Partial and complete amino acid sequences of histones

```
                 10         20         30         40         50         60
                      *    *      ****     ***           *     * *         **
F1    Ac-SEAPAETAAP APAPKSPAKT PVKAAKKKKP AGARRKASGP PVSELITKAV AASKERSGVS LAALKKALA

             *      **  *   **   **  **** * **         *  *  *        *
F2b      PQPAKSAPAP KKGSKAVTKK AQKKDGKKRK RSRKESYSYY VYKVLKQVHP DTGISSKAMG IMNSFVNDI

           * *   *  * * * *  *         *  **  **      *
F2a2  Ac-SGRGKQGGKA RAKAKTRSSR AGLQFPVGRV HRLLRKGNYA ERVGAGAPVY LAAVLEYLTA EILELAGNA

                         Ac  Me
                         |   |
           * *  *    *   ** **   *        *   **  **    **         *   *       *
F2a1  Ac-SGRGKGGKGL GKGGAKRHRK VLRDNTQGIT KPAIRRLARR GGVKRISGLI YEETRGVLKV FLENVIRDA
                                                                          I

                       Ac         Ac  Me
                       |          |   |
          * *   **     *  **     *  **        ** **  *      *   **  *       **    *
F3       ARTKQTARKS TGGKAPRKQL ATKAARKSAP ATGGVKKPHR YRPGTVALRE IRRYQKSTEL LIRKLPFQR
```

Table 10.2 contd.

```
                70          80              90              100             110             120             130
F1 contd.    A  A G Y . . . . . . .                                                                                          Ref. 42
                 *        *       *     * *         *             *                * *            *             *
F2b contd.   F  E R I  A G E A S R L  A H Y N  K R S T I T  S R E I  Q T A V R L  L L P G E L A K H A  V S E G T K A V T K  Y T S S K   Ref. 53
                *      * *   *        * *         *            *       *                          * *         * * *   *   *
F2a2 contd.  A  R D N K K T R I  I P  R H L Q  L A I R N D  E E L N K L L G K V  T I  A Q G G V L P N  I Q A V L L P K K T  E S H H K A K G K   Ref. 61
                        *
                        K1* *                    * *     *
F2a1 contd.  V  T Y T E H A  R K T  V T A M  D V V Y A L  K R Q G R T L Y G F  G G                                              Ref. 66
                        R2
                 *          *     *                                                   *  * *            *           * *  *     *
F3 contd.    L  Y R E I  A Q D F K T  D L R F  Q S S A Y M  A L Q E A C E A Y L  V G L F E D T N L C  A I  H A K R V T I  M  P K D I  E L A R R I  R G E R A   Ref. 74
```

SINGLE LETTER AMINO ACID CODE

A	Ala	H	His	Q	Gln	Y	Tyr
B	Asx	I	Ile	R	Arg	Z	Glx
C	Cys	K	Lys	S	Ser	—	Gap
D	Asp	L	Leu	T	Thr		
E	Glu	M	Met	V	Val		
F	Phe	N	Asn	W	Trp		
G	Gly	P	Pro	X	?		

1. Calf
2. Pea

Me and Ac refer to amino acid residues which are methylated and acetylated respectively.
Basic amino acid residues are denoted by asterisks.
In each sequence residue number 1 is the *N*-terminus.

10.2.2.1 Histone F1

This group consists of the largest histone molecules (molecular weight 22 000) and apart from possessing some 30 mol% of lysine they are also rich in alanine and serine. The heterogeneity of the F1 fraction has been demonstrated by electrophoretic[37] and ion-exchange chromatographic studies[38, 39]. Although the number of parent F1 molecules in rodent tissues would appear to be at least three, superimposed on this number is a second form of heterogeneity caused by postsynthetic phosphorylation which yields tissue-specific electrophoretic patterns[40, 41].

Structural studies of parental F1 components obtained by ion-exchange chromatography indicate that, although there is a considerable degree of homology between the subfractions[42], a number of differences in amino acid sequence do occur[42, 43]. Although these chromatographic studies have not detected significant tissue differences amongst the F1 components within a species, they have confirmed the species specificity of this fraction[44-46]. From such studies the extent of the F1 heterogeneity appears to depend on the physiological state of the animal[45, 47, 48], but only quantitative variations were found between F1 subfractions from tumour and normal tissues[49].

The partial sequence studies indicated above have been carried out on two purified subfractions of rabbit thymus F1 and on one component of calf thymus F1[42, 50]. The sequence of the 72-residue *N*-terminal peptide of the calf protein is shown in Table 10.2. These studies indicate that there is an asymmetrical distribution of amino acid residues, since the *C*-terminal peptide of some 100 residues in length is highly basic as it contains more than 65% of the total basic amino acids. This region also contains most of the prolyl residues indicating that it may also have an extended non-helical conformation. The *C*-terminal region would also appear to be invariant in structure, in contrast to the *N*-terminus which can vary from F1 subfraction to another and which may also be species specific[50-52]. The conservation of the structure of the *C*-terminal region may well be obligatory considering that, as it is the most basic half, it may well be the site of binding to DNA.

10.2.2.2 Histone F2b

This histone has a molecular weight of 14 000 and is rich in lysine, serine and alanine. Although pea bud histone F2b has been shown to consist of two distinct components[36], other species appear to have a single F2b fraction[34]. Complete and partial amino acid sequences have been carried out on this fraction from calf thymus[53, 54] and trout testes[55]. The comparative studies, though incomplete, indicate two regions (residues 9–10, 19–21) at which there are sequence variations between the two species. These results lend support to electrophoretic evidence[56] that the structure of histone F2b varies from species to species.

The amino acid sequence of calf histone F2b is given in Table 10.2 and shows an asymmetrical distribution of residues, such that 60% of the basic amino acids are found in the first 50% of the 125 residues of the molecule.

Since the majority of the acidic and hydrophobic residues are located in the remainder of the molecule, the *N*-terminal region would appear to be the most basic part of this histone.

Despite the apparent homogeneity of this fraction in many species, other studies have found multiple bands on electrophoresis of histone F2b in systems of high resolution[55]. This is probably due to the effects of postsynthetic modifications, since histone F2b has been found to be acetylated, phosphorylated and methylated *in vivo*[57-59].

10.2.2.3 Histone F2a2

A number of laboratories have now succeeded in isolating this leucine, alanine and lysine-rich protein (molecular weight 12 500) in purified form. The partial sequences of the calf thymus and trout testes proteins have been reported[55, 60] and recently the complete sequence of the calf histone has been obtained in Busch's laboratory[61]. The preliminary comparative studies of Candido and Dixon[55] suggest that there may be some degree of sequence variation between the trout and calf proteins, a conclusion already reached from electrophoretic studies[56].

The amino acid sequence of calf histone F2a2[61] is given in Table 10.2 and shows a clustering of basic residues between long hydrophobic regions. Thus the first 45 residues from the *N*-terminus constitute a very basic peptide since it contains 13 basic, but no acidic, amino acids. This is separated by a stretch of 28 amino acids of a slightly acidic nature from the next basic region (11 basic groups in 35 residues). A second stretch of acidic amino acids separates this region from the basic *C*-terminus.

Electrophoresis in systems of high resolution again shows that this apparently single species of histone can be separated into several components. As previously described for other histones these effects would appear to be associated with postsynthetic phosphorylation[42, 62] and acetylation[63] of the parent F2a2 molecule.

10.2.2.4 Histone F2a1

This glycine, arginine and lysine-rich histone is the smallest of the histones (molecular weight 11 000) and the calf thymus histone F2a1 was the first histone to be fully sequenced[64, 65]. In addition the sequences of histone F2a1 from pea seedling, pig thymus and rat hepatoma[66-68] have also been reported. The important feature of these studies is the finding that although the primary structure of the calf, pig and rat histones are identical, that of the more distant species, the pea, differs in only two conservative substitutions[66] (see Table 10.2). Thus in contrast to the other histones already described, the structure of histone F2a1 appears to have been rigidly conserved during evolution, perhaps because of some specific function it has to fulfil. Similar conclusions were also reached by Panyim *et al.*[56], on finding that the electrophoretic mobilities of histone F2a1 prepared from numerous species were similar.

The sequence of calf thymus histone F2a1 is given in Table 10.2 and shows that in this histone also there is an asymmetrical distribution of basic amino acids. Thus the majority of the basic residues are located in the first 45 positions from the *N*-terminus, with the rest of the protein containing most of the aliphatic, aromatic and hydrophobic residues. Hence it is highly probable that the *N*-terminal region is the DNA-binding site of the histone.

Electrophoresis has shown this fraction to be present in different chromatins either as a single species or as multiple components. As in the case of the other histones, these microheterogeneities would appear to be due to different degrees of acetylation and phosphorylation of the parent molecule[37, 64, 69, 70].

10.2.2.5 *Histone F3*

This alanine, arginine and glutamic acid-rich histone of molecular weight 14 000 is the only histone of mammals to contain cysteine. All species up to and including rodents possess one cysteine residue per molecule, whilst the F3 fraction of higher mammals contains two such residues[71, 72]. Although a number of workers[34, 36, 56, 71, 72] have reported that this histone contains only one species of protein, a recent study[73] has shown that calf thymus F3 also contains a minor (20 %) component which has only one cysteine residue per molecule and is thus similar to the single F3 component of lower species. The two forms of histone F3 are apparently present in varying amounts in other higher animals and moreover they are metabolically distinct[73].

The complete amino acid sequences of calf thymus[74] and chick erythrocyte[75] histone F3 have been determined, whilst partial sequences of the calf[76, 77] and trout testes[55] proteins have also been reported. As far as can be seen, the sequences of the calf and trout F3 histones are similar[55], but those of the chick and calf proteins differ in the region between residues 86 and 89 and in the calf the seryl residue at position 96 is replaced by cysteine[75].

The amino acid sequence of the calf histone F3 is given in Table 10.2. Here the *N*-terminal half of the histone would appear to be the more basic portion, whilst the *C*-terminal region contains most of the acidic, aliphatic and aromatic residues. An outstanding feature is the stretch of 29 acidic residues between positions 84 and 112 which is followed by a cluster of nine basic amino acids in the last 23 residues before the *C*-terminus. The *C*-terminal half also contains the two cysteinyl residues separated by only 13 residues.

The presence of free thiol groups can obviously lead to dimerisation and polymerisation of F3 molecules, a feature which has been reported by a number of workers[71-73]. Histone F3 can also be affected by postsynthetic modifications, since acetylated, methylated and phosphorylated residues have been detected[55, 63, 74, 75, 78, 79].

10.2.2.6 *Distribution and cell specificity of histones*

The characterisation of the histones of mammals such as calf and rodents as consisting of the five groups outlined above led to the investigation of the nature of the basic proteins associated with DNA in other vertebrates and in plants. The outstanding result of such comparisons is the marked similarity of the histone fractions in different tissues and species, a subject adequately reviewed by others, e.g. Johns[34]. As an illustration of this point Panyim *et al.*[56,80] showed by electrophoresis that the same five histone fractions were present in mammals, fish, amphibians, birds and reptiles. Secondly, it is interesting to note that plants, having evolved in parallel with the vertebrates, should have histones with characteristics very similar to those of mammals[36,66,81].

Other investigations have attempted to identify histone-like proteins in the chromatin of organisms more primitive than, and distant from, the vertebrates. This subject has also been reviewed in detail elsewhere[34,35]; the following are the important features. Bacteria do not appear to possess histones[82,83] but adenovirus contains DNA-associated proteins which are rich in arginine[84]. Of the algae, *Chlorella* has been reported to possess histones[85], but both blue–green[86] and dinoflagellate algae[87] contain little or no histones. Similar types of differences appear amongst the fungi. Thus *N. crassa* does not appear to contain histones[88], whilst among the yeasts, *S. pombi* has much acid-soluble but acidic protein associated with its DNA[89] and *S. cerevisiae* has been found to have a set of histones limited in types and function[90-92]. On the other hand the slime mould *P. polycephalum* has been found to contain a complement of histones very similar to those of calf thymus[93,94]. Histones have also been detected in certain protozoa, e.g. *Tetrahymena*[95], *Euplotes*[96] and *Euglena*[97] as well as in a number of arthropods, e.g. mealy bug and cricket[98] and *Drosophila* (see Oliver and Chalkley[99]). These investigations indicate a general similarity of the histones of these species with those of mammals, the major differences being due to variations in the F1 histone fraction. A similar situation has been reported with respect to the histone complement of a number of molluscs and echinoderms (see Cozcolluela and Subirana[100]; Subirana, *et al.*[101]).

Since most histone-containing organisms possess a similar group of DNA-associated basic proteins, there would appear to be little cell specificity of these proteins. Nevertheless, variations within a histone fraction do occur, suggesting some degree of cell specificity. Most of these differences have been detected in the F1 histone fraction which has already been described as being heterogeneous by virtue of both the number of parental polypeptide molecules and the effects of phosphorylation. Thus, although a new histone species has been detected in a number of cases, the true classification of such a histone as being cell specific is complicated, since it may only be a reflection of the heterogeneity of this particular group of proteins. For example, the components of the F1 histone fraction have been described as differing significantly in amino acid analyses and electrophoretic properties in numerous species (see Mohberg and Rusch[93]; Oliver and Chalkley[99]; Subirana *et al.*[101]).

The most outstanding of these cell specific F1-like components is the additional lysine, arginine, alanine and serine-rich histone found in the nucleated erythrocyte of a number of species (see Vendreley and Picaud[102]) which has been designated histone F2c[103] or FV[104]. The chick erythrocyte fraction has been found to consist of two components which structural analysis shows to be due simply to a genetically-determined amino acid substitution (glutamine for arginine) near the *N*-terminus[105,106]. Other workers have observed a certain degree of species difference of this particular fraction[107]. Other details of this cell-specific histone will be dealt with in the section dealing with histones during differentiation.

Other cell specific F1-like histone components have also been described; thus both Panyim and Chalkley[37,108] and Ozaki[109] observed the presence of such components specific to both dividing and non-dividing tissues. Very lysine-rich histones have also been found which are specific to meiotic plant cells[110] and trout tissues[111,112]. An additional lysine-rich component has also been reported in the silk worm[113] whilst Hamana and Iwai[114] found a small lysine and glutamic acid-rich histone in *Tetrahymena*. Qualitative differences were also reported in the F1 components of plant tissues of differing metabolic states[81] and during the cell cycle of a slime mould[94]. Some species, e.g. *Drosophila*, have only one F1 histone component[99], whilst others, e.g. yeast[90] and the erythrocytes of certain fish and amphibians[107], are totally lacking in this fraction.

Of the other histone fractions, the electrophoretic studies of Panyim *et al.*[56], indicate that there is little variation in fractions F2a1 and F3 but some differences would appear to be present in fractions F2b and F2a2 amongst different species. In this connection histone F2b from pea bud chromatin has been reported as consisting of two distinct components[36]. The histones of wheat germ, on the other hand, would appear to be fairly unique, since Johns and Butler[115] found that they did not contain any arginine-rich fractions but only a number of lysine-rich species.

In addition to the nucleated erythrocyte specific histone F2c other well-defined cell specificities concern the proteins associated with the DNA of sperm. The sperm of a considerable number, but not all, species contain low molecular weight arginine-rich protamines which replace the somatic type of histones (see a review by Bloch[116]). As in the case of histones the sperm proteins of fish are also heterogeneous, those from herring (clupeine), salmon (salmine), trout (iridine) and tuna (thynnin) having been separated into a number (2–4) of components (see Ando and Watanabe[117]; Bretzel [118]). The primary structures of a number of fish protamines have been determined (see a summary by Bretzel[118]) and as evidenced by the amino acid sequences of clupeine Z and thynnin Y2 given in Table 10.3 they all appear to have a similar structure. The arginine residues are distributed throughout the molecule in six or seven clusters of up to six residues each separated by one to three non-basic amino acids. The sequence of such 'spacer' regions would also appear to be similar throughout the protamine species examined so far. Protamines also have a high content of hydroxy amino acids which can be phosphorylated[119,120], a feature which may play an important part in their subsequent transport and binding to DNA (see Section 10.3.2.3). The clustering of the arginine residues at points along the entire polypeptide chain indicates that most of the protamine molecule could bind to DNA, but Iwai *et al.*[121] suggest that in the case of clupeine Z only the very basic terminal ends may perform this function.

In contrast to the fish protamines, the basic protein found in the sperm of a number of mammals is a somewhat larger single polypeptide rich in arginine and cysteine[122,123]. There would appear to be some degree of species

Table 10.3 Amino acid sequence of sperm basic proteins. Details are the same as those given in Table 10.2

Protein	Ref.	Sequence
		* * * * * * *
Clupeine Z	(Ref. 118)	R_4 S R_2 — A S R_1 P V R_4 P R_2 V S — — R_4 A — R_4
		* * * * * *
Thynnin Y2	(Ref. 118)	— P R_4 Q A S R_1 P V R_5 Y R_2 S T A A R_5 V V R_4
		* * * * * * * *
Bull sperm	(Ref. 123)	A R_1 Y R_1 C_2 L T H S G S R_1 C_1 R_7 C_1 R_6 F G R_6 V C_1 Y T V I R_1 C_1 T R_1 Q

specificity of this protein amongst the placental-bearing mammals[123] with evidence from electron microscope work that such proteins are absent from the sperm of the marsupials[124]. The amino acid sequence of the bull sperm protein is given in Table 10.3. Unlike the situation in the fish protamines, the majority of the arginine residues are located in three clusters of 6–7 residues, each near the centre of the molecule. Thus it would appear that the cysteine residues might be responsible for the cross-linking of the deoxyribonucleoprotein in the mature sperm[123].

Features distinctive of the properties of the F1 histones are present in other species whose sperm contain a histone-type of protein. For example, during spermiogenesis in the sea urchin a very lysine-rich fraction apparently replaces the somatic form of histone F1[109, 125, 126]. In addition, sea urchin sperm histones do not appear to be modified through postsynthetic phosphorylation or acetylation[127, 128]. This is an unusual property which is not directly associated with the lack of template activity as evidenced from similar studies on erythrocytes[56], but a similar situation has been found in the genetically inactive micronuclei of *Tetrahymena*[129].

10.2.2.7 Histones during embryogenesis, development and differentiation

Indications from early work showed that there was little major change in the electrophoretic pattern of histones during development in the chick[131-133], but in a similar study using chick histones separated by ion-exchange chromatography, Agrell and Christensson[133] found that the nature of these proteins varied during development, until an adult pattern emerged.

In the newt and certain sea urchins little or no histones have been found in the chromatin of the embryo till early gastrula or blastula stages[134-139] when the appearance of the arginine-rich histones precedes that of the lysine-rich fractions. In other species of sea urchin, all histone fractions are present during early embryogenesis, fraction F2b being particularly dominant. As the embryo matures, histone F2b decreases in amount whilst the F1 fraction increases[127, 140]. Johnson and Hnilica[138] interpreted their results as showing that histones were stored in the cytoplasm of early embryos of the sea urchin after their translation from maternal messengers and were only transferred to the chromatin at blastula. However, these suggestions differ considerably from the results of Kedes and Gross[141] in which they isolated histone messenger-like RNA from sea urchin embryos at similar stages. Development in the newt has been reported as being associated with the appearance of the various histone fractions in different regions of the embryo[142]. Later work[143] showed that the histones which appear at gastrula in the newt come from precursors stored in the cytoplasm, which themselves have been derived from messenger RNA molecules synthesised during the very early stages of development.

During the development of the pea cotyledon, the amount of histone F1 has been found to increase as maturation progresses[144], whilst during the life cycle of the slime mould *P. polycephalum* variations in the relative amounts of each histone fraction were observed[94]. The histone pattern of *Drosophila* larvae differs from that of the adult, mainly due to the F3 histone being acetylated in the former; also an additional basic protein is present in the larvae[145]. Although in calf thymus each histone fraction accounts for *ca.* 20% of the total[146], other workers have found considerable variation in the relative amounts of each histone fraction in other mature tissues[147-149].

Considerable interest has been shown recently in the possibility that the nucleated erythrocyte specific histone F2c may be associated with the process of erythrocyte differentiation which leads to the mature, but relatively inert, cell. In more specific terms, it is attractive to speculate that this histone could be responsible for the decreased template activity of the mature cell, especially since preliminary evidence points to its involvement with the condensation of the erythrocyte nucleus[150].

Initial experiments by Dick and Johns[151] showed that the F2c histone possibly replaced the F1 component in duck erythrocytes, but these workers could find little difference in the ratio of the two histone fractions in mature and erythroblast cells. Since one of the major problems in such studies is the purity of the cells, Sotirov and Johns[152] prepared purified cell populations and were then able to show that chick erythroblasts contained only 20–25% of the F2c content of mature erythrocytes. Other workers[149, 153, 154] were able to confirm these results. Recently a detailed study[155] of embryogenesis in the chick has shown that histone F2c does not appear in the embryo until a definitive erythroid cell line is established at 5–6 days, after which time its concentration increases, reaching 16% of the total histone after 17 days. The evidence then would point to histone F2c being associated with the initiation and process of erythrocyte differentiation but it is unknown at present whether its precise function is to cause heterochromatinisation or to inhibit transcription by some other method.

In the mature cell the ratio of total histone to DNA remains near unity (see Table 10.1) even during the cell cycle[94, 156]. Little or no quantitative or qualitative differences have been found in histones from heterochromatin and euchromatin[2, 157, 158], active (female) and inactive (male) chromatin of the mealy bug[159, 160], diploid and polyploid nuclei[161], normal and exponentially growing cells[162], during differentiation in the chick oviduct[163] or in polytene and non-polytene insect nuclei[164]. However, in *Tetrahymena*, Gorovsky[165] did find differences in the electrophoresis patterns of histones from macro- and micro-nuclei (the latter of which are genetically inactive) which appear to be related to a high degree of acetylation of histone F2a1 in the former and a complete lack of such modification in the latter[129].

10.2.2.8 Conclusions

Of the eukaryotic species so far examined the majority have been found to possess histone-like proteins bound to their DNA. Although it is feasible to conclude that these organisms containing approximately one picogram of DNA or more per cell require histones to be associated with their DNA (see MacGillivray *et al.*[35]), there are exceptions to this general proposal. Thus the dinoflagellate algae whose DNA content is greater than that of many mammals[166] contain little or no histones. Admittedly, these algae have permanently condensed chromosomes which have unusual methods of division[167], but on the other hand the chromosomes of *Euglena* have similar properties and they do possess histones[97].

The striking feature of the histones is their marked conservation both in type and structure during evolution. Thus the primary structures of histones F2a1 and F3 appear to have been maintained fairly rigidly with a lesser degree of constraint having been applied to the F2b and F2a2 fractions. Fraction F1 is the only histone type to show marked tissue and species heterogeneity, consisting of relatively few components in primitive organisms but showing marked complexity in mammalian cells. Other recent evidence from electrophoretic studies carried out in the presence of non-ionic detergents suggests that histone F2b and F2a2 have a heterogeneity similar to that of the F1 fraction[168, 169], but it is difficult to assess these results fully, since the basis of such separations would appear to be unknown at present.

The determination of the primary structure of most of the histones has thrown light on another important feature, namely the asymmetrical distribution of basic residues in the histone molecule. Thus the *N*-terminal region of histones F2b and F2a1, the *C*-terminus of histone F1 and both termini of histones F2a2 and F3 are predominantly basic and probably the binding sites for DNA. This means that a considerable portion of each histone molecule is free to form α-helical conformations and to interact with other proteins. It is also of interest to note that in cells where a high degree of condensation of the DNA is required, e.g. in certain sperm and in adenovirus, the extra basicity of arginine-rich proteins such as the protamines is utilised.

Although the five histone fractions each appear to be fairly unique proteins, a number of cross-homologues in amino acid sequence have been pointed out[32, 74], the outstanding example being the similarity of the *N*-terminal sequences of histones F2a1 and F2a2. In addition, histone F2a1 has several internal homologies[170], whilst histone F3 appears to have a duplication of a peptide near the *N*-terminus[74].

Thus at first sight the conservation of the histones throughout the species could suggest that these proteins have a specialised function which prohibits major changes in the primary structure. Since they are a somewhat simple group of proteins it is difficult to envisage them as acting as, for example, gene regulators. Hence their role may be a more passive one, perhaps associated with the maintenance of chromosomal structure. The post-synthetic modifications such as phosphorylation which have already been referred to could impart a more varied role and this will be discussed in Section 10.3.2.

10.2.3 Non-histone proteins

10.2.3.1 Introduction

As will be described later, considerable interest is now being shown in the non-histone proteins, since they would appear by some criteria to be involved in controlling DNA template function. Many workers describe them simply as those proteins remaining after the removal of histones from chromatin by dilute acid, but as has already been pointed out, methods are now available for their isolation as a discrete class of chromosomal proteins[35]. By amino acid analysis they are acidic and, in contrast to the histones, they also contain tryptophan (see MacGillivray *et al.*[35]). The ratio of non-histone protein to DNA also varies with the source of chromatin (Table 10.1) and attempts have been made to correlate this finding with the level of RNA synthesis in tissues. Thus the content of non-histone proteins of maturing sperm cells decreases as the template capacity of the chromatin DNA decreases[171], whilst, in contrast, during sea urchin embryo development the non-histone protein content of the chromatin rises as the template capacity of the chromatin increases[4, 30]. During oestrogen induced differentiation of the chick oviduct the amount of non-histone protein may be correlated with changes in the template activity of the chromatin[163]. Also the level of non-histone protein appears to be lower in mature than in immature chick cells[172], whilst in rat liver the level may be influenced by the action of carcinogens and hormones[173].

10.2.3.2 Isolation and characterisation

We have already described in detail elsewhere[35] the various procedures currently in use to isolate non-histone proteins from chromatin. In summary, these consist of the following approaches. Non-histone proteins have been separated from DNA by phenol[174, 175], detergent[176], or salt[177] after acid extraction of the chromatin to remove histones. Alternatively, chromatin has been dissociated in sodium dodecyl sulphate (SDS)[178] or in various salt solutions[29-31, 163, 179-185], so that after removal of the DNA by ultra-centrifugation[31, 179, 181, 182, 185], gel filtration[29, 30, 185] or precipitation with histones[180] or heavy metal ions[183], the non-histone proteins have been isolated as such[180] or subsequently separated from the accompanying histones by ion-exchange chromatography[29-31, 179, 181-183] or by electrophoresis[178,185]. Combinations of such techniques have been used by other workers[186-188] whilst the methodology of Wang[180] has been modified to isolate the phospho-protein components of the non-histone protein fraction[189-191]. In this laboratory, we have developed a single column procedure whereby salt–urea dissociated chromatin is chromatographed on hydroxylapatite, yielding a high recovery of non-histone proteins[28].

A major disadvantage of a number of these methods is the fact that they provide arbitrarily selected preparations of non-histone protein with no guarantee that the proteins are representative of the whole, or that the selection is constant when applied to different chromatins (see MacGillivray *et al.*[35]). Such criticisms can be applied to the salt dissociation procedures of Wang[180] and Benjamin and Gellhorn[177], the phenol procedure as used by a number of workers[174, 192-194] and the methods devised to isolate the phospho-protein species of non-histone proteins[189-191]. On the other hand, those procedures which initially cause the complete dissociation of chromatin in SDS[176,178] or high concentrations of salt–urea or guanidine hydrochloride[28, 30, 31, 185], followed by removal of the DNA and histones in bulk, usually give a high recovery of non-histone proteins representative of those present in the original chromatin. Other methods, such as removal of histones from chromatin by dilute acid, have also been criticised on grounds that such treatment can cause irreversible binding of histones to DNA[195] and radically affect the binding of the non-histone proteins to DNA, as seen from immunological studies[196].

10.2.3.3 Heterogeneity and sources of non-histone proteins

Since the non-histone proteins tend to aggregate with themselves and with histones and nucleic acids, electrophoresis of these proteins has proved difficult and even those investigators who have reported reasonable electrophoretic separations still find that considerable material will not enter the electrophoresis gel[29, 31, 182, 185, 197-199]. Hence many workers now completely dissociate their non-histone protein preparations in SDS, sometimes in combination with mercaptoethanol and urea, and apply the SDS-proteins to SDS-containing polyacrylamide gels (see MacGillivray *et al.*[28], Shelton and Neelin[175], Elgin and Bonner[176]) thus achieving a separation based on differences in the molecular weights of the denatured polypeptides[200]. The outcome of such studies has been the finding that the non-histone proteins are extremely heterogeneous since they consist of a far wider range of molecular weights than, say, the histones.

However, it has to be remembered that since these experiments have been carried out by electrophoresis in SDS, the separation is based only on differences in the molecular weights of SDS–protein complexes; differences in primary structure between peptides of the same molecular weight would not be detected. Hence in order to gain better resolution of these complex mixtures various investigators have combined ion-exchange chromatography and electrophoresis (in SDS and in other systems)[31, 181, 198, 201]. Such investigations have emphasised the limitations of the one-dimensional SDS gel system, but for purely analytical work, however, more satisfactory attempts have been made to 'fingerprint' the non-histone proteins by two-dimensional electrophoresis employing methods similar to those used to analyse ribosomal proteins[201, 202]. In our laboratory fully-reduced non-histone proteins have been separated by isoelectric focusing in polyacrylamide gels in the first dimension and then subjected to electrophoresis in SDS in the second dimension. In this way the full extent of the heterogeneity of the non-histone proteins has been demonstrated as consisting of proteins of a wide range of isoelectric points (pH 2–9) and of molecular weights, generally in excess of 40 000. Of the two non-histone protein fractions recoverable from the hydroxylapatite columns, that eluted in 0.05 M phosphate was found to be extremely heterogeneous whilst the more highly-retained material (eluted in 0.2 M phosphate) was found to be a relatively simple group of high molecular weight proteins whose isoelectric points were less than pH 6. In addition to a number of protein species whose isoelectric points were within the pH range 4–7, the non-histone proteins eluted from hydroxylapatite in 0.05 M phosphate were found to contain a number of relatively basic proteins with isoelectric points between pH 8 and 9. Some of these were recovered as material unretained by QAE-Sephadex at pH 8.3, and, by amino acid analysis, they appeared to be acidic, being rich in glutamic acid, serine and glycine[201]. The possibility exists, therefore, that their glutamic and aspartic acid residues may be amidated, thus conferring an overall basic charge on the proteins.

The evidence to date then is that the chromatin non-histone proteins are extremely complex and largely of high molecular weight. This could be explained by the presence of various enzymes and structural proteins being common to most chromatins. On the other hand, non-histone proteins are acidic, like many other cellular proteins, and this raises the question of their true origin. This aspect was first discussed by Johns and Forrester[203] who demonstrated that much acidic protein of calf thymus chromatin could be derived from the cytoplasm. Later experiments by Goodwin and Johns[204] showed that much of the chromatin protein isolated as Wang's 'chromosomal acidic protein'[180] could be accounted for in this way. However, Wilhelm *et al.*[205] were not able to confirm Johns' results; the anomaly probably lies in the fact that Johns prepares chromatin simply by repeated extraction of whole tissue with saline, whereas Wilhelm *et al.*, conducted their experiments using chromatin prepared from purified nuclei. Recently, Harlow *et al.*[206] have concluded that much of the non-histone proteins of chick erythrocyte chromatin are contaminants, since they were able to find a correlation between the amount of non-histone protein and the membrane content of the chromatin samples. Doubt about the origin of these proteins caused Hill *et al.*[30] and Bhorjee and Pederson[207] to estimate the contamination of their preparations by cytoplasmic proteins to be not greater than 30% (sea urchin) and 1.9 μg per 100 μg HeLa cell chromatin respectively.

Other studies indicate that part of the non-histone protein complement is loosely bound to chromatin. Thus Gronow[208, 209] has found that some 70% of the total protein of rat liver nuclei can be extracted with 8 M urea at pH 7.6, thus indicating that *ca.* 60% of the non-histone proteins of the chromatin have been lost in this way. Studies involving the digestion of chromatin with nucleases also indicate a loose attachment of some non-histone proteins to chromatin[210, 211]. In this laboratory we have also been interested in the origin of these chromatin proteins. By removing cytoplasmic remnants together with the outer and possibly also the inner nuclear membrane[212] using Triton X-100 we have been able to show that the non-histone proteins we isolate from mouse liver are in fact nuclear in origin[28]. In addition, we also found that much of the high molecular weight non-histone proteins could be removed by treatment of chromatin with various salt solutions or by preparation of nuclei using detergents only. Apart from the histones, such nucleoproteins contain small amounts of low molecular weight acidic proteins. Similar proteins, tightly bound to the DNA were also found by Gronow and Griffith[209] after extraction of nuclei with urea and by Patel and Thomas[213] in the nucleohistone which precipitates during the solubilisation of chromosomal acidic proteins. It would appear that much of the high molecular weight non-histone proteins can be removed from chromatin with relative ease, leaving species of non-histone protein of about the same size as histones, firmly bound to the DNA. Hence the possibility exists at present that some part of the non-histone proteins is derived from other parts of the cell, but the situation is complex since it is known from hormone localisation[214] and cell activation studies *in vivo*[215] that some proteins can migrate from the cytoplasm to the nucleus.

10.2.3.4 Distribution and cell specificity of non-histone proteins

In reviewing the distribution of non-histone proteins in different tissues and species it has to be appreciated that our state of knowledge of the proteins is perhaps similar to that of the histones a decade or so ago. Information has been obtained using a variety of methods to isolate both chromatin and proteins. As described above the procedures available for the isolation of non-histone proteins from chromatin belong to two types, those which yield a partial selection and those which give preparations representative of the total complement of proteins.

In the first category, Kostraba and Wang[197] reported tissue variations and differences in the electrophoretic patterns of 'chromosomal acidic' and 'residual' proteins of tumour and normal cells, confirming findings by others that there were tissue differences in the chromosomal acidic protein fraction[216-220]. Such studies indicate the complexity of these preparations of non-histone proteins, but the methods of preparation and electrophoresis may have led to aggregation effects making interpretation difficult. Benjamin and Gellhorn[177] prepared a fraction of non-histone proteins from dehistoned chromatin by extraction with CsCl at pH 11.6, but could find little difference between rat and mouse proteins after electrophoresis at the same pH. Teng *et al.*[174] however, found distinct tissue differences in both SDS electrophoresis and ^{32}P-labelling patterns of the proteins solubilised from dehistoned chromatin by phenol at pH 8.6. Other workers have also reported tissue differences amongst the phosphoproteins of the non-histone fraction. On a quantitative scale Kleinsmith and Allfrey[190] observed that rat chromatin contains some three times more phosphoprotein than calf thymus chromatin, whilst Platz *et al.*[221] and Wang[199] both reported qualitative differences in preparations of phosphoproteins from different tissues. However, protein aggregation effects may have affected the material in both of these investigations and in the case of the ^{32}P-labelling experiments of Platz *et al.*[221], it should be pointed out that the conditions used *in vitro* may have resulted in all available serine and threonine residues being phosphorylated, a situation which probably does not occur in the intact chromatin *in vivo*.

In contrast the methods which appear to give total extracts of non-histone proteins from chromatin have shown little cell specificity of these proteins. Thus Elgin and Bonner[176] solubilised non-histone proteins in SDS after removal of the histones from chromatin with dilute acid and as judged by electrophoresis in SDS subsequently found the proteins to be of a molecular weight range of 5000 – 100 000. However, they could find few tissue differences amongst the non-histone proteins, although the proteins from pea bud chromatin were dissimilar to those of mammalian cells. A similar lack of cell specificity of the non-histone proteins of pig tissues was reported by Shaw and Huang[185] who used polyacrylamide gel electrophoresis at pH 2.7. Shelton and Neelin[175] separated non-histone proteins into four fractions, by extraction of dehistoned chick chromatin with phenol and by gel electrophoresis in SDS they found a total of some 40 components in chick liver with erythroid cells possessing a somewhat smaller complement. They were able to show some quantitative and qualitative tissue differences but the specific variations between mature and immature erythroid cells were confined to minor components. Richter and Sekeris[181] also reported some degree of cell specificity of rat non-histone proteins after their fractionation on QAE–Sephadex followed by electrophoresis in SDS and at pH 11.6.

In this laboratory we have examined non-histone protein preparations from various tissues after their isolation by chromatography of salt–urea dissociated chromatins on hydroxylapatite[28]. As judged by electrophoresis in SDS only quantitative differences were found in these proteins from different cow and mouse tissues. The exceptions were the proteins obtained from duck erythrocyte and calf thymus chromatins which both appeared to consist predominantly of low molecular weight proteins. We did find, however, that the outcome of such tissue comparisons could be affected by the methods used to prepare nuclei. Thus our evidence indicated that nuclei made by 'sucrose' procedures contained somewhat less high molecular weight proteins that those made by the 'citric acid' method, whilst nuclei prepared in detergents alone were relatively devoid of such large species of protein. More recently we have used our methodology to examine the rapidly-labelled phosphoproteins of the non-histone fraction[222]. By incubating isolated nuclei with [γ-^{32}P]-ATP we found that the major phosphorylation product was protein and, moreover, by chromatography of the resulting chromatins on hydroxylapatite we obtained three fractions of ^{32}P-labelled protein. These were the histones, which were only lightly labelled, the non-histone proteins eluted from the column in 0.05 M phosphate and the remaining acidic proteins recovered from the column in 0.2 M phosphate, the ratio of specific activities c/min mg^{-1} of these fractions being 1:70:130. By electrophoresis in SDS many of these phosphoproteins of the non-histone species were found to be common to several mouse tissues, tissue specific proteins being relatively few. The phosphoproteins of the Landschutz ascites chromatin, however, differed markedly from those of other mouse tissues. We have now used the two-dimensional electrophoresis technique described already to examine the nature of these phosphoproteins further[201]. Such experiments have again shown the similarity of the non-histone proteins and their phosphoprotein complement amongst various mouse tissues. One dominant feature of these phosphorylation experiments was the presence in all of the non-histone protein preparations examined of a highly-labelled low molecular weight phosphoprotein, eluted from hydroxylapatite in 0.05 M phosphate[222]. Additional work has now shown that this component is also present in the saline-soluble proteins of the nucleus, indicating that this protein could belong to the proteins already described as being loosely bound to chromatin[201].

In contrast to these results Wilhelm *et al.*[158] reported tissue specificity of the SDS electrophoresis patterns of non-histone proteins obtained from rat tissues. However, certain details of the procedures used by these workers differ considerably from those of others. For example, in attempting to remove contaminating 'cytoplasmic proteins' and histones from their chromatins, Wilhelm *et al.*, also extracted variable amounts of non-histone protein at each stage of washing. In addition, analyses of the resulting chromatins indicated that a considerable amount of nucleolar protein was still present.

In summary, the considerable variations in procedures between laboratories makes full assessment of results difficult. Those methods which yield total extracts of non-histone proteins appear in general to show little tissue specificity of these proteins, whilst procedures which give only selected fractions of these proteins often show marked tissue variations. However, the disadvantage of the latter type of approach is that different fractions of proteins may be obtained from different chromatins.

10.2.3.5 Non-histone proteins during development, the cell cycle and after hormone treatment

During development in the sea urchin, varying degrees of quantitative and qualitative changes in the non-histone protein complement of the embryo chromatin have been reported between blastomere and hatching blastula stages[223, 224], but the transition from blastula to pluteus would appear to be accompanied only by an increase in the amount of all non-histone species[30]. During differentiation in the slime mould *P. polycephalum* a number of newly-synthesised high molecular weight proteins have been reported to appear in a fraction on non-histone protein soluble in phenol, whilst two low molecular weight polypeptides disappeared[193, 225].

As will be described later, histones are synthesised largely along with DNA during S phase, whereas the non-histone proteins can be synthesised throughout the entire cell cycle. In HeLa cells the greatest accumulation of non-histone protein occurs in the chromatin in late G prior to the start of DNA synthesis. Electrophoresis in SDS gels showed that a number of indidividual non-histone chromatin proteins were synthesised specifically at different stages of the cycle[226, 227]. In similar investigations of the HeLa cell cycle, Bhorjee and Pederson[207] found quantitative changes in only four out of 22 non-histone protein components separated on SDS-gels.

In contrast, neither LeStourgeon and Rusch[193] nor Becker and Stanners[228] found any variation in the pattern of non-histone proteins synthesised during the cell cycles of a slime mould and hamster fibroblasts, respectively. However, differences between stationary and proliferating cells have been reported, since the chromatin of growing cells appears to contain additional species of non-histone protein[228, 229].

A number of workers have observed changes in the non-histone protein complement of target organs after treatment with hormones. Thus Teng and Hamilton[230] found that oestrogen caused the appearance of new non-histone protein in uterine tissue, whilst Shelton and Allfrey[192] reported that cortisol caused the appearance of a newly synthesised non-histone protein of molecular weight 41 000 in liver chromatin. A similar situation was found in the polytene chromosomes of *Drosophila* during specific band changes brought about by the hormone ecdysone or by temperature shock, except that the additional protein did not result from *de novo* synthesis[188, 231].

10.2.3.6 Non-histone proteins of heterochromatin and euchromatin

The conclusion that there was little cell specificity of the non-histone proteins, as judged by criteria described above, has led to a number of comparisons of these proteins obtained from different chromatin fractions, in the hope that the active euchromatin would show distinct differences from the inactive heterochromatin. In fact the original experiments of Frenster[2] indicated that there was twice the amount of non-histone protein in euchromatin as in heterochromatin and that the former also contained some four times as much phosphoprotein. However, recent investigations have shown little or no significant differences between the SDS gel patterns of non-histone proteins of these two chromatin fractions[158, 232, 233]. On the other hand, Reeck *et al.*[234] prepared fractions similar to heterochromatin and euchromatin by chromatography of sonicated rabbit liver chromatin on ECTHAM–cellulose. In this case the euchromatin-like material was associated with a decrease in the amount of histone F1 and an increase in the amount of one high molecular weight non-histone protein.

10.2.3.7 Conclusions

Although procedural variations make direct comparisons difficult, the non-histone proteins appear to be extremely heterogeneous in terms of both size and charge distribution. Another factor to be considered is the evidence which suggest that some of these proteins shuttle between the nucleus and other parts of the cell. As yet few major differences have been found in the total non-histone complement of different cells or between active and inactive chromatin fractions. Since other evidence (to be discussed later) points to these proteins being involved in controlling chromatin template activity, this presumably means that our analytical techniques are not yet sufficiently sensitive to detect tissue differences.

From the data available there would appear to be a considerable range of chromatin proteins, stretching from the very basic histones through the modified slightly basic non-histone proteins to a group of relatively acidic non-histone proteins. It has to be remembered in this context that the distinction between histones and non-histone proteins was made originally on the basis that the former could be extracted from chromatin with dilute acid or strong salt. Such a separation may merely reflect differences in binding to DNA, since once isolated, both types of protein are soluble in these solvents. Certain low molecular weight non-histone proteins do appear to bind tightly to DNA and hence some of these species of protein may have properties similar to those of histones, e.g. structural roles.

10.2.4 Chromosomal RNA

The analyses give in Table 10.1 show that most chromatins contain a low content of RNA. This low value is generally accepted as indicating that much of the nucleolar material has been extracted by saline–Tris washes during the preparation of chromatin from nuclei. The remaining RNA must, therefore, be bound tightly to the chromatin and could represent either nascent messenger RNA adhering to transcription sites on the DNA or RNA closely associated with chromosomal protein. Huang and Bonner[235] and then Benjamin *et al.*[236] reported the finding of covalently linked RNA-histone complexes in pea chromatin and rat liver respectively but others were unable to confirm these observations[177, 237]. Later both Bonner[238, 239] and Huang[240] presented evidence for the covalent linking of a species of RNA to non-histone chromosomal protein. Other investigations (see Huang and Huang[241] and Bonner[238]) confirmed that histone–RNA complexes did exist in several tissues but only through hydrogen bonding of the non-histone protein-bound RNA to the basic proteins.

Bonner has termed this RNA 'chromosomal RNA' and his group has devoted much attention to its chemical and biological properties (see review by Holmes *et al.*[242]). This RNA has been reported as being of short chain length (40–60 nucleotides depending on the species[235, 243]) and containing a high amount (7–10 mol %) of dihydropyrimidines (dihydrouridylic acid in pea, cow and chick, dihydroribothymidylic acid in rat[239, 243, 244]). According to Huang[240] and Jacobson and Bonner[239] chromosomal RNA is linked to protein through an amide bond formed with the dihydropyrimidine nucleotides. In ascites tumour chromatin, the protein has been identified as being a single acidic polypeptide[239]. Other evidence points to the organ specificity of chromosomal RNA[245], its sequence heterogeneity within a tissue[239] and its half-life of 17 h in ascites cells indicating that it is a discrete class of rat RNA[245]. Despite such complexity all chromosomal RNA molecules apparently possess cytosine and guanine at their 5′ and 3′ ends respectively[239].

Much attention has been drawn to the biological role of chromosomal RNA as a result of attempts to reconstitute chromatin after dissociation of its components in salt–urea solutions. Both Bonner[246] and the Huangs[241] interpreted their results as indicating that the tissue specific template properties of chromatin were associated not primarily with non-histone protein but with the chromosomal RNA covalently bound to it. Following upon this it was found that, in hybridisation experiments, *in vitro* chromosomal RNA could anneal to at least 2% of denatured homologous DNA (see Holmes *et al.*[242]) and other investigations provided the interesting, if not curious, result that chromosomal RNA could also bind specifically to native homologous DNA[247]. The fraction of DNA to which chromosomal RNA hydridises has been reported to be the middle repetitive sequences[242, 248]. In chromosomes this RNA is bound to DNA in such a way as to render the complex resistant to ribonuclease[249]. Mayfield and Bonner[250] have reported that shortly after hepatectomy the production of rapidly labelled high molecular weight RNA preceded the appearance of new specific sequences of chromosomal RNA; these events are followed by a large increase in chromatin template activity.

However, despite this apparent wealth of knowledge concerning chromosomal RNA, other workers have disputed the significance and even the existence of this material. Thus Heyden and Zachau[251] found that in their experiments chromosomal RNA was an artefact derived from the degradation of transfer RNA by contaminating nucleases present in the pronase. (The original method of preparing chromosomal RNA involved pronasing the protein pellicle obtained by ultracentrifugation of CsCl dissociated chromatin). Artman and Roth[252] reported that purification of chick embryo chromatin caused the loss of chromosomal-like RNA. Other workers[253] could find little evidence in chromatin for the type of DNA–RNA hybrids suggested by Bonner's findings[242] and later showed that not only was much of the chromosomal-like RNA of rat liver not bound to protein but also it consisted of a spectrum of components with different labelling characteristics[254]. Hill *et al.*[30] prepared chromatin from sea urchin embryos which possessed conventional analyses and template properties but in which the RNA did not appear to contain dihydropyrimidines.

In a recent attempt[242] to answer some of their critics, Bonner's group have redefined the properties of chromosomal RNA, presenting evidence that it is a distinct class of nuclear RNA not related to ribosomal or transfer RNA. They also suggest that some of their critics may have been handling species of RNA other than chromosomal RNA and describe alternative improved procedures for its isolation. Hence according to Bonner's group[242] the existence of chromosomal RNA in chromatin is beyond doubt. The biological role of this RNA seems less clear, since as will be described below, much evidence now points to non-histone protein, rather than RNA, being involved in controlling the tissue specific properties of chromatin. On the other hand, another independent group of workers[255, 256] have reported the presence of a low molecular weight RNA in chromatin which stimulates the transcription of chick liver chromatin by bacterial polymerase. Although these studies differ in detail and effect from those of Bonner[246] and Huang[241] the possibility that some species of RNA may regulate transcription cannot be entirely dismissed at this time.

10.3 METABOLIC PROPERTIES OF CHROMATIN PROTEINS

10.3.1 Synthesis and turnover

10.3.1.1 Introduction

It would be reasonable to expect that if a chromatin protein fraction possessed gene-regulatory properties, then its rate of turnover should be distinct from that of other types of chromosomal proteins. Thus proteins which show some degree of stability and which have a turnover rate perhaps similar to that of DNA might be expected to be structural elements of chromatin or permanent repressors. A considerable amount of work has been carried out on the biosynthesis and turnover of chromatin proteins and this will be summarised below.

10.3.1.2 Histones

There is now considerable evidence to indicate that, like other cell proteins, histones are synthesised on cytoplasmic ribosomes. Firstly, histone-like polypeptides were observed on functional ribosomes[257-259] and a class of slowly sedimenting polysomes, which were probably the site of histone synthesis, were detected in sea urchin embyros[260, 261]. Secondly, a 7–9S RNA has been found to be synthesised in the nucleus prior to the beginning of DNA and histone synthesis in both HeLa cells[257, 258, 262] and sea urchin embryos[141]. More recently RNA of this type has been identified as the messenger for histones since it has been found to code for such proteins in cell-free systems[263, 264]. In sea urchins these histone messengers appear to be transcribed from repetitious sequences of DNA[265] and unlike the other mammalian messengers studied so far, they do not contain polyA sequences[266]. Hence histones would appear to be synthesised by conventional mechanisms, except that the absence of polyA sequences suggests that the original transcript may be processed in the nucleus in a special manner.

Despite the fact that there is little translational control of histone synthesis during the cell cycle[267] these proteins are synthesised largely near the time of DNA synthesis. The exact timings of these events are controversial points. Thus Butler and Cohn[268], Orlova and Rodionov[269] and Gutierrez-Cernosek and Hnilica[270] reported that histones were synthesised prior to DNA, but others[156, 271-274] have claimed concomitant synthesis of both types of molecules during S phase. In contrast, Cross[275] found that in mast cells, histone F1 was synthesised in early S phase, the other histone fractions being produced later in this period. Other workers have found evidence for histone synthesis occurring at other periods of the cell cycle[156, 276-278].

The rate of biosynthesis appears to vary from histone to histone. Thus Hnilica *et al.*[279] found differential rates of labelling of histones with [^{14}C]-lysine in normal and regenerating liver, but not in hepatoma. Similar results were found in a number of tissues[280], including liver[270], HeLa[281] and Chinese Hamster ovary cells[282]. On the other hand, Dick and Johns[283] found similar rates of synthesis of all histone fractions in rat thymus.

A number of groups have investigated the turnover of histones in relation to that of DNA. For example, Byvoet[284], Hancock[285] and Chalkley's group[286] showed that histones and DNA had similar turnover rates in rat tissues, mouse mastocytes and cultured hepatoma cells respectively. Induction and repression of enzymes had no effect on such turnover properties of histones in liver[287]. In developing chick brain, however, Bondy[288] found that all histone fractions turned over at a similar rate, which was faster than that of DNA. In the mature chick erythrocyte only the cell specific F2c histone turns over[289], whereas in continuously dividing hamster ovary cells only histone F1 has been found to turnover, the other fractions being stable for several cell cycles. Since part of the histone F1 of these cells appeared to be synthesised at least an hour before its attachment to DNA, the existence of non-chromatin pools of histones was postulated[282, 290].

It could be concluded from such varying results that the histones of different tissues have different degrees of turnover and rates of biosynthesis. However, other factors, such as contamination of histone fractions with highly labelled non-histone proteins[291] and the lack of true synchrony in cultured cells[285] could also explain such variations.

10.3.1.3 Non-histone proteins

Work by Patel and Wang[292, 293] and more recently by Fleischer-Lambropoulos and Reinsch[294] indicated that non-histone proteins could be synthesised within the nucleus itself. However, Stein and Baserga[295] found that in [^{14}C]-leucine labelling experiments there was a transfer of labelled non-histone proteins from the cytoplasm to the nucleus of HeLa cells, indicating a cytoplasmic site of synthesis of these proteins.

In stationary cells the rate of synthesis of non-histone proteins is sensitive to various treatments, independent of DNA and histone synthesis, e.g. by cortisol in liver[192, 296], by oestrogens in the uterus[297-299], by phenobarbitone in liver[300], by isoproterenol in mouse salivary glands[274] and after hepatectomy[301]. However, synthesis of non-histone proteins at the same time as that of DNA has been reported in foetal liver cells after treatment with erythropoietin[302] and in explanted mammary cells[303]. Baserga's group have studied the stimulation of growth of density-inhibited fibroblasts induced by a change of medium. They found that synthesis of non-histone proteins started to increase shortly after the medium change and continued for some hours reaching a maximum at the same time as that of DNA synthesis[194, 304]. Only a limited number of non-histone proteins were involved in this effect[194]. Infection of cells with SV40 virus also causes an early increase in non-histone protein synthesis, although at the same time there is a reduction in the rate of cytoplasmic protein synthesis[305]. Inhibition of DNA synthesis has been reported to follow inhibition of non-histone protein synthesis in salivary glands and fibroblasts[274], a situation which has been interpreted as showing that in fibroblasts the early synthesis of non-histone protein, which is sensitive to actinomycin, activates genes which are responsible for the control of DNA synthesis[304]. Two fractions of non-histone proteins have been studied in rat ovaries after treatment with gonadotrophin. The synthesis of the more loosely bound fraction started soon after hormone treatment and was maintained at a high level until a second increase occurred along with a period of DNA synthesis. On the other hand, synthesis of a protein fraction tightly bound to DNA occurred only at the same time as DNA synthesis[278].

In a number of continuously dividing cell systems, non-histone proteins have been found to be synthesised throughout the whole of the cell cycle[226, 275, 306-309] at a rate which is similar to or higher than that of histones (similar effects have also been reported in a number of solid tissues[279, 288, 310]). Synthesis of these proteins also occurs during mitosis[307], more labelled non-histone proteins having been found in mitotic compared with interphase cells[285]. However, the relationship between the period of maximal synthesis of these proteins and that of DNA varies. Halliburton and Mueller[308] found that in HeLa cells the synthesis of non-histone proteins came after DNA synthesis, but Ingles[309] observed concomitant synthesis of DNA, histones and non-histone proteins in ascites cells. In contrast to these results, Stein and Borun[226], also using HeLa cells, found that non-histone proteins were synthesised and accumulated in chromatin late in G1 phase before DNA synthesis began. Unlike histones the synthesis of these HeLa cell proteins was insensitive to inhibitors of DNA synthesis and other experiments showed that maximal and minimal turnover of the non-histone proteins occurred during mitosis and S phase respectively. In addition, individual non-histone proteins appeared to have different turnover kinetics at different times of the cell cycle[227]. A reduction in the rate of non-histone protein synthesis during S phase has also been reported in hamster ovary cells[306] and in mast cells[275].

During the cell cycle of the slime mould *P.polycephalum*, a group of phenol-soluble non-histone proteins were found to be synthesised after DNA and histones[193] and during development in the sea urchin a number of non-histone proteins are synthesised at a constant rate, whilst the synthesis of histones and DNA gradually diminish[311]. However, in the latter case, the phenol-insoluble non-histone proteins were exceptional in that their specific activities increased rapidly during development[311].

Halliburton and Mueller[308] found that in HeLa cells, the non-histone proteins turned over more rapidly than the histones but in ascites cells Ingles[309] demonstrated that they had a turnover rate similar to that of the histones. Bondy[288] found that, in contrast to the uniform decay rate of histones, the non-histone proteins of developing chick brain decayed in a heterogeneous manner.

Many of these studies indicate that non-histone proteins have metabolic properties quite distinct from those of the histones, but a number of discrepancies and differences are found amongst the reported results. Since numerous methods are presently in use for the preparation of chromatin and its proteins, it is possible that such differences are due to different workers handling different species of non-histone proteins.

10.3.1.4 Conclusions

On the basis of turnover rates, histones would not be expected to fulfil a regulatory role since they decay at about the same rate as DNA and during the cell cycle their actual synthesis is closely tied to that of DNA. These facts would suggest that the histones are permanently associated with the DNA and may have a more structural role. It would appear, on the other hand, that the non-histone proteins are synthesised and turnover in a heterogeneous manner and at all times of the cell cycle including mitosis. In addition they appear to be preferentially synthesised during the early stages of stimulation of cells by hormones and growth responses. Such properties would be compatible with the non-histone proteins possessing a role in the regulation of gene activity.

10.3.2 Postsynthetic modifications

10.3.2.1 Introduction

In describing the chemical nature of both histone and non-histone proteins, we have already mentioned that these proteins can undergo post-translational modifications at a number of sites. Thus acetylated, phosphorylated and methylated amino acids have been detected in chromosomal proteins. When it was found that some of these groups (e.g. phosphate[312]) were metabolically unstable, considerable interest developed in these modifications, since it was possible that they could impart some degree of specificity upon proteins which appeared similar not only in whole chromatin but also in heterochromatin and euchromatin fractions. Hence if such modifications were selective, proteins which otherwise appeared ubiquitous could possess some degree of cell specificity. Moreover, since these modifications involved altering or introducing charged groups (e.g. positive for methylation, negative for phosphorylation) into a protein molecule, this could change the charge of a protein, either uniformly or at localised sites. For example in this way, the ionic bonds between histones and DNA could be weakened, thus allowing a previously repressed portion of DNA to be made available for transcription. Alternatively, the attraction between protein and DNA could be enhanced by the addition of extra basic groups leading perhaps to a more permanent repression of the DNA template.

The next sections will deal with investigations designed to test such hypotheses. As well as acetylation, phosphorylation and methylation, the effects of reversible oxidation of thiol groups will also be described.

10.3.2.2 Acetylation of histones

Histones contain two forms of acetyl group, one which is stable and the other which is not. The former concerns the acetyls of the *N*-terminal serines of histone F1 and F2a1, 2 (Table 10.2) which are incorporated at synthesis[313, 314]. The latter type of acetyl group, which is unstable, is added enzymatically to the amino groups of histones F2a1, 2 and F3[315-317] and possibly also to *O*-seryl groups of histone F3[317, 318] after they have been synthesised. Histone F2b has been reported by some to contain no acetyl groups[316, 319] but since others have found incorporation of labelled acetate into this histone, there would appear to be a species difference in this effect[320]. Little is known concerning the acetylation of the non-histone proteins. Some studies indicate that this modification process does occur in these proteins, but the biological effect may not be so important as that of the histones[321].

The donor of these histone acetyl groups has been found to be acetyl CoA[322]; the reaction is carried out by transferase enzymes, a number of which have been identified[323-327] and which appear to be tissue specific[328]. Turnover of these acetyl groups is accomplished by deacetylase enzymes, a number of which have also been identified[329, 330], while calf thymus appears to contain at least two species of deacetylase[331]. In contrast to these enzyme systems, non-enzymatic acetylation of histones *in vitro* by acetyl CoA itself has also been reported[332].

Early studies by Mirsky[333] showed that chemically acetylated histone F3 was less efficient in inhibiting the transcription of DNA *in vitro* than the unmodified protein. This result suggested that acetylation of histones might well play a role in the control of chromatin template activity *in vivo* and many investigations have now been carried out to test this association. One of the first biological systems to be studied was the early effect of the mitogen phytohaemagglutinin in stimulating RNA synthesis in small lymphocytes. Mirsky's group[334] reported that this response was preceded by an increased acetylation of the arginine-rich histones. Later, other workers[335, 336] were able to confirm this apparent association, whilst other investigations suggested that the effect may be an artefact of the lymphocyte system. Thus hydrocortisone which inhibits the response of the lymphocyte to phytohaemagglutinin, did not affect the incorporation of labelled acetate into histones[337]. In addition, preparations of phytohaemagglutinin, devoid of mitogenic activity, were found to cause histone acetylation, whilst purified forms of the mitogen which still retained their stimulatory activity were found to have lost the ability to cause histone acetylation[338]. The interpretation of such findings suggested that phytohaemagglutinin, a biologically-complex mixture of proteins[339], also altered the membrane permeability of the lymphocyte and thus enhanced the uptake of labelled acetate from the medium into the acetate pool of the cell and thence into histone. Support for such explanations came from Pogo[340] who confirmed the increased permeability of the phytohaemagglutinin treated cell and showed the increased specific activity of the acetate pool in such cells.

More clear cut associations between histone acetylation and RNA synthesis have been reported in a number of other systems. In the rat, partial hepatectomy is followed by an increase in histone acetylation at the same time as the synthesis of DNA-like RNA and the stimulation of RNA polymerase activity[317]. In the sea urchin embryo histone F2a1 is acetylated at various sites during a period of intense RNA synthesis, while in the inactive sperm cell this histone is unmodified[128]. The genetically-active chromosomes of the female mealy bug are acetylated some seven times more than the inactive male chromosomes thus suggesting a direct link between gene activity and acetylation[341]. In addition, an early response of the immature rat uterus to oestrogens, appears to be the increased acetylation (140% after 10 min) of histone F2a1[321]. Another hormone, erythropoietin, has been found to cause increased acetylation of mouse spleen histones[342]. In contrast to its apparent effects on lymphocytes, phytohaemagglutinin has been reported as inhibiting RNA synthesis in granulocytes with a simultaneous deacetylation of histones[343].

In a number of systems, however, there would appear to be little or no association between histone acetylation and RNA synthesis. Thus, the RNA synthesis which follows puff formation in dipteran polytene chromosomes is not accompanied by the increased acetylation of histones[344-346], although Allfrey[347] has reported a correlation between these two events. Of the two periods of RNA synthesis found following thyroxine-induced metamorphosis in amphibia, one was followed by, and another coincided with, histone acetylation[348]. Different growth conditions have also been found to cause histone acetylation in HeLa cells, but this response could not be linked directly with changes in RNA synthesis[349].

Attempts have also been made to determine changes in histone acetylation during the cell cycle. Thus in hamster ovary cells, it was found that the acetylation of histones F2a1 2, F2b and F3 rose to a maximum at the same time as DNA synthesis and then decreased, leaving the arginine-rich histones partially acetylated and histones F1 and F2b completely deacetylated during the rest of the cycle[57]. However, in further experiments these authors found little turnover of histone acetyl groups in these cells and concluded that their previous observations were due to the acetylation in early S phase of portions of each histone fraction which were then diluted by non-acetylated histones synthesised during the rest of this phase[350]. The reported lack of turnover of histone acetyl groups in these cells is somewhat surprising in view of the evidence presented above for distinct variations in histone acetylation in some cells. In addition, the studies of Byvoet[351] have shown differences in the turnover rate of histone acetyl groups in a number of systems. Shepherd *et al.*[350] suggest that, since their ovary cells are dedifferentiated, histone acetyl group turnover is a function only of differentiated cells.

Thus, although the structural studies referred to in Sections 10.2.1–10.2.4 indicated the existence of acetylated amino acids in histones, the association between this modification process and gene activity is complicated, since a temporal correlation occurs in only some cells. In addition, there is some evidence for histone acetylation near the time of DNA synthesis in the S phase of the cell cycle. This is at a time when modifications to histone structure may be required for transport purposes or the correct binding to DNA (see Dixon[352]). However, there is little indication from the studies so far reported if the acetylation occurs on 'old' pre-existing histones or only on newly-synthesised proteins. Further consideration will be given to this point in Section 10.2.2.

10.3.2.3 Phosphorylation of chromatin proteins

(a) *Phosphokinases*—Phosphorylation of chromatin proteins takes place at serine and to some extent at threonine residues (see Kleinsmith, *et al.*[312]; Benjamin and Gellhorn[177]; Balhorn *et al.*[41]) and very often the incorporated groups are unstable since they have a high turnover (see Kleinsmith *et al.*[312, 353]). Both histones and non-histone proteins are phosphorylated, the latter being the major phosphoproteins of chromatin as shown by labelling studies[222, 354-357]. These phosphorylation processes are energy dependent, the donor of the phosphate group being the γ-phosphate group of ATP.

A number of kinases have been studied which can utilise histones or protamines as substrates[358-360] and a phosphatase with a similar specificity has also been described[361]. Both nuclei and chromatin have been shown to contain endogenous phosphokinases which will phosphorylate both histones and non-histone proteins *in situ*[354-357]. A number of laboratories have obtained partially purified forms of these enzymes which appear to be heterogeneous[360, 363-365]. Kish and Kleinsmith[366] have reported the presence of 11 distinct phosphokinase activities in the non-histone phosphoprotein fraction of calf liver chromatin, only five of which could phosphorylate non-histone proteins *in vitro*. Phosphokinases of cytoplasmic origin, together with protamine kinase, can be stimulated by cyclic AMP[359, 367, 368], but in general nuclear kinases do not appear to be influenced in this way[363, 365, 369]. On the other hand, Kish and Kleinsmith[366] found that five kinase activities in calf liver chromatin were cyclic AMP dependent while six kinases were inhibited by the nucleotide.

(b) *Histones*—The various histone fractions are phosphorylated to different extents in a number of tissues and cells. Thus histones F1 and F3 are phosphorylated in rat liver[370, 371], F1 and F2a2 in mouse tissues[40], while other workers have reported that all of the histone fractions of rat tissues and trout sperm are phosphorylated[69, 120, 372]. The phosphate content of histones, particularly F1, is dependent on the physiological state of the tissue[371, 373, 374] such that the phosphate contents of histones F1 and F3 increase during DNA synthesis[375, 376]. In addition, the phosphorylation of most histones of rat liver has been reported to be subject to a circadian rhythm[377]. Histone phosphorylation is partly responsible for the apparent electrophoretic heterogeneity of some of the histone fractions (see Sections 10.2.2.1–10.2.2.8) whilst phosphorylation studies *in vitro* have shown different sites to be available for modification in F1 fractions from different tissues[378].

The relationship between histone phosphorylation and RNA synthesis has been investigated in a number of cells. The induction of RNA synthesis in mammary tissue explants by hormones such as insulin, prolactin and glucagon has been found to be accompanied by increased histone phosphorylation[379]. The *in vivo* response of the liver in glucagon-treated rats is associated with the phosphorylation of a specific seryl residue of histone F1[380]. A rapid phosphorylation of histones F1 and F2a1 was also observed in ovary cells after treatment with gonadotrophin[278].

In regenerating liver, increased phosphorylation and increased phosphate content was reported for histones, other than F1, during the period of increased RNA synthesis[371, 374], whilst a similar effect was found for histone F1 during the later period of DNA synthesis[370, 373]. In contrast, other experiments showed intense phosphorylation of all histones apart from F3 within the first 6 h after hepatectomy, with all histones being phosphorylated during the period of DNA synthesis at a rate proportional to their rates of synthesis[270]. Chalkley's group[41] have reinvestigated the phosphorylation of histone F1, in particular, during liver regeneration. Using purified fractions, they found that the rate of phosphorylation of this fraction was some ten times greater in treated than in control animals. These phosphorylated F1 components appeared at the start of DNA synthesis and were subsequently followed by two other peaks of F1 phosphorylation which coincided with the first wave of mitosis and a second period of DNA synthesis.

The phytohaemagglutinin stimulated lymphocyte system has also been used to study histone phosphorylation. Initial work by Kleinsmith *et al.*[353] showed an early increased phosphorylation of total nuclear phosphoproteins in these cells. Further work by Cross and Ord[381] showed that the stimulation of RNA synthesis in such lymphocytes was accompanied by an increased turnover of histone F1 phosphate groups, but not by a net increase in phosphate content. During the subsequent period of DNA synthesis in these cells both the phosphate content and the rate of phosphorylation of histone F1 increased.

The investigations using regenerating liver and stimulated lymphocytes showed that there were differences in the modes of phosphorylation of histones at times of RNA and DNA synthesis. Further studies were carried out using synchronised cells in culture. Thus Gurley and Walters[382] and Chalkley's group[383] found that after the removal of non-specifically bound phosphate, histones F1 and F2a2 were the only histones to be phosphorylated and that this occurred at the time of DNA synthesis in hamster ovary and HTC cells respectively. Later evidence[384] has now shown that hamster ovary cell histone F2a2 is phosphorylated at other times of the cell cycle but it is unknown as yet whether this represents a genuine difference between these cell types or whether it is due to different methods of synchronisation.

However, Cross[275] found that during the cell cycle of mast cells the phosphorylation of histones followed the timing of their synthesis, so that histone F1 was phosphorylated in early S phase, whilst phosphorylation was associated with the remaining histones, which were unfractionated, later in the same period. Other workers[385] have studied metaphase chromosomes of a number of cell lines and reported that the F1 histones were in a highly phosphorylated state. Since on entering G1 phase the histones became dephosphorylated, this phosphorylation process would appear to be associated in some way with chromosome condensation.

In a series of investigations Chalkley's group[386-388] has reported that significant phosphorylation of histone F1 is found only in dividing cells, non-replicating cells being practically devoid of phosphorylated F1 components. Further studies using cultured hepatoma cells have shown that during these events both existing and newly-synthesised histone F1 undergo phosphorylation, the evidence pointing to both types of histones being modified only when bound to DNA. In addition most of the 'new' F1 histones are phosphorylated only once in the cell cycle, namely, during S phase, but surprisingly this phosphorylation does not appear to take place until after a lag of up to 1 h after these histones have been synthesised[389]. Cross[275] has also supplied evidence that both 'old' and 'new' histones are phosphorylated during the S phase of the cell cycle. These results have led Chalkley[383, 386-388] to conclude that F1 phosphorylation is associated with neither the control of gene activity nor with transport devices, but that it is somehow an integral part of chromosome replication. Recent investigations by Gurley's group[384] agree with those of Chalkley that phosphorylation of histone F1 occurs only during S phase, but their present evidence shows that histone F2a2 can be phosphorylated independently of the cell cycle stage, DNA synthesis or F1 phosphorylation. Thus its modification occurs at the same high rate during both G1 and S phases in hamster ovary cells and at a somewhat decreased rate in G1 arrested cells. Hence these authors feel that phosphorylation of this histone may be connected with processes other than those involved with DNA replication.

Of the histone fractions which are phosphorylated during these cell cycle studies, only the phosphorylation of histone F1 is sensitive to x-ray irradiation[382]. It is interesting to note, in relation to Chalkley's conclusion[388] concerning histone F1 phosphorylation and its correlation with DNA synthesis, that Gurley and Walters[382] interpret their results as indicating that phosphorylation of this histone is not necessarily associated with DNA replication *per se* but with the incorporation of a radiosensitive precursor of DNA. In addition, in hamster ovary cells turnover of histone F1 does not appear to require phosphorylation of the protein[384].

(c) *Histones in spermatogenesis*—The data described above show that phosphorylation of histones can occur under two sets of circumstances. The first involves the loose correlation between histone phosphorylation and changes in chromatin template activity and the second is the high level of histone phosphorylation near the time of synthesis of these proteins. However, it is extremely difficult to find a biological system whose characteristics are simple enough to allow a thorough investigation of these phenomena. One such system which has given a wealth of information about changes in chromatin during development concerns the process of spermatogenesis in rainbow trout.

Dixon's group have studied a number of aspects of this system with particular reference to chromatin protein modification (for a review see Dixon[352]). Since a reasonable degree of synchrony of these cells is obtained by the action of pituitary gonadotrophin, it was found that a major feature of the process of spermatogenesis was the replacement of the somatic type of histone by arginine-rich protamines[171, 390]. Recent studies have shown that of the sperm line cells, the spermatocytes possess and actively produce histones, whereas during the maturation of the middle spermatids the histones are gradually replaced until the mature sperm contains only the metabolically stable protamines[391].

As in the case of the histones described above, protamines are phosphorylated after they have been synthesised and are transported to the nucleus in this form[119, 120]. Although all the seryl residues are originally phosphorylated[392], the protamines become dephosphorylated as the spermatids age[393]. Dixon[352] has suggested that this cycle of events is necessary for the correct binding of protamine to DNA. Since both molecules are of high but opposite charge incorrect binding could take place unless some form of control was applied. Phosphorylated protamine has been found to have a reduced ability to bind to DNA[120] and hence it is possible that the phosphorylation process allows the correct alignment of the protein with DNA. Once the correct conformations have been established dephosphorylation by phosphatases could then allow the final complexing to take place. It is of interest to note that protamine kinase is activated by cyclic AMP, and that the adenyl cyclase enzyme is in turn stimulated by pituitary gonadotrophin (see Dixon[352]).

Spermatocyte histones are also phosphorylated[120] near the time when they (and DNA) are synthesised[393]. Other studies have shown that the histones of these cells are also acetylated shortly after synthesis[394]. In the case of histone F2a1, the newly-synthesised product passes through pools of varying degrees of acetylation before being complexed to DNA in the unmodified form[70]. Dixon[352] suggests that these histone modifications are also required to ensure correct binding to the DNA. For example, space filling models show that the *N*-terminal region of histone F2a1 could bind in the major groove of DNA provided it was in an α-helical form. Dixon's view is that the modifications described above could cancel out the high positive charges and allow such a conformation to take place[69, 352]. As in the case of protamines, the modifying group could be removed by enzymatic action to allow the final complexing to DNA to take place.

Histones of spermatids have also been found to be acetylated when there is no DNA synthesis, but at a time when active replacement of histones by protamines is occurring[359]. Dixon[352, 395] suggests that in contrast to the situations described above, this particular effect causes a weakening of the ionic bonds between histones and DNA. In this way the histone could be detached completely from the DNA and then possibly degraded by proteases[390].

Hence these cells have given considerable information concerning the association between postsynthetic modifications of histones and protamines and their binding to DNA. Several features of the sperm system are biologically extreme but it gives an indication of the possible processes which might be involved in regulation during changes in gene activity.

(d) *Non-histone proteins*—It has already been described that, as judged by ^{32}P-labelling experiments, the non-histone proteins are phosphorylated to a greater extent than the histones. In addition, Frenster[2] showed that euchromatic fractions contained more phosphoprotein than heterochromatin. Hence major changes in phosphorylation of the non-histone proteins would not be unexpected during changes in gene activity. Unfortunately, the systems studied to date are often complicated by increased turnover and other chemical modifications of both histones and non-histone proteins (see Turkington and Riddle[379], Jungmann and Schweppe[278]). Thus it appears that insulin and hydrocortisone caused the phosphorylation of non-histone proteins in explanted mammary tissue[379], whilst a similar effect was found in gonadotrophin-stimulated ovary cells[278]. Evidence has also been obtained from apparently less complex sources. For example, the early work of Kleinsmith *et al.*[353] showed that there was an increased phosphorylation of total chromatin proteins in phytohaemagglutinin-stimulated lymphocytes. Testosterone has been found to cause increased phosphorylation of ventral prostrate non-histone proteins, as judged by the incubation of isolated nuclei with [γ-^{32}P]-ATP *in vitro*[356]. A similar effect was observed when rabbit endometrial tissue was incubated *in vitro* with oestradiol[396].

Relatively few studies have been carried out on the phosphorylation of non-histone proteins during the cell cycle. Cross[275] found that phosphorylation of these proteins occurred during G1, S and G2 phases of the mast cell cycle, the highest level being found during S phase. A phenol-soluble fraction of slime mould non-histone protein was also found to be phosphorylated in S phase, immediately after its synthesis[193].

Thus like the histones, the non-histones also appear to undergo considerable phosphorylation during S phase, when presumably some of their complement is synthesised prior to attachment to DNA. The evidence from both stationary and dividing tissues is that phosphorylation of these proteins can also occur independently of DNA synthesis. It is possible, moreover, that their overall rate of phosphorylation may be an important factor involved in regulating RNA synthesis. For example, studies carried out in conjunction with our laboratory have shown that short-term phosphorylation of chromatin protein varies from tissue to tissue, for example a non-dividing tissue such as brain has a higher phosphorylation rate than liver. Of more importance, however, is the finding that when exponentially growing tumour cells become stationary the phosphorylation rate of their chromatin proteins doubles. Since the histones are relatively minor phosphoproteins, much of this effect probably resides in the non-histone protein fraction[397].

In addition, a parallel situation exists in relation to histone modification, since mitotically-active cells have a lower ability to phosphorylate these proteins than non-dividing tissues[372]. It is also of importance to note that in differentiated or stationary cells, more of the genome is available for transcription than in dividing cells[5, 398]. Hence differentiated cells may possess a larger functional genome, which in turn is associated with an increase in the phosphorylation of chromatin proteins.

10.3.2.4 Methylation of histones

The postsynthetic methylation of proteins has recently gained much prominence (for a review see Paik and Kim[399]). Of the nuclear proteins only the histones are methylated at high rates[400, 401], although non-histone proteins do appear to be modified in this way to some extent[400]. In fact in one cell line the 'residual' non-histone proteins contain methylarginine of higher specific activity than that of the histones[401].

S-Adenosyl methionine has been found to be the methyl group donor[333, 402, 403] and a number of methylase enzymes have been identified. These have been described as consisting of methylase I (protein arginine methyl transferase) found in both nucleus and cytosol, methylase II (protein carboxyl methyl transferase) found in the cytosol and methylase III (protein lysine methyl transferase) found in the nucleus. The substrate for both methylases I and III appears to be histones, whilst methylase II can modify other proteins[399]. Histone-methylating enzymes have been reported by others[404, 405] including Gallwitz[406] who partially purified two such activities from rat thymus which preferentially methylate histones F2a1 and F3. Although the arginine-rich fractions tend to be the major methylated histones, the lysine-rich fractions can also be methylated[404, 405, 407, 408]. In contrast to these enzymatic reactions methylation of rat histones at lysine, arginine, cysteine and methionine residues through direct chemical alkylation by carcinogens such as dimethylnitrosamine has also been reported[409].

The role of protein methylation is largely unknown[399]. Methylation of a protein amino group to yield a mono- or tri-substituted derivative will increase the basicity of the polypeptide, although disubstitution has the opposite effect. The modification involves the introduction of non-polar groups to the polypeptide which could lead to changes in conformation[399].

The biological significance of histone methylation has been studied in a number of systems. Early work by Mirsky's group[410] showed that the ε-amino groups of the arginine-rich histones were methylated after DNA and histone synthesis had occurred in regenerating liver. Later Benjamin[411] confirmed the increased methylation of histones F2a1 and F3 in these cells. Shepherd *et al.*[59] examined the situation during the cell cycle of hamster ovary cells and found that the [^{14}C]-methyllysine content of histones F2a1, 2 and F3 increased throughout the cycle and reached a maximum after the end of DNA synthesis in S phase near the start of mitosis. On the other hand, fraction F2b was methylated at a lower level reaching a maximum in early S phase, whilst histone F1 was totally unaffected at any stage. Similar results were obtained using HeLa cells, in which histone methylation was also found to continue at a significant level during the entire cell cycle with mitotic cells containing most methylated histone molecules. The methylarginine and methylhistidine residues of histone F3 in these cells were methylated at a maximal rate during S phase[400].

These results have been interpreted as indicating that this modification of histones, which could lead to their increased basicity, takes place at a time when conformational changes are occurring in the nucleus prior to the formation of the condensed inactive mitotic chromosomes. However, not all investigations agree with this hypothesis. Lee and Paik[412] examined histone methylation in regenerating liver in some detail and concluded that, relative to DNA synthesis, histone methylation was not a late event. In fact they found that the major incorporation of [^{14}C]-methyl groups into mono- and di-substituted lysines occurred along with the peak of DNA synthesis and at a time which did not coincide with the activities of methylases I and III. The reason why these results differ from those of Tidwell *et al.*[410] is unknown at present.

Other aspects of histone methylation are also of importance. For example, although methylation of arginine and lysine residues occurs throughout the cell cycle, present evidence indicates that once incorporated the methyl group remains metabolically stable, decaying with a half-life similar to that of the histone itself[400, 401]. Unlike similar aspects of histone acetylation and phosphorylation this does not appear to be a feature of dedifferentiation, since the same process has been found in a wide range of cells and tissues[401, 413, 414]. Histone methylation does, however, possess some degree of cell specificity. Thus in hamster cells only histone F2b contains methylated arginine residues[401], whilst in human (HeLa) cells fraction F3 contains both modified arginine and histidine[400]. Cow, but not pea, histone F2a1 has been found to have a methylated lysine at residue 20[66]. In addition, the ratio of di- to mono-substituted lysine is higher in normal than in tumour or embryonic cells[59, 400, 415]. The function of these features is largely unknown but it is of interest that Byvoet *et al.*[401] should find evidence for methylation occurring mostly on newly synthesised histones. This result taken together with the apparent irreversibility of the amino acid modification tends to implicate histone methylation in the postsynthetic processing of histones prior to them becoming part of chromatin.

10.3.2.5 Reversible oxidation of thiol groups

We have already described in Section 10.2.2.5. that fraction F3 is the only mammalian histone to contain cysteine and that, in addition, species up to and including rodents possess one such residue per F3 molecule, whereas two such amino acids are found in the histones of higher animals. Hence under suitable conditions polymerisation of F3 molecules could occur through the formation of disulphide bonds and evidence for such cross-linking has been reported (see Section 10.2.2.5). However, unlike the amino acid modification processes described in the preceding sections there is no indication as yet for such disulphide bond formation being due to other than direct chemical interaction of thiols.

In the sea urchin embryo the F3 thiol content increases after fertilisation and during the first cell cycle[375, 416, 417]. In this system nucleic acid synthesis would appear to require the F3 thiols to be in the reduced state[371]. In liver cells the F3 thiols appear to be mostly reduced, even during regeneration[371], but in the mature avian erythrocyte such groups may be oxidised[104].

After stimulation by phytohaemagglutinin, the thiol content of lymphocyte histone F3 remained unchanged during the early period of RNA synthesis, but increased almost three times during the later phase of DNA synthesis[336]. A similar effect was also evident during DNA synthesis in mast cells[275]. In contrast, metaphase chromosomes of HeLa cells have been found to contain more F3 histones in the disulphide form than interphase chromatin[276, 277]. Also, during interphase, Ord and Stocken[370] observed that heterochromatin contained more disulphide bonds than euchromatin.

Hence it would appear that in condensed states, e.g. in heterochromatin or metaphase chromosomes, the F3 histone molecules are crosslinked by disulphide bonds. In interphase chromatin these bonds may be partially reduced, whereas DNA synthesis is accompanied by a large increase in histone thiol groups.

Although non-histone proteins contain thiol groups, little is known regarding their properties. Evidence has been presented that these groups could be linked by disulphide bridges to F3 histones in metaphase chromosomes[276,277]. The increase in RNA synthesis after cortisol treatment of liver is accompanied by an increased thiol content of nuclear proteins, most of which appear to be associated with the non-histone fraction[418].

10.3.2.6 Conclusions

The common feature of histone acetylation, phosphorylation and methylation is their occurrence at or near the time of DNA synthesis. The data which are available show that both 'old' and 'new' histones are acetylated and phosphorylated, indicating that these modifications are reversible. Hence it is possible that histones already bound to DNA require modification in order that they can be removed from the DNA prior to its replication, whereas newly synthesised histones appear to be modified in preparation for their final complexing to DNA. In contrast, methylation appears to take place only on 'new' histones and, moreover, the modification is irreversible. Hence this may be a process which is mandatory for some histone fractions after synthesis, but its function is as yet unknown.

In some cell systems acetylation and phosphorylation, but not methylation, of histones takes place before increases in RNA synthesis. Non-histone proteins can also undergo phosphorylation at times of increased template activity. The relationship between these effects strongly suggests that this form of reversible modification of chromatin proteins is implicated in gene control. However it has to be stressed that the correlation is temporal only, since it has proved difficult to date to locate and study such changes in chromatin during the transcription of single genetic loci, at the time of the induction of specific proteins. Recent evidence suggests that dedifferentiated cells do not possess the high turnover rates of acetylation and phosphorylation seen in differentiated cells and this correlates with the lack of template activity in the former.

The oxidation and reduction of thiol and disulphide bonds appears to be involved in more structural changes in chromatin. Thus disulphide bond formation might assist in the condensation of the chromosomes at mitosis, whilst these bonds appear to be broken in the looser structure of active euchromatin.

10.4 CHROMATIN TEMPLATE STUDIES

10.4.1 Studies employing intact chromatin

An important step towards an understanding of the roles of the various chromosomal components in selective gene repression has been the use of bacterial and eukaryotic RNA polymerases as probes for determining the amount of transcribable DNA in native and reconstituted chromatin. The experiments were first carried out by Huang and Bonner[419], Bonner and Huang[420] and Frenster *et al.*[421]. They found that the template activity of the DNA in chromatin was much less than that of deproteinised DNA. These early studies were criticised by Sonnenberg and Zubay[422] who demonstrated that solubilisation of chromatin by shearing increased its template capacity. Further results from other workers also suggested that the decreased template activity was a reflection of the lower solubility and physical state of the chromatin as compared with DNA in the assay system used[423-427]. However, other workers have been unable to confirm these results[428], moreover, further studies[429, 430] on the binding affinity of RNA polymerase and the melting curves of DNA and nucleohistone reaffirmed the earlier results and indicated that certain parts of the DNA were stabilised against heat denaturation and were repressed such that they could not be transcribed by RNA polymerase.

In addition to the rate studies, the RNA synthesised *in vitro* from chromatin has also been studied by means of RNA–DNA hybridisation. Georgiev *et al.*[431] and Paul and Gilmour[433] found that the RNA synthesised using chromatin as a template hybridised to a smaller fraction of the DNA than did RNA synthesised using DNA as a template. Therefore, it was concluded that much of the DNA in chromatin is not available for transcription, i.e. there was selective repression of specific DNA sequences. Further, Paul and Gilmour[432] showed, by competition hybridisation experiments, that the RNA synthesised using chromatin as a template was similar to that synthesised *in vivo;* they also claimed that the species of RNA synthesised were related to the tissue from which the chromatin was obtained. Subsequently other workers were able to show, by a variety of hybridisation techniques, that the synthesis of RNA was organ specific[434-437]. Moreover, changes in the species of RNA synthesised during embryogenesis could also be found[438].

The conclusions from the experiments described above have been criticised on two separate counts. Firstly, most of the rate studies, and the hybridisation experiments with *in vitro* synthesised RNA, have utilised bacterial RNA polymerase. However, although one group has indicated that different sequences may be transcribed by the prokaryotic and eukaryotic RNA polymerases[439], these differences are not apparently detectable by hybridisation experiments[436]; moreover, recent work suggests that while the homologous RNA polymerase can only bind to, and transcribe certain defined sequences, the bacterial polymerase binds in a random fashion and is able to transcribe all of the 'exposed DNA' in the chromatin[440]. The hybridisation results have also been criticised on the grounds that, as shown by the work of Britten and Kohne[1], the conditions of low RNA concentrations and relatively short incubation times (low C_0t conditions) permit only the RNA transcribed from the repetitious DNA to form a hybrid. Therefore, the above results obtained using low C_0t conditions only show that the RNA transcribed from a particular fraction of the DNA, i.e. the repetitive DNA, is organ specific. However, in more recent work in which RNA was hybridised to 'unique DNA' sequences, the RNA synthesised also appeared to be organ specific[5].

In view of the organ specific repression of DNA transcription, much work has been directed towards the elucidation of the constituents of chromatin which are responsible for this effect. The DNA as isolated from eukaryotic interphase cells is complexed with approximately an equal amount of histone, together with variable amounts of non-histone protein and RNA (see Section 10.2.1). Different approaches have been used to determine the roles of the various chromosomal components in the control of DNA transcription. Early experiments looked at the effect of binding total histone and histone fractions to DNA, and this technique was then refined so that the properties of reconstituted chromatin could also be studied. The other approach has been that of studying the effect of depleting the chromatin of certain chromosomal constituents, particularly the histones.

10.4.2 Histone–DNA binding experiments

In 1950 Stedman and Stedman[14] first proposed that histones could be specific gene repressors, and this led to an intensive investigation of the effect of histones on the transcription of DNA by RNA polymerase. It was found that the addition of histones to DNA reduced the template capacity of the DNA[419, 441]; however, the reported inhibitory effects of each of the histone fractions varied from one group to another[430, 442, 443]. Also other results suggested that the observed variability of the apparent decrease in DNA template activity in the presence of histones merely reflected the precipitation of the nucleohistone complex[422, 424, 426, 427]. Johns and Hoare[444] also demonstrated that variable amounts of DNA could be precipitated by each of the histone fractions by changing the ratio of histone to DNA. On the other hand, Hindley[443] was able to show that precipitation of the DNA by itself was not responsible for the decreased template activity; also, in the experiments of Butler and Chipperfield[441], the decrease of template activity did not precisely parallel the amount of DNA precipitated. More recently, Johns[445] has studied the binding of histone F2b fragments to DNA and has found that while the *N*-terminal fragment and to a lesser extent the *C*-terminal fragment can both bind to and precipitate DNA, only the *N*-terminal fragment is able to repress the template activity of the DNA.

The histones are bound to the DNA by both ionic and hydrophobic bonds[446, 447] possibly in the major groove of the DNA helix[448, 449]. Some evidence has been presented which suggests that the lysine-rich histones preferentially bind to A–T rich regions[450-453] while the arginine-rich histones preferentially bind to G–C rich regions of the DNA[453, 454]. The above results have also been supported by the observation that in a model system, the histones appear to bind to bacteriophage DNA in a cooperative manner and electron microscopy of the complexes further showed that the histones are bound non-uniformly along the DNA[455]. These binding studies may have some relevance to the situation *in vivo*, since Varshavsky and Georgiev[454] have reported that in chromatin the arginine-rich histones are distributed in clusters along the DNA helix.

The observed specificity of histone binding to DNA described above would not appear to be sufficient to account for the tissue specific transcription of DNA. Thus although F1 is claimed to inhibit the transcription of certain classes of DNA[456] this may only be a reflection of the base composition of these parts of the genome. Also the results from low C_0t hybridisation experiments carried out by Gilmour and Paul[457] revealed that the histone repression of the DNA template in reconstituted chromatin was non-specific; they further showed that the non-histone fraction was required for accurate reconstitution of the chromatin.

10.4.3 Chromatin reconstitution studies

The histone binding studies described above generally involved mixing the components at high ionic strength (e.g. 1–2 M NaCl) followed by stepwise removal of the salt, thus enabling the histone to bind onto the DNA. The original reconstitution studies of Paul and Gilmour[458] involved a similar procedure; in their experiments they showed that if the non-histone fraction was present during the reconstitution process, then the reconstituted chromatin had a template activity similar to that of native chromatin. In later, more refined experiments, Gilmour and Paul[457] showed that RNA synthesised *in vitro* from chromatins reconstituted from 2 M NaCl, 5 M urea were indistinguishable from those synthesised *in vivo*, as judged by low C_0t hybridisation experiments. The evidence from Paul and Gilmour's work suggested that proteins complexed with the DNA, and in particular the non-histone proteins, were able to permit organ specific transcription of the nucleohistone complex. Further work[459] showed that the template properties of the reconstituted chromatin were completely dependent on the source of the non-histone fraction present in the reconstitution mixture.

The non-histone proteins used by Paul and Gilmour contained variable amounts of nucleic acid. In parallel with the experiments of Paul and Gilmour, Bekhor *et al.*[246] and Huang and Huang[241] carried out chromatin reconstitution studies, which claimed to show that a particular species of RNA, termed 'chromosomal RNA' was responsible for organ specific transcription (see Section 10.2.4). In general, however, the work of many groups has indicated that it is the non-histone proteins which are responsible for organ specific transcription of the DNA. Recent work has also shown that the phosphoproteins in particular may play an important role in the control of DNA transcription, since the dephosphorylation of calf uterine non-histone proteins prior to reconstitution inhibits the template activity of the reconstituted chromatin[396]. The putative roles of the different types of non-histone protein are further discussed in Section 10.5.2.

10.4.4 Effects of partial deproteinisation of chromatin

The other approach used to study the roles of the various chromosomal components has been the use of procedures which can specifically deplete the chromatin of certain types of macromolecules. In particular it has been found that the histones, which are an extremely basic group of proteins, may be removed specifically from the chromatin by acid or salt washes[146, 460-468]. More recently the relatively harsh acid extraction procedures to remove histones have been superseded by the use of salt solutions, both in the absence[469] and presence of urea[470], or low concentrations of sodium deoxycholate[471] or by treating chromatin with transfer RNA in the presence of low salt and divalent cations[454]. However, it must be appreciated that the results obtained require cautious interpretation, since all such procedures generally remove variable amounts of non-histone protein from the chromatin and may indeed give a partial redistribution of the chromatin proteins.

The very lysine-rich histone F1, can be removed selectively from the chromatin by several of the procedures described above. The reported effects of the removal of F1 from the chromatin has varied considerably from group to group. Thus Georgiev *et al.*[431] and Bonner *et al.*[472] found that the removal of F1 greatly stimulated DNA transcription by bacterial RNA polymerase. Seligy and Neelin[473] also found that removal of the lysine-rich erythrocyte-specific histone F2c stimulated the transcription of the previously inactive chromatin; however, no such stimulation was observed after the removal of histone F1. Other groups also found that the removal of F1 had little effect on template activity, and that only after removal of F2b and F2a was the chromatin template capacity significantly increased[460, 473, 474]. These conflicting results may, in part, be due to differences in technique and material used.

Early results suggested that removal of the histones from the chromatin resulted in a complete derepression of the DNA template[429, 431]; however, later work has shown that the template activity of the dehistonised chromatin, as measured by bacterial RNA polymerase, is still significantly less than that of pure DNA[458, 469, 473, 475]. On the other hand, using homologous RNA polymerase it appears that dehistonised chromatin is a better template than pure DNA[476], which shows that the non-histone proteins which remain bound to the DNA have the ability to stimulate DNA transcription. Spelsberg and co-workers have investigated the immunological properties of dehistonised chromatin and its constituents and have found that in oviduct tissue the non-histone fractions can be arbitrarily divided into those that stimulate DNA transcription and those which can form tissue-specific antigenic complexes with DNA[196]; the latter fraction also contains the tissue-specific protein which can bind the progesterone nuclear receptor complexes[477]. Therefore these results raise the interesting possibility that it may be possible to characterise the tissue specific non-histone proteins by immunological means.

10.5 PROBABLE MODE OF ACTION OF CHROMATIN PROTEINS

The details of the structure of chromosomes outlined in Section 10.2.1 indicate that the strands of the DNA double helix of several metres in length are folded and foreshortened into chromosomes of only μmetres in dimension. Two main points arise from these observations. The first relates to the mechanism whereby such a degree of condensation is achieved and the second concerns the processes which allow the transcription of tissue-specific portions of the genome under these conditions. The next two sections (10.5.1 and 10.5.2) describe the possible functions of the chromatin proteins in the organisation of the chromosomes, whilst section 10.6 summarises some current models of gene regulation.

10.5.1 Histones

Various spectroscopic and physical chemical techniques have been used to probe the structure of chromatin and nucleohistones. Although circular dichroism studies of nucleoproteins are complicated by solvent conditions, e.g. ionic strength[478], they indicate that the conformation of DNA is changed in nucleoprotein[479-483]. This effect is generally accepted as being due to the interaction of DNA with chromosomal proteins, partial deproteinisation[482,483] and reconstitution experiments[484-486] both point to the histones as being the proteins responsible for this change. However, neither histone F1[483] nor the erythrocyte-specific F2c histone[484] appear to be associated with these conformational changes. Optical rotatory dispersion spectra have also shown changes in the conformation of DNA in chromatin[487,488]. In addition, Henson and Walker[489] have found that DNA is more asymmetrical than nucleohistone, histones F2b, F2a2 and F3 being responsible for this effect.

These observations are interpreted as indicating that there is a slight conformational change in DNA when it is part of chromatin[485,488]. Other spectroscopic and analytical data suggest that the DNA in chromatin consists of two forms, a portion in which the DNA is in the native B form and is exposed to the solvent environment and the remainder in which the DNA is in the deformed C form and is neutralised completely by histones[481,490].

Various physical chemical techniques have shown that isolated histones are capable of possessing α-helical and random coil conformations[486,491-493]. Histones, e.g. F2a1 when bound to DNA, still possess α-helical structures according to some[486,494], whilst other circular dichroism studies have indicated atypical structures[478]. In contrast, histone F1, when bound to DNA, appears to have a structure similar to that of denatured or random coil proteins[494]. Physical studies, e.g. using nuclear magnetic resonance, have shown that the non-basic regions of a number of histones can undergo conformational changes which could be responsible for inter- and intra-chain interactions. Such studies have also shown that both the *N*-terminal and *C*-terminal of histone F1 could react with DNA, since both ends of the polypeptide possess high positive charges. The *N*-terminal half of histone F2b appears to be the primary site of interaction with DNA, although some binding may take place at the *C*-terminus. In contrast, only the *N*-terminal third of histone F2a1 can bind to DNA[495-498]. These results confirm the impressions already gained from the primary structure of these histones described in Sections 10.2.2.1–10.2.2.8.

Since histones appear to be involved in forming the structure of chromatin various attempts have been made to locate the position of the various fractions on the DNA. Both the lysine-rich F1[499] and the erythrocyte specific F2c histones[500] have been reported as lying in the large groove of DNA, but suggestions have also been made that histone F1 is associated with neither groove[500]. Dye binding studies have indicated that the other histones (and non-histone proteins) may be distributed along both the large and small grooves[500], whilst Bradbury and Crane-Robinson[486] regard the arginine-rich histones as being in an extended form lying parallel to the helix itself. Other aspects of this topic have been dealt with in Section 10.3.2.

The most popular concept of the structure of nucleohistone which could explain the conformational properties of DNA and histones already described is that of the 'supercoil'. Based on x-ray diffraction work carried out in Wilkin's laboratory[18,501,502], the organisation of DNA and its histones is regarded as being one in which the complex folds upon itself to form a coil of diameter 100 Å and pitch 120 Å. (For a discussion of the evidence see Bradbury and Crane-Robinson[491].) Other investigators, e.g. those using the electron microscope, also favour the supercoil structure[19,503], but others suggest that the supercoil is restricted to the portion of DNA in chromatin which is in the *C* conformation[490]. Reconstitution of nucleohistones show that histones are indeed required for such a structure[504,505] whilst deproteinisation experiments show that neither the F1 histones nor the erythrocyte specific F2c histone are associated with the supercoil[506,507]. Since nucleohistones prepared from DNA and histones from widely different sources still show evidence of supercoil, there would appear to be little evidence for specific base-sequence associations between DNA and histones[508]. Factors such as divalent cations[490,508] have been considered necessary for the formation of supercoil and the stabilisation of its structure has been suggested as being due to histone bridges[509]. Nucleohistone reconstitution experiments have confirmed that the very lysine-rich histone F1 fraction is not required for supercoil formation and they have further shown that nucleoproteins containing DNA and a mixture of either histones F2a1, F2a2 and F3 or histones F2a2 and F3 do show supercoiling properties[491,501]. Other experiments have shown that the specific removal of histones F2a1 and F2a2 from nucleohistone causes the loss of supercoil, an effect which has been reversed in some cases by replacement of these fractions[491]. Hence supercoiling may be the result of the co-operative action of a number of histone fractions.

The circular dichroism and x-ray diffraction data described show that of the histone fractions, the F1 and F2c proteins are not involved in producing the altered conformation of DNA in chromatin. In fact their removal causes chromatin to adopt a more regular structure[507]. The F1 histones have a high content of proline (9 mol %) (see Johns[34]), suggesting that each molecule of this histone could have a 'non-α' helical extended conformation. Such a structure has been implied from physical studies of the isolated protein[510,511] and recent studies of DNA–F1 complexes show that the histone exists in an extended conformation even when bound to DNA[494]. Furthermore, work carried out on both interphase and metaphase chromosomes indicates that histone F1 molecules link DNA-containing fibrils together[466,512,513]. Hence, as regards the structure of chromatin, the function of the F1 histones appears to be markedly different from that of the other histones.

The question remains as to how histones cause the DNA to supercoil. Bradbury and Crane-Robinson[491] suggest that the structure results from histone-induced constraints applied at regular intervals along the DNA helix. Such constraints could be caused by the basic regions of the histones straining the DNA, or they could be due to interactions between the portions of the histone polypeptide chain which do not bind to DNA. Further degrees of condensation could then be brought about through additional histone–histone interactions or by cross-linking by F1 histones.

Hence the histones appear to be responsible for the structure of chromatin through condensation effects based on the supercoil principle. It has to be admitted, however, at this stage that we do not know precisely what role, if any, the non-histone proteins play in these conformations. Since the complexing of histones with DNA can occur in part through ionic links between the basic amino acid residues of the protein and the phosphate groups of the nucleic acid, alteration of these forces by processes such as the postsynthetic modifications of histones described in Sections 10.3.2.1.–10.3.2.6 could provide a means for changing the structure of chromatin. In this way transcription of a portion of the genome or replication of the whole of the DNA could be initiated. Thus it is interesting to note that phosphorylated F1 histone is less effective than the unmodified protein in producing conformational changes in DNA[514]. The role of the non-histone proteins in regulating the transcription of chromatin DNA will be discussed in the following section, but it is relevant to point out here that in forming condensed chromatin structures, histones may indirectly play a role of template repressors. Thus Johns[515] has suggested that the dimensions of the supercoil would prevent the passage of RNA polymerase molecules along the DNA. Evidence has also been presented indicating that histone F2a1 is mainly responsible for the conformational features of chromatin which inhibit transcription[482].

10.5.2 Non-histone proteins

As described in Section 10.4, chromatin reconstitution studies have indicated that the non-histone fraction is responsible for the tissue-specific repression of genes in eukaryotic cells. Moreover, these early experiments indicated that the non-histone proteins in particular fulfil this role.

The non-histone fractions used in the early experiments were, however, generally contaminated with variable amounts of nucleic acid material; also many of the methods of preparation involved exposure of the proteins to extremes of temperature, pH and solvent environment (see MacGillivray *et al.*[35]). However, as described below, even non-histone proteins which have been isolated by rigorous procedures still appear to possess some activity.

The mechanisms by which the non-histone proteins might influence transcription are at present unknown. Most of the current theories (see Section 10.6) envisage that some of these proteins could be analogous to the regulatory proteins present in prokaryotic systems, e.g. the *lac* repressor and CAP factor in *E.coli*, or that they might be able to form complexes with the histones and thus derepress certain parts of the genome. Since one of the difficulties encountered in determining the activities of the non-histone proteins has been the absence of suitable methods of assaying their biological activities, this has been tackled using several different types of approach.

10.5.2.1 Studies of DNA transcription

The work of Wang and his co-workers[197, 199, 516, 517] has shown that the addition of chromosomal acidic proteins to heterochromatin results in a stimulation of template activity. This stimulation was proportional to the quantity of non-histone proteins, while the addition of this non-histone fraction to euchromatin had a much smaller stimulatory effect. In these experiments the template activity was assayed in the presence of bacterial RNA polymerase, thus it would appear likely that the stimulation may result from an opening up of certain regions of the condensed chromatin. In agreement with this, low C_0t hybridisation experiments revealed that, in the presence of added non-histone proteins, new parts of the genome became available for transcription and that the sequences of DNA exposed were dependent on the source of the non-histone proteins. It is of interest to note that rate studies also revealed that the addition of non-histone proteins to homologous heterochromatin stimulated the template more than if the proteins were derived from a different tissue. This effect could reflect the presence of a tissue specific factor in the heterochromatin which interacts with tissue specific components in the non-histone fraction, resulting in an activation of the DNA template. On the other hand, the hybridisation experiments showed that only the addition of heterologous proteins exposed fresh DNA sequences for transcription.

The components of the non-histone fraction responsible for the template stimulation of heterochromatin have been partially characterised. The *in vitro* phosphorylation of heterochromatin by kinases present in the chromosomal acidic fraction stimulates the template activity[362]. However, further work showed that it was the phosphoproteins rather than the kinases which were responsible for template activation, since only the former were able partially to derepress the template activity of nucleohistone complexes[360, 364]. In contrast to these results, Spelsberg and Hnilica[518] have reported previously that while non-histone proteins could complex with histones and thereby prevent the histones binding to DNA, the non-histone proteins were unable to remove histones from nucleohistone complexes. Therefore the order in which the reactants are mixed is obviously very important.

10.5.2.2 Studies of DNA binding proteins

As previously described in Section 10.4.4, the removal of histones from chromatin does not result in a complete derepression of the DNA template, and these results indicated that up to 25% of the DNA may be non-transcribable by bacterial RNA polymerase, by virtue of the fact that it is complexed with non-histone proteins[458, 469, 470, 473, 475]. However, in the presence of homologous RNA polymerase the proteins that remain bound to the DNA can stimulate the template activity to above that of DNA alone[476]. Studies have also shown that some species of non-histone proteins can stimulate the bacterial RNA polymerase-mediated transcription of DNA[174,519]. These proteins appear to bind to the DNA in a species specific manner and thereby stimulate the efficiency of transcription, apparently by increasing the number of RNA chains initiated. Much effort has been directed towards studies of those proteins which can bind to DNA in a species-specific manner. Teng *et al.*[174] have claimed that certain species of phenol-soluble proteins could bind specifically to DNA, however, binding only took place at low ionic strength (0.01 M NaCl), at which concentration the proteins become aggregated and precipitate. More recent work suggests that non-specific binding is avoided only if the final concentration of salt during binding is 0.2–0.25 M NaCl[520,521]. Kleinsmith *et al.*[522] have studied the binding of chromosomal acidic proteins to DNA-cellulose columns in 0.14 M NaCl. In these experiments, non-specific binding was minimised by first exposing the proteins to heterologous DNA, and under these conditions only a very small fraction of the total proteins bound. The effect of these proteins on DNA transcription was not reported.

10.5.2.3 Immunological studies

A detailed description of some of this work will be found elsewhere in this volume[214]. Briefly, it has been found that antibodies may be raised against dehistonised chromatin[523], but only some species of the total non-histone complement are antigenic[196]. However, it is of great interest to note that to date it has not been possible to raise antibodies to purified non-histone proteins alone, only to DNA–protein conjugates[524]. The reason for their apparent lack of antigenicity is not clear, since, in contrast to the histones, the number and heterogeneity of these proteins would lead one to expect that many immunologically different proteins should be present in different tissues and species.

10.5.2.4 Factors which affect RNA polymerase

The discovery of the role of σ-factor in determining the initiation sites of DNA transcription in *E. coli* led to a parallel search for similar factors in eukaryotic cells. Thus Mondal and his co-workers[525-527] found that the chromosomal acidic protein fraction from coconut endosperm contains a species of protein which can bind to both the nucleolar and nucleoplasmic RNA polymerases resulting in an increased efficiency of transcription of native homologous DNA *in vitro*. This stimulation of transcription appeared to be due to an increase both in the rate of initiation and in the size of RNA synthesised. Another protein isolated from the same fraction appears to facilitate the release of RNA from the DNA template, but, unlike the bacterial ρ-factor, it does not affect the size of RNA synthesised *in vitro*. In mammals, rat liver nucleolar extracts contain a protein fraction, distinct from the RNA polymerase activity, which when added to the purified RNA polymerase can stimulate the *in vitro* transcription of native homologous DNA[528]. Other workers have shown that fractions of whole tissue extracts can stimulate both the initiation[529] and rate of transcription[530] by homologous RNA polymerases; however, their exact relationship to the chromosomal proteins has not been determined.

In conclusion, many of the studies on the biological activity of the non-histone proteins have used proteins which have been exposed previously to extreme conditions of pH and/or solvent environment. Therefore it is likely that the use of the less rigorous procedures will enable many more of the biological activities of the non-histone proteins to be characterised, which should in turn lead to a better understanding of the control of DNA transcription in eukaryotic cells.

10.6 GENERAL THEORIES OF GENE REGULATION IN EUKARYOTES

The evidence detailed in the preceding sections has pointed to the involvement of the non-histone proteins in the genetic regulation of eukaryotic cells. However, at the present time, when our ideas concerning gene regulation in prokaryotic systems have yet to be finalised, the current ideas concerning eukaryotic gene regulation are necessarily speculative. Thus, while many of the current ideas are based on extensions of the prokaryotic models of transcription, the genomes of eukaryotic cells possess several unique characteristics. In particular, much of the DNA of eukaryotic cells consists of reiterated sequences[1,531] many of which are not transcribed[6]; moreover, these reiterated sequences are intimately dispersed amongst the unique sequences of DNA[5], which probably represent the bulk of the 'informational DNA'. Genetic evidence has also revealed that some structural genes which are transcribed coordinately are not linked[532,533]. Evidence has also been accumulating which suggests that in eukaryotic cells most species of RNA are synthesised as larger precursor molecules (see a review by Darnell[534]). These precursors then undergo specific modifications prior to their transport to the cytoplasm, while other species of RNA appear to be restricted to the nucleus[535,536].

One of the simplest models suggested was that of Georgiev[537]. He has suggested that the reiterated sequences found amongst the unique DNA could have a regulatory function, and that the repetitious nature of the regulatory DNA results from the presence of a series of linked similar regulatory protein binding sites. Thus each structural gene would be preceded by a large number of repressor sites (operator loci) which could be transcribed. Binding of a repressor to any one of these sites would prevent transcription of the gene. In agreement with this; low C_0t hybridisation experiments have shown that the 5′ end of nuclear RNA hybridises more readily than the rest of the molecule[538]. Another model, similar to that of Georgiev's, has been proposed by Britten and Davidson[539], who concluded that the coordinate transcription of non-linked genes might be carried out by the transcription products of 'integrator genes', either RNA itself or specific DNA-binding proteins.

Both of the theories described above assume that the portions of the genome required for transcription consists essentially of 'free DNA'. Several methods have been devised to determine the amount of free DNA in chromatin, mainly by titrating the free phosphate groups with either cationic dyes[540-542] or polylysine[543,544]. These studies indicated that 30–60% of the DNA phosphate groups are not complexed with protein. However, it has since been pointed out that, as most of the chromosomal proteins appear to be bound in the major groove of the DNA[448], the results obtained from studies of the binding of polylysine and dyes to DNA could be a reflection of their interaction with the phosphate groups in the minor groove of the DNA. Indeed some evidence suggests that polylysine preferentially binds to the minor groove[545].

On the other hand, extensive electron microscopic studies have revealed that chromatin contains no long stretches of free DNA. In agreement with this, titration of chromatin with specific anti-DNA sera also suggests that only 1–5% of the total DNA is exposed[546], while titration studies with *E. coli* RNA polymerase suggests that in most tissues only *ca.* 10% or less of the DNA can be transcribed. Thus it seems that most of the DNA in the chromatin is complexed with protein such that it is not available for transcription.

As described previously, the difference in transcriptional patterns in eukaryotic and prokaryotic cells has resulted in large differences in the organisation of chromosomes. Thus the more recent models of gene regulation proposed by Crick[547] and Paul[548] take into account the morphology of the chromosomes, as deduced from studies of the giant interphase chromosomes of diptera. These giant chromosomes consist of densely-coiled band regions and less dense interband regions[549], each band plus interband region apparently corresponding to a complementation group[550]. Interband regions actively involved in DNA transcription appear as 'puffs'; moreover the distribution of these puffs depends on the cell type[551]. Crick's model supposes that the interband DNA is informational, while the compact band regions contain the regulatory sequences, thus transcription proceeds from the band to the interband regions. Crick has also suggested that these compact regions might contain hair-pin loop structures which have single-stranded recognition sites to which the regulatory proteins could bind. However, Levy and Simpson[552] were unable to detect such regions in isolated chromatin. Paul's theory varies from that of Crick in that he suggests transcription may proceed from the interband regions to the band regions. Paul proposes that the interband regions are reiterated and may consist of a large number of similar protein binding sites ('address sites'). It is visualised that transcription initiated at these sites could destabilise the condensed chromatin in the banded regions.

It is apparent that we are still some way from understanding the roles of the various chromosomal components in the control of gene expression in eukarytoic cells. The evidence quoted in this review suggests that the non-histone proteins may act as gene regulators, while the histones appear to fulfil a more structural function. Definitive conclusions, however, can only be reached when the biological activities of the individual non-histone proteins, in relation to the organisation of the chromatin, have been clearly defined.

References

1. Britten, R. J. and Kohne, D. E. (1968). *Science*, **161,** 529
2. Frenster, J. H. (1965). *Nature* (*London*), **206,** 680
3. Littau, V. C., Allfrey, V. G., Frenster, J. H. and Mirsky, A. E. (1964). *Proc. Nat. Acad. Sci.* (*U.S.A.*), **52,** 93
4. Marushige, K. and Ozaki, H. (1967). *Develop. Biol.*, **16,** 474
5. Grouse, L., Chilton, M. and McCarthy, B. J. (1972). *Biochemistry*, **11,** 798
6. Southern, E. C. (1970). *Nature* (*London*), **227,** 794
7. Pardue, M. L. and Gall, J. G. (1970). *Science*, **168,** 1356
8. Yasmineh, W. G. and Yunis, J. J. (1970). *Expl. Cell Res.*, **59,** 69
9. Gurdon, J. B. and Laskey, R. A. (1970). *J. Embryol. Exp. Morphol.*, **24,** 227
10. Coon, H. G. (1966). *Proc. Nat. Acad. Sci.* (*U.S.A.*), **55,** 66
11. Cahn, R. D. and Cahn, M. B. (1966). *Proc. Nat. Acad. Sci.* (*U.S.A.*), **55,** 106
12. Hadorn, E. (1965). *Brookhaven Symp. Biol.*, **18,** 148
13. Königsberg, I. R. (1963). *Science*, **140,** 1273
14. Stedman, E. and Stedman, E. (1950). *Nature* (*London*), **166,** 780
15. Ris, H. (1961). *Can. J. Genet.*, **3,** 95
16. Dupraw, E. J. (1965). *Nature* (*London*), **206,** 338
17. Wilkins, M. H. F., Zubay, G. and Wilson, H. R. (1959). *J. Molec. Biol.*, **1,** 179
18. Pardon, J. F., Wilkins, M. H. F. and Richards, B. M. (1967). *Nature* (*London*), **215,** 508
19. Bram, S. and Ris, H. (1971). *J. Molec. Biol.*, **55,** 325
20. Paul, J. and More, I. A. R. (1972). *Nature New Biol.*, **239,** 134
21. Solari, A. J. (1968). *Expl. Cell Res.*, **53,** 567
22. Abuelo, J. G. and Moore, D. E. (1969). *J. Cell Biol.*, **41,** 73
23. Davies, H. G. (1968). *J. Cell Sci.*, **3,** 129
24. Davies, H. G. and Small, J. V. (1968). *Nature* (*London*), **217,** 1122
25. Miller, O. L. and Bakken, A. H. (1972). *Karolinska Symp. on Research Methods in Reproductive Endocrinology, 5th Symposium Gene Transcription in Reproductive Tissue*, 155 (E. Diczfalusy, editor)
26. Landel, A. M., Aloni, Y., Raftery, M. A. and Attardi, G. (1972). *Biochemistry*, **11,** 1654
27. Bonner, J., Chalkley, G. R., Dahmus, M., Fambrough, D., Fujimura, F., Huang, R. C., Huberman, J., Jensen, R., Marushige, K., Ohlenbusch, H., Olivera, B. and Widholm, J. (1968). *Methods in Enzymology, Nucleic Acids*, Vol. 12, part B, 3 (L. Grossman and K. Moldave, editors) (New York: Academic Press)
28. MacGillivray, A. J., Cameron, A., Krauze, R. J., Rickwood, D. and Paul, J. (1972). *Biochim. Biophys. Acta*, **277,** 384
29. Graziano, S. L. and Huang, R. C. C. (1971). *Biochemistry*, **10,** 4770
30. Hill, R. J., Poccia, D. C. and Doty, P. (1971). *J. Molec. Biol.*, **61,** 445
31. Levy, S., Simpson, R. T. and Sober, H. A. (1972). *Biochemistry*, **11,** 1547
32. Delange, R. J. and Smith, E. C. (1971). *Annu. Rev. Biochem.*, **40,** 279
33. Elgin, S. C. R., Froehner, S. C., Smart, J. E. and Bonner, J. (1971). *Advances in Cell and Molecular Biology* (E. J. DuPraw, editor) (New York: Academic Press)
34. Johns, E. W. (1971). *Histones and Nucleohistones* (D. M. P. Phillips, editor) (London: Pelham Press)
35. MacGillivray, A. J., Rickwood, D. and Paul, J. (1974). *Advan. Molec. Genetics*, **1,** in the press
36. Fambrough, D. M. and Bonner, J. (1969). *Biochim. Biophys. Acta*, **175,** 113
37. Panyim, S. and Chalkley, R. (1969). *Biochemistry*, **8,** 3972
38. Kinkade, J. M. and Cole, R. D. (1966). *J. Biol. Chem.*, **241,** 5790
39. Bustin, M. and Cole, R. D. (1969). *J. Biol. Chem.*, **244,** 5286
40. Sherod, D., Johnson, G. and Chalkley, R. (1970). *Biochemistry*, **9,** 4611
41. Balhorn, R., Rieke, W. O. and Chalkley, R. (1971). *Biochemistry*, **10,** 3952
42. Rall, S. C. and Cole, R. D. (1971). Quoted in Delange and Smith (1971). *Annu. Rev. Biochem.*, **40,** 279
43. Langan, T. A., Rall, S. C. and Cole, R. D. (1971). *J. Biol. Chem.*, **246,** 1942
44. Bustin, M. and Cole, R. D. (1968). *J. Biol. Chem.*, **243,** 4500
45. Stellwagen, R. H. and Cole, R. D. (1968). *J. Biol. Chem.*, **243,** 4456

46. Kinkade, J. M. (1969). *J. Biol. Chem.*, **244,** 3375
47. Hohmann, P. and Cole, R. D. (1969). *Nature* (*London*), **223,** 1064
48. Hohmann, P. and Cole, R. D. (1971). *J. Molec. Biol.*, **58,** 533
49. Hohmann, P., Cole, R. D. and Bern, H. A. (1971). *J. Nat. Cancer Inst.*, **47,** 337
50. Bustin, M. and Cole, R. D. (1970). *J. Biol. Chem.*, **245,** 1458
51. Kinkade, J. M. and Cole, R. D. (1966). *J. Biol. Chem.*, **241,** 5798
52. Bustin, M. and Cole, R. D. (1969). *J. Biol. Chem.*, **244,** 5291
53. Iwai, K., Ishikawa, K. and Hayashi, H. (1970). *Nature* (*London*), **226,** 1056
54. Hnilica, L. S., Kappler, H. A. and Jordan, J. J. (1970). *Experientia*, **26,** 353
55. Candido, E. P. M. and Dixon, G. H. (1972). *Proc. Nat. Acad. Sci.* (*U.S.A.*), **69,** 2015
56. Panyim, S., Bilek, D. and Chalkley, R. (1971). *J. Biol. Chem.*, **246,** 4206
57. Shepherd, G. R., Noland, B. J. and Hardin, J. M. (1971). *Biochim. Biophys. Acta*, **228,** 544
58. Shepherd, G. R., Noland, B. J. and Hardin, J. M. (1971). *Arch. Biochem. Biophys.*, **142,** 299
59. Shepherd, G. R., Hardin, J. M. and Noland, B. J. (1971). *Arch. Biochem. Biophys.*, **143,** 1
60. Sautiere, P., Tyrou, D., Laine, B., Mizon, J., Lambelin-Breynaert, M. D., Ruffin, P. and Biserte, G. (1972). *C. R. Acad. Sci. Paris*, **274,** 1422
61. Yeoman, L. C., Olson, M. O. J., Sugano, N., Jordan, J. J., Taylor, C. W., Starbuck, W. C. and Busch, H. (1972). *J. Biol. Chem.*, **247,** 6018
62. Sung, M. T., Dixon, G. H. and Smithies, O. (1971). *J. Biol. Chem.*, **246,** 1358
63. Candido, E. P. M. and Dixon, G. H. (1972). *J. Biol. Chem.*, **247,** 3868
64. Delange, R. J., Smith, E. C., Fambrough, D. M. and Bonner, J. (1968). *Proc. Nat. Acad. Sci.* (*U.S.A.*), **61,** 1145
65. Ogawa, Y., Puagliarotti, G., Jordan, J., Taylor, C. W., Starbuck, W. C. and Busch, H. (1969). *J. Biol. Chem.*, **244,** 4387
66. Delange, R. J., Fambrough, D. M., Smith, E. C. and Bonner, J. (1969). *J. Biol. Chem.*, **244,** 5669
67. Sautiere, P., Breynaert, M. D., Moschetto, Y. and Biserte, G. (1970). *C. R. Acad. Sci. Paris*, **271,** 364
68. Wilson, R. K., Starbuck, W. C., Taylor, C. W., Jordan, J. and Busch, H. (1970). *Cancer Res.*, **30,** 2942
69. Sung, M. T. and Dixon, G. H. (1970). *Proc. Nat. Acad. Sci.* (*U.S.A.*), **67,** 1616
70. Louie, A. J. and Dixon, G. M. (1972). *Proc. Nat. Acad. Sci.* (*U.S.A.*), **69,** 1975
71. Fambrough, D. M. and Bonner, J. (1968). *J. Biol. Chem.*, **243,** 4434
72. Panyim, S., Sommer, K. R. and Chalkley, R. (1971). *Biochemistry*, **10,** 3911
73. Marzluff, W. F., Sanders, L. A., Miller, D. M. and McCarty, K. S. (1972). *J. Biol. Chem.*, **247,** 2026
74. Delange, R. J., Hooper, J. A. and Smith, E. C. (1972). *Proc. Nat. Acad. Sci.* (*U.S.A.*), **69,** 882
75. Brandt, W. F. and von Holt, C. (1972). *FEBS Lett.*, **23,** 357
76. Olson, M. O. J., Jordan, J. and Busch, H. (1972). *Biochem. Biophys. Res. Commun.*, **46,** 50
77. Yokotsuka, K., Kikuchi, A. and Shimura, K. (1972). *J. Biochem.* (Tokyo), **71,** 133
78. Marzluff, W. F. and McCarty, K. S. (1972). *Biochemistry*, **11,** 2672
79. Marzluff, W. F. and McCarty, K. S. (1972). *Biochemistry*, **11,** 2677
80. Panyim, S., Chalkley, R., Spiker, S. and Oliver, D. (1970). *Biochim. Biophys. Acta*, **214,** 216
81. Pipkin, J. C. and Larson, D. A. (1972). *Exp. Cell Res.*, **71,** 249
82. Leaver, J. L. and Cruft, H. J. (1966). *Biochem. J.*, **101,** 665
83. Raaf, J. and Bonner, J. (1968). *Arch. Biochem. Biophys.*, **125,** 567
84. Laver, W. G. (1970). *Virology*, **41,** 488
85. Iwai, K. (1964). *The Nucleohistones*, 59 (J. Bonner and P. Ts'o, editors) (San Francisco, London, Amsterdam: Holden-Day)
86. Makino, F. and Tsuzuki, J. (1971). *Nature* (*London*), **231,** 446
87. Rizzo, P. J. and Nooden, L. D. (1972). *Science*, **176,** 796
88. Leighton, J. J., Dill, B. C., Stock, J. J. and Phillips, C. (1971). *Proc. Nat. Acad. Sci.* (*U.S.A.*), **68,** 677
89. Duffus, J. H. (1971). *Biochim. Biophys. Acta*, **228,** 627

90. Tonino, G. J. M and Rozijn, T. H. (1966). *Biochim. Biophys. Acta*, **124,** 427
91. Van der Vliet, P. C., Tonino, G. J. M. and Rozijn, T. H. (1969). *Biochim. Biophys. Acta*, **195,** 473
92. Van der Vliet, P. C., Zandvliet, G. M. and Rozijn, T. H. (1971). *Biochim Biophys. Acta.*, **247,** 373
93. Mohberg, J. and Rusch, H. (1969). *Arch. Biochem. Biophys.*, **134,** 577
94. Mohberg, J. and Rusch, H. (1970). *Arch. Biochem. Biophys.*, **138,** 418
95. Iwai, K., Hamana, K. and Yabuki, H. (1970). *J. Biochem.*, **68,** 597
96. Prescott, D. M. (1966). *J. Cell Biol.*, **31,** 1
97. Lynch, M. J., Leake, R. E. and Buetow, D. E. (1972). *J. Cell. Biol.*, **55,** 160A
98. Pallotta, D. and Berlowitz, L. (1970). *Biochim. Biophys. Acta*, **200,** 538
99. Oliver, D. R. and Chalkley, R. (1972). *Exp. Cell Res.*, **73,** 295
100. Cozcolluela, C. and Subirana, J. A. (1968). *Biochim. Biophys. Acta*, **154,** 242
101. Subirana, J. A., Palau, J., Cozcolluela, C. and Ruiz-Carrillo, A. (1970). *Nature*, (*London*), **228,** 992
102. Vendrely, R. and Picaud, M. (1968). *Exp. Cell Res.*, **49,** 13
103. Hnilica, L. S. (1964). *Experientia*, **20,** 13
104. Vidali, G. and Neelin, J. M. (1968). *Europ. J. Biochem.*, **5,** 330
105. Greenaway, P. and Murray, K. (1971). *Nature New Biol.*, **229,** 233
106. Garel, A., Burckard, J., Mazen, A. and Champagne, M. (1972). *Biochimie*, **54,** 451
107. Edwards, L. J. and Hnilica, L. S. (1968). *Experientia*, **24,** 228
108. Panyim, S. and Chalkley, R. (1969). *Biochem. Biophys. Res. Commun.*, **37,** 1042
109. Ozaki, H. (1971). *Develop. Biol.*, **26,** 209
110. Sheridan, W. F. and Stern, H. (1967). *Exp. Cell Res.*, **45,** 323
111. Wigle, D. T. and Dixon, G. H. (1971). *J. Biol., Chem.*, **246,** 5636
112. Huntley, G. H. and Dixon, G. H. (1972). *J. Biol., Chem.*, **247,** 4916
113. Yoshida, M., Yokotsuka, K. and Shimura, K. (1966). *J. Biochem.* (*Tokyo*), **60,** 586
114. Hamana, K. and Iwai, K. (1971). *J. Biochem.* (*Tokyo*), **69,** 1097
115. Johns, E. W. and Butler, J. A. V. (1962). *Biochem. J.*, **84,** 436
116. Bloch, D. (1969). *Genetics Suppl.*, **61,** 1
117. Ando, T. and Watanabe, S. (1969). *Int. J. Protein Res.*, **1,** 221
118. Bretzel, G. (1972). *Hoppe-Seyler's Z. Physiol. Chem.*, **353,** 209
119. Ingles, C. J. and Dixon, G. H. (1967). *Proc. Nat. Acad. Sci.* (*U.S.A.*), **58,** 191
120. Marushige, K., Ling, Y. and Dixon, G. H. (1969). *J. Biol., Chem.*, **244,** 5953
121. Iwai, K., Nakahara, C. and Ando, T. (1971). *J. Biochem.* (*Tokyo*), **69,** 493
122. Coelingh, J. P., Rozijn, T. H. and Montfoort, C. H. (1967). *Biochim. Biophys. Acta*, **188,** 353
123. Coelingh, J. P., Montfoort, C. H., Rozijn, T. H., Gevers-Leuven, J. A., Schiphof, R., Steyn-Parve, E. P., Braunitzer, G., Schrank, B. and Ruhfus, A. (1972). *Biochim. Biophys. Acta*, **285,** 1
124. Calvin, H. I. and Bedford, J. M. (1971). *J. Reprod. Fert. Suppl.*, **13,** 65
125. Palau, J., Ruiz-Carrillo, A. and Subirana, J. A. (1969). *Europ. J. Biochem.* **7,** 209
126. Paoletti, R. A. and Huang, R. C. C. (1969). *Biochemistry*, **8,** 1615
127. Easton, D. and Chalkley, R. (1972). *Exp. Cell Res.*, **72,** 502
128. Wangh, L., Ruiz-Carrillo, A. and Allfrey, V. G. (1972). *Arch. Biochem. Biophys.*, **150** 44
129. Gerovsky, M. A., Pleger, G. L., and Kevert, J. (1972). *J. Cell Biol.*, **55,** 92a
130. Lindsay, D. T. (1964), *Science*, **144,** 420
131. Kischer, C. W. and Hnilica, L. S. (1967). *Exp. Cell Res.*, **48,** 424
132. Kischer, C. W., Gurley, L. R. and Shepherd, G. R. (1966). *Nature* (*London*), **212,** 304
133. Agrell, I. P. S. and Christensson, E. G. (1965). *Nature* (*London*) **207,** 638
134. Asao, T. (1969). *Exp. Cell Res.*, **58,** 243
135. Orengo, A. and Hnilica, L. S. (1970). *Exp. Cell Res.*, **62,** 331
136. Hnilica, L. S. and Johnson, A. W. (1970). *Exp. Cell Res.*, **63,** 261
137. Johnson, A. W. and Hnilica, L. S. (1970). *Biochim. Biophys. Acta*, **224,** 518
138. Johnson, A. W. and Hnilica, L. S. (1971). *Biochim. Biophys. Acta*, **246,** 141
139. Thaler, M. M., Cox, M. C. L. and Villee, C. A. (1970). *J. Biol. Chem.*, **245,** 1479
140. Benttinen, L. C. and Comb, D. G. (1972). *J. Molec. Biol.*, **57,** 355
141. Kedes, L. H. and Gross, P. R. (1969). *Nature* (*London*), **223,** 1335
142. Asao, T. (1970). *Exp. Cell Res.*, **61,** 255

143. Asao, T. (1972). *Exp. Cell Res.*, **73,** 73
144. Fambrough, D. M., Fujimura, F. and Bonner, J. (1968). *Biochemistry*, **7,** 575
145. Oliver, D. R. and Chalkley, R. (1972). *Exp. Cell Res.*, **73,** 303
146. Johns, E. W. (1964). *Biochem. J.*, **92,** 55
147. Hnilica, L. S., Edwards, L. J. and Hey, A. E. (1966). *Biochim. Biophys. Acta*, **124,** 109
148. MacGillivray, A. J. (1968). *Biochem. J.*, **110,** 181
149. Appels, R., Wells, J. R. E. and Williams, A. F. (1972). *J. Cell Sci.*, **10,** 47
150. Billett, M. A. and Barry, J. M. (1972). *Biochem. J.*, **130,** 10P
151. Dick, C. and Johns, E. W. (1969). *Biochim. Biophys. Acta*, **175,** 414
152. Sotirov, N. and Johns, E. W. (1972). *Exp. Cell Res.*, **73,** 13
153. Mazen A. and Champagne, M. (1969). *FEBS Lett.*, **2,** 248
154. Billett, M. A. and Hindley, J. (1972). *Europ. J. Biochim.*, **28,** 451
155. Moss, B. A., Joyce, W. G. and Ingram, V. M. (1972). *Fed. Proc.* (*Fed. Amer. Soc. Exp. Biol.*), **31,** 417
156. Gurley, L. R. and Hardin, J. M. (1968). *Arch. Biochem. Biophys.*, **128,** 258
157. Comings, D. E. (1967). *J. Cell Biol.*, **35,** 25a
158. Wilhelm, J. A., Ansevin, A. T., Johnson, A. W. and Hnilica, L. S. (1972). *Biochim. Biophys. Acta*, **272,** 220
159. Loewus, M. W. (1968). *Nature* (*London*), **218,** 474
160. Pallotta, D., Berlowitz, L. and Rodriguez, L. (1970). *Exp. Cell Res.*, **60,** 474
161. Chanda, S. K., and Dounce, A. L. (1971). *Arch. Biochem. Biophys.*, **145,** 211
162. Johns, E. W., Davies, A. J. S. and Barton, M. E. (1970). *Biochim. Biophys. Acta*, **213,** 537
163. Spelsberg, T. C., Mitchell, W. M., Chytil, F., Wilson, E. M. and O'Malley, B. W. (1973). *Biochim. Biophys. Acta*, **312,** 765
164. Cohen, L. H. and Gotchel, B. V. (1971). *J. Biol. Chem.*, **246,** 1841
165. Gorovsky, M. A. (1970). *J. Cell Biol.*, **47,** 631
166. Rizzo, P. J. (1971). *Ph.D. Thesis*, University of Michigan
167. DuPraw, E. J. (1970). *Molecular and Cellular Biology Series, DNA and Chromosomes* (H. Koffler, editor) (New York: Holt Reinehart and Winston, Inc.)
168. Zweidler, A. and Cohen, L. H. (1972). *Fed. Proc.* (*Fed. Amer. Soc. Exp. Biol.*), **31,** 4051
169. Zweidler, A. (1972). *J. Cell Biol.*, **55,** 297a
170. Delange, R. J., Fambrough, D. M., Smith, E. L. and Bonner, J. (1969). *J. Biol. Chem.*, **244,** 319
171. Marushige, K. and Dixon, G. H. (1969). *Develop. Biol.*, **19,** 397
172. Dingman, C. W. and Sporn, M. B. (1964). *J. Biol. Chem.*, **239,** 3483
173. Sporn, M. B. and Dingman, C. W. (1966). *Cancer Res.*, **26,** 2488
174. Teng, C. S., Teng, C. T. and Allfrey, V. G. (1971). *J. Biol. Chem.*, **246,** 3597
175. Shelton, K. R. and Neelin, J. M. (1971). *Biochemistry*, **10,** 2342
176. Elgin, S. C. R. and Bonner, J. (1970). *Biochemistry*, **9,** 4440
177. Benjamin, W. and Gellhorn, A. (1968). *Proc. Nat. Acad. Sci.* (*U.S.A.*), **59,** 262
178. Shirey, T. and Huang, R. C. (1969). *Biochemistry*, **8,** 4138
179. Umansky, S. R., Tokarskaya V. I., Zotova, R. N. and Migushina, V. C. (1971). *Molec. Biol. U.S.S.R.*, **5,** 270
180. Wang, T. Y. (1967). *J. Biol. Chem.*, **242,** 1220
181. Richter, K. H. and Sekeris, C. E. (1972). *Arch. Biochem. Biophys.*, **148,** 44
182. Arnold, E. A. and Young, K. E. (1972). *Biochim. Biophys. Acta*, **257,** 482
183. Yoshida, M. and Shimura, K. (1972). *Biochim. Biophys. Acta*, **263,** 690
184. Spelsberg, T. C., Steggles, A. W., Chytil, F. and O'Malley, B. W. (1972). *J. Biol. Chem.*, **247,** 1368
185. Shaw, L. M. J. and Huang, R. C. C. (1970). *Biochemistry*, **9,** 4530
186. Steele, W. J. and Busch, H. (1963). *Cancer Res.*, **23,** 1153
187. Patel, G., Patel, V., Wang, T. Y. and Zobel, C. R. (1968). *Arch. Biochem. Biophys.*, **128,** 654
188. Helmsing, P. J. and Berendes, H. D. (1971). *J. Cell Biol.*, **50,** 893
189. Langan, T. A. (1967). *Biochim. Biophys. Acta Library*, Vol. 10, 233 (V. V. Koningsberger and L. Bosch, editors) (Amsterdam: Elsevier Publishing Company)
190. Kleinsmith, L. J. and Allfrey, V. G. (1969). *Biochim. Biophys. Acta*, **175,** 123
191. Gershey, E. L. and Kleinsmith, L. J. (1969). *Biochim. Biophys. Acta*, **194,** 331
192. Shelton, K. R. and Allfrey, V. G. (1970). *Nature* (*London*), **228,** 132

193. LeStourgeon, W. M. and Rusch, H. P. (1971). *Science*, **174,** 1233
194. Tsuboi, A. and Baserga, R. (1972). *J. Cell Physiol.*, **80,** 107
195. Sonnenbichler, J. and Nobis, P. (1970). *Europ. J. Biochem.*, **16,** 60
196. Spelsberg, T. C. (1973). Mitchell, W. M. and Chytic, F. (1973). *Molec. Cell. Biochem.*, **1,** 243
197. Kostraba, N. C. and Wang, T. Y. (1970). *Int. J. Biochem.* **1,** 327
198. Elgin, S. C. R. and Bonner, J. (1972). *Biochemistry*, **11,** 772
199. Wang, T. Y. (1971). *Exp. Cell Res.*, **69,** 217
200. Weber, K. and Osborn, M. (1969). *J. Biol. Chem.*, **244,** 4406
201. MacGillivray, A. J. and Rickwood, D. (1973). *Biochem. Soc. Trans.*, **1,** 686
202. Barrett, T. and Gould, H. J. (1973). *Biochim. Biophys. Acta*, **294,** 165
203. Johns, E. W. and Forrester, S. (1969). *Europ. J. Biochem.*, **8,** 547
204. Goodwin, G. H. and Johns, E. W. (1972). *FEBS Lett.*, **21,** 103
205. Wilhelm, J. A., Groves, C. M. and Hnilica, L. S. (1972). *Experientia*, **28,** 514
206. Harlow, R., Tolstoshev, P. and Wells, J. R. E. (1972). *Cell Differentiation*, **1,** 341
207. Bhorjee, J. S. and Pederson, T. (1972). *Proc. Nat. Acad. Sci.* (*U.S.A.*), **69,** 3345
208. Gronow, M. (1969). *Europ. J. Cancer*, **5,** 497
209. Gronow, M. and Griffiths, G. (1971). *FEBS Lett.*, **15,** 340
210. Itzhaki, R. F. (1971). *Biochem. J.*, **125,** 221
211. Wood, J. (1972). *Ph.D. Thesis*, University of Edinburgh
212. Barton, A. D., Kisieleski, W. E., Wassermann, F. and Mackevicius, F. (1971). *Z. Zellforsch. Mikrosk. Anat.*, **115,** 299
213. Patel, G. L. and Thomas, T. L. (1972). *J. Cell Biol.*, **55,** 200a
214. Means, A. R., Spelsberg, T. C. and O'Malley, B. W. (1973). This volume, Chapter 6
215. Carlsson, S. A., Moore, G. M. and Ringertz, N. (1973). *Exp. Cell Res.*, **76,** 234
216. Kruh, J., Tichonicky, L. and Wajcman, H. (1969). *Biochim. Biophys. Acta*, **195,** 549
217. Kruh, J., Tichonicky, L. and Dastugue, B. (1970). *Bull. Soc. Chim. Biol.*, **52,** 1287
218. Loeb, J. and Creuzet, C. (1969). *FEBS Lett.*, **5,** 37
219. Loeb, J. and Creuzet, C. (1970). *Bull. Soc. Chim. Biol.*, **52,** 1007
220. Dastugue, B., Tichonicky, L., Penit-Soria, J. and Kruh, J. (1970). *Bull. Soc. Chim. Biol.*, **52,** 391
221. Platz, R. D., Kish, V. M. and Kleinsmith, L. J. (1970). *FEBS Lett.*, **12,** 38
222. Rickwood, D., Riches, P. G. and MacGillivray, A. J. (1973). *Biochim. Biophys. Acta*, **299,** 162
223. Cognetti, G., Settineri, D. and Spinelli, G. (1972). *Exp. Cell Res.*, **71,** 465
224. Connor, B. J. and Patel, G. L. (1972). *J. Cell Biol.*, **55,** 49a
225. LeStourgeon, W. M. and Rusch, H. P. (1972). *J. Cell Biol.*, **55,** 153a
226. Stein, G. and Borun, T. W. (1972). *J. Cell Biol.*, **52,** 292
227. Borun, T. W. and Stein, G. S. (1972). *J. Cell Biol.*, **52,** 308
228. Becker, H. and Stanners, C. P. (1972). *J. Cell Physiol.*, **80,** 51
229. Jeter, J. R., Pavlat, W. A. and Cameron, I. L. (1972). *J. Cell Biol.*, **55,** 125a
230. Teng, C. S. and Hamilton, T. H. (1970). *Biochem. Biophys. Res. Commun.*, **40,** 1231
231. Helmsing, P. J. (1972). *Cell Differentiation*, **1,** 19
232. Gronow, M. (1972). *Biochem. J.*, **130,** 11P
233. Rickwood, D. and MacGillivray, A. J. Unpublished results
234. Reeck, G. R., Simpson, R. T. and Sober, H. A. (1972). *Proc. Nat. Acad. Sci.* (*U.S.A.*), **69,** 2317
235. Huang, R. C. and Bonner, J. (1965). *Proc. Nat. Acad. Sci.* (*U.S.A.*), **54,** 960
236. Benjamin, W., Levander, O. A., Gellhorn, A. and Debellis, R. H. (1966). *Proc. Nat. Acad. Sci.* (*U.S.A.*), **55,** 858
237. Commerford, S. L. and Delihas, N. (1966). *Proc. Nat. Acad. Sci.* (*U.S.A.*), **56,** 1759
238. Bonner J. (1967). *Biochim. Biophys. Acta Library*, Vol. 10 (V. V. Koningsberger and L. Bosch, editors) (Amsterdam: Elsevier Publishing Company)
239. Jacobson, R. N. and Bonner, J. (1971). *Arch. Biochem. Biophys.*, **146,** 557
240. Huang, R. C. (1967). *Fed. Proc.* (*Fed. Amer. Soc. Exp. Biol.*), **26,** 1933
241. Huang, R. C. and Huang, P. C. (1969). *J. Molec. Biol.*, **39,** 365
242. Holmes, D. S., Mayfield, J. E., Sander, G. and Bonner, J. (1972). *Science*, **177,** 72
243. Dahmus, M. E. and McConnell, D. J. (1969). *Biochemistry*, **8,** 1524
244. Shih, T. Y. and Bonner, J. (1969). *Biochim. Biophys. Acta*, **182,** 30
245. Mayfield, J. E. and Bonner, J. (1971). *Proc. Nat. Acad. Sci.* (*U.S.A.*), **68,** 2652

246. Bekhor, I., Kung, G. M. and Bonner, J. (1969). *J. Molec. Biol.*, **39,** 351
247. Bekhor, I., Bonner, J. and Dahmus, G. K. (1969). *Proc. Nat. Acad. Sci.* (*U.S.A.*), **62** 271
248. Sivolap, Y. M. and Bonner, J. (1971). *Proc. Nat. Acad. Sci.* (*U.S.A.*), **68,** 387
249. Bonner, J., Huang, R. C. C. and Maheshwari, N. (1961). *Proc. Nat. Acad. Sci.* (*U.S.A.*), **47,** 1548
250. Mayfield, J. E. and Bonner, J. (1972). *Proc. Nat. Acad. Sci.* (*U.S.A.*), **69,** 7
251. Heyden, H. W. and Zachau, H. G. (1971). *Biochim. Biophys. Acta*, **232,** 651
252. Artman, M. and Roth, J. S. (1971). *J. Molec. Biol.*, **60,** 291
253. Szeszak, F. and Pihl, A. (1971). *Biochim. Biophys. Acta*, **247,** 363
254. Szeszak, F. and Pihl, A. (1972). *FEBS Lett.*, **20,** 177
255. Kanehisa, T., Fujitani, H., Sano, M. and Tanaka, T. (1971). *Biochim. Biophys. Acta*, **240,** 46
256. Kanehisa, T., Tanaka, T. and Kano, Y. (1972). *Biochim. Biophys. Acta*, **277,** 584
257. Borun, T. W., Scharff, M. D. and Robbins, E. (1967). *Proc. Nat. Acad. Sci.* (*U.S.A.*), **58,** 1977
258. Gallwitz, D. and Mueller, G. C. (1969). *J. Biol. Chem.*, **244,** 5947
259. Gurley, L. R., Walters, R. A. and Enger, M. D. (1970). *Biochem. Biophys. Res. Commun.*, **40,** 428
260. Nemer, N. and Lindsay, D. T. (1969). *Biochem. Biophys. Res. Commun.*, **35,** 156
261. Moav, B. and Nemer, M. (1971). *Biochemistry*, **10,** 881
262. Gallwitz, D. and Mueller, G. C. (1970). *FEBS Lett.*, **6,** 83
263. Gallwitz, D. and Breindl, M. (1972). *Biochem. Biophys. Res. Commun.*, **47,** 1106
264. Jacobs-Lorena, M., Baglioni, C. and Borun, T. W. (1972). *Proc. Nat. Acad. Sci.* (*U.S.A.*) **69,** 2095
265. Kedes, A. and Birnsteil, M. (1971). *Nature New Biol.*, **230,** 165
266. Adesnik, M. and Darnell, J. E. (1972). *J. Molec. Biol.*, **67,** 397
267. Pederson, T. and Robbins, E. (1970). *J. Cell Biol.*, **45,** 509
268. Butler, J. A. V. and Cohn, P. (1963). *Biochem. J.*, **87,** 330
269. Orlova, L. V. and Rodionov, V. M. (1970). *Exp. Cell Res.*, **59,** 329
270. Gutierrez-Cernosek, R. M. and Hnilica, L. S. (1971). *Biochim. Biophys. Acta*, **247,** 348
271. Yarbo, J. W. (1967). *Biochim. Biophys. Acta*, **145,** 531
272. Robbins, E. and Borun, T. W. (1967). *Proc. Nat. Acad. Sci.* (*U.S.A.*), **57,** 409
273. Takai, S., Borun, T. W., Muchmore, J. and Leiberman, I. (1968). *Nature* (*London*), **219,** 860
274. Stein, G. and Baserga, R. (1970). *J. Biol. Chem.*, **245,** 6097
275. Cross, M. E. (1972). *Biochem. J.*, **128,** 1213
276. Sadgopal, A. and Bonner, J. (1970). *Biochim. Biophys. Acta*, **207,** 206
277. Sadgopal, A. and Bonner, J. (1970). *Bochim. Biophys. Acta*, **207,** 227
278. Jungmann, R. A. and Schweppe, J. S. (1972). *J. Biol. Chem.*, **247,** 5535
279. Hnilica, L. S., Kappler, H. A. and Hnilica, V. S. (1965). *Science*, **150,** 1470
280. Chalkley, G. R. and Maurer, H. R. (1965). *Proc. Nat. Acad. Sci.* (*U.S.A.*), **54,** 498
281. Spalding, J., Kajiwara, K. and Mueller, G. C. (1966). *Proc. Nat. Acad. Sci.* (*U.S.A.*) **56,** 1535
282. Gurley, L. R. and Hardin, J. M. (1970). *Arch. Biochem. Biophys.*, **136,** 392
283. Dick, C. and Johns, E. W. (1969). *Biochim. Biophys. Acta*, **174,** 380
284. Byvoet, P. (1966). *J. Molec. Biol.*, **17,** 311
285. Hancock, R. (1969). *J. Molec. Biol.*, **40,** 457
286. Balhorn, R., Oliver, D., Hohmann, P., Chalkley, R. and Granner, D. (1972). *Biochemistry*, **11,** 3915
287. Byvoet, P. (1967). *Molec. Pharmacol.*, **3,** 303
288. Bondy, S. C. (1971). *Biochem. J.*, **123,** 465
289. Appels, R. and Wells, J. R. E. (1972). *J. Molec. Biol.*, **70,** 425
290. Gurley, L. R., Hardin, J. M. and Walters, R. A. (1970). *Biochem. Biophys. Res. Commun.*, **38,** 290
291. Stellwagen, R. H. and Cole, R. O. (1969). *J. Biol. Chem.*, **244,** 4878
292. Patel, G. and Wang, T. Y. (1965). *Biochim. Biophys. Acta*, **95,** 314
293. Patel, G. and Wang, T. Y. (1965). *Life Sciences*, **4,** 1481
294. Fleischer-Lambropoulos, H. and Reinsch, I. (1971). *Hoppe-Seyler's Z. Physiol. Chem.*, **252,** 593

295. Stein, G. and Baserga, R. (1971). *Biochem. Biophys. Res. Commun.*, **44**, 218
296. Buck, M. D. and Schauder, P. (1970). *Biochim. Biophys. Acta*, **224**, 644
297. Teng, C. S. and Hamilton, T. H. (1969). *Proc. Nat. Acad. Sci.* (*U.S.A.*), **63**, 465
298. Smith, J. A., Martin, L., King, R. J. B. and Vertes, M. (1970). *Biochem. J.*, **119**, 773
299. Barker, K. L. (1971). *Biochemistry*, **10**, 284
300. Ruddon, R. W. and Rainey, C. H. (1970). *Biochem. Biophys. Res. Commun.*, **40**, 152
301. Dastugue, B., Hanoune, J. and Kruh, J. (1971). *FEBS Lett.* **19**, 65
302. Malpoix, P. J. (1971). *Exp. Cell Res.*, **65**, 393
303. Stellwagen, R. H. and Cole, R. D. (1969). *J. Biol. Chem.*, **244**, 4878
304. Rovera, G. and Baserga, R. (1971). *J. Cell Physiol.*, **77**, 201
305. Rovera, G., Baserga, R. and Defendi, V. (1972). *Nature New Biol.*, **237**, 240
306. McClure, M. E. and Hnilica, L. S. (1970). *J. Cell Biol.*, **47**, 133a
307. Stein, G. and Baserga, R. (1970). *Biochem. Biophys. Res. Commun.*, **41**, 715
308. Halliburton, I. W. and Mueller, G. C. (1971). Personal communication
309. Ingles, C. J. (1971). Personal communication
310. Holoubek, V. and Crocker, T. T. (1968). *Biochim. Biophys. Acta*, **157**, 352
311. McClure, M. E. and Hnilica, L. S. (1972). *J. Cell Biol.*, **55**, 169a
312. Kleinsmith, L. J., Allfrey, V. G. and Mirsky, A. E. (1966). *Proc. Nat. Acad. Sci.*, (*U.S.A.*), **55**, 1182
313. Liew, C. C., Haslett, G. W. and Allfrey, V. G. (1970). *Nature* (*London*), **226**, 414
314. Marzluff, W. F. and McCarty, K. S. (1970). *J. Biol. Chem.*, **245**, 5635
315. Gershey, E. L., Vidali, G. and Allfrey, V. G. (1968). *J. Biol. Chem.*, **243**, 5018
316. Vidali, G., Gershey, E. L. and Allfrey, V. G. (1968). *J. Biol. Chem.*, **243**, 6361
317. Pogo, B. G. T., Pogo, A. O., Allfrey, V. G. and Mirsky, A. E. (1968). *Proc. Nat. Acad. Sci.* (*U.S.A.*), **59**, 1337
318. Nohara, H., Takahashi, T. and Ogata, K. (1968). *Biochim. Biophys. Acta*, **154**, 529
319. Phillips, D. M. P. (1963). *Biochem. J.*, **87**, 258
320. Marzluff, W. F., Miller, D. M. and McCarty, K. S. (1972). *Arch. Biochem. Biophys.*, **152**, 472
321. Libby, P. R. (1972). *Biochem. J.*, **130**, 663
322. Allfrey, V. G. (1970). *Fed. Proc.* (*Fed. Amer. Soc. Exp. Biol.*), **29**, 1447
323. Nohara, H., Takahashi, T. and Ogata, K. (1966). *Biochim. Biophys. Acta*, **127**, 282
324. Bondy, S. C., Roberts, S. and Morelos, B. S. (1970). *Biochem. J.*, **119**, 665
325. Gallwitz, D. (1970). *Biochem. Biophys. Res. Commun.*, **40**, 236
326. Gallwitz, D. (1970). *Hoppe-Seyler's Z. Physiol. Chem.*, **351**, 1050
327. Racey, L. A. and Byvoet, P. (1971). *Exp. Cell Res.*, **64**, 366
328. Gallwitz, D. (1971). *FEBS Lett.*, **13**, 306
329. Inoue, A. and Fujimoto, D. (1969). *Biochem. Biophys. Res. Commun.*, **36**, 146
330. Libby, P. R. (1970). *Biochim. Biophys. Acta*, **213**, 234
331. Kikuchi, H. and Fujimoto, D. (1973). *FEBS Lett.*, **29**, 280
332. Paik, W. K., Pearson, D., Lee, H. W. and Kim, S. (1970). *Biochim. Biophys. Acta*, **213**, 513
333. Allfrey, V. G., Faulkner, R. and Mirsky, A. E. (1964). *Proc. Nat. Acad. Sci.* (*U.S.A.*), **51**, 786
334. Pogo, B. G. T., Allfrey, V. G. and Mirsky, A. E. (1966). *Proc. Nat. Acad. Sci.* (*U.S.A.*), **55**, 805
335. Mukherjee, A. B. and Cohen, M. M. (1969). *Exp. Cell Res.*, **54**, 257
336. Cross, M. E. and Ord, M. G. (1970). *Biochem. J.*, **118**, 191
337. Ono, T., Terayama, H., Takaku, F. and Nakao, K. (1968). *Biochim. Biophys. Acta*, **161**, 361
338. Monjardino, J. P. P. V. and MacGillivray, A. J. (1970). *Exp. Cell Res.*, **60**, 1
339. Haber, J., Rosenau, W. and Goldberg, M. (1972). *Nature New Biol.*, **238**, 60
340. Pogo, B. G. T. (1969). *J. Cell Biol.*, **40**, 571
341. Berlowitz, L. and Pallotta, D. (1972). *Exp. Cell Res.*, **71**, 45
342. Takaku, F., Nakao, H., Ono, T. and Terayama, H. (1969). *Biochim. Biophys. Acta*, **195**, 396
343. Pogo, B. G. T , Allfrey, V. G. and Mirsky, A. E. (1967). *J. Cell Biol.*, **35**, 477
344. Clever, U. (1967). *The Control of Nuclear Activity*, 161 (L. Goldstein, editor) (New Jersey, Englewood Cliffs: Prentice-Hall)
345. Clever, V. and Ellgaard, E. G. (1970). *Science*, **169**, 373

346. Ellgaard, E. G. (1967). *Science*, **157,** 1070
347. Allfrey, V. G., Pogo, B. G. T., Littau, V. C., Gershey, E. L., and Mirsky, A. E. (1968). *Science*, **159,** 314
348. Pearson, D. B. and Paik, W. K. (1972). *Exp. Cell Res.*, **73,** 208
349. Wilhelm, J. A. and McCarty, K. S. (1970). *Cancer Res.*, **30,** 418
350. Shepherd, G. R., Noland, B. J. and Hardin, J. M. (1972). *Exp. Cell Res.*, **75,** 397
351. Byvoet, P. (1968). *Biochim. Biophys. Acta*, **160,** 217
352. Dixon, G. H. (1972). *Karolinska Symp. on Research Methods in Reproductive Endocrinology, 5th Symposium Gene Transcription in Reproductive Tissue*, 128 (E. Diczfalusy, editor)
353. Kleinsmith, L. J., Allfrey, V. G. and Mirsky, A. E. (1966). *Science*, **154,** 780
354. Schiltz, E. and Sekeris, C. E. (1969). *Hoppe-Seyler's Z. Physiol. Chem.*, **350,** 317
355. Schiltz, E. and Sekeris, C. E. (1971). *Experientia*, **27,** 30
356. Ahmed, K. and Ishida, H. (1971). *Molec. Pharmacol.*, **7,** 323
357. Lurquin, P. F., Seligy, V. L. and Neelin, J. M. (1972). *Arch. Int. Physiol. Biochim.*, **80,** 202
358. Langan, T. A. and Smith, L. K. (1967). *Fed. Proc. (Fed. Amer. Soc. Exp. Biol.)*, **26,** 603
359. Jergil, B. and Dixon, G. H. (1970). *J. Biol. Chem.*, **245,** 425
360. Kamiyama, M., Dastugue, B. and Kruh, J. (1971). *Biochem. Biophys. Res. Commun.*, **44,** 1345
361. Meisler, M. H. and Langan, T. A. (1969). *J. Biol. Chem.*, **244,** 4961
362. Kamiyama, M. and Dastugue, B. (1971). *Biochem. Biophys. Res. Commun.*, **44,** 29
363. Takeda, M., Yamamura, H. and Ohga, Y. (1971). *Biochem. Biophys. Res. Commun.*, **42,** 103
364. Kamiyama, M., Dastugue, B., Defer, N. and Kruh, J. (1972). *Biochim. Biophys. Acta*, **277,** 576
365. Ruddon, R. W. and Anderson, S. C. (1972). *Biochem. Biophys. Res. Commun.*, **46,** 1499
366. Kish, V. M. and Kleinsmith, L. J. (1972). *J. Cell Biol.*, **55,** 138a
367. Langan, T. A. (1968). *Science*, **162,** 579
368. Granner, D. K. (1972). *Biochem. Biophys. Res. Commun.*, **46,** 1516
369. Siebert, G., Ord, M. G. and Stocken, L. A. (1971). *Biochem. J.*, **122,** 721
370. Ord, M. G. and Stocken, L. A. (1966). *Biochem. J.*, **98,** 888
371. Ord, M. G. and Stocken, L. A. (1969). *Biochem. J.*, **112,** 81
372. Gutierrez, R. M. and Hnilica, L. S. (1967). *Science*, **157,** 1324
373. Stevely, W. S. and Stocken, L. A. (1968). *Biochem. J.*, **110,** 187
374. Fitzgerald, P. J., Marsh, W. H., Ord, M. G. and Stocken, L. A. (1970). *Biochem. J.*, **117,** 711
375. Ord, M. G. and Stocken, L. A. (1968). *Biochem. J.*, **107,** 403
376. Adams, G. H. M., Vidali, G. and Neelin, J. M. (1970). *Can. J. Biochem.*, **48,** 33
377. Letnansky, K., and Reisinger, L. (1972). *Biochem. Biophys. Res. Commun.*, **49,** 312
378. Jergil, B., Sung, M. and Dixon, G. H. (1970). *J. Biol. Chem.*, **245,** 5867
379. Turkington, R. W. and Riddle, M. (1969). *J. Biol. Chem.*, **244,** 6040
380. Langan, T. A. (1969). *Proc. Nat. Acad. Sci. (U.S.A.)*, **64,** 1276
381. Cross, M. E. and Ord, M. G. (1971). *Biochem. J.*, **124,** 241
382. Gurley, L. R. and Walters, R. A. (1971). *Biochemistry*, **10,** 1588
383. Balhorn, R., Bordwell, J., Sellers, L., Granner, D. and Chalkley, R. (1972). *Biochem. Biophys. Res. Commun.*, **46,** 1326
384. Gurley, L. R., Walters, R. A. and Tobey, R. A. (1972). *Arch. Biochem. Biophys.*, **154,** 212
385. Lake, R. S., Goidl, J. A. and Salzman, N. P. (1972). *Exp. Cell Res.*, **73,** 113
386. Balhorn, R., Chalkley, R. and Granner, D. (1972). *Biochemistry*, **11,** 1094
387. Balhorn, R., Balhorn, M., Morris, H. P. and Chalkley, R. (1972). *Cancer Res.*, **32,** 1775
388. Balhorn, R., Balhorn, M. and Chalkley, R. (1972). *Develop. Biol.*, **29,** 199
389. Oliver, D., Balhorn, R., Granner, D. and Chalkley, R. (1972). *Biochemistry*, **11,** 3921
390. Marushige, K. and Dixon, G. H. (1971). *J. Biol. Chem.*, **246,** 5799
391. Louie, A. J. and Dixon, G. H. (1972). *J. Biol. Chem.*, **247,** 5490
392. Sanders, M. M. and Dixon, G. H. (1972). *J. Biol. Chem.*, **247,** 851
393. Louie, A. J. and Dixon, G. H. (1972). *J. Biol. Chem.*, **247,** 5498

394. Louie, A. J., Candido, E. P. M. and Dixon, G. H. (1972). *Fed. Proc.* (*Fed. Amer. Soc. Exp. Biol.*), **31,** 1121
395. Candido, E. P. M. and Dixon, G. H. (1972). *J. Biol. Chem.*, **247,** 5506
396. Andress, D., Mousseron-Canet, M., Borgna, J. C. and Beziat, Y. (1972). *C. R. Acad. Sci. Paris*, **274D,** 2606
397. Riches, P. G., Harrap, K. R., Sellwood, S. M., Rickwood, D. and MacGillivray, A. J. (1973). *Biochem. Soc. Trans.*, **1,** 684
398. Birnie, G. D. (1973). Personal communication
399. Paik, W. K. and Kim, S. (1971). *Science*, **174,** 114
400. Borun, T. W., Pearson, D. and Paik, W. K. (1972). *J. Biol. Chem.*, **247,** 4288
401. Byvoet, P., Shepherd, G. R., Hardin, J. M. and Noland, B. J. (1972). *Arch. Biochem. Biophys.*, **148,** 558
402. Kim, S. and Paik, W. K. (1965). *J. Biol. Chem.*, **240,** 4629
403. Murray, K. (1964). *Biochemistry*, **3,** 10
404. Comb, D. G., Sarkar, N. and Pinzino, C. J. (1966). *J. Biol. Chem.*, **241,** 1857
405. Burdon, R. H. and Garvin, E. V. (1971). *Biochim. Biophys. Acta*, **232,** 371
406. Gallwitz, D. (1971). *Arch. Biochem. Biophys.*, **145,** 650
407. Orenstein, J. M. and Marsh, W. H. (1968). *Biochem. J.*, **109,** 697
408. Paik, W. K. and Kim, S. (1969). *Arch. Biochem. Biophys.*, **134,** 632
409. Turberville, C. and Craddck, V. M. (1971). *Biochem. J.*, **124,** 725
410. Tidwell, T., Allfrey, V. G. and Mirsky, A. E. (1968). *J. Biol. Chem.*, **243,** 707
411. Benjamin, W. (1971). *Nature New Biol.*, **234,** 18
412. Lee, H. W. and Paik, W. K. (1972). *Biochim. Biophys. Acta*, **277,** 107
413. Thomas, G., Lange, H. W. and Hempel, K. (1972). *Hoppe-Seylers Z. Physiol., Chem.*, **353,** 1423
414. Byvoet, P. (1972). *Arch. Biochem. Biophys.*, **152,** 887
415. Desai, L. S. and Foley, G. E. (1970). *Biochem. J.*, **119,** 165
416. Ord, M.G. and Stocken, L. A. (1970). *Biochem. J.*, **120,** 671
417. Ord, M. G. and Stocken, L. A. (1970). *Biochem. J.*, **116,** 415
418. Doenecke, D., Beato, M., Congote, L. F. and Sekeris, C. E. (1972). *Biochem. J.*, **126,** 1171
419. Huang, R. C. and Bonner, J. (1962). *Proc. Nat. Acad. Sci.* (*U.S.A.*), **48,** 1216
420. Bonner, J. and Huang, R. C. C. (1963). *J. Molec. Biol.*, **6,** 169
421. Frenster, J. H., Allfrey, V. G. and Mirsky, A. E. (1963). *Proc. Nat. Acad. Sci.* (*U.S.A.*), **50,** 1026
422. Sonnenberg, B. P. and Zubay, G. (1965). *Proc. Nat. Acad. Sci.* (*U.S.A.*), **54,** 415
423. Roy, A. K. and Zubay, G. (1966). *Biochim. Biophys. Acta*, **129,** 403
424. Clark, P. R. and Byvoet, P. (1970). *Experientia*, **26,** 725
425. Hoare, T. A. and Johns, E. W. (1970). *Biochem. J.*, **119,** 931
426. Moskowitz, G., Ogawa, Y., Starbuck, W. C. and Busch, H. (1969). *Biochem. Biophys. Res. Commun.*, **35,** 741
427. Moskowitz, G. J., Wilson, R. K., Starbuck, W. C. and Busch, H. (1970). *Physiol., Chem. and Phys.*, **2,** 217
428. Seligy, V. and Miyagi, M. (1969). *Exp. Cell Res.*, **58,** 27
429. Marushige, K. and Bonner, J. (1966). *J. Molec. Biol.*, **15,** 160
430. Huang, R. C. C., Bonner, J. and Murray, K. (1964). *J. Molec., Biol.*, **8,** 54
431. Georgiev, G. P., Ananieva, L. N. and Kozlov, J. V. (1966). *J. Molec. Biol.*, **22,** 365
432. Paul, J. and Gilmour, R. S. (1966). *Nature* (*London*), **210,** 992
433. Paul, J. and Gilmour, R. S. (1966). *J. Molec. Biol.*, **16,** 241
434. Sullivan, D. T. (1968). *Proc. Nat. Acad. Sci.* (*U.S.A.*), **59,** 846
435. Ursprung, H., Smith, K. D., Sofer, W. H. and Sullivan, D. T. (1968). *Science*, **160,** 1075
436. Smith, K. D., Church, R. B. and McCarthy, B. J. (1969). *Biochemistry*, **8,** 4271
437. Tan, C. H. and Miyagi, M. (1970). *J. Molec. Biol.*, **50,** 641
438. Daniel, J. C. and Flickinger, R. A. (1971). *Exp. Cell Res.*, **64,** 285
439. Butterworth, P. H. W., Cox, R. F. and Chesterton, C. J. (1971). *Europ. J. Biochem*, **23,** 229
440. Keshgegian, A. A. and Furth, J. J. (1972). *Biochem. Biophys. Res. Commun.*, **48,** 757
441. Butler, J. A. V. and Chipperfield, A. R. (1967). *Nature* (*London*), **215,** 1188
442. Barr, G. C. and Butler, J. A. V. (1963). *Nature* (*London*), **199,** 1170
443. Hindley, J. (1963). *Biochem. Biophys. Res. Commun.*, **12,** 175

444. Johns, E. W. and Hoare, T. A. (1970). *Nature* (*London*), **226**, 650
445. Johns, E. W. (1972). *Nature New Biol.*, **237**, 87
446. Bartley, J. A. and Chalkley, R. (1972). *J. Biol. Chem.*, **247**, 3647
447. Kleiman, L. and Huang, R. C. C. (1972). *J. Molec. Biol.*, **64**, 1
448. Simpson, R. T. (1970). *Biochemistry*, **9**, 4814
449. Farber, J., Baserga, R. and Gabray, E. J. (1971). *Biochem. Biophys Res. Commun.*, **43**, 675
450. Skalka, A., Fowler, A. V. and Hurwitz, J. (1966). *J. Biol. Chem.*, **241**, 588
451. Sponar, J. and Vatavova, H. (1970). *Studia Biophys.*, **24**, 351
452. Sponar, J. and Sormova, Z. (1972). *Europ. J. Biochem.* **29**, 99
453. Clark, R. J. and Felsenfeld, G. (1972). *Nature New Biol.*, **240**, 226
454. Varshavsky, A. J. and Georgiev, G. P. (1972). *Biochim. Biophys. Acta*, **281**, 669
455. Rubin, R. C. and Moudrianakis, E. N. (1972). *J. Molec. Biol.*, **67**, 361
456. Limborska, S. A. and Georgiev, G. P. (1972). *Cell Differentiation.*, **1**, 245
457. Gilmour, R. S. and Paul, J. (1969). *J. Molec. Biol.*, **40**, 137
458. Paul, J. and Gilmour, R. S. (1968). *J. Molec. Biol.*, **34**, 305
459. Gilmour, R. S. and Paul, J. (1970). *FEBS Lett.*, **9**, 242
460. Hindley, J. (1964). *Abstr. Int. Congr. Biochem. 6th New York*, 1
461. Johns, E. W. (1967). *Biochem. J.*, **105**, 611
462. Murray, K. (1966). *J. Molec. Biol.*, **15**, 409
463. Murray, K. (1969). *J. Molec. Biol.*, **39**, 125
464. Ohlenbusch, H. H., Olivera, B. M., Tuan, D. and Davidson, N. (1967). *J. Molec. Biol.*, **25**, 299
465. Fambrough, D. and Bonner, J. (1968). *Biochim. Biophys. Acta*, **154**, 601
466. Mirsky, A. E., Burdick, C. J., Davidson, E. H. and Littau, V. C. (1968). *Proc. Nat. Acad. Sci.* (*U.S.A.*), **61**, 592
467. Murray, K., Vidali, G. and Neelin, J. M. (1968). *Biochem. J.*, **107**, 207
468. Johns, E. W. and Diggle, J. H. (1969). *Europ. J. Biochem.*, **11**, 495
469. Spelsberg, T. C. and Hnilica, L. S. (1970). *Biochem. J.*, **120**, 435
470. Spelsberg, T. C., Hnilica, L. S. and Ansevin, A. T. (1971). *Biochim. Biophys. Acta*, **228**, 550
471. Smart, J. E. and Bonner, J. (1971). *J. Molec. Biol.*, **58**, 661
472. Bonner, J., Dahmus, M. E., Fambrough, D., Huang, R. C. C., Marushige, K. and Tuan, D. Y. H. (1968). *Science*, **159**, 47
473. Seligy, V. C. and Neelin, J. M. (1970). *Biochim. Biophys. Acta*, **213**, 380
474. Spelsberg, T. C. and Hnilica, L. S. (1971). *Biochim. Biophys. Acta*, **228**, 202
475. Spelsberg, T. C. and Hnilica, L. S. (1971). *Biochim. Biophys. Acta*, **228**, 212
476. Singh, V. K. and Sung, S. C. (1972). *Biochem. J.*, **130**, 1095
477. Schrader, W. T., Toft, D. O. and O'Malley, B. W. (1972). *J. Biol. Chem.*, **247**, 2401
478. Li, H. J., Isenberg, J. and Johnson, W. C. (1971). *Biochemistry*, **10**, 2587
479. Fasman, G. D., Schaffhausen, B., Goldsmith, L. and Adler, A. J. (1970). *Biochemistry*, **9**, 2814
480. Simpson, R. T. and Sober, H. A. (1970). *Biochemistry*, **9**, 3103
481. Matsuyama, A., Tagashira, Y. and Nagata, C. (1971). *Biochim. Biophys. Acta*, **240**, 184
482. Wagner, T. and Spelsberg, T. C. (1971). *Biochemistry*, **10**, 2599
483. Ramm, E. I., Vorob'ev, V. I., Birshtein, T. M., Bolotina, I. A. and Volkenshtein, M. V. (1972). *Europ. J. Biochem.*, **25**, 245
484. Boffa, L., Saccomani, G., Tachburro, A. M., Scatturin, A. and Vidali, G. (1971). *Int. J. Protein Res.*, **3**, 357
485. Fric, J. and Sponar, J. (1971). *Biopolymers*, **10**, 5125
486. Shih, T. Y. and Fasman, G. D. (1971). *Biochemistry*, **10**, 1675
487. Tuan, D. Y. H. and Bonner, J. (1969). *J. Molec. Biol.*, **45**, 59
488. Sponar, J., Boublik, M., Fric, I. and Sormova, Z. (1970). *Biochim. Biophys. Acta*, **209**, 532
489. Henson, P. and Walker, I. O. (1970). *Europ. J. Biochem.*, **14**, 345
490. Johnson, R. S., Chan, A. and Hanlon, S. (1972). *Biochemistry*, **11**, 4347
491. Bradbury, E. M. and Crane-Robinson, C. (1971). *Histones and Nucleohistones* (D. M. P. Phillips, editor) (London: Pelham Press)
492. D'Anna, J. A. and Isenberg, I. (1972). *Bochemistry*, **11**, 4017
493. Wickett, R. R., Li, H. J. and Isenberg, I. (1972). *Biochemistry*, **11**, 2952

494. Shih, T. Y. and Fasman, G. D. (1972). *Biochemistry*, **11,** 398
495. Boublik, M., Bradbury, E. M., Crane-Robinson, C. and Rattle, H. W. E. (1971). *Nature New Biol.*, **229,** 149
496. Bradbury, E. M., Cary, P. D., Crane-Robinson, C., Riches, P. L. and Johns, E. W. (1971). *Nature New Biol.*, **233,** 265
497. Bradbury, E. M., Cary, P. D., Crane-Robinson, C., Riches, P. L. and Johns, E. W. (1972). *Europ. J. Biochem.*, **26,** 482
498. Bradbury, E. M. and Rattle, H. W. E. (1972). *Europ. J. Biochem.*, **27,** 270
499. Olins, D. E. (1969). *J. Molec. Biol.*, **43,** 439
500. Lurquin, V. F. and Seligy, V. L. (1972). *Arch. Int. Physiol. Biochem.*, **80,** 606
501. Richards, B. M. and Pardon, J. F. (1970). *Exp. Cell Res.*, **62,** 184
502. Pardon, J. F. and Wilkins, M. H. F. (1972). *J. Molec. Biol.*, **68,** 115
503. More, I. A. R. (1971). *Ph.D. Thesis, University of Glasgow*
504. Zubay, G. and Wilkins, M. H. F. (1964). *J. Molec. Biol.*, **9,** 246
505. Palau, J., Pardon, J. F. and Richards, B. M. (1967). *Biochim. Biophys. Acta*, **138,** 633
506. Murray, K., Bradbury, E. M., Crane-Robinson, C., Stephens, R. M., Haydon, A. J. and Peacocke, A. R. (1970). *Biochem. J.*, **120,** 859
507. Bradbury, E. M., Molgaard, H. V., Stephens, R. M., Bolund, L. A. and Johns, E. W. (1972). *Europ. J. Biochem.*, **31,** 474
508. Garrett, R. A. (1968). *J. Molec. Biol.*, **38,** 249
509. Simpson, R. T. (1972). *Biochemistry*, **11,** 2003
510. Bradbury, E. M., Crane-Robinson, C., Goldman, H., Rattle, H. W. E. and Stephens, R. M. (1967). *J. Molec. Biol.*, **29,** 507
511. Haydon, A. J. and Peacocke, A. R. (1968). *Biochem. J.*, **110,** 243
512. Littau, V. C., Burdick, C. J., Allfrey, V. G. and Mirsky, A. E. (1965). *Proc. Nat. Acad. Sci.* (*U.S.A.*), **54,** 1204
513. Sluyser, M. and Snellen-Jurgens, N. H. (1970). *Biochim. Biophys. Acta*, **199,** 490
514. Adler, A. J., Schaffhausen, B., Langan, T. A. and Fasman, G. D. (1971). *Biochemistry*, **10,** 909
515. Johns, E. W. (1969). *Homostatic Regul., Ciba Foundation Symp.*, 128 (London: J. and A. Churchill, Ltd.).
516. Wang, T. Y. (1970). *Exp. Cell Res.*, **61,** 455
517. Kamiyama, M. and Wang, T. Y. (1971). *Biochim. Biophys. Acta*, **228,** 563
518. Spelsberg, T. C. and Hnilica, L. S. (1969). *Biochim. Biophys. Acta*, **195,** 63
519. Rickwood, D., Threlfall, G., MacGillivray, A. J., Paul, J. and Riches, P. G. (1972). *Biochem. J.*, **129,** 50P
520. Yamamoto, K. R. and Alberts, B. M. (1972). *Proc. Nat. Acad. Sci.* (*U.S.A.*), **69,** 2105
521. O'Malley, B. W. (1972). Personal communication
522. Kleinsmith, L. J., Heidema, J. and Carroll, A. (1970). *Nature* (*London*), **226,** 1025
523. Chytil, F. and Spelsberg, T. C. (1971). *Nature New Biol.*, **233,** 215.
524. Wakabayashi, K. and Hnilica, L. S. (1972). *J. Cell Biol.*, **55,** 271a
525. Mondal, H., Mandel, R. K. and Biswas, B. B. (1970). *Biochem. Biophys. Res. Commun.*, **40,** 1194
526. Mondal, H., Ganguly, A., Das, A., Mandal, R. K. and Biswas, B. B. (1972). *Europ. J. Biochem.*, **28,** 143
527. Mondal, H., Mandal, R. K. and Biswas, B. B. (1972). *Europ. J. Biochem.*, **25,** 463
528. Higashinakagawa, T., Ohnishi, T. and Muramatsu, M. (1972). *Biochem. Biophys. Res. Commun.*, **48,** 937
529. Stein, H. and Hausen, P. (1970). *Europ. J. Biochem.*, **14,** 270
530. Lentfer, D. and Lezius, A. G. (1972). *Europ. J. Biochem.*, **30,** 278
531. Flamm, W. G., Bond, H. E. and Burr, H. E. (1966). *Biochim. Biophys. Acta*, **129,** 310
532. Epstein, C. J. and Motulsky, A. G. (1966). *Prog. Med. Genets.*, **4,** 85
533. Nabholz, M., Miggiano, V. and Bodmer, V. (1969). *Nature* (*London*), **223,** 358
534. Darnell, J. E. (1968). *Bacterial Rev.*, **32,** 262
535. Shearer, R. W. and McCarthy, B. J. (1967). *Biochemistry*, **6,** 283
536. Drews, J., Brawerman, G. and Morris, H. P. (1968). *Europ. J. Biochem.*, **3,** 284
537. Georgiev, G. P. (1969). *J. Theoret. Biol.*, **25,** 473
538. Ryskov, A. P., Mantieva, V. L., Avakian, E. R. and Georgiev, G. P. (1971). *FEBS Lett.*, **12,** 141
539. Britten, R. J. and Davidson, E. H. (1969). *Science*, **165,** 349

540. Miura, A. and Ohba, Y. (1967). *Biochim. Biophys. Acta*, **145,** 436
541. Salganik, R. I., Morozova, T. M. and Zakharov, M. A. (1969). *Biochim. Biophys. Acta*, **174,** 755
542. Kurashina, Y., Ohba, Y. and Mizuno, D. (1970). *J. Biochem.* (*Tokyo*), **67,** 661
543. Itzhaki, R. F. (1970). *Biochem. Biophys. Res. Commun.*, **41,** 25
544. Clark, R. J. and Felsenfeld, G. (1971). *Nature New Biol.*, **229,** 101
545. Carroll, D. and Botchan, M. (1972). *Biochem. Biophys. Res. Commun.*, **46,** 1681
546. Stollar, B. D. (1970). *Biochim. Biophys. Acta*, **209,** 541
547. Crick, F. H. C. (1971). *Nature* (*London*), **234,** 25
548. Paul, J. (1972). *Nature* (*London*), **238,** 444
549. Dupraw, E. J. and Rae, P. M. M. (1965). *Nature* (*London*), **212,** 598
550. Judd, B. H., Shen, M. W. and Kaufman, T. W. (1972). *Genetics*, **71,** 139
551. Berendes, H. D. and Beerman, W. (1969). *Handbook of Molecular Biology*, 51 (A. Lima-de-Faria, editor) (Amsterdam: North-Holland Publishing Co.)
552. Levy, S. and Simpson, R. T. (1973). *Nature New Biol.*, **241,** 139

Index